Mathematics in Victorian Britain

Queen Victoria (1819–1901).

# MATHEMATICS IN VICTORIAN BRITAIN

*Edited by*

RAYMOND FLOOD

ADRIAN RICE

*and*

ROBIN WILSON

OXFORD

UNIVERSITY PRESS

# OXFORD
## UNIVERSITY PRESS

Great Clarendon Street, Oxford OX2 6DP

Oxford University Press is a department of the University of Oxford.
It furthers the University's objective of excellence in research, scholarship,
and education by publishing worldwide in

Oxford  New York

Auckland  Cape Town  Dar es Salaam  Hong Kong  Karachi
Kuala Lumpur  Madrid  Melbourne  Mexico City  Nairobi
New Delhi  Shanghai  Taipei  Toronto

With offices in

Argentina  Austria  Brazil  Chile  Czech Republic  France  Greece
Guatemala  Hungary  Italy  Japan  Poland  Portugal  Singapore
South Korea  Switzerland  Thailand  Turkey  Ukraine  Vietnam

Oxford is a registered trade mark of Oxford University Press
in the UK and in certain other countries

Published in the United States
by Oxford University Press Inc., New York

© Oxford University Press 2011

The moral rights of the authors have been asserted
Database right Oxford University Press (maker)

First published 2011

British Library Cataloguing in Publication Data

Data available

Library of Congress Cataloging in Publication Data

Data available

Typeset by SPI Publisher Services, Pondicherry, India
Printed and bound by
CPI Group (UK) Ltd, Croydon, CR0 4YY

ISBN 978-0-19-960139-4

1 3 5 7 9 10 8 6 4 2

# FOREWORD

The Industrial Revolution began in the 18th century, and by the time Victoria came to the throne in 1837 Britain led the world in the production of iron, steel, and steam engines. The British navy ruled the world, and the atlas was rapidly being coloured pink as the Empire grew.

This power led to tremendous wealth and tremendous self-confidence. Invention took off: in 1837 about five hundred patents were applied for; by the end of the century there were about two hundred and fifty thousand patent applications every year. Engineers thought they could do anything: they built bridges, railways, and steam-powered machines of every kind.

The electric telegraph started as a warning system to be strung along the new-fangled railway lines in order to prevent two trains from using the same single track simultaneously from opposite ends. Within thirty years cables had been laid across the Atlantic Ocean and around the world. From her equipment in Buckingham Palace the queen could in principle send a message to her subjects in a dozen countries and receive a reply within minutes.

Manchester and its outlying towns were built almost entirely on the wealth from the cotton trade—Oldham alone boasted more than three hundred mills—and the amount of money sloshing about was extraordinary. At one point it was claimed that ninety per cent of all the world's cotton passed through the port of Liverpool.

This spirit of enthusiasm and the 'can-do' attitude invaded not only engineering and commerce, but the academic world too. There was more time and support for research. Mathematics was not merely a question of reciting Euclid in order to become a vicar; instead, people began to peel back the intricate petals of the subject, both for its own sake and for practical ends.

The result was a splendid flowering, in areas ranging from algebra and geometry to electromagnetism and mathematical machinery—and this book will take you on a splendid walk through the garden of Victorian mathematics.

Adam Hart-Davis

# PREFACE

Today, the history of 19th-century mathematics is a well-studied and vibrant area of research, with books, articles, and scholarly papers published frequently on various aspects of the subject. One of these aspects is the development of mathematics in Victorian Britain, a period of tremendous vitality and change—social, political, and scientific. In recent years, numerous scholars worldwide have undertaken research into facets of the history of Victorian mathematics, resulting in a variety of published works. However, most of this research has appeared in specialized and hard-to-find scholarly journals, practically inaccessible to all but the most devoted historian of mathematics. Moreover, since these publications have largely focused on specific themes or mathematical areas, no single volume has explicitly considered the broad topic of mathematics in Victorian Britain, until now.

This book thus serves two purposes:

- it constitutes what is perhaps the first general survey of the mathematics of the Victorian period;
- it assembles, for the first time in a single source, researches on the history of mathematics in Victorian Britain that would otherwise be out of the reach of the general reader.

The chapters have been contributed by a group of sixteen authors, with topics corresponding to their individual areas of research. Some of this research receives its first publication on these pages; other chapters are based on older work that has appeared elsewhere but is all but unavailable to those without access to research journals or large university libraries. Consequently, the chapters feature a variety of styles and levels of technical difficulty, depending on their area of coverage. Thus, while they no doubt differ in their level of accessibility, we hope that they will nevertheless be found to be interesting and informative.

Whereas some books on the history of mathematics concentrate on the contributions of particular individuals, and while the following pages will feature details of a number of mathematicians and their work, we have chosen a different approach for this volume, focussing on two main themes.

The first of these is geographical, with six chapters detailing mathematical life in London, Oxford, and Cambridge, as well as Scotland, Ireland, and the British Empire. The motivation for this theme is three-fold. First, these chapters chart the growth and institutional development of mathematics as a profession through the course of the 19th century. Second, they document changes in the teaching of the subject at the principal British centres of higher mathematics education. And third, the chapters provide a contextual background for the discussion of the mathematics and mathematicians featured in the second part of the book.

Following a chapter highlighting the dissemination of mathematics in Victorian Britain via journals and learned societies, the second thematic part of the book focuses on developments in specific mathematical areas, with ten chapters ranging from pure mathematics (such as geometry, algebra, and logic) to the applied sciences (including statistics, calculating machines, and astronomy). Finally, lest the reader assume that the mathematical developments in Victorian Britain are beyond criticism, we close with Jeremy Gray's somewhat subversive epilogue, which argues against viewing the contributions of Victorian British mathematicians too indulgently.

Naturally, as the first book of this kind, this volume cannot hope to be comprehensive, and the motivated reader will no doubt find topics that are missing or only minimally discussed. Such topics include school-level mathematics education; engineering, military, or actuarial mathematics; interactions and correspondence with European and American mathematical contemporaries; and larger sociological issues, such as the role of women in Victorian intellectual society. While all of these subjects (and more) were considered for inclusion, we felt that, under the necessary constraints of a volume of manageable scope and size, there was simply not enough room. Indeed, any one of these topics could provide sufficient material for several books of this size. These limitations notwithstanding, we hope that as an overview of recent and ongoing research this book will provide both information and inspiration for future work.

As with many edited volumes, this book is not necessarily intended to be read from cover to cover, although it can be. Rather, it is intended to serve as a valuable resource to a variety of audiences. It is our hope that mathematicians with little or no knowledge about the development of their field will find many of the historical details stimulating and intriguing, and that the specialist historian of mathematics will view its assorted surveys as a useful resource and an impetus towards further research in the area. For the more general reader, we hope that it will provide an introduction to a fascinating and little-known subject that continues to stimulate and inspire the work of scholars today: mathematics in Victorian Britain.

The Editors
June 2011

# CONTENTS

Oxford's neo-gothic University Museum, seen here shortly after its opening in 1860, was specifically designed to improve the provision of mathematics and science in mid-Victorian Oxford.

# Introduction

ADRIAN RICE

Although the Victorian era was characterized by the dominance of Great Britain, both militarily and economically, the accomplishments of the British in mathematics seem, at first sight, rather less spectacular. However, significant contributions were made by British mathematicians throughout the period, with vectors, matrices, and histograms being just a few of the mathematical innovations for which they were responsible. In this preliminary survey of some of the achievements of Victorian mathematics, we highlight not just those mathematical scholars such as Charles Babbage, James Clerk Maxwell, and Bertrand Russell who are still well known today, but also draw attention to the lesser known mathematical contributions of other prominent Victorians such as Florence Nightingale and Lewis Carroll.

On 2 November 1839, the British literary magazine the *Athenæum* published an article in which was contained the following, apparently unremarkable, sentence:

Perhaps the Annean authors, though inferior to the Elizabethans, are, on a general summation of merits, no less superior to the latter-Georgian and Victorian.[1]

The anonymous author of this article thus distinguished his otherwise run-of-the-mill essay by providing the first appearance (in print at least) of a new contribution to the English language: the word *Victorian*.

This word has been used ever since to denote something typical of, or belonging to, the sixty-four-year reign of Queen Victoria, and provides, at least in part, the title of this book. Since the First World War—and, in particular, assaults on the preceding generation by the outspoken writer, critic, and Bloomsbury-group leading light Lytton Strachey—the word 'Victorian' has often been used in a somewhat derogatory manner to mean 'old-fashioned', 'out-dated',

Queen Victoria at the beginning of her reign.

'backward-looking', and 'reactionary', among other things. And, while some of the mathematics practised in Victorian Britain might appear somewhat outmoded today, much also seems recognizably contemporary. Indeed, to the modern-day mathematician, Victorian mathematics is a curious mixture of both the antiquated and progressive!

The Victorian era, which lasted from 1837 to 1901, was the age of Peel and Palmerston, Gladstone and Disraeli, Tennyson and Thackeray, Dickens and Wilde, Stanley and Livingstone, Gilbert and Sullivan. It was the age of the Great Exhibition, the Crimean War, the Suez Canal, the Indian Mutiny, and the American Civil War. It was characterized by rapid industrialization and urban growth, far-reaching social and political reforms, vast colonial expansion overseas, and impressive scientific development: increased urbanization resulted in Britain moving from being a primarily agrarian to a largely industrial society; parliamentary reforms enfranchised whole new sections of the population; innovations such as the telegraph and the railways revolutionized daily life; while Darwin's theory of evolution challenged the very basis of people's beliefs.

Not surprisingly, the Britain of 1837 was very different from the nation that entered the 20th century in 1901. In 1837 there were no telephones, no light bulbs, no motor cars. X-rays and electrons had still to be discovered, as had the planet Neptune. There was no income tax, evolution was not a mainstream scientific theory, and the Pope was still fallible. Slavery was still legal in the southern-most members of the twenty-six United States of America. And, despite being recognized geographical entities, the countries of Italy, Germany, Canada, and Australia had yet to be created as distinct unified sovereign nations. All this would change in the sixty-four years from 1837 to 1901.

The Victorian age is remembered, in the UK at least, as the period when Britain rose from being a major military and economic force to the hub of the most powerful and extensive empire the world had ever seen. It was an age of pride and 'jingoism' (a very Victorian word)—a time when Britain quite literally 'ruled the waves'.

But whereas Britain dominated the 19th century militarily and economically, we have to search harder to uncover its Victorian mathematical heritage. Were we to list the most influential mathematicians who were active during this period (among them Gauss, Cauchy, Jacobi, Riemann, Weierstrass, Cayley, Maxwell, Cantor, Klein, Poincaré, and Hilbert), we would find that the British names are far outnumbered by their French and German counterparts. Indeed, mathematically, this period was characterized by the dominance of the French giving way to the rise, in particular, of German mathematics, with British mathematics being relatively peripheral. Nevertheless, there are still compelling reasons to devote an entire book to the fascinating story of mathematics in Victorian Britain.

After a long period of stagnation from about 1750 to 1830, British mathematics experienced a dramatic renaissance during Queen Victoria's reign, with a resurgence of interest and achievement in the subject. British Victorian mathematicians made numerous significant contributions to mathematics in this period and, as we will see, were innovators in several respects. Indeed, anyone who has

Oxford Circus in London in Victorian times.

studied mathematics at school or university will have come across at least some of the fruits of this remarkably fertile period: matrices, vectors, Maxwell's equations, Boolean algebra, histograms, and even the concept of standard deviation, were all invented by British mathematicians of the Victorian era. To provide a backdrop for the chapters that follow, this introduction highlights just a few of the many mathematical developments made in Britain during this time.

## Sixty-four years of invention and discovery

The Victorian period coincided with a revival of British mathematics from its mid-18th-century slump. Despite the strong impetus given to British mathematics by the prestige and achievements of Isaac Newton and the ability of his immediate successors (such as Roger Cotes, Brook Taylor, Colin Maclaurin, and Thomas Simpson), British mathematics had entered a period of stagnation from the middle of the 1700s.[2] Indeed, no really first-rate mathematicians were produced from about 1750 until around 1830.

By that time, Charles Babbage was lamenting what he termed 'the decline of science in England' in the title of a widely-read book. In it he wrote:[3]

...in England, particularly with respect to the more difficult and abstract sciences, we are much below other nations, not merely of equal rank, but below several even of inferior power.

Quoting fellow mathematician and friend John Herschel, he added:[4]

In mathematics we have long since drawn the rein, and given over a hopeless race.

The good news was that, by the time these words were published, Britain was emerging from what has been described as 'its mathematical fog'.[5] By the early 1800s, mathematicians at the University of Edinburgh, the University of Cambridge, and Trinity College, Dublin, were initiating an interest in Continental results, a trend furthered by subsequent scholars, including George Peacock, William Whewell, and Babbage and Herschel themselves.

There were several reasons why 18th-century British mathematics had lagged so far behind that of mainland Europe, and these factors remained well into the 19th century. The country's peripheral geographical position, together with the effects of a prolonged war with France, resulted in the slight delay and limited availability of Continental publications reaching its shores. Thus, while still in touch with developments in mainland Europe, mathematicians in Britain were inevitably slower in assimilating results from the continent than might otherwise have been the case. The most obvious example was their patriotic insistence on sticking to (and failing to develop adequately) Newton's fluxional version of the calculus in the face of more versatile Continental equivalents.[6]

Another was the famous distrust of complex numbers, imaginary quantities, and even negative numbers, by several high-ranking British scholars.[7] For a variety of mathematical and philosophical reasons, the meaning, interpretation, and legitimacy of negative (and, by extension, imaginary and complex) numbers were called into question. The most vocal opponents of the use of such 'impossible' quantities were the Cambridge-trained mathematicians Francis Maseres and William Frend, who in various publications argued vigorously for their rejection, stimulating a fierce debate on the veracity of their use in mathematics. Consequently, the last half of the 18th century and the opening third of the 19th saw the question of negative and imaginary numbers occupy a major place in the discussions of British mathematicians, philosophers, and men of science.[8]

Indeed, as the Victorian era began, British algebraists were only just beginning to move on. In his *Treatise on Algebra* of 1830 and his book-length 'Report on the recent progress and present state of certain branches of analysis' of 1833, George Peacock had initiated a different, and more abstract, algebraic methodology that fundamentally altered the way the subject was perceived in Britain. His emphasis was not on what negatives and imaginaries actually meant, but rather on the laws under which they operated. Thus, for Peacock, algebra was a subject based not on generalizations of arithmetical concepts, but on a series of formal assumptions or axioms. In short, in order to legitimize the use of negative and imaginary quantities in algebra, Peacock had answered, not by attempting to clarify the meaning of such entities, but by redefining what was meant by algebra itself.[9]

Among those following Peacock's more axiomatic approach to algebra was the Cambridge-educated Scottish mathematician Duncan Gregory. Having been heavily influenced by his reading of recent French mathematics, he published the first explicit mention of the commutative and distributive laws in an English work in 1838.[10] The commutative law, which for addition may be stated as

$$a + b = b + a,$$

George Peacock (1791–1858).

and the distributive law,

$$a(b + c) = ab + ac,$$

are simple rules from everyday arithmetic, which at that time were believed to be universal truths in algebra too. Indeed, to those today who have studied algebra only at school, the idea that these rules may not always hold may still seem vaguely unnerving! However, by arguing that such laws 'may be said to exist by convention only',[11] Peacock and Gregory questioned this universality for the first time, raising the possibility of constructing new algebras that are independent of the usual rules of arithmetic.

The first such algebra was created by the Irish mathematician, Sir William Rowan Hamilton in 1843: his creation (or discovery) of quaternions was rightly regarded as ground-breaking. Here, for the first time, was an algebraic system in which, although $a + b$ is the same as $b + a$, the quantity $a \times b$ does not necessarily equal $b \times a$. It was thus the first fully consistent algebraic system to break one of the previously inviolable laws of arithmetic: commutativity.

It also enlarged the world's mathematical vocabulary: the terms *quaternion*, *scalar*, and *vector* (in its modern sense) were all coined by Hamilton in 1843. Whereas normal complex numbers, $a + bi$, were composed of a real and imaginary part, quaternions were four-part complex numbers of the form

$$a + bi + cj + dk,$$

where $a$ is the real, or 'scalar', quantity and the $i$, $j$, and $k$ components form the imaginary part. Hamilton called this latter quantity the 'vector' part, with the three constituents representing three perpendicular lines; this explains why, to this day, when studying vectors in three dimensions we label their components $i$, $j$, and $k$. Hamilton and his like-minded mathematical followers worked vigorously to apply quaternions to geometrical and physical problems; indeed, when James Clerk Maxwell published his revolutionary *Treatise on Electricity and Magnetism* in 1873, the now-famous 'Maxwell's equations' were stated there in terms of quaternions. But it was not long before they received their current form as vector equations. Thus, a consequence of Hamilton's quaternions was the development throughout the Victorian era (mainly by British applied mathematicians) of the algebra and calculus of vectors, essential to modern-day mathematical physics.[12]

William Rowan Hamilton (1805–65).

With the law of commutativity breached, it was not long before other algebras were developed, and once again British mathematicians were in the vanguard. Within months of Hamilton's discovery of quaternions, his friend John Graves and the Englishman Arthur Cayley had independently created a consistent algebra of complex numbers with eight components, one real and seven imaginary, called *octonions*. These numbers turned out to be even more unusual because, in addition to violating the commutative law of multiplication, they also break the associative law: for octonions, we have

$$(a \times b) \times c \neq a \times (b \times c).$$

In 1878 this idea was generalized still further by William Kingdon Clifford to comprise systems with $2^n$ components, an area now known as *Clifford algebras*.[13]

Another very different form of algebra invented in Victorian Britain, and one with which many are familiar today, is that of *matrices*. The term 'matrix' had first been given in 1850 in an article by James Joseph Sylvester. But in 1858, in a landmark paper, Cayley used this definition as the basis of a brand new area of mathematics: the algebra of matrices.[14] Cayley quickly found that, just like quaternions, matrices do not always give the same results when multiplied together in different orders: matrix multiplication is also non-commutative. Over the next quarter-century, Cayley and Sylvester, together with other British mathematicians such as the Oxford professor Henry Smith, developed this new mathematical subject, helping to lay the foundations for much of what we today call *linear algebra*.

Cayley and Sylvester were also largely responsible for the creation of a totally new area of algebra, which quickly became almost an obsession for several British Victorian mathematicians. The subject of *invariant theory* involved, among other things, the search for algebraic expressions known as *invariants* (a word coined by Sylvester in 1851), whose basic form is preserved when they undergo a particular kind of transformation. These searches became increasingly complicated and time-consuming as higher-degree expressions were investigated. This time, although the British could be considered the originators of the theory, it was the Germans who produced the deeper results. Their more theoretical and abstract approach contrasted sharply with the British reliance on computation.[15]

Indeed, Victorian British mathematics was often characterized by a far greater concern for computational techniques and mastery of symbolic manipulation, than for rigorous proof. For example, in his inaugural paper on matrices of 1858, Cayley proved a fundamental result (now known as the *Cayley–Hamilton theorem*) for $2 \times 2$ matrices, noting that he had verified it for the $3 \times 3$ case, but saying that:[16]

I have not thought it necessary to undertake the labour of formal proof of the theorem in the general case of a matrix of any degree.

Statements like this, and others such as 'similar proofs may be given for equations of higher degree', occur regularly in British work of this time.

Meanwhile, at Oxford in the mid-1860s, a mathematics lecturer by the name of Charles Dodgson was researching an area very closely related to matrices, that of determinants. A *determinant* is a particular number or expression related to a matrix, from which important algebraic and geometrical characteristics can be deduced. For $2 \times 2$ matrices, determinants are very easy to find, but things quickly get more complicated, and by the time one gets to $5 \times 5$ matrices, the computations are very laborious indeed. But Dodgson created an ingenious method that only ever required the calculation of $2 \times 2$ determinants; this was a considerable achievement, especially considering that, at roughly the same time

Arthur Cayley (1821–95).

James Joseph Sylvester (1814–97).

under his pen-name Lewis Carroll, he had completed his first children's book, *Alice's Adventures in Wonderland*.[17]

Another area in which Dodgson worked was logic. Although logic was a subject with a long and varied history, it was a relative newcomer as a mathematical discipline. In fact, at the beginning of our period it was not even considered a part of mathematics at all: to study logic was to study a branch of philosophy. But it was not long before a disciplinary shift began. In 1839, stimulated by its use in teaching geometry from Euclid's *Elements*, the mathematician Augustus De Morgan began researching into the study of logic as a means of facilitating geometrical proof. This led him into full-blown publications in the 1840s, in which he developed a symbolic notation for making logical deductions. Unfortunately for De Morgan, this work on logic resulted in his being drawn into a dispute with the Scottish philosopher Sir William Hamilton, a firm believer in preserving the disciplinary boundaries between logic and mathematics.[18]

In 1847, Hamilton erroneously accused De Morgan of plagiarizing his ideas on quantification of the predicate, since both men had been working on (very different) modifications to quantified logical statements, such as 'all men are mortal' and 'no dogs are cats'. This charge turned out to be unfounded, but De Morgan's work marked the birth of algebraic logic, wherein logical premises are manipulated purely symbolically in order to derive consequent conclusions. A further consequence of this lengthy battle with Hamilton was that it prompted De Morgan's friend, the Lincoln-based mathematician George Boole, to enter the subject, resulting in the creation of what was to become *Boolean algebra*.[19]

Among those to take up the new algebraic approach to logic was the Cambridge academic John Venn in the 1880s. Venn's current fame rests principally on his introduction in 1880 of diagrammatic representations of logical relationships, his so-called *Venn diagrams*. Actually, the pictures we now refer to as 'Venn diagrams' might more accurately be called 'Euler circles', since they were first popularized by the Swiss mathematician Leonhard Euler in his *Letters to a German Princess* (1768), and even these first occurred in the work of Leibniz.[20] But Venn did something different: first he generalized Euler's system considerably, and secondly he based his logical diagrams firmly on the algebraic traditions initiated by De Morgan and Boole a generation before. He thus established a powerful and general method by means of which logical problems can be solved purely symbolically.[21]

Through the work of Venn and others, including Charles Dodgson, symbolic logic continued to develop and thrive until, by 1900, it had attracted the attention of perhaps its most famous British practitioner, Bertrand Russell. However, in contrast to preceding British logicians, the influences on Russell came almost exclusively from overseas—in particular the work of Georg Cantor and especially Giuseppe Peano and Gottlob Frege. Moreover, whereas Boole, De Morgan, and other algebraic logicians had based their logic on mathematics, Russell and his fellow logicians were trying to show that mathematics could be grounded on logic.[22] The culmination of their efforts came in the publication of Russell's monumental

A self-portrait by the Revd. Charles Dodgson: 'what I look like when lecturing'.

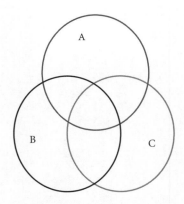

A 'Venn diagram' with three circles.

**∗110·632.** $\vdash : \mu \, \epsilon \, \mathrm{NC} \, . \supset . \, \mu +_c 1 = \hat{\xi} \, \{ (\exists y) \, . \, y \, \epsilon \, \xi \, . \, \xi - \iota\text{'}y \, \epsilon \, \mathrm{sm}\text{''}\mu \}$

> *Dem.*
>
> $\vdash . \, \ast 110 \cdot 631 \, . \, \ast 51 \cdot 211 \cdot 22 \, . \supset$
>
> $\vdash : \mathrm{Hp} \, . \supset . \, \mu +_c 1 = \hat{\xi} \, \{ (\exists \gamma, y) \, . \, \gamma \, \epsilon \, \mathrm{sm}\text{''}\mu \, . \, y \, \epsilon \, \xi \, . \, \gamma = \xi - \iota\text{'}y \}$
>
> $[\ast 13 \cdot 195] \qquad = \hat{\xi} \, \{ (\exists y) \, . \, y \, \epsilon \, \xi \, . \, \xi - \iota\text{'}y \, \epsilon \, \mathrm{sm}\text{''}\mu \} : \supset \vdash . \, \mathrm{Prop}$

**∗110·64.** $\vdash . \, 0 +_c 0 = 0 \qquad\qquad [\ast 110 \cdot 62]$

**∗110·641.** $\vdash . \, 1 +_c 0 = 0 +_c 1 = 1 \quad [\ast 110 \cdot 51 \cdot 61 \, . \, \ast 101 \cdot 2]$

**∗110·642.** $\vdash . \, 2 +_c 0 = 0 +_c 2 = 2 \quad [\ast 110 \cdot 51 \cdot 61 \, . \, \ast 101 \cdot 31]$

**∗110·643.** $\vdash . \, 1 +_c 1 = 2$

> *Dem.*
>
> $\vdash . \, \ast 110 \cdot 632 \, . \, \ast 101 \cdot 21 \cdot 28 \, . \supset$
>
> $\vdash . \, 1 +_c 1 = \hat{\xi} \{ (\exists y) \, . \, y \, \epsilon \, \xi \, . \, \xi - \iota\text{'}y \, \epsilon \, 1 \}$
>
> $[\ast 54 \cdot 3] \quad = 2 \, . \supset \vdash . \, \mathrm{Prop}$

The above proposition is occasionally useful. It is used at least three times, in ∗113·66 and ∗120·123·472.

Russell and Whitehead's proof that 1 + 1 = 2.

three-volume collaboration with his former Cambridge tutor Alfred North White-head, *Principia Mathematica* (1910–13), which today is perhaps most famous for taking several hundred pages to build up to a proof that $1 + 1 = 2$—a result that they described as being 'occasionally useful'![23]

Of course, the most well known application today of symbolic logic is computing, and one of the forefathers of today's programmable computer was that most famous of Victorian technological pioneers, Charles Babbage. He is often remembered for designing precursors to the modern computer that were never built.[24] The first, his Difference Engine no. 1, was designed in the 1820s, but it was his analytical engine of the 1830s that was particularly remarkable. Designed as a more general-purpose calculating machine, it also featured aspects of data storage and programmability.[25]

This feature was famously explored by Ada King, Countess of Lovelace, the only legitimate child of Lord Byron, and a long-time friend of Babbage. Gifted with an ability and intense interest in science and mathematics, she had been tutored privately by a number of family friends, including William Frend and Augustus De Morgan. In 1843, in a series of commentaries on Babbage's machine, she gave a table of execution for an algorithm that Babbage wrote to calculate the series of coefficients known as the Bernoulli numbers.[26] For this reason, she is often credited with being the world's first computer programmer, and it was in her honour that the programming language Ada was so named in 1979.

Augusta Ada King (Countess of Lovelace) (1815–52).

Babbage's Difference Engine No. 1, demonstration assembly, 1832.

Naturally, due to Babbage's failure to see any of his projects to completion, his and Lovelace's achievements in this area of applied mathematics can only be appreciated in retrospect. But quite the opposite is true of many other areas of mathematical applications. Indeed, if there was one area of Victorian mathematics of which the British could be uniformly proud, it was mathematical physics.

In this field the Scotsman William Thomson—better known today as Lord Kelvin—acquired a high reputation, both at home and overseas, while still fresh out of Cambridge. Well known and respected abroad, he published widely in many areas of applied mathematics. His work on thermodynamics led him to propose the well-known absolute scale of temperature in 1848, and his *Treatise on Natural Philosophy*, co-authored with P. G. Tait, was one of the most influential textbooks on physics of the entire Victorian era.[27] Most significantly, with respect to furthering the work of British mathematicians, he famously promoted the hitherto obscure *Essay on the Application of Mathematical Analysis to Electricity and Magnetism* by the recently deceased George Green.[28]

William Thomson—a sketch at age 16 by his sister Elizabeth.

Green's windmill in Sneinton, Nottinghamshire.

James Clerk Maxwell (1831–79).

Green had spent most of his life working in the family mill just outside Nottingham.[29] A self-taught mathematician, he published his *Essay* at his own expense in 1828. The work was remarkable—introducing the concept of *potential*, as now used in physics, and containing an important result now known as *Green's theorem*—but initially went unnoticed by the mathematical community at large. Nevertheless, Green was encouraged to develop his scientific credentials, and in 1833 he entered Cambridge as a mature student, graduating in 1837, and continued to publish papers on mathematical physics. Sadly, his health deteriorated and he died in 1841 at the age of just 47. But his mathematics was rescued from obscurity by Thomson; indeed, the fact that we call his key result *Green's theorem* is due to Thomson's tireless promotion of the *Essay*, particularly in France, where papers arising from it appeared in *Liouville's Journal* in the 1840s, and Germany, where Thomson arranged for the whole work to be reprinted with an introduction in *Crelle's Journal* in the 1850s.[30]

Thomson was also responsible for an important generalization of Green's theorem, which first appeared in a letter from him to fellow applied mathematician George Stokes on 2 July 1850. Three and a half years later, in January 1854, Stokes set the result (now named after him) as a question in a Cambridge University prize exam paper.[31] The student who won the prize that year was a certain James Clerk Maxwell, so we may infer that he was presumably one of the first to prove Stokes' theorem! This result later appeared in the opening chapter of Maxwell's ground-breaking *Treatise on Electricity and Magnetism*, which revolutionized the subject and featured the first appearance of Maxwell's equations (albeit in a much more involved quaternion form, as mentioned above).

---

8. If $X$, $Y$, $Z$ be functions of the rectangular co-ordinates $x$, $y$, $z$, $dS$ an element of any limited surface, $l$, $m$, $n$ the cosines of the inclinations of the normal at $dS$ to the axes, $ds$ an element of the bounding line, shew that

$$\iint \left\{ l\left(\frac{dZ}{dy} - \frac{dY}{dz}\right) + m\left(\frac{dX}{dz} - \frac{dZ}{dx}\right) + n\left(\frac{dY}{dx} - \frac{dX}{dy}\right) \right\} dS$$
$$= \int \left( X\frac{dx}{ds} + Y\frac{dy}{ds} + Z\frac{dz}{ds} \right) ds,$$

the differential coefficients of $X$, $Y$, $Z$ being partial, and the single integral being taken all round the perimeter of the surface.

9. Explain the geometrical relation between the curves, referred to the rectangular co-ordinates $x$, $y$, $z$, whose differential equations are

$$\frac{dx}{P} = \frac{dy}{Q} = \frac{dz}{R},$$

and the family of surfaces represented by the partial differential equation

$$P\frac{dz}{dx} + Q\frac{dz}{dy} = R.$$

Stokes's theorem: Question 8 in the 1854 Smith's prize examination paper.

These twenty equations were later simplified down to four in their vector form, by Oliver Heaviside in the 1880s.[32]

Among several other significant British Victorian contributors to mathematical physics, we mention John Strutt, the 3rd Lord Rayleigh, who contributed to wave theory, acoustics, and the theory of gases. Horace Lamb made important contributions to fluid dynamics, and Joseph Larmor worked on electricity, dynamics, and thermodynamics, being instrumental in restructuring the subject of electromagnetism around the theory of the electron. The first physicist to use what are now known as 'Lorentz transformations', Larmor's research was backed up by J. J. Thomson's experimental identification of the electron in 1897.[33] Valuable contributions by applied mathematicians such as these and others made mathematical physics the indisputable forte of British Victorian mathematics.

But the Victorians did not just apply mathematics to physics. Mathematical methods were also introduced into a variety of areas in the social sciences, medicine, and public health. One of the leaders in this field was one of the most famous of all Victorians, Florence Nightingale. Although best remembered as a humanitarian and founder of the modern nursing profession, Nightingale also pioneered the introduction of statistical methods into medicine, following her studies of the effects of disease on soldiers in the Crimean War published in 1858.[34] In this work, she popularized the then novel use of diagrams to represent statistical information, particularly by means of her *polar area graph*, similar to the modern pie chart. Her graphical representations were among the first visual presentations of statistics to be accessible to the general reader, and highlighted the extent of the unnecessary deaths during the war most effectively. Moreover, their user-friendly format served as an effective tool to persuade parliament and the medical profession of the advisability of sanitation reforms.

Oliver Heaviside (1850–1925).

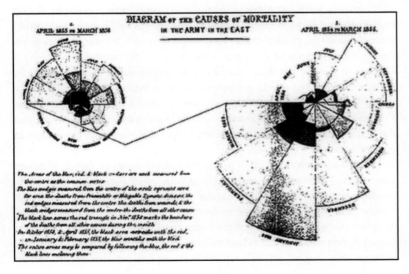

Florence Nightingale's polar area graph.

Karl Pearson (1857–1936).

Many of Nightingale's ideas were championed by Francis Galton, who also tried to mathematize Darwin's theory of evolution by looking at the inheritance of variation,[35] originating the statistical study of regression. The related correlation coefficient was defined soon afterwards by Karl Pearson, who studied how to analyse relationships between several variables.[36] Pearson introduced several now-commonplace statistical tools. One of these was the *histogram*, a diagram similar to a bar chart, but which represents a set of continuous, rather than discrete, data. For this reason, Pearson explained that it could be employed as a tool in the study of history, for example to chart historical time periods, and coined the name 'histogram' in 1891 to convey its use as a 'historical diagram'. In the following year, he devised a sophisticated new statistical function to measure the spread of data sets around their mean. This function, which he called the *standard deviation*, is now a fundamental concept in mathematical statistics. One of Pearson's most celebrated contributions to the subject came at the very end of the Victorian period. His *chi-square test* for goodness of fit between observed data and values predicted from a hypothesis was published in 1900. It is now a standard and widely used method in modern statistics.[37]

For over fifty years from 1884, Pearson worked at University College London, initially as its professor of applied mathematics. But in 1911, he founded a new department of applied statistics—the first of its kind in Britain. It was there that Pearson, and the school of students that grew up around him, went on to lay the groundwork that was in many ways responsible for the creation and establishment of the theory of statistics as a mathematical discipline in the early decades of the 20th century.[38]

## Conclusion

This overview gives just a flavour of the many contributions made by British mathematicians between 1837 and 1901. Of course, it is by no means exhaustive: we could go on to mention British contributions to further areas, such as graph theory, combinatorics, mathematical economics, and astronomy, to name but a few, but in view of the many mathematical developments that took place in Victorian Britain, one should not be tempted into a Whiggish historical perspective of 'onwards and upwards'. Indeed, when studying the mathematics of the Victorian era, we should be aware of certain peculiarities and differences that distinguished British mathematical practice from that elsewhere, particularly in mainland Europe.

The first concerns the lack of a genuine research ethos at British Victorian universities. This contrasts considerably with the situation in Germany, where mentors such as Klein in Göttingen and Weierstrass in Berlin actively created an environment to encourage their students to emulate their research interests. Compare this with Britain, where the leading mathematicians at this time were solitary figures who did little to lead students into research. Among students and lecturers alike, original research was never an explicit requirement, or even an implicit expectation, at Cambridge or anywhere else in Victorian academia. A true research ethos came into being in British universities only around the First World War.

That said, Britain was essentially the first to establish what became a national learned society devoted purely to research in mathematics.[39] Founded in 1865, the London Mathematical Society (LMS) quickly developed from a local body to the *de facto* national society for British mathematicians.[40] The fact that France followed suit from Britain eight years later was seen by some French savants (such as the geometer Michel Chasles) as further evidence of the decline in their mathematical fortunes.[41] But the LMS also provided a role model for up-and-coming mathematical nations, such as the United States, where the New York Mathematical Society was founded in emulation of its British precursor in 1888, evolving six years later into the American Mathematical Society.[42] But Britain did not just lead the way for emerging mathematical powers in this regard: even Germany—perhaps the world's leading mathematical nation by the end of the 19th century—had no national mathematical society until 1890.[43]

A second difference concerns the range of mathematical subjects studied by the British at this time. Although the interests of Victorian British mathematicians were less specialized than today, there were very few 'universal mathematicians' in the sense of a Gauss, Cauchy, Riemann, Weierstrass, or Poincaré. Although this is due in part to the massive expansion of the subject during this time period, some people, like William Rowan Hamilton and William Kingdon Clifford, were equally comfortable in both pure and applied mathematics. But Cayley was probably the only true British mathematics 'all-rounder' of the period.

Even then there were several areas of the subject missing or largely absent from British mathematics. For example, despite major British contributions to mathematical statistics, there was little original work in probability. Few Victorian British mathematicians were interested in group theory, with one of its (few) practitioners noting that the subject had 'failed, so far, to arouse the interest of any but a very small number of English mathematicians'.[44] There was also very little number theory. However, Henry Smith's monumental *Report on the Theory of Numbers* from the 1860s was highly regarded by European mathematicians—so much so that three decades later, when David Hilbert and Hermann Minkowski were asked to prepare a similar report for the German Mathematical Society, one of the reasons Minkowski gave for his failure to produce his half was the comprehensiveness of Smith's *Report*.[45]

There was also little in the way of analysis (real or complex)—this is ironic, since this area is now such a staple of British university mathematics courses. The change came with the publication of Andrew Forsyth's *Theory of Functions of a Complex Variable* in 1893, which, according to his former student E. T. Whittaker, 'had a greater influence on British mathematics than any work since Newton's *Principia*'.[46] However, the book was not highly regarded outside Britain and in the opening years of the 20th century its standard of rigour was quickly surpassed by younger up-and-coming British mathematicians, particularly Whittaker, W. H. Young, and (most famously) G. H. Hardy and J. E. Littlewood. Together with the reform of the Cambridge mathematics syllabus

The London Mathematical Society's De Morgan medal.

Queen Victoria at the end of her reign.

and its exam system in 1909, these works marked a turning point in the way that pure mathematics was taught and researched in Britain: from then on, analysis would be a fundamental component.

But the very absence of certain mathematical areas gives an indication of the distinctiveness of British mathematics at this time, for its character and style were very different from those of its Continental counterparts. Indeed, there were periods when the mathematical work that was being done was virtually unique to Britain. Consider, for example, the creation of algebraic logic during the 1840s and 1850s: virtually the only people involved (or interested) in this area at that time were George Boole and Augustus De Morgan.[47] Consider also the development of vector algebra in the 19th century: with the principal exceptions of Hermann Grassmann in Germany and Josiah Willard Gibbs in America, a substantial proportion of this work was carried out by British mathematicians.[48] Similarly in the last decades of the period, while scholars such as Galton and Pearson were laying the foundations of modern mathematical statistics, the only comparable work that was happening elsewhere was around Chebyshev in St. Petersburg.[49] Thus, while aware of mathematics from overseas, the British retained their own characteristic style and individualism.

By 1901, British mathematics was certainly less insular than it had been in 1837, but it still had some way to go before becoming fully international. Even though work by British mathematicians was known and appreciated abroad, there were some surprising exceptions. For example, the French organizers of the 1882 prize competition of the Académie des Sciences were embarrassed to learn that the solution to the problem they set had actually been published in Britain by Henry Smith fifteen years earlier, resulting in prizes eventually being awarded to Smith and the young Hermann Minkowski. Interestingly, when Hilbert was preparing his famous address on open mathematical problems for the Second International Congress of Mathematicians (ICM) in Paris in 1900, Minkowski advised him to consult Henry Smith's valedictory presidential address to the London Mathematical Society in which Smith had set out a similar list of topics that he felt warranted further study.[50]

Indeed, the history of the early International Congresses shows how relatively quickly British mathematicians began to participate at the international level after 1900. At the first International Congress of Mathematicians, held in Zürich in 1897, only three British delegates attended and none presented papers. Yet in 1912, just over a decade after the close of the Victorian era, the fifth International Congress of Mathematicians was held at Cambridge, perhaps the surest sign of the growing maturity of British mathematicians in the international arena. As the Oxford algebraist Edwin Elliott observed:[51]

There was a time not long ago when British Mathematicians may have been thought too self-centred. If the judgment were ever correct, it is no longer. We are alive to what is being done elsewhere, and now aim at cooperation.

In fact, the Victorian age ended just as another renaissance of British (pure) mathematics was on the horizon. The real coming of age of British mathematics happened around the time of the First World War. These were the years that saw the reform of the mathematical syllabus at Cambridge, the foundation of the country's first statistics department by Pearson at University College London, the publication of Russell and Whitehead's *Principia Mathematica*, the beginning of the Hardy–Littlewood collaboration, and the ICM meeting at Cambridge. These events would have profound and long-lasting consequences for the future of British mathematics. But they did not come from nowhere: the foundations for all of them were laid in the Victorian period. The chapters that follow will focus in greater detail on some of these developments and the environments that produced them.

The Hall at Trinity College, Cambridge, 1838.

## CHAPTER I

# Cambridge

*The rise and fall of the mathematical tripos*

TONY CRILLY

When Queen Victoria came to the throne in 1837, virtually all Cambridge students were obliged to receive a mathematically based education. The students who studied the famed honours degree, the mathematical tripos, faced a challenging course. Even the ordinary degree students received a basic training in the subject, as mathematics was viewed as the key to all further study. During the century, the educational fare at Cambridge became more diverse and later students were freed from this obligation. The traditions of the mathematical tripos came under increasing scrutiny, and by 1901 it was on the brink of a major reform that would leave it barely recognizable from its former self.

On 20 June 1837 King William IV died and the young Victoria became Queen. As a war leader the Duke of Wellington was still a living hero, but the Battle of Waterloo was becoming a historical artefact. It was now permissible to make contact with the mainland of Europe without being admonished for consorting with the enemy.

At home, industrialization gathered pace. In the same year that Victoria began her long reign, Charles Wheatstone and William Cooke took out patents on their electric telegraph, and Isambard Kingdom Brunel's steam-driven *Great Western* crossed the Atlantic in fifteen days. The Great Western Railway's first passengers left Paddington for Bristol Temple Meads, and 'God's Wonderful Railway', the great GWR, joined the capital of the West Country to the London metropolis. Although provincial Bristol was eleven minutes behind London, the railway moved the clocks into line. Cambridge managed to remain isolated for a little longer, but

View of Cambridge from
Castle Hill, 1841.

The Market Place in early
Victorian Cambridge.

after a town-and-gown debate of some length, the railways came to Cambridge in 1845.

At the beginning of the Victorian period, the market town of Cambridge with its ancient university was the most important place for mathematics in the entire country—the 'holy city of mathematics'. There were two reasons for this. Cambridge University housed the famed mathematical tripos as the mainstream course of study for its students. This was the jewel in the University's crown, and in a religion-dominated society it was likened by many to the Ark of the Covenant. Alongside this venerated institution a new generation of mathematicians was learning to do mathematical research and transmit its work internationally. Indeed, the position of Cambridge as the embryonic national centre for mathematical research was the second reason that placed Cambridge at the fulcrum of mathematics in Victorian Britain.

## Old Cambridge reforming: the 1840s

Since the Reformation, one of the principal functions of Cambridge University had been to fill out the ranks of the Anglican Church. But by the 1840s, the choices of future career for its graduates had widened. If students discovered that the Church was not for them, they could become lawyers, schoolmasters, engineers, or men of business. Mathematics had already assumed a central place in the curriculum when the mathematical tripos was instituted in the middle of the 18th century. The mathematics of the tripos was held to be the basis of all study, and it was widely believed that it would hold a man in good stead whatever his future calling. It is a sobering thought that a large proportion of the clergymen, lawyers, and schoolmasters of Victorian England were well grounded in mathematics. Women did not feature in these plans, coming onto the Cambridge scene only in the 1880s.

At the beginning of the Victorian period, the examination for the mathematical tripos was still *the* 'Senate House examination'.[1] Mathematics at Cambridge was regarded as the basis for a 'liberal education'. It was a sort of pre-knowledge and it was held to be important to teach it to the young. Taught too late and it would be as useless as trying to teach 'the violin to a grown man'. The knowledge of mathematics was not claimed to be useful in itself (except for those destined to become tutors or schoolteachers), but it was believed that the study of mathematics would develop and strengthen the faculties of the mind, and *after* the completion of this study one could go on to other fields and be more effective in them. In short, mathematics gave the 'art of acquiring all arts', and like physical games that prepared the body, mathematics toned the intellectual muscle.

This mathematics-based liberal education was aimed at the middle and upper classes, and for some it was a bitter pill to swallow. But swallow it they must, for students of the 1840s were required to acquit themselves in the mathematical tripos before sitting the examinations for the classics degree, a newcomer course of study established only in 1824. Many a fine classicist was stymied by this requirement; for example, three out of four Kennedy brothers starred as the 'senior classic' (the first in the entire class of classics students), but W. J. Kennedy (the youngest brother) was 'gulfed' in the mathematical tripos examination and so debarred from competing for the classics degree. There were few exceptions to this rule, but in an unreformed Cambridge noblemen were exempted by virtue of their aristocratic birth.

While generations of students passed through the Cambridge system, the mathematical tripos and its examination evolved over the course of Victoria's reign. The tripos examination sat by Arthur Cayley in the 1840s was not the tripos sat by James Clerk Maxwell in the 1850s. This in turn differed from the one sat by

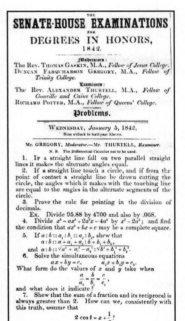

Senate House examination paper, 1842.

Karl Pearson in the 1870s, and still more from that of Bertrand Russell in the 1890s which was barely recognizable as the same exam of half-a-century earlier. In the 1840s the mathematical tripos was a wide-ranging course covering most aspects of mathematics in some depth, but by the time Russell sat down in the Senate House in 1894 it had become specialized, and the products of the tripos at that time were constricted by a knowledge in a branch of the subject that was acquired towards the end of their studies.

Some institutional practices were central to the mathematical tripos, and though they were vigorously debated at various times they were resistant to change. In this category were:

- The once-and-for-all nature of the mathematical tripos final examination.
- The order of merit—the listing of students according to their marks obtained in the final examination; this was divided into three distinct classes: wranglers, senior optimes, and junior optimes (corresponding to first, second, and third class honours in later British honours degrees).
- The acknowledgement of the 'senior wrangler'—the top student in the order of merit.

In January 1837, at the top of the list were William Griffin (senior wrangler), J. J. Sylvester (2nd wrangler), Edward Brummell (3rd wrangler), George Green (4th wrangler), Duncan Gregory (5th wrangler) and A. J. Ellis (6th wrangler). This list is typical of those that followed on throughout the Victorian era, where important mathematicians and unknowns appear side by side. The list is unimportant as a mathematical guide, but it was important socially. To be a 'wrangler' in the top class of the order of merit was a distinction that continued for life. If a student was the '14th wrangler' (say), then this would be as widely known as if it were branded across his forehead.

In the 1837 list we have a highly creative mathematician like J. J. Sylvester 'beaten' by someone unknown to the mathematical community. The order of merit—the grandfather of all league tables—did not indicate any research potential, and it was not supposed to. The mathematical tripos examination was primarily a contest for clever schoolboys who could jump through hoops at speed.

At the pinnacle of the order of merit was the champion student, the senior wrangler, a young man who held a fascination in the Victorian mind. He signified all that was good about the mathematical tripos and acted as a focus for the whole system. One contemporary remarked:[2]

In my opinion it is this continuance of solving problems, this general course of not only acquiring principles but applying them, that at last makes the senior wrangler, who perhaps at the time is one of the most expert mathematicians in existence.

The senior wrangler was an early example of a 'celebrity' and became someone believed to have superior powers quite beyond the competence tested in the mathematical examination. Occasionally the system did identify future mathematicians and scientists. In the early 1840s the senior wranglers were R. L. Ellis (1840), George Gabriel Stokes (1841), Arthur Cayley (1842), and John Couch Adams (1843), a group who achieved fame in mathematical subjects, and who were still talked about by the end of the century as that 'illustrious quartet'. They were unusual, and a present-day mathematician would find it difficult to name other senior wranglers of the period. Ellis detested the Cambridge system, Cayley seemed unaffected by it, while Adams was caught up in the excitement of it all and thought that he had little chance of the top spot.

One thing was clear. The serious students coming up to Cambridge were rapidly moved into examination mode. There were examinations at every turn—on arrival and at the annual college examinations—and the whole procession culminated in the university-wide mathematical tripos examination at the end of ten terms, an examination held in icy rooms in the January of each year. The skills of solving problems and working quickly under pressure were all part of the Cambridge package for its students. If they succeeded, there would be a week-long set of examinations for the two Smith's prizes, and then there were the College fellowships to strive for. A fellowship gained a person status, income, and freedom from earning a living for at least seven years. If they were at Trinity College, the high flyers with a fellowship in prospect would face another batch of examinations nine months after the exertions of the mathematical tripos and the Smith's prize contest.

The tripos examinations of the 1840s spread over six days covering equal proportions of 'pure' and 'applied' questions (then called 'mixed mathematics'). The exam schedule was typically from Wednesday to Saturday with Sunday off, followed by Monday and Tuesday—five and a half hours each day, but six hours on Saturday. The examination papers contained questions on such subjects as astronomy, algebraic topics, the theory of integration, differential equations, mechanics, and the application of mathematics to such questions as the shape of the rotating earth. The examinations were of varying length, from nine to twenty-four questions, and in all, one hundred and seventy-seven questions were put before the students (see Box 1.1).

What is not conveyed by citing individual questions is the time dimension. To answer *all* questions completely on a paper would require superhuman qualities. This would not be expected, but the subtext generated by the competition was still 'speed'. While the mathematical tripos of the 19th century was described as a 'great writing race' by Augustus De Morgan, it was also remarkable in its wide coverage of mathematical topics.

To make success a reality, the private mathematical coaches came into their own by supplying tuition and examination tips. They stressed the practical acquisition of examination technique geared towards examination performance and, in that sole aim, they inculcated hard work. One of the last students to experience this regime wrote:[3]

not a day, not an hour was wasted; the perfect candidate should be able to write the bookwork automatically while his thoughts were busy with the rider, and the

# Box 1.1: Some Cambridge examination questions

The questions at the start of examination week for honours were fairly straightforward. The first paper carried the rubric 'The differential calculus not to be used' and consisted of sixteen geometrical and algebraic questions of varying length, with the candidates being invited to answer as many or few questions as they wished. A quarter of the questions on the examination paper were allocated to 'problems', with the bulk of questions of bookwork type. One problem question was to:

> Shew that the sum of a fraction and its reciprocal is always greater than 2.
> How can we, consistent with this truth, assume that $2 \cos \theta = x + 1/x$?

A question that might have caused many to scratch their heads was a bookwork question once set by Duncan Gregory on the final day. Many of the questions on these examination papers were written in a prose style:

> If a rigid body be struck by any couple, the axis of the couple can never be the corresponding instantaneous axis of rotation, unless it coincide with a principal axis of the body. Shew also that, when this condition is fulfilled, the angular velocity round the principal axis is equal to the moment of the couple divided by the moment of inertia round the axis.

At the end of this paper, the end of the examination period, was a long question on the 'shape of the earth' to wish them on their way.

The level of the honours examination lay somewhere between the standard expected of the ordinary (poll) degree student and the Smith's prize candidate. The poll student might be asked practical questions in the mathematics section such as:

> Find the sum of $3\frac{5}{12}$, $7\frac{1}{2}$, $8\frac{2}{3}$, 4, and $2\frac{1}{4}$.

The questions set for the Smith's prize examination, a competition for the top students, were technically demanding. The opportunity to set them was an occasion for the professors to show off their erudition, and the questions demanded both bookwork knowledge and technical skill. They were challenging, even allowing for students being coached to answer them:

> The problem: 'To find the path of a body, acted on by gravity and moving in a medium of which the resistance is as the *2n* power of the velocity': was solved by Bernoulli as follows:
> Assumptâ indeterminatâ *z*, construatur area $\int (aa + zz)^{n-1/2} dz$ quae vocetur *Z*; sint autem co-ordinatæ curvæ quæsitæ *x* et *y*. Fiat $x = \int (zdz. Z^{1/4})$ et $y = \int (adz. Z^{1/4})$. Dico curvam quæ inde oritur esse quæsitam. Prove this construction.

fingers could be trained even when the brain was weary; above all, curiosity about unscheduled mathematics was depravity.

The Victorian era was spanned by the two great exemplars of the coaching trade, William Hopkins and Edward John Routh (see Box 1.2).

William Hopkins (1793–1866), early Victorian mathematical coach.

## Box 1.2: Students of the coaches William Hopkins and E. J. Routh

**HOPKINS' PUPILS INCLUDED:**

J. H. Pratt, P. Kelland, A. Smith, J. J. Sylvester, M. O'Brien, R. L. Ellis, G. G. Stokes, A. Cayley, F. Fuller, F. Galton, W. Thomson (Lord Kelvin), W. F. L. Fischer, H. Blackburn, I. Todhunter, N. M. Ferrers, P. G. Tait, W. J. Steele, E. J. Routh, and J. C. Maxwell.

**ROUTH'S PUPILS INCLUDED:**

J. W. Strutt (Lord Rayleigh), W. D. Niven, J. Stuart, C. Niven, W. K. Clifford, G. H. Darwin, J. W. L. Glaisher, H. Lamb, W. W. Rouse Ball, G. Chrystal, J. H. Poynting, K. Pearson, J. Larmor, J. J. Thomson, A. R. Forsyth, H. F. Baker, and W. McF. Orr.

Hopkins took a paternal interest in his pupils and achieved success after success during the early part of the Victorian period. For a handful of students the mathematical tripos held no terrors, and under the watchful eye of a Hopkins they succeeded and were enthused by mathematics. But, in taking thoroughness and efficiency to new levels, it was Hopkins' pupil E. J. Routh who dominated the coaching scene in the latter half of the century. The successful coaches taught their students by giving lectures to them in their rooms, marking their work, and above all focusing on those topics likely to appear in the examination. Routh knew what would 'pay' and what would not, and the best students opted to be under his guidance. In effect, the private coaches ran private colleges within the university.

George Peacock criticized the 'unhappy system' of private tuition. He challenged the notion that mathematics was good medicine for all students and the rule that students had first to acquit themselves in the mathematical tripos examination before taking the classics degree. He saw all too plainly that the mathematical tripos was crammed full of subjects that resulted in an indigestible course of study, and a reduction in the previous wide coverage was proposed and accepted. His contemporary William Whewell argued strongly for student attendance at the lectures given by the professors who, he observed, had little input in the workings of the mathematical tripos, but otherwise he opposed changes to it. In 1848, the Board of Mathematical Studies was set up and reforms made, but Peacock could do little

E. J. Routh (1831–1907), later Victorian mathematical coach.

about the issue of private coaching.[4] It was recommended that the mathematical tripos examinations be in two parts, thus reducing the pressure brought about by a continuous battery of examinations one after the other—now there would be a few days between the two parts.

New 'technologies' came to the aid of the student. The Macmillan book empire, set up originally in Cambridge, was in its ascendancy in the 1840s, and expanded from one of pure bookselling to one of publishing as well. One of their first publishing ventures in mathematics was George Boole's *Mathematical Analysis of Logic* (1847), but they made most money from textbooks, and one of the first of these, Isaac Todhunter's *Differential and Integral Calculus* (1852) was a bestseller. Publishing textbooks became easier and more available to the student population. Todhunter exceeded all expectations with his edition of Euclid which sold over half a million copies.[5]

The requirements for a tripos student were really quite basic, as can be judged from Lord Rayleigh's student booklist of the 1860s: Todhunter's textbooks on differential calculus, integral calculus, algebra, conic sections, trigonometry, and geometrical conics. Also included were Ferrers' *Trilinear Coordinates*, Drew's *Analytical Statics* and *Dynamics of a Particle*, Herschel's *Outlines of Astronomy*, Salmon's *Conic Sections*, and Routh's *Rigid Dynamics*.[6]

The early Victorian period did see some collective movement in the direction of research and scholarship in mathematics. One outcome was the founding of *The Cambridge Mathematical Journal* in 1837 and its successor, *The Cambridge and Dublin Mathematical Journal*, in 1846. As described in Chapter 7, these gave the young mathematicians an opportunity to see their work in print and get them used to publication.

While the Cambridge journals had an international dimension and enjoyed the support of a few continental mathematicians, they also brought together students and fellows from the different colleges of Cambridge. In the 1840s teaching was college-based and a man had no necessity to mix with students from other colleges. The Cambridge journals performed the useful function of removing this insularity and, when *Dublin* was added to the title, of enlarging the research base in Britain. In 1845 the British Association for the Advancement of Science came to Cambridge, and Peacock handed over the reins to the incoming president, Sir John Herschel, who praised the 'green shoots' of mathematical research that Britain was beginning to produce.

The importance of mathematics was brought home to the Victorian public in the 1840s by the discovery of the planet Neptune. This scientific triumph was claimed by both England and France and was given wide coverage in the media, but it was the method of discovery that brought mathematics to centre stage. The recently graduated senior wrangler John Couch Adams discovered the planet not by searching for it in the night sky, but by mathematics. Using Newton's theory of gravitation, he discovered it at the 'tip of his pen', and it was this power of mathematics that was proclaimed in the newspapers and pulpits around the country.

John Couch Adams (1819–92), Lowndean professor.

# The settled Victorians: the 1850s and 1860s

In 1850, a Royal Commission was appointed to look into the workings of both Oxford and Cambridge Universities. The resulting Cambridge University Act (1856) gave a new impetus to the creation of the University as something more than a collection of autonomous colleges. A new form of governance was given to the University and the powers of the individual colleges were reduced. The need to declare allegiance to the Church of England for admission to a degree was abolished for all degrees except divinity. New chairs were created, such as the one gained by converting the previously existing Sadleirian college lecturerships in algebra into the Sadleirian chair of pure mathematics.

The road to research proved bumpy. Through financial problems, *The Cambridge and Dublin Mathematical Journal*, which William Thomson had launched with such brio in 1846, collapsed in 1854. It was rebranded in the following year as *The Quarterly Journal of Pure and Applied Mathematics* (see Chapter 7). After a rocky start, when it was even doubtful if this would continue, it made a long run until 1927, for many years under the editorship of the Cambridge don J. W. L. Glaisher. During its first years it provided a vital link in the establishment of a research ethos in British mathematics, but this ethos was fragile. W. P. Turnbull, who gained a Trinity College fellowship in 1865, published a text on analytical geometry in 1867, but said that the 'book was produced at a time when abstract thought was rather at a discount, for physical research was in the ascendant'.[7]

It was the mathematical tripos that mattered most at Cambridge, and not least the order of merit and the senior wranglership: it was a Cambridge affair that in hindsight now seems somewhat parochial. Some admired the stability of the system, as did the theologian F. J. A. Hort in the 1850s, who bemoaned that there had been no Trinity College senior wrangler in eleven years: 'I feel a proper pride in the mathematical tripos and senior wranglership as great existing institutions'.[8] Hort was a man who sat more than his fair share of Cambridge examinations and held the system in great respect. But what are we to make of the rebellious Leslie Stephen, a mathematics tutor in the 1850s who ridiculed the mathematical tripos and wrote, 'To this day I do not realise—though on purely intellectual grounds I accept—the fact that even a senior Wrangler is made of flesh and blood'?[9] In the Cambridge *Student's Guide* of 1863, J. R. Seeley summed up its position:[10]

The Mathematical Examination of Cambridge is widely celebrated, and has given to this University its character of the Mathematical University *par excellence*.

The 'reading men' who took the honours degree represented about one-fifth of each year's intake of students. The ordinary or 'poll' degree students were required to study mathematics as well, but their standard of mathematics was undoubtedly low. This was a reflection on the public schools, where mathematics

J. W. L. Glaisher (1848–1928), editor of *The Quarterly Journal* and *The Messenger of Mathematics*.

occupied a very small part of the curriculum and where the emphasis was on the teaching of classics. One veteran mathematics tutor at Cambridge reported at the beginning of the 1860s on 'Euclid got by heart and not understood; arithmetic worked by rule of thumb, without any understanding of the simplest first principles; algebra, a chaos of confusion'.[11] As to the working of the system, the student culture offered another viewpoint. It was said about poll examinations that:

the most common method [of cheating] of all was to take in many of the most likely answers, especially propositions in mechanics, Hydrostatics, in Euclid and the like, ready written out, and to produce them from the pocket…Twenty examiners, all parading the Senate House, and all looking sharp too, would not be too many to detect the numerous tricks implied to hoodwink an examiner.[12]

Notwithstanding this, the Cambridge system of education was held in the highest regard, especially as many products of it were included in the 'Upper Ten Thousand' who ran the country. While Lord Palmerston naturally preferred patronage to examinations as a basis for sound appointments, he did acknowledge the value of mathematics, and especially Euclid, which he thought excellent training for a diplomat.

William Everett, an American student who spent three years at Cambridge in the 1860s, noted the characteristics of the mathematical education he received and its continued reliance on Newton and Euclid:[13]

Englishmen hate going back to first principles, and mathematics allows them to accept a few axiomatic statements laid down by their two gods, Euclid and Newton, and then go on and on, very seldom reverting to them. This system of mathematics developed in England, is exceedingly different from that either of the Germans or the French, and though at different times it has borrowed much from both these countries, it has redistilled it through its own alembic, till it is all English of the English.

Whewell's idea of bringing the professors' lectures into contact with undergraduates went unfulfilled, except for ordinary degree students who for a term attended a professor's lectures by compulsion. Professors Cayley and Adams regularly gave their erudite lectures to audiences of two or three—a number that would occasionally include each other.

When reform of the mathematical tripos was considered in the 1860s, the newly installed Sadleirian professor, Arthur Cayley, engaged in debate with George Airy, the Astronomer Royal and former Lucasian professor of mathematics at Cambridge. Cayley thought of his subject independently of any students, while Airy's thinking was shaped by the ideals of the university as a teaching institution. It was Cayley's ill-advised sentence: 'I do not think everything should be subordinated to the educational element',[14] which caused Airy the greatest consternation. Airy wrote back:[15]

Arthur Cayley (1821–95), Sadleirian professor.

I cannot conceal my surprise at this sentiment, assuredly the founders of the Colleges intended them for education (so far as they apply to persons in *statu pupillari*), the statutes of the University and the Colleges are framed for education, and fathers send their sons to the University for education. If I had not your words before me, I should have said that it is impossible to doubt this.

When Airy was a student at Cambridge in the 1820s, the University had been solely a teaching institution. With the new professors appointed in the 1860s, the seeds of change were planted, but a research culture in mathematics had yet to grow and mature.

## Box 1.3: The Victorian mathematical professors

### LUCASIAN PROFESSORS (MATHEMATICS, 1663)

| | |
|---|---|
| 1828–39 | C. Babbage |
| 1839–49 | J. King |
| 1849–1903 | G. G. Stokes |

### PLUMIAN PROFESSORS (ASTRONOMY AND EXPERIMENTAL PHILOSOPHY, 1704)

| | |
|---|---|
| 1836–83 | J. Challis |
| 1883–1912 | G. H. Darwin |

### LOWNDEAN PROFESSORS (ASTRONOMY AND GEOMETRY, 1749)

| | |
|---|---|
| 1837–58 | G. Peacock |
| 1859–92 | J. C. Adams |
| 1892–1914 | R. S. Ball |

### SADLEIRIAN PROFESSORS (PURE MATHEMATICS, 1860)

| | |
|---|---|
| 1863–95 | A. Cayley |
| 1895–1910 | A. R. Forsyth |

### (MECHANISM AND APPLIED MATHEMATICS, 1875)

| | |
|---|---|
| 1875–90 | J. Stuart |
| 1890–1903 | J. A. Ewing |

# The high-Victorians unsettled: 1870–1901

By the beginning of the 1870s, Britain was facing increased economic competition from Germany and the United States of America. There was a concern that the scientific base of new developments could be eroded and that Britain would lose her leadership in the world. The Devonshire Commission on Scientific Instruction and the Advancement of Science sat during 1872–75 and produced a voluminous report. The Oxford and Cambridge Commission of 1877 resulted in a University of Oxford and Cambridge Act, which enforced further changes in the governance of the university following the first modern reorganization in 1856. The first steps towards the higher education of women began in the 1870s, and a decade later a woman was recognized as the equivalent of a 'wrangler', though the formal admittance to a degree was still a long way off.

Major reforms in the mathematical tripos came into operation in 1873. The syllabus now included the introduction (and reintroduction) of such topics as the mathematical theory of elasticity, heat, electricity, waves, and tides, with these new specialisms arranged in divisions that students could select for their study. Karl Pearson praised the mathematical tripos examination of the 1870s. He liked it for its broad sweep, for its being '*not* specialised, but [giving] a general review of the principia of many branches of mathematical science', and he valued the 'problems', which forced the private coaches to deal with such questions in their classes. He observed that this essence of mathematical research was missing in the much-heralded German system, which he saw as laying the emphasis on the teaching of theory.[16]

But overall, these reforms of the mathematical tripos were not a success and even in their first year of operation this failure was apparent. Drilled in examination technique by their coaches, students quickly learned that the art of cherry-picking across the subject divisions was an effective way of amassing marks. This led to a superficial knowledge of a wide range of subjects, rather than knowledge of particular subjects to any depth. Drastic action was required. In May 1877 a large and influential University syndicate was appointed.[17] High on its agenda for discussion were:

- Whether the order of merit should be retained.

- The status of the senior wrangler.

- How to cope with the increase in mathematical knowledge, and whether the mathematical tripos could, or should, cover the whole of mathematics.

- Whether the honours students should be allowed to sit the mathematical tripos examinations in June, or keep to the traditional January examinations.

Reaching an agreed radical solution was impossible. Syndicate members were successful products of the very system they were investigating, and there would inevitably be a strong tendency to preserve their 'golden age'. Of course, the private coaches also had a powerful incentive for maintaining a system which benefitted them financially.

The 1873 increase in subjects caused problems in the running of the order of merit, for there was little common ground on which to compare the performance of individual students. Some dons were in favour of the system whereby students were divided into classes and then listed alphabetically. But this impinged on the awarding of college fellowships, where a high position in the order of merit was a traditional criterion for election in most colleges.

An elaborate scheme for reform was accepted towards the end of 1878, but it represented a compromise which pleased few members of the University Senate and resulted in a most complicated structure.[18] The order of merit and the position of senior wrangler were both retained and would be decided on the results obtained in an examination held in June, and not in January as tradition had demanded. After this examination, only the wranglers would go on to an advanced part in the following January, so there was to be a January examination. In their number would be the 'professed mathematicians', or those going on to a career in mathematics. In the advanced part the results were given by alphabetical surname order in class divisions. It was too much for some of the diehards, and especially the coaches. Routh was said to have exclaimed: 'They will want to run the Derby in alphabetical order next'.[19]

In 1882 two tripos lists were issued, one in January for the old system, and one in June for the new system. The senior wrangler under the old regulations was R. A. Herman, one of the last private coaches, who later taught G. H. Hardy.

Attention was turning towards research being part of the university's mission. Five university lectureships in mathematics were created in 1883: the first appointed were J. J. Thomson, A. R. Forsyth, W. H. Macaulay, R. T. Glazebrook, and E. W. Hobson, and each received a stipend of £50. But how was research to be done? Thomas Muir wrote of the Scottish universities, that while:

we recognise two of the functions of a University—*instruction* and *research*, we ignore, so far as mathematics is concerned, a third and equally important function—*instruction in research*.[20]

What could be said of Scotland applied equally to Cambridge. Instruction in research was not provided in anything like the way it was in Germany, where the driving force in pure mathematics was Felix Klein. The most obvious candidate for leading research at Cambridge was Arthur Cayley, the Sadleirian professor of mathematics. Cayley did have his protégés in J. W. L. Glaisher, W. K. Clifford, A. R. Forsyth, and H. F. Baker, and he gave assistance to a number of promising students, including women, who were beginning to arrive on the scene in the 1880s. His main interest was invariant theory, but the leadership in this area had passed to Germany, and the particular way he approached it—by collecting specimens in the manner of the natural scientist—was falling out of favour. Baker started on this tack but dropped it fairly quickly, while Forsyth, a loyal apostle in the 1880s, took up other paths. By the 1880s Glaisher was forced to admit that Britain had leaders but no followers. Cayley was, in the end, a 'General without Armies'.[21]

G. H. Hardy identified the 1880s as the time when the mathematical tripos was at the zenith of its reputation in the public eye, but one that coincided with research in pure mathematics in England being at its lowest ebb.[22] The lone star in pure mathematics at Cambridge was the ageing Cayley.

Quite the opposite was the case with applied mathematics and theoretical physics. A product of the mathematical tripos of the 1840s, George Gabriel Stokes distinguished himself in mathematics generally, but he is really noted for his contributions to mathematical physics. Apart from Stokes, who was actually born in Ireland, it is notable that many of those who arrived at Cambridge to sit the tripos had already distinguished themselves in the educational system in their native Scotland. From a remarkable social family group, William Thomson and James Clerk Maxwell made their way to Cambridge to top up their education in the 'Holy City'. Thomson was a child prodigy, writing original papers on Fourier series as a 16-year old, but he quickly turned his attention to mathematical physics. In fact he was actually disdainful of pure mathematics, as shown by the frustration he expressed to Hermann von Helmholtz in 1864 that such a talent as Cayley's should not apply itself to the mathematical problems in the theory of electricity:[23]

Oh! that the CAYLEYS would devote what skill they have to such things [Kirchhoff's theory of electrical conduction] instead of to pieces of algebra which possibly interest four people in the world, certainly not more, and possibly also only the one person who works. It is really too bad that they don't take their part in the advancement of the world.

Besides Thomson, there was the genius of Maxwell, also a mathematician who devoted himself to physical theories. On the experimental side the Cavendish laboratory was created at Cambridge in the early 1870s. This vigorous enterprise had Maxwell as its first director; he was followed by Lord Rayleigh and then J. J. Thomson. Such luminaries as Rayleigh (on the physics of sound),

A student at Newnham College in the 1880s.

Thomson (on mathematical physics, and specifically the discovery of the electron), G. H. Darwin, the second son of Charles Darwin (on geodesy), and J. Larmor (on the theory of relativity) helped to gain Cambridge a worldwide reputation for excellence in mathematical physics in the late-Victorian period.

In 1886 a newer two-part mathematical tripos was created. A two-year Part I decided the order of merit, and a Part II taken a year later offered the advanced subjects.[24] Both exams were scheduled for the summer, and the famous January examination became a thing of the past. The new Part I was a technical course; indeed, the whole idea of mathematics as the underpinning of a liberal education for all students had proved impossible to maintain. It was a very different mathematical tripos from that of the beginning of the Victorian period, when mathematics had no competing subjects and students had little choice of what to study. Towards the end of Victoria's reign the number of students opting for the mathematical tripos course fell rapidly. In the 1840s, an average of one hundred and twenty-four mathematical tripos students graduated each year with an honours degree, but by the 1890s there were only ninety-two.

In 1890 G. T. Bennett graduated as senior wrangler and winner of the first Smith's prize. These prizes were now awarded for dissertations, and so impressed was Cayley by Bennett's paper on number theory that he communicated it to the Royal Society for publication.[25] But while Bennett was the *official* male senior wrangler, it was Philippa Fawcett's performance that electrified the student population when she graduated 'above the Senior Wrangler' and, both celebrated and disparaged, she was eulogized in doggerel:

Philippa Fawcett (1868–1948).

> Hail the triumph of the corset,
> Hail the fair Philippa Fawcett.

The year-on-year procession of males in the order of merit had at last been topped by a female scholar. Women could study at Cambridge, but they could not be admitted to a degree and membership of the university until 1948, when the first to be admitted was the Queen Mother, with an honorary doctorate. By Philippa Fawcett's time the state of the mathematical tripos was again under attack, and the attack would continue until the system eventually changed.

In George Bernard Shaw's play *Mrs Warren's Profession*, Vivie gave voice to the curious phenomenon that the mathematical tripos had become by the end of the 19th century:[26]

do you know what the mathematical tripos means? It means grind, grind, grind for six to eight hours a day at mathematics, and nothing but mathematics. I'm supposed to know something about science; but I know nothing except the mathematics it involves. I can make calculations for engineers, electricians, insurance companies, and so on; but I know next to nothing about engineering or electricity or insurance. I don't even know arithmetic well.

Tensions existed between the teachers of mathematics and 'active mathematicians' who researched the subject. The latter could not believe in teaching a system that was dominated by an examination consisting of artificial problems which could only be justified by being good mathematical tripos examination questions. G. H. Hardy believed that, since it dominated Cambridge mathematics, and was in turn dominated by an examination which consisted of artificial questions about subjects of no interest to professional mathematicians, the mathematical tripos stifled real mathematical research. He thus concluded that the whole thing should be abolished.[27]

## Aftermath

Queen Victoria died at Osborne on the Isle of Wight on 22 January 1901, but 'Victorian mathematics' at Cambridge continued a little longer. The big fight in the cause of mathematical tripos reform took place in 1907. The majority of active mathematicians at Cambridge were in favour of the abolition of the order of merit and the coveted title of the senior wrangler, but there was a minority who opposed the reforms. One private coach, a protégé of E. J. Routh, thought that the proposed reforms would mean the end of mathematics at Cambridge.[28]

The vote was put to the whole electorate of Cambridge M.A.s. The voting took place in February 1907, and about 55 per cent were in favour of reform. The turnout was only 20 per cent, no doubt reduced by the requirement that voters had to attend Cambridge in person. It was a close call, but in a first-past-the-post voting system 'one is enough'. The last examination under the old regulations was held in 1909.[29] It was truly the end of an era. The institution of the private coach melted away, and in the tumultuous events of 1914–1918 the mathematical tripos, which had reigned supreme during the Victorian period, became a distant memory.[30]

A caricature of Henry Smith, Savilian professor of geometry from 1861 to
1883.

CHAPTER 2

# Mathematics in Victorian Oxford

*A tale of three professors*

KEITH HANNABUSS

Whereas Cambridge was uniformly regarded as the centre of both British mathematical education and research throughout the 19th century, the curriculum at the University of Oxford was geared more towards the study of classics and the humanities. Thus, while mathematics was not unimportant at Oxford, it did not have the same dominance over the course of study as at its rival institution in the Fens. Indeed, throughout the Victorian period, Oxford mathematicians had to strive to establish a mathematical culture as prevalent as that which existed at Cambridge. The key figures in this endeavour during this period were the three holders of Oxford's most prestigious mathematical chair, the Savilian professorship of geometry, which had been founded by Sir Henry Savile in 1619. This chapter documents the changing fortunes of mathematics during the Victorian period at Britain's oldest university, and the principal mathematicians involved in this process.

In 1837 Oxford was a small and relatively isolated city, dominated by the University. Its north-eastern boundary followed roughly the line leading from Magdalen College and Wadham College through St Giles Church to the Radcliffe Observatory. (Little changed for almost twenty years, Magdalen Bridge is at the bottom right-hand corner of the figure overleaf, and the Radcliffe Observatory is near the middle of the horizon, with St Giles Church to its right and slightly lower. Beyond the visible buildings there is open country. From 1855 the

University Museum was built roughly at the right-hand edge of the horizon, and the Norham estate beyond that.) The University was still largely a mix of a finishing school for gentlemen and a seminary for the Anglican clergy. Students had to sign the 39 articles on matriculation, so that Jews, Catholics, Nonconformists, and other dissenters were effectively excluded. Moreover, this requirement could not be postponed until graduation as at Cambridge, a possibility that allowed non-Anglicans like Sylvester to study there, as long as they did not take their degree.

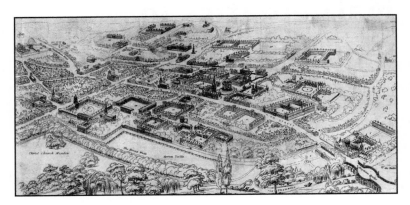

A bird's-eye view of Oxford University and city, 1850.

With a few exceptions—professors, heads of houses, and a small number of lay fellows—dons were required to take holy orders within around seven years, first as a deacon and then as a priest, and they were not allowed to marry.[1] The typical career for a bright undergraduate would be to obtain a college fellowship soon after graduation, proceed to ordination and then, assuming he wished to marry, move to a suitable living when one fell vacant. These livings were usually in remote villages where the parish demands were small and they could continue their scholarship and research. This system ensured that there were generally fellowships for the promising young graduates to start a career—although the opportunities were rather greater in classics than in mathematics—but its disadvantage was that it was difficult to develop an academic profession in the modern sense.

During Victoria's reign all this changed. The opening of the Great Western Railway line in 1844 eased the isolation. (It is a measure of Oxford's previous seclusion that one of the Oxford–London express coachmen turned out to have a wife and family at each end of his itinerary.[2]) In 1855 the University commenced construction of the University Museum as a centre for the sciences, including mathematics, on a site beyond the old northern boundary. By the time of the Museum's opening in 1860, the city too had started to extend further north, with developments like Park Town, and plans for the Norham Manor estate between Norham Gardens and Norham Road.

Two Royal Commissions that reported in 1854 and 1877, plus the ensuing acts of Parliament, abolished the religious tests and celibacy requirements; this parliamentary intervention came although the University received no public money at all until further reforms in 1923. With the relaxation of the celibacy

The new University Museum in 1860.

rule, and consequently a larger number of academic families, the city continued to expand. Where in 1860 the planned Norham Manor estate formed no more than an island of houses in the open countryside, it soon became the portal to a much larger development to the north and west.

The honour schools had started in 1800 as a more demanding alternative to the pass schools, though only three colleges (Balliol, Christ Church, and Oriel) insisted that all students must read for honours. Initially the honours were awarded in classics (*In Literis Humanioribus*) and mathematics (*In Disciplinis Mathematicis et Physicis*), which covered pure and applied mathematics. In 1849 the University decided to add honour schools of natural science and modern history with jurisprudence to the existing schools, and with the introduction of the new examinations in 1853, the mathematical honours were renamed *In Scientiis Mathematicis et Physicis*; the new mathematical honour moderations, first examined in Easter term 1852, were just *In Disciplinis Mathematicis*. In the course of the century the numbers taking honours in all subjects rose from under one hundred to nearer five hundred.

The dominance of classical studies was evident not only in student numbers, but in elections to fellowships: in 1853 it was reported that three-quarters of those with first class honours in classics got fellowships, compared with only one-third of those with first class honours in mathematics.[3] Although mathematics was a much smaller school than classics, its numbers initially rose more or less in line with the total until 1850, when they levelled off and most of the expansion went into the new natural science school. The total number of honours degrees more than trebled during the second half of the century, and the number of honours gained in the two schools of mathematics and natural science remained at a fairly steady 15 per cent of the total, but of those the number of honours awarded in mathematics was much the same at the end of the century as it had been in the 1830s.

In the 1860s, Eleanor Smith, sister of the Savilian professor of geometry Henry Smith, set up a series of 'women's lectures', primarily for the wives and families of the Oxford professors.[4] A decade later, the first women's colleges were founded: Lady Margaret Hall in 1878, Somerville College (named after the mathematician Mary Somerville) and St Anne's College in 1879, followed by St Hugh's College in 1886 and St Hilda's College in 1893. From 1875 women were allowed to take special examinations, and from 1882 they were officially permitted to take the University examinations, although they could not graduate until 1920.

## The Savilian professors

Baden Powell, Savilian professor of geometry from 1827 to 1860.

Within Oxford, much of the credit for reviving mathematics during the Victorian period goes to three successive Savilian professors of geometry: Baden Powell, Henry Smith, and James Joseph Sylvester. Not only did each introduce important changes within the University, but they were also active in the national debates about higher education. Powell and Sylvester were amongst those presenting the petition for the repeal of the religious tests to Parliament in 1855, in the Oxford and Cambridge delegations, respectively.[5] Smith served on, and was secretary of, the 1877 Royal Commission that created the Combined University Fund—now used to support lectureships in mathematics and the humanities—and new statutory professorships such as the Waynflete chair in pure mathematics, to which Edwin Elliott was elected as first professor in 1892.

### Baden Powell

Baden Powell's career was typical of the older generations.[6] After graduation and a fellowship at Oriel, he moved to a living where he carried out his researches into the infra-red end of the solar spectrum, before being elected in 1827 to the Savilian chair (being preferred, according to gossip,[7] to the computer pioneer Charles Babbage, whose work is described in Chapter 10). Baden Powell's work is no longer well known, and his name is more readily associated with his sixth son by his third wife, Robert Stephenson Baden Powell, the founder of the boy scout movement.

Just as Babbage lamented the state of mathematics at Cambridge, so Baden Powell was horrified by the low esteem enjoyed by mathematics and the sciences within the University of Oxford, and he set about trying to improve matters. He was advised not to give an inaugural lecture, since that would have been unlikely to attract an audience, but a few years later his privately printed lecture, *The Present State and Future Prospects of the Mathematical and Physical Sciences at the University of Oxford*, laid out his ideas of the importance of mathematics both within the University and to the nation.

On the practical side, Baden Powell decided to try to raise the standards in mathematics by printing the examination papers, as was already done at Cambridge, so that students could practice solving examination questions. The first question on the first paper of the honour school examination of Easter 1828 gives some idea of the huge task which he faced:

What decimal of a week is 1 hour 7 minutes and 14 seconds?

Oxford students taking their examinations.

The Oxford syllabus was somewhat smaller than that at Cambridge, and even in the 1890s the gap in standards is illustrated by the story of Grace Chisholm. Possibly as the result of a dare, she sat the Oxford examination after completing her Cambridge tripos. She became the first woman to achieve first class honours at Oxford, coming top of the list, whereas she had been placed between the 22nd and 23rd wranglers at Cambridge (where women's names did not appear on the official list).[8]

An Oxford *viva voce* examination.

Another difference between the two universities was that private coaches were probably less prominent at Oxford than at Cambridge. Henry Wall, the professor of logic, had a good track record coaching for the logic component of the classics degree,[9] but perhaps the nearest mathematical equivalent was provided by the mathematical reading parties of Bartholomew Price, Sedleian professor of natural philosophy.[10] Probably the lower status of mathematics at Oxford, and the fact that the class list was not ordered by merit, played a part in this, but also, following the lead of the more academic colleges, the tutors began to take their teaching duties quite seriously and may have rendered coaching unnecessary for most students. In the late 1860s, the mathematics lecturers of a group of colleges decided to collaborate in giving the Combined College Lectures. They were soon joined by others, and this eventually turned into the later system of departmental lectures.[11]

Part of an Oxford finals paper from 1854.

Another of Powell's practical steps was to raise money for a mathematical scholarship, to enable the brightest students to stay longer in Oxford and wait until a fellowship fell vacant. Among the early scholars were William Donkin (1837) and Bartholomew Price (1842), who later became Powell's professorial colleagues as Savilian professor of astronomy and Sedleian professor of natural philosophy, respectively. Besides his lunar observations and an interesting early paper on the least squares method, Donkin's main mathematical work was in acoustics, although his great monograph on the subject was still unfinished at the time of his death. Price wrote a four-volume treatise on the calculus, and for many years he also ran the University Press, where he transformed the academic side by commissioning new monographs such as James Clerk Maxwell's *Treatise on Electricity and Magnetism*.[12]

In 1854, Charles Dodgson (better known today as Lewis Carroll) attended one of Price's reading parties in Whitby, and some believe that Price, whose nickname was 'Bat', was the 'little bat' of *Alice's Adventures in Wonderland*. The two developed a close friendship for many years as Dodgson's rooms in Christ Church were just across the road from Pembroke College, where Price was a fellow and later Master. It was Price who, as a Fellow of the Royal Society, communicated Dodgson's paper on the condensation of determinants for the Society's *Proceedings* in 1866.

By 1844 there was enough money in the scholarship fund to provide Junior and Senior Mathematical Scholarships (nowadays, the Junior and Senior mathematical prizes). In an era when there were no graduate courses or doctorates in mathematics, the senior prizes enabled promising students to get started on some independent research of their own. Early senior scholars included two Balliol students, Henry Smith (1851), later Savilian professor of geometry, and William Spottiswoode (1846), later printer to the Queen. Spottiswoode continued his research in mathematics and physics privately, attempting in 1861 a statistical analysis of whether the mountain ranges in Asia were formed by a single cause or by more. It was to this paper that Francis Galton later attributed his interest in applications of statistics to the social sciences. Spottiswoode was eventually elected President of the Royal Society in 1878.[13]

Henry Smith received one of only two firsts awarded in the 1849 Easter term examinations, the other going to Robert Faussett of Christ Church. Faussett was almost immediately elected to a mathematical lecturership at Christ Church, and when he resigned in 1855 he was replaced by his pupil Charles Dodgson. Dodgson's mathematical prowess has been rather overshadowed by the success of the *Alice* books, but he was probably one of the most original mathematicians in Oxford. His *Elementary Treatise on Determinants* contains some interesting insights, and had he not died while producing his advanced volume on *Symbolic Logic*, he might well have had a stronger influence on the field, as might his work on voting and elections, had it been published in a book rather than in pamphlet form.

As readers of *Alice's Adventures in Wonderland*, *Alice through the Looking-Glass*, and *The Hunting of the Snark* well know, Dodgson's mathematical interests spilled over into his children's books. The less well known *Sylvie and Bruno* books include an interesting passage on a tea party in free fall, which prefigures ideas in Einstein's principle of equivalence, and also a description of a projective plane constructed out of three handkerchiefs—the 'Purse of Fortunatus', which, having neither inside nor outside, may be said to contain the entire wealth of the world.

Baden Powell, an enthusiastic popularizer of science, had joined other Oxford scientists in supporting the newly formed British Association for the Advancement of Science in 1831, and invited it to hold its second meeting in Oxford in

Bartholomew Price, in a photograph by Charles Dodgson.

Mein Herr demonstrating the 'Purse of Fortunatus', from Lewis Carroll's *Sylvie and Bruno Concluded.*

1832. This simultaneously raised the profile of the Association and that of science in Oxford. The British Association returned in 1847 (when the co-discoverers of Neptune, John Couch Adams and Urbain Leverrier, met in Powell's house) and again in 1860. By that time the University Museum was complete enough to host the meeting, and it was there that the famous discussion on evolution took place between Bishop Samuel Wilberforce and Thomas Huxley. The debate would no doubt have interested Powell, who had himself written on evolution, but he had died a couple of months earlier at the start of that summer.

## Henry Smith

Powell's death came in the middle of a huge furore about a set of Oxford *Essays and Reviews* published earlier in the year, and his contribution which questioned miraculous evidences for Christianity earned its share of the general opprobrium. Perhaps because of this, the only candidates for the vacant Savilian chair were, apparently, George Boole and the Balliol mathematical lecturer, Henry Smith. Moreover, it seems that this religious controversy made Boole hesitate to apply for the position as he did not submit a proper formal application,[14] and the electors may have decided that his candidacy was not serious. Although Smith was beginning to make a name for himself with his *Report on the Theory of Numbers* for the British Association, a survey welcomed by Continental mathematicians such as Leopold Kronecker, he did not yet have anything like the solid achievement of Boole in mathematical logic and other areas which are described in Chapters 13 and 16. Nonetheless, the electors chose Smith, who soon proved himself worthy of the chair.

Henry Smith (1826–83).

Of the British pure mathematicians of the mid-19th century, Henry Smith was one of the few whose career followed a relatively straight path through academic life. He did not go into the law like Cayley, or suffer the fragmented career of Sylvester, but was elected to a lectureship soon after graduating, and then in his mid-30s to the Savilian professorship.

In 1861 he published a paper 'On systems of linear equations and congruences', which finally gave a complete method for determining which linear Diophantine equations have solutions and for finding all solutions when they exist. It relied on applying the new theory of matrices, but with integer entries, and essentially gave the existence and uniqueness of what is now called the *Smith normal form* (see Box 2.1).

## Box 2.1: The Smith normal form of a matrix

Henry Smith showed that by row and column operations any matrix with integer entries can be put into a diagonal form in which each diagonal entry divides its successors, and that these diagonal entries are, up to sign, unique:

The transformation to which we have referred...is obtained by employing simultaneously a premultiplying and postmultiplying unit matrix. It is expressed by the equation

$$\|a\| = \|\alpha\| \times \mathrm{diag}\ (\nabla_n/\nabla_{n-1}, \nabla_{n-1}/\nabla_{n-2}, \dots, \nabla_1/\nabla_0) \times \|\beta\|,$$

in which $\|a\|$ is a given square matrix of the type $n \times n$, $\|\alpha\|$ and $\|\beta\|$ are unit matrices, and $\nabla_n$, $\nabla_{n-1}$, $\nabla_{n-2}, \dots, \nabla_1$, $\nabla_0$ are the determinant and greatest common divisors of the minor determinants of $\|a\|$, so that, in particular, $\nabla_n$ is the determinant of $\|a\|$, $\nabla_{n-1}$ the greatest common divisor of its minor determinants of order $n-1$, $\nabla_1$ the greatest common divisor of its constituents, and $\nabla_0 = 1$. The units $\|\alpha\|$ and $\|\beta\|$ are not absolutely determined, but admit, when $n > 1$, of an infinite number of different values...

It thus appears that in the series of numbers

$$\nabla_n/\nabla_{n-1}, \nabla_{n-1}/\nabla_{n-2}, \dots, \nabla_2/\nabla_1, \nabla_1/\nabla_0$$

each term is divisible by that which comes after it.

The concise clarity of this passage is typical of Smith's writing and contrasts with the vague prolixity of some of his contemporaries. In Chapter 18 we see that Smith's work seems to present a possible exception to the general lack of rigour found in British mathematics, and it is perhaps worth mentioning some possible reasons for this. Shortly after starting as a student, Smith had suffered from smallpox and then malaria, and had been forced to spend a couple of years recuperating on the Continent.[15] He had put his convalescence to good use by attending lectures at the Sorbonne, and had forged links with Continental mathematicians which he retained and which few if any of his colleagues could match.

Smith was elected to the Mathematical Lectureship at Balliol in 1849, and was asked by the college to take instruction in chemistry with Nevil Story-Maskelyne at the Ashmolean Laboratory, as well as with August Hofmann in London, so that he could run the college's teaching laboratory for the new school of natural science.[16] This provided further contact with Continental ideas, so that Smith was much less insular than most of the other Oxford-trained mathematicians. Having taken a double first in classics and mathematics, and won the top prizes in both, it also meant that Smith could never be outranked in University debates, since nobody else could match his expertise across such a wide range of subjects.

Smith kept up his contacts with Continental mathematicians, both by frequent visits to Europe and through his habit of entertaining them in Oxford. Felix Klein visited, and in 1876 Smith entertained the young German mathematician, Ferdinand Lindemann; according to Smith's Cambridge colleague J. W. L. Glaisher,[17] they discussed the problem of proving that the number $\pi$ is transcendental, presumably inspired by Charles Hermite's recent work on $e$. Although we know less about Smith's European visits, Hermann von Helmholtz's diaries show that he dined in Paris with Hermite and Smith in 1865.[18] There Smith revealed his hopes that Helmholtz might have been elected as Oxford's professor of experimental philosophy and head of the Clarendon Laboratory, but unfortunately this initiative, which could have reinvigorated Oxford science, came to nothing.

In 1874 Henry Smith was appointed Keeper of the University Museum, following the death of the geologist John Phillips. He moved with his sister, who kept house for him, into the Keeper's residence, which stood behind the Museum until its demolition in 1952 to make way for the new inorganic chemistry laboratory. He was also elected to a professorial fellowship at Corpus Christi College, enabling him to give up the arduous undergraduate teaching that he had continued to do at Balliol. In the following months he wrote more papers than in any other comparable period, confirming comments in his sister's letters that he at last had the leisure to catch up on writing. He was probably also encouraged by Glaisher, who had just taken over as editor of the *Messenger of Mathematics*, and who actively solicited papers covering the topics of Smith's lectures at various meetings. These papers were by no means restricted to number theory, or the theory of elliptic functions, which became his main interest in later years.

In an interesting paper of 1876, 'On the integration of discontinuous functions', Smith introduced a set equivalent to the Cantor set some seven years before Cantor (see Box 2.2).

In 1868 Smith had shared (with Hermann Kortum) the Steiner prize of the Prussian Academy of Sciences for solving a problem in projective geometry. In 1883, under rather contentious circumstances, he was also awarded the Grand Prix of the Paris Academy. In the previous year the Académie had set as a prize problem the task of developing some work of the German number-theorist Ferdinand Eisenstein, who had died at the age of 29 before completing it; this

# Box 2.2:  Henry Smith and the 'Cantor set'

Henry Smith was studying Riemann's new theory of integration, and was investigating just how discontinuous a function could be whilst still being integrable.

The Cantor set is obtained by taking a unit interval consisting of the points from 0 to 1, removing the middle third of points between $\frac{1}{2}$ and $\frac{2}{3}$, removing the middle thirds of what are left (the points between $\frac{1}{9}$ and $\frac{2}{9}$, and between $\frac{7}{9}$ and $\frac{8}{9}$), and so on, for ever. The remaining set of points is an infinite set of points with total length 0. Cantor showed that a function that is discontinuous at the points left after this process of subdivision can still be integrated.

Seven years earlier, Smith had constructed an example of such a set by taking a unit interval divided into $n$ parts (here, three) and removing the last subinterval, then subdividing each remaining subinterval into $n$ parts and removing each last subinterval, and so on. He showed that a function discontinuous at the points left after this process of subdivision can be integrated, whereas a function with discontinuities at the points given by successive subdivisions into $n$, $n^2$, $n^3$ parts is not integrable.

The resulting set is a 'fractal' pattern now known as the Smith–Volterra–Cantor set.

The Smith set and the Cantor set.

problem asked for the number of ways of expressing a given positive integer as a sum of five or seven squares. The Paris Academy were apparently unaware that Smith had already published the solution of this problem in 1867, in a paper entitled 'On the orders and genera of quadratic forms containing more than three indeterminates'.[19]

After consulting Glaisher, Smith wrote to Hermite pointing this out. Hermite apologized profusely, but asked Smith to submit his work for the prize, presumably hoping that Smith's would be the only entry. Despite ill health, Smith translated his work into French and complied, but another almost identical solution was submitted by the young Prussian student Hermann Minkowski. In the end Smith died a couple of months before the award, and one of his obituary notices mentioned his prior publication. This was quickly picked up by the French press, who concluded that Minkowski had plagiarized Smith's work, and there was soon a major controversy which died down only when the Académie announced that it would award separate prizes to Smith and Minkowski, rather than sharing the prize between them.[20]

James Joseph Sylvester (1814–97), Savilian professor of geometry from 1883 to 1894.

## James Joseph Sylvester

Smith's sudden death in February 1883, followed just a few months later by the death of William Spottiswoode (an elector), meant that the Savilian chair of geometry remained vacant for some months, during which time Cayley sounded out Felix Klein to see whether he might be interested, and also his friend and colleague, James Joseph Sylvester. Despite having received an honorary doctorate from the University in 1880, Sylvester was hesitant about applying, but at the age of 69 he was duly elected and started as professor in January 1884. Interestingly, the electoral board included Thomas Huxley, who had received a public rebuke from Sylvester for his rather ill-informed comments on mathematics some fifteen years earlier.

On his arrival, Sylvester set about trying to build up mathematical research in Oxford, as he had at the new Johns Hopkins University in Baltimore, where from 1876 to 1883 he had established a flourishing research school, as well as founding the *American Journal of Mathematics*. Unfortunately, he discovered that, despite the international eminence of Henry Smith, the status of mathematics within Oxford was nowhere near as great as at Cambridge, and he lacked any powerful local ally to replace Daniel Coit Gilman, the influential president of Johns Hopkins, who had given him such crucial support there.

Within Oxford, Sylvester continued Smith's tradition of inviting eminent foreign mathematicians to visit, such as Leopold Kronecker in 1884 and Henri Poincaré in 1891, and, in the face of the difficulties preventing the establishment of a research group, he set up the Oxford Mathematical Society. This latter organization remained for several years the only forum in Oxford at which mathematicians could present and discuss recent research. Failing health dogged Sylvester's later years at Oxford, but he worked a little in graph theory, carrying out an extended correspondence with the Danish mathematician Julius Petersen.[21] By 1894, however, his sight was too precarious for him to continue giving his lectures, and the Merton tutor William Esson was appointed to serve as his deputy.

## Other Oxford mathematicians

Although Cambridge dominated British mathematics at the time, and certainly produced many more graduates in the subject, by the time of Sylvester's professorship in the 1880s, Oxford did not lack talented mathematicians. These included several who successively held both the Junior and Senior Scholarships: Leonard James Rogers in 1881 and 1885, Percy Heawood in 1882 and 1886, and John Edward Campbell in 1885 and 1888.

In 1894 Rogers discovered a more general form of what are now called the *Rogers–Ramanujan identities*[22] (see Box 2.3), some twenty years before they were conjectured by the Indian mathematician Srinivasa Ramanujan. Ramanujan had asked

## Box 2.3:  The Rogers–Ramanujan identities

These formulas are combinatorial identities relating to hypergeometric series, in which certain infinite sums are expressed as infinite products. Two of the best-known ones are:

$$1 + \sum_{n=1}^{\infty} \frac{x^{n^2}}{(1-x)(1-x^2)\ldots(1-x^n)} = \frac{1}{\prod_n (1-x^{5n+1})(1-x^{5n+4})},$$

and

$$1 + \sum_{n=1}^{\infty} \frac{x^{n(n+1)}}{(1-x)(1-x^2)\ldots(1-x^n)} = \frac{1}{\prod_n (1-x^{5n+2})(1-x^{5n+3})}.$$

Leonard James Rogers (1862–1933).

most of the leading mathematicians about his conjecture, including the English combinatorialist Percy MacMahon, but none could supply a proof. When eventually Ramanujan made contact with Rogers, after spotting his old paper, Rogers, who was by now at Leeds, promptly supplied several more proofs.[23]

One of the most well-known problems in mathematics at this time was the *four colour problem*, which asks whether every planar map can be coloured with just four colours (see Chapter 17). In 1879, the problem had apparently been solved in the affirmative by Alfred Kempe, who published a proof in Sylvester's *American Journal of Mathematics*. Inspired by one of Smith's Oxford lectures on geometrical problems in the following year, in which the four colour problem was mentioned, Heawood became fascinated with it. He read Kempe's purported proof and eventually identified an error—one of the most infamous mistakes in the history of mathematics—and after moving to Durham University published his well-known refutation of Kempe's proof in 1890, as well as a variety of other useful related work on the problem. He was to continue working in this area until the early 1950s when his final paper on the subject was published in his ninetieth year.[24]

Campbell published the first text in English on Lie groups, and his contributions are commemorated in the 'Campbell–Baker–Hausdorff formula' (see Box 2.4).[25] Campbell was content to remain a mathematics tutor at Hertford College, despite his family's ambition that he should apply to become a head of house or professor.[26] Nonetheless, the esteem in which his contributions were held is clear from his Fellowship of the Royal Society in 1905 and his election as President of the London Mathematical Society in 1918.

Another Senior Mathematical Scholar (1875) and founding member of the Oxford Mathematical Society was Edwin Bailey Elliott, who was elected to the newly created Waynflete professorship of mathematics in 1892, one of the new chairs established by the 1877 Royal Commission. Unfortunately, in the very year

John Edward Campbell (1862–1924).

## Box 2.4: The Campbell–Baker–Hausdorff formula

This formula solves the equation $Z = \log(e^X . e^Y)$, when $X$ and $Y$ do not commute. It shows that the image of the exponential map from a Lie algebra to a Lie group is closed under the group composition.

Campbell's papers of 1897 and 1898 'On a law of combination of operators bearing on the theory of continuous transformation groups' in the *Proceedings of the London Mathematical Society* were, like those of Henry Frederick Baker (1905) and Felix Hausdorff (1906), mainly concerned with showing that, for elements $X$ and $Y$ of a Lie algebra,

$$\exp(X)\exp(Y) = \exp(Z),$$

where the Lie algebra element $Z$ is constructed from $X$ and $Y$ using only linear combinations and Lie brackets. The first terms show that

$$Z = X + Y + \tfrac{1}{2}[X, Y] + \ldots,$$

but the full series emerged only in the work of Eugene Dynkin in 1947.

---

of his election, David Hilbert's finite basis theorem answered the central questions of invariant theory so effectively that the subject lost much of its interest as a research topic and Elliott's monograph *Algebra of Quantics* (1895), with its chapter on Hilbert's proof of Gordan's finiteness theorem for binary invariants, now looks more like a eulogy for a defunct research agenda.

On Sylvester's death in 1897, his deputy William Esson succeeded him in the Savilian professorship, despite the fact that there were clearly better candidates for the chair, including Percy MacMahon, who applied unsuccessfully for it. As a young college tutor Esson, with Henry Smith and a few others, had set up the Combined College Lectures, the forerunner of the current lecturing system. He had also done some interesting work with the experimental chemist Vernon Harcourt (one of Smith's first chemistry pupils), trying to understand and model chemical reactions, although they were hampered by having unwittingly chosen to study an autocatalytic reaction with non-linear dynamics.

Alongside the professors of mathematics at the close of the century, Francis Edgeworth, the Drummond professor of political economy, is counted amongst the founders of mathematical statistics as a result of his work on multi-dimensional normal distributions and the analysis of variations. Edgeworth had studied classics at Balliol College in the late 1860s, and whilst there learned some economics from Benjamin Jowett, Regius professor of Greek. This, according to John Maynard Keynes, so caught Edgeworth's interest that he started to study the subject and its mathematical foundations further, even whilst studying for the Bar, to which he was called in 1877.[27] In that same year he published *New and Old Methods of Ethics*, and just four years later this was followed by *Mathematical Psychics: An Essay on the Application of Mathematics to the Moral Sciences*.

The fact that Oxford at this point had no professor of statistics leads us to the interesting story of a failed initiative to establish such a chair (see also Chapter 11). An imaginative plan to create a chair in statistics had been hatched in the 1870s by Florence Nightingale, the nursing pioneer and passionate statistician, and Benjamin Jowett, Master of Balliol from 1870 and the University's Vice-Chancellor from 1882 to 1886. Jowett had been introduced to Nightingale by his former pupil, the poet Arthur Hugh Clough, and at her invitation came to celebrate holy communion with her once a month.[28] Nightingale had used statistical techniques to analyse the mortality data that she had collected in the Crimea, and had introduced new methods such as polar diagrams, a forerunner of pie charts.[29]

Nightingale had been advised by William Farr of the General Registry, and it was he who first suggested the idea of a professorship of statistics, an idea that took particular hold when she heard of the death of the Belgian statistician Adolphe Quetelet in 1874. Jowett, though preferring things not to be 'overloaded ... with mathematics',[30] had his own interest in statistics, and quickly seized on the idea, suggesting that they each 'give or bequeath 2000£ for the endowment [of a chair] ... and then we might go about begging of the rich people of the world'.[31]

Unfortunately, Francis Galton, to whom Florence Nightingale had turned for advice, alarmed her by his insistence that a professor of statistics must be active in research as well as teaching. She feared that her gift would 'only end in endowing some bacillus or microbe and I do not wish that'.[32] Eventually, just a month before his death in 1893, Jowett admitted defeat and changed his will to withdraw the legacy of £2000 to Miss Nightingale 'as I fear that there is no possibility of realizing the scheme to which it was originally to have been applied'.[33]

Benjamin Jowett, Regius professor of Greek and Master of Balliol College.

## Conclusion

Despite the failure of the scheme to establish a chair of statistics, Oxford mathematics had come a long way during the reign of Queen Victoria. With the creation of the mathematical scholarships in the 1830s, Baden Powell had laid the foundations for a more serious approach to research; a generation later, Henry Smith raised the subject to international level with his own work and fostered links with Continental mathematicians, while under J. J. Sylvester, mathematics at Oxford took the first steps towards the foundation of a research school.

During the early years of the 20th century there was a brief period of relative dormancy, with pure mathematics in particular losing some of its momentum. Having been appointed Savilian professor of geometry rather late in his career, Esson was apparently not able to offer the same leadership in research as his two predecessors in the chair. Campbell's work on Lie groups and differential geometry provided a better pointer to future research directions, but as a college tutor he was not in a position to offer research leadership himself. Meanwhile, Elliott's work on invariant theory lost its edge after Hilbert's comprehensive solution of the central problem of the subject. In any case, none of these three was sufficiently motivated to stimulate a research department. As Elliott later put it:[34]

But how about research and original work under this famous system of yours, I can fancy someone saying. You do not seem to have promoted it much. Perhaps not! It had not yet occurred to people that systematic training for it was possible.

It was with the arrival of Esson's successor G. H. Hardy as Savilian professor in 1920 that the modern era of mathematics in Oxford really began. In many ways Hardy was a natural heir of Henry Smith in number theory, and he arrived in Oxford at the height of his mathematical powers, picking up essentially where Sylvester had left off by creating a fully fledged research school in mathematics for the first time in Oxford. One needs only to look at the list of subsequent professors, including E. C. Titchmarsh, Sydney Chapman, E. A. Milne, Henry Whitehead, C. A. Coulson, Graham Higman, Sir Michael Atiyah, Sir Roger Penrose, Daniel Quillen, and Simon Donaldson, to see that Oxford has maintained its research momentum in both pure and applied mathematics ever since.

London in the 1860s.

## CHAPTER 3

# Mathematics in the metropolis

*A survey of Victorian London*

ADRIAN RICE

In this chapter we survey the teaching of university-level mathematics in various London institutions during the reign of Queen Victoria. We highlight some of the famous mathematicians who were involved for many years as teachers, including Augustus De Morgan, James Joseph Sylvester, and Karl Pearson. We also investigate the wide variety of teaching establishments, from mainly academic institutions (University College, King's College) to the military colleges (the Royal Military Academy at Woolwich, and the Royal Naval College, Greenwich), women's colleges (Bedford and Queen's), and the technical colleges and polytechnics (such as the Central Technical Institute) that began to appear during the latter part of this period. Comparing the teaching styles and courses provides a fruitful way of exploring the rich development of university-level mathematics in London between 1837 and 1901.

Discussing any aspect of Victorian London is necessarily complex. In particular, it must cover a vast area, for the London of 1837 was very different, in both size and character, from the huge metropolis that evolved by the turn of the century. To consider mathematics in the capital at this time, we must therefore limit our attention to some particular aspects so as to avoid producing an unduly prolonged, and possibly less informed, account. Consequently, we omit two related aspects of London mathematics: school- or elementary-level mathematics and, at the other end of the spectrum, research-level mathematics. Instead, we provide an overview of higher mathematical education in London at this time, concentrating solely on the teaching of university-level mathematics in London between the years 1837 and 1901 and the mathematicians who taught it.

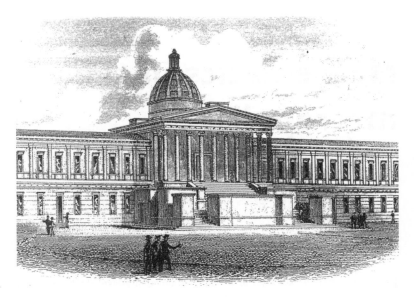

University College.

## University College

In 1837, London was served by only three institutions for the teaching of higher-level mathematics. The first was University College. Founded in 1826 as 'The London University', it was the first university to be established in England since the middle ages, opening for lectures on Gower Street, then on the northern edge of London, in 1828. Its first professor of mathematics was Augustus De Morgan, best known to mathematicians today for his research in algebra and logic, and especially for De Morgan's laws of set theory. Educated at Trinity College, Cambridge, he was appointed professor at University College at the age of only 21. A man of high principle, he resigned his post three years later over the dismissal of a colleague, only to return in 1836 following the accidental death of his successor. After another thirty years he resigned again, this time over a question of the college's adherence to its policy of religious neutrality.

De Morgan's mathematics courses were designed to be studied over two or three years, his students being divided into four groups: a junior and senior class (or 'year', as we would now term it), each with a lower and a higher division. The level at which students entered depended on their previous attainments; for example, to enter the lower junior class one needed the four rules of arithmetic and some experience of vulgar fractions. By the higher senior class, students were studying advanced applications of the differential and integral calculus, differential equations, the calculus of variations, and some probability theory. His pupils' memoirs also include references to number theory, the theory of equations, and complex numbers. We may therefore conclude that De Morgan was offering an extensive mathematical programme at University College; for a student to progress from elementary arithmetic to the calculus of variations in two to three years was quite an accomplishment.

Mathematics under De Morgan was stimulating, but never easy: even his brightest pupils had to struggle to keep up. Walter Bagehot, later to find fame as a political and constitutional writer, was a student in the 1840s, and wrote in 1843 that:[1]

Augustus De Morgan (1806–71), a caricature by a student.

De Morgan has been taking us through a perfect labyrinth lately; he was quite lost by the whole class for one lecture, but we are, I hope, getting better, and more gleg [astute] at the uptake. We have been discussing the properties of infinite series, which are very perplexing.

Seventeen years later, the economist and logician Stanley Jevons recounted:[2]

We were delighted the other day when, in the higher senior, he at last appeared conscious that a demonstration about differential equations, which extended through the lecture, was difficult; he promised, indeed, to repeat it. But then one is disappointed to find that the hardest thing he gives in any of his classes is still to him a trifle, and that the bounds of mathematical knowledge are yet out of sight.

One of De Morgan's earliest, and most accomplished, students was the algebraist James Joseph Sylvester, who attended his lectures in 1828 at the age of only 14, returning as professor of natural philosophy in 1837. He did not particularly enjoy teaching physics, however, especially the experimental side, and deliberately kept his course as mathematical as possible. The topics he covered included statics, dynamics, hydrostatics, elliptic motion, gravitation, optics, and astronomy, with little or no reference to heat, electricity, or magnetism. An illustration of how mathematically inclined Sylvester's three-year course was can be gleaned from its prerequisites. For entry into the first year, students needed a knowledge of algebraic notation, proportion, and trigonometric functions. A familiarity with conic sections, quadratic equations, and spherical trigonometry was necessary for the second year, and for the third year, the student needed analytical geometry and the differential and integral calculus.

Despite the intellectual freedom offered by University College and the support and goodwill of the other professors (especially De Morgan), Sylvester became increasingly restless and dissatisfied with having to teach applied mathematics, longing for a pure mathematics chair of his own. In 1841, he was appointed to a professorship at the University of Virginia. But all did not go well, and the following decade saw Sylvester back as a teacher of mathematics in London, as we shall see.

Sylvester's successor in the natural philosophy chair was Richard Potter, who held the professorship until his retirement in 1865. In contrast to Sylvester, Potter's natural philosophy course was far more experimental, his chief interests lying in that area, as is indicated by his fifty-nine papers, chiefly in connection with optics. He also published a few textbooks on optics, hydrostatics, and mechanics, which were well respected at the time. However, he turned out to be quite incompetent as a lecturer. As a former student later recalled, by the 1860s Potter had become a laughing-stock:[3]

The professor was the dearest of old gentlemen with long, silky, silver grey hair, a winning smile, and a very gentle deprecatory manner... But as a teacher in my day, he had one fatal defect. He was worn out, he had lost his memory and not a few of his wits... In his mathematical class the professor was dependent upon his book. Sometimes, ashamed of copying, he would attempt a few lines on his own, and get hopelessly involved. In despair he would return to his book and copy the conclusion at the bottom. Some unkind student would point out a *non sequitur* in the middle. The dear old man, with a puzzled look, would glance from the blackboard to his book and from his book to the blackboard, and then turn to his class with an air of triumph and say, 'But, gentlemen, you see the conclusion is correct. It is a case of compensation of errors.'

Potter's retirement in 1865, together with De Morgan's departure two years later, afforded University College an opportunity to reorganize its mathematics teaching. The chair of natural philosophy was divided into two professorships: experimental physics and mathematical physics. The former chair was filled by George Carey Foster, a former student of the college, and the latter by Thomas Archer Hirst, a Marburg graduate, Fellow of the Royal Society, and a highly respected geometer. After De Morgan left the college in 1867, Hirst took over as professor of pure mathematics.

Although his tenure as professor was fairly brief, Hirst instituted two notable changes to the pure mathematics syllabus. The introduction of the theory of determinants was one, but the other was even more significant: the removal of Euclid from his geometrical classes. As we see in Chapter 14, the 19th century witnessed an intense debate among British mathematicians about the value of Euclid as a didactical tool. Hirst sided firmly with the modernizers who urged for its abandonment, concurring fully with Sylvester that it should be 'honourably shelved or buried "deeper than did ever plummet sound"'.[4] He pursued his belief with considerable effect, becoming in 1871 the first president of the body established to achieve such a purpose, the Association for the Improvement of Geometrical Teaching (or AIGT).

Hirst resigned the chair of pure mathematics in 1870, becoming assistant registrar of the University of London. His successor was Olaus Magnus Friedrich Erdmann Henrici, an intriguing and capable mathematician whose name, like Hirst's, is largely forgotten today. Born in Denmark, Henrici came to London in 1865, becoming acquainted with many of the foremost British mathematicians of the day. In 1867 he became Hirst's assistant at University College and in 1870, following Hirst's resignation, was appointed the new professor of pure

Thomas Archer Hirst (1830–92).

mathematics. He held the chair for ten years, before transferring to the applied mathematics professorship, which he held for a further four years.

One of the changes that Henrici instituted was the introduction of projective geometry and graphical statics into the mathematics syllabus—a radical departure from the analytically biased Cambridge-style course previously taught. To help his students with this new geometry, Henrici published a small book, *Congruent Figures* (1878), 'with the object of familiarising students from the very first with those modern methods'.[5] Henrici can also be given the credit for introducing vector analysis into English mathematical teaching, making much use of it in his classes. It was as a mathematical teacher that Henrici was primarily remembered, his success corroborated by student accounts of 'the singular lucidity of his teaching'[6] and the 'masterly ease and freedom'[7] of his exposition. He continued to teach projective geometry, vector analysis, and graphical statics, before being lured away in 1884 to the new Central Technical Institute in South Kensington.

In the accompanying position of applied mathematics, 1871 saw the appointment of possibly the most promising young British mathematician of the time, William Kingdon Clifford. A graduate of the rival King's College, Clifford's appointment to the applied chair at University College was his first and, as it turned out, his last academic position. In addition to his outstanding contributions to the fields of geometry and philosophy, Clifford was an excellent lecturer, giving enjoyable and (apparently) intelligible talks on abstruse topics with only a few brief notes. He left a profound impression on those who heard him lecture, and also on the scientific world at large:[8]

Olaus Henrici (1840–1918).

the word 'fascinating' could truly be applied to his oral communications... So much, however, depended on Clifford's manner and his imagery, his gentle voice, rapid diction, and clever way of putting familiar ideas, that it was afterwards difficult to recall what it was that had made so much impression at the time.

Clifford was one of the first to protest against the analytical bias of the Cambridge mathematical syllabus. Like De Morgan before him, he aimed to teach students not the analytical solution of a problem, but how to think for themselves. His applied mathematics course at University College (like Henrici's pure mathematics course, with which it ran parallel) was far more geometrical than those of his predecessors; his lectures introduced to England the graphical and geometrical methods of Möbius, Culmann, and other German geometers. Clifford thus shares the credit with Henrici for introducing graphical statics to English university education.

In 1884, University College appointed Micaiah John Muller Hill as the new professor of pure mathematics, a position he would hold until 1923. Hill had been a student of Henrici and Clifford at University College in the early 1870s, taking his B.A. degree in the University of London and coming first in the mathematical honours list. Described as 'one of the most commanding personalities'[9] of the college, as a teacher Hill was skilful, methodical, and extremely popular. Infinitely patient,

William Kingdon Clifford (1845–79).

he possessed that rare quality, which students so keenly appreciate, of never slurring over difficulties: time spent on making a demonstration perfect was always to him time well spent.[10]

For the chair of applied mathematics, the college managed to obtain yet another outstanding figure, Karl Pearson. Educated at University College School, he had proceeded to Cambridge and then to Heidelberg, where he studied law. Finally, in 1884, he was appointed professor of applied mathematics at University College, a position he held for twenty-seven years before becoming the country's first professor of eugenics in 1911. Pearson has been described as one of 'the most influential university teachers of his time'.[11] In his lectures he took 'great pains to be intelligible and could hold a large audience either of students or of merely casual hearers who were without special interest in his topics'.[12]

Not content with his duties at University College, in 1890 Pearson was appointed to the professorship of geometry at Gresham College, which had been providing free public lectures for Londoners since 1597, long before the appearance of any university in the city.[13] This post, which he held until 1894, required him to give popular lectures on subjects of his choosing. One of these subjects was probability theory, on which he lectured 'with that wealth of illustration, diagrammatic and arithmetical, which characterized all his popular lectures'.[14] It was in these lectures that he first introduced the terms *histogram* and *standard deviation*, now standard terms in statistics.

Pearson had become increasingly interested in mathematical statistics and in 1898 he introduced the subject into the University College curriculum. This comprised an elementary course, in which topics included the general theory of statistics, normal and skew variation, normal correlation, and an advanced course on the quantitative theory of heredity. His teaching involved the use of actual statistical data to calculate various types of statistical measurements and coefficients using tables and mechanical calculators. It constituted the first undergraduate course in mathematical statistics in Britain. (For more on Pearson's statistical work, see Chapter 12.)

Pearson's success in the applied mathematics chair is illustrated by the growth of his department, from an average number of nineteen students in the 1870s, rising to forty-three in the 1880s, and to seventy-seven by the late 1890s. By 1896 the department was employing one assistant professor and two demonstrators. The mathematics course and its teaching had changed considerably since the days of De Morgan and Sylvester, but throughout the Victorian period University College had remained the prime source of advanced mathematical tuition in London.

## King's College

But University College was far from being the only place to study university-level mathematics in the city. The second main establishment for higher education in the capital was set up as a direct consequence of the first. Disturbed by the secular nature of the 'godless institution of Gower Street',[15] several leading political and religious figures proposed a rival body, King's College, in 1829. It opened on the Strand two years later in 1831, offering tuition similar to that of its rival, but with the addition of compulsory lectures in theology. Throughout the 19th century, King's College remained the only major academic rival to University College in London, but as far as its mathematics was concerned, the overall calibre on the Strand was far more modest.

King's College.

The rivalry between the two colleges was at least partially resolved in 1836 with the Whig government's creation of what is now called the University of London. This was founded purely for the purpose of examining students from the two London colleges and awarding degrees. Tuition was deliberately excluded from its functions, being still the province of the colleges—a situation that remained unaltered throughout the century. However, as we shall see, the number of teaching bodies affiliated to the University of London would change dramatically by the end of our period, as would the character of the University itself.

In 1827, King's first professor of mathematics, the Reverend Thomas Grainger Hall, had been a candidate for the chair at University College to which De Morgan was subsequently elected.[16] A fellow of Magdalene College, Cambridge, Hall was elected to the mathematics chair at King's in 1830, 'which he continued modestly, faithfully and inconspicuously to occupy (rather than fill) for the next thirty-nine years'.[17]

During the 1840s and early 1850s, Hall's classes bore notable fruit: between 1840 and 1844, twenty-five of the Cambridge wranglers had attended King's College, including one of the most outstanding mathematicians of the 19th century, Arthur Cayley. By the mid-1850s, however, Hall's interest in mathematics and its teaching had languished. He had been appointed a prebendary of St Paul's Cathedral in 1845, a position he held until his death, and it seems that he became far more concerned with church matters than the mundane instruction of undergraduates. Indeed, by the time he resigned his chair in 1869, 'he had long been apathetic and devoid of active interest in either his subject or his pupils'.[18]

His colleague in the chair of natural philosophy, the Revd. Henry Moseley, was a more active mathematician. Unfortunately his class sizes were small, and since a professor's income was determined by the number of his students, Moseley's income was never comparable to that of his colleague in the mathematical chair. Eventually, possibly lured by a higher (and more reliable) salary, Moseley resigned his professorship to become a school inspector. His two successors, the Revd. Matthew O'Brien and Thomas Minchin Goodeve, were both eventually lured to professorships at the Royal Military Academy in Woolwich 'by the superior emoluments which the government could offer'.[19]

In 1860 the college acquired a man of outstanding scientific skill as its new professor. James Clerk Maxwell was arguably the foremost British mathematical physicist of the 19th century. Yet, remarkable though his scientific credentials may have been,

as a teacher of raw youths . . . he did not prove to be a success. "He was," says one who knew him, "a quiet and rather silent man, and it seems not unlikely that the students were too much for him."[20]

He resigned his professorship in 1865.

After Hall finally retired as professor of pure mathematics in 1869, he was succeeded first by the Revd. William Henry Drew, and then in 1882 by William Henry Hoar Hudson, a King's alumnus. Both were apparently Cambridge graduates of 'immense vivacity and energy',[21] though not, it would seem, of particular pedagogical originality. The same can also be said of subsequent mathematics professors following Hudson's retirement in 1903; in fact, for the first third of the 20th century, King's continued in much the same spirit as before, offering no serious mathematical opposition to its Gower Street competitor.

Throughout the 19th century, in both pure and applied mathematics, tuition at King's was adequate though hardly innovative. With the obvious exception of Maxwell, King's was also noticeably bereft of first-rate mathematical researchers, especially in pure mathematics. The combined skill in research and teaching, evident in so many of the staff at University College (such as De Morgan, Clifford, Henrici, and Pearson), was curiously absent from King's mathematical personnel, while Maxwell's first-class research abilities were accompanied by a disappointing performance as a lecturer. After him, one observes no mathematical professors of note in the Strand until the appointment of G. B. Jeffery in 1922, and it was not until the arrival of George Temple and J. G. Semple in the 1930s that the superiority of University College mathematics was significantly challenged by its long-term rival.

# Mathematics for women

The educational situation for London's women in 1837 was highly deficient, a fact that remained unchanged during the next decade, despite the installation of a female monarch. King's College statutes denied membership to non-Anglican men, let alone women (Anglican or otherwise), while University College, notwithstanding its doctrinal liberality, also remained an exclusively male domain.

Yet it was the staff of the apparently more conservative college on the Strand who were to be instrumental in the establishment of a college for London's women. Chief among them was Frederick Denison Maurice, a deeply committed Christian Socialist clergyman and professor of divinity at King's. Largely through his efforts, the first college in the country expressly for the education of women was founded in Harley Street. Queen's College, as it was called, opened in 1848 as a branch of the Governesses' Benevolent Institution. Maurice and several other professors from King's lectured there in its opening months, including Hall in mathematics and O'Brien in natural philosophy.[22]

Like its parent institution, Queen's College was operated on explicitly Anglican lines, a fact that soon led to the inauguration of a second women's college, on a Non-conformist basis. The principal figure this time was Elizabeth Jesser Reid, a widowed lady of property, whose dissenting background had acquainted her with many liberal educationalists of the day, including some of the professors at University College. With their support, and her money, Mrs Reid opened The Ladies' College at 47 Bedford Square in 1849.[23] Many of its early teachers were drawn from University College, including its first, Augustus De Morgan, who gave classes on arithmetic and algebra during its first year of operation.[24]

For twenty years the two ladies' colleges remained the sole teaching establishments for young women in London, until they were supplemented by the London Ladies Educational Association. In 1868, this body began to organize lectures in the vicinity of University College, though outside its premises. During the academic session of 1871–72, gradual moves towards mixed classes were made in the college, with the first integrated classes being given in art and political economy. Several professors followed suit; for instance, in 1876 Henrici admitted ladies to his higher senior mathematics class. Finally, in 1878, University College became the country's first coeducational institution, with 288 women admitted as undergraduates. Simultaneously, the University of London opened its examinations to women, who could now compete for degrees on an equal basis with men.

The integration of higher education in the capital did not signal the end of the ladies' colleges, or even a decline in the number of students. On the contrary, due to increased demand, a higher mathematics class was introduced in 1879 at Bedford College.[25] This also reflected the increased proficiency of its students: now that women were examined equally with men, it was reasonable for them to expect their tuition to reach the same standard. However, it seems that the same attitude did not prevail at Queen's College, which became, and remains, a private school for girls.[26]

Sophie Bryant (1850–1922).

It was not long before women began to graduate with distinction in mathematics from the University of London. Sophie Bryant, later headmistress of the North London Collegiate School and an early female member of the University Senate, was one of the first women to take the B.Sc. examination in 1881. She was also the first woman to attain a doctorate when she received a D.Sc. in 1884.[27] Other early female graduates were Philippa Fawcett, an alumna of both Bedford and University Colleges, who, as we saw in Chapter 1, gained the distinction of being placed above the senior wrangler in the Cambridge Tripos of 1890 (although she could not actually *graduate* from there), and Alice Lee, later to become a lecturer in physics at Bedford College.[28]

The 1880s also saw the opening of three new ladies' colleges in the vicinity of the rapidly expanding capital. In 1882, The College for Ladies at Westfield was founded in Hampstead,[29] to be followed four years later by the opening of Royal Holloway College in Egham, Surrey.[30] King's College opened a Ladies' Department in 1885, its location in Kensington rendering it a distinct entity from its parent college and ensuring continued separation of male and female students. It was finally incorporated in the University of London in 1910 as King's College for Women.[31]

Thus, by 1901, the situation for women's higher education in the capital was beyond any comparison with that of sixty-four years earlier. Not only was university-level instruction in mathematics now available to women, it was almost as accessible to them as it was to men. One's ability to graduate was no longer contingent on one's gender—a very different situation to that in military mathematics, to which we now turn.

## The military academies

Throughout the 19th century the British Army excluded women from entry into any of its branches. This policy was also followed at the prestigious Royal Military Academy in Woolwich. Although, strictly speaking, it lay outside London during the Victorian period, its adjacency to the capital and the many institutional and personal links between it and the rest of London make it an important constituent of this chapter. Founded by George II in 1741, the Academy's reputation in the mathematical world had been swelled in its first century by the distinguished professors that its generous salaries attracted—notably, Thomas Simpson and Charles Hutton.

As for the curriculum, Niccolò Guicciardini has pointed out that the teaching staff at Woolwich 'could not introduce any sophisticated innovations into the curriculum for the "raw and inexperienced" cadets'.[32] Thus, favourably disposed as Woolwich professors may have been to progressive new European methodology, they were 'unable to use it in research and, in reality, never even attempted to teach it in written works'.[33] Consequently, Hutton's *Course of Mathematics*, upon which the Woolwich mathematical programme had been based since 1798, still employed the Newtonian fluxional calculus in preference to more recent

The Royal Military Academy at Woolwich.

continental methods. Indeed, when we recall that differential and integral calculus were taught to a substantially high level at both of the Academy's scholastic London counterparts, we can only view the content of the Woolwich mathematical course of 1837 as embarrassingly behind the times. Moreover, says Guicciardini:[34]

we suspect that even the very elementary level required was not reached: from the *Records of the Royal Military Academy* . . . one gets the strong impression that the discipline of both the masters and the cadets was not exemplary.

The professor of mathematics at the Royal Military Academy in 1837 was Olinthus Gregory. A protégé of Hutton, Gregory had been appointed a master at Woolwich in 1803. More of an engineer than a mathematician, his most noteworthy contributions to science were in the form of his *Treatise of Mechanics* (1806) and his experiments to determine the speed of sound in 1823. A possible reason why the course taught to cadets at Woolwich changed so little was that Gregory was far more conservative than Hutton, especially in matters concerning the calculus. Indeed he went on record as saying that 'in point of intellectual conviction and certainty, the fluxional calculus is decidedly superior to the differential and integral calculus'.[35]

The chances of reforming the Woolwich course were substantially increased with Gregory's retirement in 1838. The Academy's governing body selected Samuel Hunter Christie as his successor, the first Cambridge man to hold the position. Christie was keenly aware of the need to reform all academic aspects of the Academy, and one of his first decisions as professor was to abandon Hutton's *Course* as the foundation of the Woolwich curriculum and replace it by his own *Elementary Course of Mathematics for the Use of the Royal Military Academy, and for Students in General*, published in two volumes in 1845 and 1847.

Olinthus Gregory (1774–1841).

Although it attempted to remedy the weaknesses inherent in Hutton's syllabus by introducing a programme more appropriate to the capabilities of the cadets, the new course had considerable flaws. The fluxional calculus had at last been banished, but in its place were only the 'elements' of differential and integral calculus, with no problems more taxing than finding maxima and minima or the areas under curves. Due, no doubt, to the shortcomings both of masters and cadets, it was apologetically noted that 'at present this subject cannot be much dwelt on'.[36] Even more extraordinary, however, is the virtual omission of applied mathematics, with the instruction in elementary mechanics proceeding no further than motions of projectiles *in vacuo*.

If the notion of a prospective artillery officer or engineer taking up his commission with such a trifling mathematical training seems absurd today, it was considered scandalous in certain contemporary quarters. Christie's new course was deemed far too elementary and was quickly rejected in favour of a new one, drawn up by three mathematical masters at the Academy: Stephen Fenwick, William Rutherford, and Thomas Stephens Davies. Christie retired in 1854 and was succeeded by the Revd. Matthew O'Brien, who, while at King's College, had been lecturing on astronomy at the Academy since 1849. He was awarded the Woolwich professorship in August 1854, but died a few months later.

At this point, another character reappears: James Joseph Sylvester. Following his premature departure from Virginia, he was back in England far sooner than he had anticipated. Being Jewish, he was barred from seeking employment at Oxford, Cambridge, or King's College in London, and with De Morgan and Potter firmly ensconced at the secular University College, Sylvester recognized the vacancy at Woolwich as the only opportunity for him to re-enter the academic world, since the Academy imposed no religious restrictions on its staff or cadets. Unsuccessful in 1854, he re-applied for the position and this time his application was successful. He was finally appointed to the chair in September 1855.

Sylvester's term of office saw several new developments in the administration of the Academy, chief of which was the introduction of a system of open competitive examinations. The chief examiner was Henry Moseley, another erstwhile professor of natural philosophy at King's College. His report reveals that the level of the examination was hardly severe: 'Only 31 out of 151 candidates afforded evidence of mathematical knowledge to which the designation "moderate" was applied by the examiners'[37]: this they defined as the 'power to work an easy sum in arithmetic, demonstrate a proposition in the first book of Euclid, and solve a simple equation'.[38] But this new system of exams was far from perfect: in June 1869 a cadet 'who sent in blank papers and wrote no fair notes was allotted 120 marks in Practical Mechanics'.[39]

Unfortunately, as with his former teaching posts, Sylvester's term as professor of mathematics did not improve his already erratic teaching skills. Indeed his reputation among the gentlemen cadets as an irritable and absent-minded eccentric was well earned, if the following anecdote is to be believed:[40]

on one occasion he suddenly looked up from a paper in the hall of study and demanded of the corporal on duty, 'What year is it?' An explosion of laughter in the room led to a 'scene', and the subsequent infliction of many punishments upon the cadets.

Yet despite occasional wrangles and his general dissatisfaction with the standard of mathematics he was obliged to teach, he remained at Woolwich until 1870, and even then left under duress when changes in Academy regulations decreed that all members of staff over the age of 55 had to retire.

Sylvester's replacement was the applied mathematician Morgan William Crofton. Educated at Trinity College, Dublin, he had taught mathematics at Woolwich since 1864, where he was appointed on the recommendation of Sylvester. We are told that his method of teaching was the antithesis of Sylvester's, being 'terse and lucid', and his mechanics relying on a 'direct geometrical presentation'.[41] This method was a great improvement on the efforts of his predecessor, and far more appropriate to the needs of trainee engineers or artillery officers. It seems to have been successful too, both militarily and mathematically, since at least two of Crofton's students went on to achieve fame: Lord Kitchener in the army, and Major Percy MacMahon in combinatorics, partitions, and invariant theory.

Crofton retired in 1884, to be succeeded by Harry Hart, a Cambridge graduate who had been a mathematical instructor at Woolwich since 1873. The principal event of his period as professor was the unveiling of a new mathematical syllabus in 1892. This course, divided into four classes and designed to take two years in total, built on alterations already initiated by Crofton.[42] The most noticeable feature of the new syllabus was the prevalence of applied mathematical subjects, indispensable to an apprentice engineer. Statics, dynamics, hydrostatics, and mechanisms were all taught to a considerable level, although the subject of hydrodynamics was curiously omitted. There was also a shift from a primarily analytical course to a more graphical and geometrically inclined one, perhaps influenced by a similar bias at University College. The standard to which the Woolwich course of 1892 aspired was thus considerably higher than that of a few years before. Now, at last, the Academy had a mathematics course comparable to its continental rivals.

Comparison with its counterpart at the start of our period is instructive. Mathematics at Woolwich in 1837 had been old-fashioned and irrelevant to the needs of most of the cadets. By the end of the Victorian era the course had changed almost beyond recognition, in the level of acquirement and applicability to the objectives of the institution. This was a clear rejection of the sloppy methods and over-simplification that had largely dominated instruction in mathematics at Woolwich for much of the 19th century. Hart's syllabus was the most progressive that the Academy had implemented, consolidating the improvements begun by his predecessor. While there was certainly still room for improvement, both in terms of content and quality of tuition, it was with this new curriculum that the Royal Military Academy entered the 20th century.

The British Army was not alone in teaching mathematics to its cadets. The subject was also an important ingredient in the curriculum of an establishment run by the Royal Navy, which began to rank as a London institution during the second half of our period. Founded as the Royal Naval Academy at Portsmouth in 1722, 'for instructing young gentlemen in the sciences useful for navigation',[43] it served as a naval counterpart to Woolwich, but the age of its cadets and the standard of their instruction were equally low. However, reforms had been under way since 1806 (when the school was renamed the Royal Naval College), which reduced its resemblance to the Woolwich Academy. Since 1829 it had also been training some commissioned officers and from 1839 had been an institution for adult education. In 1873, the College transferred from Portsmouth to Greenwich, reopening that autumn with Thomas Archer Hirst as its first Director of Naval Studies and upwards of two hundred students.

Cadets at the Royal Naval College at Greenwich, with their director of studies, Thomas Hirst.

To the modern mathematician, the most famous holder of a mathematics chair at the Naval College during this period was William Burnside. He had taught mathematics at Pembroke College, Cambridge, since 1875, and was appointed professor at the Naval College in 1885, where he remained until his retirement in 1919. Burnside is best remembered today for his work in group theory (see Chapter 15), and in particular for *The Theory of Groups of Finite Order* (1897) which was a standard work for many years. However, his mathematical research ranged over an extensive area: he wrote over one hundred and fifty papers on topics ranging from automorphic functions and complex analysis to probability theory and hydrodynamics.

As professor of mathematics at Greenwich, Burnside was engaged in the teaching of three main topics: ballistics for gunnery and torpedo officers, mechanics and heat for engineer officers, and dynamics for naval constructors, where his expertise in kinematics, kinetics, and hydrodynamics was invaluable. But, like any good teacher of mathematics, his success did not rest solely upon his mathematical proficiency. In an obituary of Burnside, the mathematician Andrew Forsyth wrote:[44]

Records and remembrance declare that he was a fine and stimulating teacher, patient with students in their difficulties and their questions—although elsewhere, as in discussions with equals, his manner could have a directness that, to some, might appear abrupt.

Thus, by 1901, London was twice as well served for instruction in higher-level mathematics for military use as it had been in 1837, and this does not refer only to the number of such institutions. At the start of Victoria's reign, neither the Military Academy nor the Naval College could be accurately described as university-level teaching establishments. At Woolwich, in particular, neither the course offered nor the tuition given were comparable to their scholastic counterparts. Yet by the turn of the century, we see in the teaching of Burnside at Greenwich and Hart at Woolwich, consideration of topics that would not have been out of place in the advanced mathematical courses of any contemporaneous high-level academic institution. More importantly, the move towards applied mathematics at both schools reflects the growing awareness of the need for instruction in the utilization of mathematics. This realization was not unique to the military, as we now see.

# Technical education

The Great Exhibition of 1851.

By the mid-19th century, British industry was fully aware of the need for a thorough technical education of the working population. The Great Exhibition of 1851, while providing a showcase for Britain's impressive industrial prowess, had also highlighted growing competition from new rivals such as Germany and the United States, where technical education was of major importance: Germany already had several Technische Hochschulen in cities such as Munich, Hanover, Stuttgart, and (most famously) Charlottenburg in Berlin, while in America, the Massachusetts Institute of Technology opened in 1865. Other European countries were also amply equipped with technical institutions, such as the Federal Technische Hochschule of Zürich and the École Centrale des Arts et Manufactures in Paris. It was realized that the technical deficiency in the training of British artisans, if not quickly remedied, would soon result in Britain losing her place as the world's foremost industrial power.

A start had been made earlier in the century by the Scotsman George Birkbeck, who, with other educational reformers, established the London Mechanics' Institute in 1823; this was renamed the Birkbeck Literary and Scientific Institution in 1866, and then Birkbeck College in 1907. Birkbeck and his associates had recognized early on that Britain, 'though the first manufacturing country in the world, is singularly deficient in schools for instructing the people in the Mechanical Arts'.[45] The new Mechanics' Institute was designed to redress this state of affairs, offering tuition in the physical sciences to working men. The lectures were popular, prompting the inauguration of similar institutes across the country until, by 1850, there were six hundred literary and mechanics' institutes nationwide. However, it quickly transpired that these institutes catered more to the lower middle classes than the workman, providing more in the way of general elementary education and social facilities than vocational training for the artisan.

The first moves towards constituting a thorough technical education at university level began around the time of the Great Exhibition. In 1845, the Royal College of Chemistry was founded in South Kensington. This was followed six years later by the establishment of the Government School of Mines and of Science Applied to the Arts. In 1853, these two schools were combined, with the latter renamed the Royal School of Mines ten years later. The next change occurred in 1881, when the schools moved to Exhibition Road in South Kensington and reopened as the Normal School of Science and Royal School of Mines, with Thomas Huxley as its first Dean. The former school's title soon proved unpopular and was changed in 1890 to the Royal College of Science.

The education provided by the College of Science was of a general scientific nature (physics, chemistry, and biology), whereas at the School of Mines instruction was more specialized (mining, metallurgy, and geology). The intention of both bodies to provide a high standard of instruction is reflected by the professors they appointed to teach, most notably the professors of physics at the College of Science. The first such professor was George Gabriel Stokes, supplementing his income as Lucasian professor of mathematics at Cambridge by lecturing part-time at the college between 1853 and 1859. He was followed by John Tyndall, Frederick Guthrie, and Arthur Rücker.

Initially, at least, training in mathematics was not a high priority at the Royal College, the emphasis being on practical, rather than theoretical, science. The chair of mathematics eventually grew out of the professorship of mechanics which, between 1869 and 1896, was ineffectively occupied by the former King's College professor Thomas Goodeve. His replacement in the chair (renamed mechanics and mathematics in 1881) was a far more successful mathematical educator. John Perry was an engineer who had previously taught at the Imperial College in Tokyo from 1875 to 1878 and later at the Technical College, Finsbury (see below). As we shall see in Chapter 14, he is primarily remembered today as the leader of the 'Perry Movement', a large body of technical and applied mathematical teachers, chiefly responsible for the complete divorce of Euclid and university education at the end of the Victorian era.[46]

Perry's success in increasing the mathematical reputation of his college is evidenced by the fact that, in 1913, his successor was the first professional mathematician to teach there—Andrew Russell Forsyth. Moreover, the following year they were able to procure the services of another equally distinguished practitioner, Alfred North Whitehead. The recruitment of such eminent figures demonstrated that, in mathematics, the Royal College of Science could now rival the previously unchallenged academic prestige of University College. However, the college at which Forsyth and Whitehead found themselves had nominally ceased to exist in 1907 when it was incorporated into the newly-formed Imperial College of Science and Technology.[47] This had been created from the amalgamation of three South Kensington colleges specializing in scientific education, the other principal constituents being the Royal School of Mines and a more recent creation, the Central Technical College.

The City and Guilds of London Institute for the Advancement of Technical Education was founded in 1878 by the various Guilds (such as the Mercers', Drapers', and Clothworkers' Companies) of the City of London. Teaching began in Cowper Street, in Finsbury, an area slightly north of the city. It was officially inaugurated in 1883 as The Technical College, Finsbury, with professorships in electrical engineering, chemistry, and mechanical engineering, the last of which was held by Perry. As intended, the majority of students there were artisans, such as engineers, engravers, electricians, brewers, instrument makers, and printers, numbering one hundred in 1882 but increasing to two hundred and ten by 1894.[48] The college filled two complementary roles, serving as a finishing technical school for those about to enter industrial life, and operating as an intermediate college for those intending to progress to the proposed central technical college in South Kensington.

This opened in 1884 as the Central Technical Institute (renamed the Central Technical College in 1893). The college was essentially a school of engineering with four professorships: chemistry, physics (later electrical engineering), civil and mechanical engineering, and mechanics and mathematics.[49] The founding professor in this fourth chair was Olaus Henrici, who had been enticed from his post at University College. At South Kensington, he continued his teaching of projective geometry and vector analysis, also exploiting his new purpose-built premises to establish an innovative laboratory of mechanics upon which many later versions were based. Here he continued his research, developing (among other things) a harmonic analyser to calculate Fourier coefficients mechanically, following a similar machine by Lord Kelvin. He finally retired from the college in 1911.

The Central Technical College soon established a high reputation. When its first courses began in January 1885, the number of full-time students had been only six, this number rapidly increasing to two hundred and eight just ten years later.[50] By 1900, the college's premises, designed to accommodate two hundred students, were considerably overcrowded. Indeed, so wide had its standing grown that, by 1902, students were coming from India, South Africa, Japan, Italy, and even Germany, and paying the substantial sum of £35 per annum for the privilege. It was the Central Technical College that was to form the third component of the new Imperial College upon its foundation in 1907, evolving into what is today its Faculty of Engineering.

The Central Technical College, later incorporated into Imperial College.

We come, finally, to the provision of technical education for London's working population. In the 1830s, there had briefly existed in London an institution called 'The Adelaide Gallery', after the wife of King William IV. While ostensibly an educational establishment, it was devoted more to the exhibition of new scientific instruments and curiosities than to scientific research or teaching. In 1838, an imitation was set up on Regent Street in central London. Entitled the 'Polytechnic' it functioned along similar lines, but with the addition of occasional popular lectures. Both institutions enjoyed periods of evanescent popularity and prosperity, but after a few years eventually went bankrupt. In 1880, Quintin Hogg, a wealthy philanthropist, bought the Polytechnic's disused premises on Regent Street and reopened it under the same name but with a different agenda.

The new Regent Street Polytechnic now operated as a centre for the improvement of the working man, with classes in science, art, and literature; physics and chemistry laboratories; and a library, gymnasium, and various sporting, religious,

Contemporary drawing of the Regent Street Polytechnic.

and educational clubs. Over six thousand students enrolled in its first year, rising to fifteen thousand by 1900.[51] Like Birkbeck's Mechanics' Institute half a century earlier, the success of Hogg's Polytechnic inspired the foundation of similar polytechnics for the working population of London, such as the East London Technical College founded at Mile End in 1884,[52] the Northern Polytechnic at Holloway, and The Goldsmith's Institute at New Cross in south-east London, founded in 1894.[53] The initial purpose of most of these polytechnics was to provide basic mechanical and manual instruction for the working classes, but before long, more academic studies had been brought in to supplement the technical training.

This increase in the general range and quality of polytechnic courses coincided with significant alterations in the constitution of the University of London around the turn of the century, establishing it as a teaching as well as an examining university. These changes, resulting from the 1898 University of London Act, permitted it to admit educational institutions of a certain standard as 'Schools of the University'. Naturally, University and King's Colleges were included, together with many others, such as the Bedford College for Women. The Central Technical College was also admitted as a University school in its Faculty of Engineering. But perhaps the most remarkable consequence of the University's new constitution was the admission of three polytechnics as schools of the University by 1907—namely, Birkbeck College, the East London College (now Queen Mary's College), and Goldsmith's College. This move was all the more desirable since by 1907 there were eighty-six recognized 'teachers of the University' working in the London polytechnics, and more than seven hundred and fifty polytechnic students studying for University of London degrees.[54]

# Conclusion

In this chapter we have surveyed the immense changes undergone by Britain's capital during the Victorian period. We have alluded to the great contrast in size between 1837 and 1901, with the population growing from one and a half million to four and a half million people: this was reflected in the huge increase in the number of institutions relevant to our subject during the intervening period. London and its environs had begun the Victorian era with a mere three institutions offering higher-level mathematical tuition. By the beginning of the 20th century, this number had increased to more than twenty, providing courses in mathematics no longer solely for purely academic or military purposes, but also for other facets of society such as industry and commerce.

We have also seen that a study of mathematics education at university level can shed some light on social developments in the capital, particularly with respect to women and the working class. The majority of the new institutions created in Victorian London were designed to improve the education of at least one of these two groups, and these improvements to a certain extent mirrored the social and political fluctuations that occurred during the period. The changes that took place with regard to the mathematical education of women and the working classes both reflected and participated in the alteration of both groups' political status between the beginning and end of our period. In 1837, both parties were politically impotent, having no right to vote, but by 1901 much of the working class population had received the franchise and even women could vote in local government elections. However, it was the 20th century that would witness the final progression (political and educational) that would aim to place women and workers on an equal footing with the rest of the population.

If we now turn our attention to the general characteristics of advanced mathematical tuition in Victorian London, several distinguishing features become apparent. One of the most striking is the number of prominent mathematical researchers who earned a living by teaching the subject. Certainly, the capital had more than its fair share of less academically distinguished professors (such as Hall, Drew, and Goodeve), but a remarkable number of top-rank mathematicians were also involved. Little explanation is required for this phenomenon. Throughout the 19th century, mathematicians could not support themselves by research alone: academics were paid solely to teach, and research constituted no part of a professor's duties.

It is therefore hardly surprising that a considerable number of high-calibre London-based mathematicians chose to earn their living by teaching mathematics. In many cases, strong researchers (such as De Morgan, Clifford, Henrici, and Pearson) also proved to be equally successful teachers. However, just as effective lecturing does not imply profound research, not all skilled mathematicians made good teachers (as witness Maxwell and Sylvester). This is not to say that all mathematicians supported themselves by tuition: teaching appealed little enough to many of those engaged in it! Perhaps the best example of a London-based mathematical researcher who preferred not to teach is Arthur Cayley: he subsidized his research by working for twenty years as a lawyer. Even when appointed Sadleirian professor at Cambridge in 1863, he kept his lecturing duties to an absolute minimum.

We also note the connections between the various institutions provided by the migration of pupils or professors from institution to institution. For example, Sylvester was both a pupil and professor of natural philosophy at University College, and later professor of mathematics at Woolwich; similarly, Clifford was a pupil of King's and a professor at University College. But not all connections were professorial: Hirst was linked with both University College and the Greenwich Naval College, having been a professor of mathematics at the former and director of studies at the latter. While links with other locations would also be interesting, it remains a testament to the growth of university-level mathematical instruction in this period that so many mathematicians were able to spend so much of their careers teaching within the same geographical area.

If one had to compare one locality outside London with the capital, the obvious place to pick would be Cambridge. Most of the principal characters in this chapter were associated with the Cambridge mathematical community at some time in their careers, either as staff, students, or both. Indeed, it would be quicker to mention those involved in London mathematics who were *not* Cambridge men (such as Hirst, Henrici, Crofton, and Perry) than it would to list those who were. All the major London institutions of this period had at least one Cambridge graduate on their staff. Moreover, at King's College every professor of mathematics or natural philosophy appointed throughout the entire Victorian period was a Cambridge man. So the prevalence of Cambridge-trained mathematicians is one more characteristic of 19th-century London mathematics.

The dominance of University College mathematics has been stressed throughout this chapter, and is another distinguishing feature of higher mathematical education in London during this period. It is no coincidence, therefore, that the great majority of eminent scholars who also happened to be good teachers were associated at one time or another with that institution. But one further tendency, prevalent not only at University College but also in the other London institutions, was an increased inclination towards applied mathematics. In 1837, before receiving tuition in 'mixed mathematics' at either of the two academic London colleges, it would have been necessary to pass through much of the grounding in pure mathematics before one could begin to deal with its applications. At Woolwich, most of the mathematics course was pure, and the standard of the applied mathematics was scarcely adequate.

As the century progressed, the availability of advanced classes in applied mathematics rose sharply, especially with the inauguration of the technical colleges and polytechnics towards the latter part of the period. Thanks to the progressive methodology of such professors as John Perry, students in these new institutions were taught mathematics to facilitate construction, design, engineering, and other related disciplines, without reference to many of the abstract notions previously considered as prerequisites for the study of applied mathematics. In the older establishments, the trend towards the applied side can also be detected. Most of the course innovations at University College after the 1870s took place in the applied department, while at Woolwich, the syllabus that had evolved by the 1890s was strikingly more applied than its predecessors. Thus, by the death of Queen Victoria in 1901, both the standard and the availability of tuition in the applied branches of mathematics had increased dramatically. Consequently, as Britain entered the 20th century, mathematics in London had become more accessible and of more service to its population than ever before.

The University of Glasgow in 1870, as depicted in *The Graphic*.

CHAPTER 4

# Scotland

*Land of opportunity but few rewards*

### A. J. S. MANN AND A. D. D. CRAIK

The Scottish universities were very different from Oxford and Cambridge. They were far less well endowed, had fewer staff, suffered recurrent financial crises, and the student intake was much more egalitarian. The full M.A. degree was broadly based, but many students took only a few courses. There was just one professor of mathematics and one of natural philosophy at each university, and few assistants; accordingly, the professors spent much time on elementary teaching. Nevertheless, several also carried out notable research, mainly in applied mathematics and physics: the most illustrious were William Thomson, Peter Guthrie Tait, and James Clerk Maxwell, but others made substantial contributions. From around 1830, ambitious Scottish graduates often proceeded to Cambridge, where they did well in the mathematical tripos: several later returned to Scotland as professors and promoted changes in the curriculum. The Royal Society of Edinburgh and, later, the Edinburgh Mathematical Society provided opportunities for discussion and dissemination of ideas and research.

The situation of mathematics in Victorian Scotland is well illustrated by the case of one young Scot. Alexander Bain was born in 1818, one of eight children of a poor Aberdeen handloom weaver, who struggled to make ends meet as mechanization threatened his trade.[1] Bain's early education was from a succession of schoolmasters with whom he studied arithmetic and algebra, becoming proficient 'as far as equations, simple and quadratic'. Aged 9 or 10 he attended a school attached to Gilcomston Church in Aberdeen, where, he recalled:[2]

I got a "Bonnycastle" for my own special use, and went on to the cubic and higher equations, and a number of miscellaneous topics,—which I understood generally, but the exercises under them had often a peculiar trick that I could not work out: no more could the master, until he took them to his own rooms in the evening, and consulted a Bonnycastle Key. With such a master, my progress in algebra, at least during the three years, was but small.

His attempt to learn Euclidean geometry was even less successful: it soon became evident that

I was working by force of memory alone, and had not the smallest comprehension of the geometrical processes . . . [but] What I could not do at nine or ten, was found perfectly easy at fourteen, by mere brain growth.[3]

Alexander Bain (1818–1903).

When he was 11, Bain left school since his father could not afford the grammar school fee. But, while working at the loom, he continued to educate himself, particularly in science, drawing on the library of the Aberdeen mechanics' institution. At age 16 he was learning fluxions from Charles Hutton's *A Course of Mathematics*, and reading Thomas Simpson's *Geometry*, Hutton's *Recreations*, and various volumes of the *Library of Useful Knowledge*. He also befriended George Innes, a watchmaker with an interest in mathematics and astronomy, who 'knew a great deal of the gossip of Mathematics . . . It was from him I learned that Fluxions had now given place in England to the Calculus'.[4]

About this time, Bain attended the evening mathematical school of William Eigen, with whom he studied differential calculus from the English translation of Lacroix, and from whom he borrowed Laplace's *Système du Monde* and, to his joy, Andrew Motte's English translation of Newton's *Principia Mathematica*.[5] He was also reading metaphysics and theology, and attending the mechanics' mutual instruction class. Resolved to study Latin, Bain bought a copy of 'the Jesuits' edition' of the *Principia*, and by the end of 1835 he was well through the second book.

A minister who met Bain encouraged him to go to university and helped him to prepare for the Latin element of the bursary competition. After three months' attendance at the grammar school, for which the teacher took no fee, Bain sat the bursary competition for Aberdeen's Marischal College. Although placed just outside the bursary list, he was advised to join the classes and a vacant bursary was found for him.

Thus, at the age of 18, Bain went to university, while initially also teaching mathematics at the Aberdeen mechanics' institution two evenings a week, and working at the loom between sessions. No mathematics was taught in his first year, but in his second year he enrolled in Professor John Cruickshank's course;[6] this comprised Euclid, plane trigonometry, and algebra up to quadratics. All this was already familiar, but Bain recalled that 'what I gained from Cruickshank was the correction of many slovenly ways of dealing with the propositions of Euclid, and an improvement in precise handling generally—just what the self-taught student is deficient in'.[7]

In his third year, Bain studied natural philosophy and more advanced mathematics—higher algebra, conic sections, spherics, and geometrical astronomy—and towards the end of his studies he won the competition for a mathematical bursary of £30 a year for two years, which 'helped to give me a maintenance... independent of private teaching or other drudgery'.[8] With his bursary Bain was required to take the otherwise optional third mathematics course involving differential calculus and analytical trigonometry, but by this stage 'my exclusive interest in mathematics had long since faded... For purposes of general scientific culture, I had as much as was at all needed'.[9]

On completion of his studies, Bain became assistant to the professor of moral philosophy, George Glennie, and he also gave Cruickshank's mathematics classes while the latter was indisposed. At the end of the 1841–42 session, he visited London to seek other opportunities, turning down a situation at Greenwich Hospital, which he considered 'very much of the nature of a common school'.[10] He also applied, without success, for a chair in natural philosophy and astronomy at Bombay (now Mumbai) in India.

Bain had hopes of eventually replacing Glennie as professor of moral philosophy, but, on learning of this, Glennie took offence and in 1844 dismissed Bain as his assistant. Bain then applied for the newly vacant chair of natural philosophy, which however went to David Gray, head of Inverness Academy, Bain's religious skepticism perhaps being the decisive factor. In March 1845 Bain applied, also unsuccessfully, for the chair of logic at St Andrews, but a month later he was appointed professor of mathematics and natural philosophy at the 'Andersonian University' of Glasgow; Anderson's College provided scientific and technical instruction mainly for trainee shipbuilders and engineers.[11] There he taught mathematics to Irish medical students, among others, and gave popular evening lectures in natural philosophy. He made later unsuccessful applications for the chair of moral philosophy at St Andrews, and as substitute for the ailing professor of natural philosophy at the University of Glasgow.[12]

When Glennie died in November 1845, Bain applied for his chair. His main rival, William Robinson Pirie, had influential support and Bain's religious views again counted against him. However, he believed that, had Lord John Russell been able to form a Whig government when Robert Peel resigned in 1846, he would have been appointed. In the end, the Lord Advocate, perhaps unwilling to upset supporters of either Pirie or Bain by preferring the other, appointed a third candidate, William Martin of Madras College, St Andrews. Bain next considered applying for positions at the new Queen's Colleges opening in Ireland (see Chapter 5), but he did not pursue these since his inquiries suggested that native Irishmen would be preferred. Late in 1846 he was an unsuccessful candidate for the chair of natural philosophy at St Andrews, which went to the Cambridge-educated German, W. F. L. Fischer.

Bain then moved to London to a post as assistant secretary to the Metropolitan Sanitary Commission. He subsequently taught moral science and geography at the all-women Bedford College, and examined for the University of London and the Indian civil service. His interest in two further chairs, those of philosophy at Queen's College, Belfast in 1855, and of logic at St Andrews in 1860, again came to nothing. At last, in 1860, Bain was appointed by the Home Secretary to the new chair of logic at Aberdeen, against the wishes of the principal, following the merger of King's College and Marischal College. He is best remembered today for his pioneering books on psychology, *The Senses and the Intellect* (1855) and *The Emotions and the Will* (1859), as well as being the founding editor of the journal *Mind*.

Bain's experience was not untypical for a young Scot considering a university career. Long before, the talented John West, an assistant to Nicolas Vilant, the professor of mathematics at St Andrews, had emigrated to Jamaica through lack of prospects in his native land. And James Ivory, perhaps the most able British applied mathematician of his day, managed a spinning mill before joining the Royal Military College in Marlow (later Sandhurst) after his business failed. Others, such as Adam Anderson and Thomas Duncan, were school rectors for many years before securing university chairs.[13]

We have dwelt at some length on Bain's mathematical education, and on his early attempts to forge an academic career, because they illustrate key aspects of mathematical life in Scotland in the 19th century. Bain took advantage of a variety of educational opportunities that were available to the son of impoverished parents. Without major financial outlay, he was able to attend and do well at his local university. But his prospects of a university career were dependent entirely on his being able to obtain a chair that chanced to become vacant at the right time, and he had to be flexible about which discipline to follow. His unwillingness to join the established church, and so gain the influential support of one or other of its moderate or evangelical factions, also significantly diminished his prospects.[14]

In contrast, top-ranking Cambridge and Oxford graduates had almost automatic access to college fellowships that provided financial security and stability, and time for private study and research, until an attractive opportunity came along. Many used this time to prepare for the Anglican Church, and were rewarded by a church living in the gift of their college; others studied law at one of the London Inns of Court, and a few became university professors, sometimes in Scotland.

## Scottish universities in Victorian times

University reform was hotly debated throughout the Victorian period, and the Scottish universities underwent significant and controversial change in the last years of Victoria's reign: as we shall see, mathematics and mathematicians featured significantly in the controversy. We begin by describing the situation that prevailed for most of the 19th century: this account draws heavily on the work of R. D. Anderson,[15] and also acknowledges G. E. Davie's provocative analysis in his book *The Democratic Intellect*.[16] Davie views this period as 'the tortuous, dark revolution whereby a nation noted educationally both for social mobility and for fixity of first principle gradually reconciled itself to an alien system in which principles traditionally did not matter and a rigid social immobilism was the accepted thing'.[17]

The Scottish system was very different from the English one. As Anderson noted:[18]

The belief that Scottish education was peculiarly 'democratic', and that it helped to sustain certain corresponding democratic features of Scottish life, formed a powerful historical myth, using that word to indicate not something false, but an idealization and distillation of a complex reality, a belief which influences history...shaping the form in which institutions inherited from the past are allowed to change.

In every rural parish there was a school, inspected by the Church of Scotland, and the towns had burgh schools, so some form of education was available to all, although Samuel Johnson noted:[19]

How knowledge was divided among the Scots, like bread in a besieged town, to every man a mouthful, to no man a bellyful.

In the early 19th century Scotland had five universities—Edinburgh, Glasgow, St Andrews, and the two Aberdeen Universities, King's College and Marischal College (which merged in 1860). The separate, mainly technological, Anderson's College in Glasgow had been founded in 1796, and University College, Dundee, was founded in 1881, becoming soon afterwards a college of St Andrews University. All these universities attracted a genuine social mix, although women were not admitted until the 1880s and 1890s, except at Anderson's College which admitted a few women from the outset.

Marischal College in Aberdeen.

Not all university students aimed to take the complete four-year Master of Arts course. Some attended lectures only in those subjects considered relevant to their work, sometimes attending early morning classes before their day's labour. An informal survey of the junior humanity (Latin) class in Glasgow in 1876 found that, out of two hundred and eighty-three students, only about sixty did not have jobs that interfered with their attendance during at least part of the academic year.[20] Some were clerks in the city, while others, such as Highland school-teachers and farm workers, attended seasonally when their labours permitted. P. G. Tait estimated that only a quarter of the two hundred and twenty-nine students in his 1879 natural philosophy class at Edinburgh were taking the full M.A. course. Even those who took the full course rarely bothered to graduate: with no honours classifications, personal testimonials from individual professors, rather than degree certificates, carried weight with employers. Students lived at home if they could, or in private lodgings: an environment very different from the collegiate (and expensive) Oxford and Cambridge.

The curriculum was designed for this mixed intake. George Chrystal, professor of mathematics at Edinburgh at the end of our period, describes his situation at Aberdeen University, where he went with a scholarship in 1867, having won several prizes at the grammar school in Aberdeen:[21]

When I entered the University of Aberdeen some eighteen years ago I was a moderate classical scholar, but I had learned practically no mathematics. We used to read the first book of Euclid as far as *pons asinorum*;[22] but regularly as we reached the dreadful pass we were turned back for a revisal. Algebra I had none, not to speak of other mathematical furniture. Yet large demands were made upon me during my second session under Professor Fuller, and I had to work hard during the spare time of my first year to be able to take his junior class with advantage. The fact that mathematical students from Aberdeen had been doing well in the world long before the time I allude to, was due to no exertions on the parts of the schools, but simply to the presence in the Faculty of Arts of two teachers, Professors Fuller and Thomson, of exceptional energy and ability, whose efforts were ably seconded by a private tutor, Mr Rennet, well known and much beloved by all Aberdeen graduates, who combined in a way more happy than common the power of dealing at once with the best and with the worst material that came up to the university.[23]

However, the use of private tutors in Scotland was far less widespread than at Cambridge, where they were indispensable for advanced mathematical education (see Chapter 1).[24]

In the Scottish universities, Greek was taught from scratch, though some knowledge of Latin was assumed, and mathematics began from the start of Euclid's *Elements*. Some students were very young, often 13 or 14 (and William Thomson attended classes at Glasgow from the age of 10); others like Alexander Bain were adult. The overlap between school and university mathematics was demonstrated by an episode in 1836 when the St Andrews professor of mathematics, Thomas Duncan, wrote to the Trustees of Madras College, the popular local school, asking that university students be barred from Thomas Miller's mathematics classes, which they were attending in preference to his own.[25]

Students, then as now, did not always behave perfectly in class, as Robert Louis Stevenson indicates in his memoir of Fleeming Jenkin, his erstwhile professor of engineering at the University of Edinburgh:[26]

Fleeming Jenkin (1833–85).

At the least sign of unrest his eye would fall on me and I was quelled. Such a feat is comparatively easy in a small class, but I have misbehaved in smaller classes and under eyes more Olympian than Fleeming Jenkin's…I was not able to follow his lectures; I somehow dared not misconduct myself, as was my customary solace; and I refrained from attending.

But Stevenson needed a certificate of attendance from Jenkin:[27]

During the year, bad student as I was, he had shown a certain leaning to my society; I had been to his house, he had asked me to take a humble part in his theatricals; I was a master in the art of extracting a certificate even at the cannon's mouth; and I was under no apprehension. But when I approached Fleeming, I found myself in another world; he would have naught of me. 'It is quite useless for *you* to come to me, Mr. Stevenson. There may be doubtful cases, there is no doubt about yours. You have simply *not* attended my class…You see, Mr. Stevenson, these are the laws and I am here to apply them,' said he. I could not but say that this view was tenable, though it was new to me…

Happily, when Stevenson was able to prove that he had already accumulated enough certificates for his degree and needed Jenkin's only to satisfy his family, it was provided.

From the beginning of Victoria's reign, problems in Scottish higher education were becoming increasingly apparent. The system was seriously underfunded and the professors' meagre salaries were augmented by the class fees from pupils, even as late as Chrystal's day:[28]

The position of the Professor of Mathematics is this—he draws the main part of his income from the fees, the larger part of this comes from the Junior Class; for his higher work he receives practically nothing. Every step that he takes in improving the teaching of his subject, every schoolmaster that he helps to train to teach mathematics better in Scotland, aids in diminishing the number of students attending the junior class, the effect of which is to diminish the Professor's income, and to bring down upon him abuse in the newspapers regarding the fall of numbers in his department, and unpopularity with parents because the standard for the pass degree shows a tendency to rise in sympathy with improvement in the learning both in and outside the university. The unfortunate Professor may be accused in one and the same day of teaching too low in order to secure fees, and of examining too high for the same base purpose. (laughter and applause).

Obviously, earning tuition fees was of great concern to professors. For example, in the 1830s William Wallace, Edinburgh's professor of mathematics, wrote

William Wallace (1768–1843).

that he was materially affected by rival establishments offering mathematics teaching, and was worried that the appointment of a professor of astronomy would provide further competition. Extra sources of income for professors included writing for encyclopedias and, of course, textbooks for students: indeed, when Wallace reverted from using his predecessor John Leslie's textbook to that of John Playfair, the financial loss to Leslie may have contributed to his subsequent poor relations with Wallace.[29]

Funding was one issue. There was also an increasing awareness that the general nature of the degree, and its accessibility to all, meant that the Scottish universities were not vital centres of research—although natural philosophy under William Thomson in Glasgow and P. G. Tait in Edinburgh were notable exceptions. But there was no unanimity on whether degrees should remain general or should permit some specialization, or on whether universities should be primarily teaching institutions accessible to all or should raise their standards, and so foster research, by introducing entrance requirements. (Like the Scottish universities, neither Oxford nor Cambridge had any formal entry requirements at this time.) But university reform was on the agenda throughout the 19th century, with no fewer than four commissions (in 1826, 1858, 1876, and 1889) making recommendations for change in Scotland.

Advocates for reform put forward proposals for increasing the number of professors and subjects taught, for an entrance examination to allow more advanced teaching, and for specialist honours degrees. These proposals were stoutly resisted by those who supported the democratic nature of Scottish education and the general arts degree with metaphysics at its heart. They argued that learned Scots like John Leslie, James D. Forbes, Sir William Hamilton, and Thomas Chalmers (all foreign associates of the Institute of France) showed that the Scottish system could produce internationally recognized scholars.[30]

The Scots were proud of their universities. The social mix that resulted from their system was seen as a benefit. In 1878, the Revd. John Macleod claimed in a school inspector's report that:[31]

The communist of Paris or the republican of London becomes the clergyman, the doctor, the lawyer, or the schoolmaster of Scotland,

and that the Scots were peaceful by nature because of:

the absence of these sharp lines of demarcation which separate the several grades of society in other countries; and this again is largely, if not entirely, the result of the bridge across which so many youths passed yearly from the common schools to the universities.

And in 1889, the parliamentarian Lyon Playfair, a former professor of chemistry at Edinburgh who had represented the Universities of Edinburgh and St Andrews as an MP, said in Parliament that:

The English Universities … teach men how to spend £1,000 a year with dignity and intelligence while the Scotch Universities teach them how to make £1,000 a year with dignity and intelligence.[32]

On the other hand, reformers pointed to the merits of the research-intensive German universities, as well as to Oxford and Cambridge. The debate continued for most of the century. Davie, in his controversial account of the period, sees the campaign over the mathematics chair in Edinburgh in 1838 as one of the early rounds of the battle. The leading candidates were the Scot Duncan Farquharson Gregory, from the famous mathematical family, and the Englishman Philip Kelland, both then based in Cambridge. Sir David Brewster (who favoured a third candidate, the Scottish-educated John Scott Russell) argued strongly against Kelland in a public letter to the Lord Provost (the decision being taken by the town council):[33]

After I had the honour of an interview with your Lordship on the subject of the Mathematical Chair, I found among some of the members of the Council a strange delusion prevailing respecting the superiority of Cambridge Mathematics to that which is taught in Scotland, and the necessity of creating a higher standard of mathematical learning.

If this is the case, then we must consider our Maclaurins, Stewarts, Playfairs, and Leslies, as men that ought to have been replaced by the first Wranglers of Cambridge who flourished at the time of their election, although the very names of such Wranglers have fallen into utter oblivion. A First Wrangler of Cambridge, as your Lordship knows, is not a person of *inventive genius* who has distinguished himself by positive discoveries; but a mere hard-working individual, who, aspiring after University honours and Church preferment, has crammed himself with mathematical learning, which is often never digested either for his own use, or that of the public. Is Mr Kelland a greater mathematician than Dealtry or Woodhouse, or are any of these persons to be compared with Playfair or Leslie—men who were eloquent writers, eminent philosophers, as well as mathematicians?

Kelland was appointed, with the strong support of Edinburgh's professor of natural philosophy, J. D. Forbes. Davie sees this as a milestone in the Anglicization of the Scottish universities, arguing that Kelland was to break decisively with the metaphysical approach to mathematics that had been traditional in the Scottish universities (exemplified in the notes to Robert Simson's 1756 edition of Euclid's *Elements*, where two pages of commentary are attached to the very first sentence, Euclid's definition of a point).[34] Yet Kelland was to be a strong defender of the Scottish system, arguing its appropriateness in two lectures, later published, that were less than complimentary about the Cambridge professorial arrangements. But Kelland was not the sole Cambridge wrangler amongst Scottish professors for long: of the twenty-eight professors of mathematics and natural philosophy who held office at Scottish universities between 1846 and 1900, sixteen had taken the mathematical tripos at Cambridge.[35]

While the number of students taking advanced mathematical classes in Scotland remained small, the competitive examinations of the Cambridge mathematical tripos encouraged ever higher standards, with considerable rewards available to the best performers.[36] From around 1830 it became increasingly popular for the best students from Scottish universities subsequently to take the Cambridge mathematical tripos, and often Scots (among them D. F. Gregory, W. Thomson, Tait, and Maxwell) achieved high positions in the list of wranglers. That this was a double-edged national achievement was noted by Thomas Muir, the second President of the Edinburgh Mathematical Society, in his presidential address of 1884:[37]

Probably in the newspapers we observe that Mr. Donald Scott of a certain northern university has gained an open scholarship at Johnshouse, and the competition having been between him and a number of young men fresh from the English public schools, we are gratified accordingly with his startling success. Gentlemen, I put it to you, if this is a thing for us as Scotsmen to be altogether proud of. When in these cases a young Scotch student competing with English students *of the same age* gains a scholarship, there may be cause for gratulation: but the Scotsman who glories in the part his Universities play in the matter glories in his own shame. Is it really past hoping for that all this may yet be changed?

The increasing number of gifted Scots progressing to the Cambridge tripos, together with the Cambridge background of many of the Scottish professors, meant that the influences between the two educational systems were by no means in one direction only. The reasons to study in Cambridge were practical as well as academic. The high prestige accorded to top wranglers in the mathematical tripos, the availability of college fellowships, and the growing demand for Cambridge-trained mathematicians to fill academic posts elsewhere (including Scotland) were strong incentives for ambitious young Scots. Cambridge also benefited by the presence of several able Scots who became fellows or professors, and who contributed much to the research ethos. But the status of the Scottish universities gradually diminished as that of Cambridge grew.[38]

Despite the stir that they caused, the Royal Commission of 1826 and the Executive Commission of 1858 had relatively little immediate effect. The 1826 commission produced its report in 1831: it suggested an entrance examination for Scottish universities, and a four-year M.A. programme for students wishing to graduate, with the following curriculum (Box 4.1).[39]

Throughout the 19th century (and beyond), Westminster was never quick to make time for purely Scottish matters. Although a Bill to implement the proposals was introduced in 1836, vocal opposition from many quarters led to its withdrawal. But pressure for reform was maintained by the Association for the

## Box 4.1 Suggested M.A. curriculum

| Year 1: | Latin 1 | Greek 1 | |
| Year 2: | Latin 2 | Greek 2 | Maths 1 |
| Year 3: | Maths 2 | Logic (including rhetoric) | |
| Year 4: | Natural philosophy | Moral philosophy | |

Improvement and Extension of the Scottish Universities, led by the classical scholar John Stuart Blackie and the advocate James Lorimer. Factors such as the introduction in 1855 of an examination for entry to the Indian Civil Service, with an adverse effect on the numbers of Scots being accepted, reinforced the arguments for university entrance examinations and higher levels of study. Finally, the Universities (Scotland) Act of 1858 set up an Executive Commission under John Inglis, which modernized university management and made some changes to the curriculum, but did not greatly expand the opportunities for specialization in science.

The Royal Commission of 1878 finally addressed the matter of specialist degrees. It dealt with the difficult issue of an entrance examination (to which many were still hostile, believing that publicly-funded institutions should have open access) by proposing a first examination that could be taken at school or university, so that those, like Bain, who were not qualified on leaving school could still have access to university education. After this first examination, three further years of study would lead to the M.A. The commission proposed six 'lines of study': the existing arts curriculum; literature and philology; philosophy; law and history; mathematical science; and natural science. This proposal, which was regarded by the universities as being weighted towards the sciences, had no immediate effect: parliamentary bills in 1883, 1884, and 1885 did not get as far as second readings, but finally in 1889 the Universities (Scotland) Act was passed and an Executive Commission established.

The unsatisfactory funding arrangements were never fully addressed. The government remained unwilling to provide generous, or indeed adequate, financial support, and this long hindered the development of the Scottish universities, while the better-endowed Cambridge and Oxford colleges thrived on their private wealth. One outspoken Liberal MP, Robert Wallace, stated in 1889 that

Parliament has been a cruel stepmother to the Scotch Universities. If we had had a Home Rule Parliament, we should not have starved our higher education and stunted its growth.[40]

Furthermore, when the fixing of an annual university grant was debated in 1883, the idea of privatization was in the air. James Donaldson, Principal of the United College at St Andrews (facing the alternatives of closure or amalgamation with the new University College in Dundee), was just one who objected, writing that the proposals for privatization

run counter to the entire history of these institutions and to the ideas of the Scottish people in regard to them. The Scottish Universities are not private corporations—they are national seats of learning, existing for the nation, and controlled by the Parliament of the nation.[41]

Although privatization was averted, the 1889 Act of Parliament set up a Commission for the Scottish Universities that brought many changes. Faculties of science were established to oversee the B.Sc. degrees, new subjects such as engineering and agriculture were introduced, and provision was made for the award of research degrees.

Another controversy during the early part of the Victorian period was over the requirement that professors should declare allegiance to the Church of Scotland.[42] In practice, this was not usually an obstacle for mathematicians. In 1805, John Leslie, who had been accused of religious unorthodoxy, was elected professor of mathematics at Edinburgh. His opponents took the matter to the General Assembly of the Church of Scotland, but Leslie was confirmed in his position in a close vote of the Assembly.[43] Later, Philip Kelland, an ordained priest of the Church of England, regularly preached in Edinburgh, and William Thomson habitually attended both Episcopalian and Presbyterian services. The religious test did however deter the devout Anglican, George Gabriel Stokes, whom Thomson had hoped to attract to Glasgow as professor of mathematics in 1849.

The so-called Disruption of 1843, when over four hundred and fifty Church of Scotland ministers walked out of the General Assembly to found the breakaway Free Church of Scotland, brought profound changes that eventually weakened the position of the Church of Scotland in public affairs. (The leader of the Disruption, Thomas Chalmers, had been a mathematical assistant at St Andrews at the beginning of the century.[44]) An unsuccessful attempt was made to unseat Sir David Brewster, Principal of the United College, St Andrews, who had joined the Free Church, but this case was rather *ad hominem*. In 1853 the religious test was altered so that professors (except for professors of theology) had only to declare that they would do nothing opposed to the doctrine of the Church of Scotland; the test was abolished altogether in 1889.

A painting by David Octavius Hill (1802–70) of the Disruption of 1843 which formed the Free Church of Scotland.

One other change should also be noted. Before the 1880s, only Anderson's College admitted women students. From 1876, St Andrews operated a lucrative scheme of worldwide examinations in many subjects for women, with a sufficient number of passes leading to a diploma. In 1880 this diploma, previously called the L.A. (Literate in Arts), was renamed the L.L.A. (Lady Literate in Arts). Between 1877 and its withdrawal in 1931, five thousand, one hundred and seventeen L.L.A.s were awarded, of which about one thousand six hundred were during the period 1877–97.[45] Despite this venture, women were first admitted to classes in St Andrews only in 1892, whereas they had been able to study at University College,

Dundee, since its foundation in 1881. By the end of Victoria's reign, women were taking mathematics degrees in all of the Scottish universities: by 1900 six women (and eighty-four men) had graduated with the new M.A. with honours in mathematics and natural philosophy.[46] George Chrystal's address to the Edinburgh graduates of 1892 welcomed this development:[47]

Meanwhile the Faculty of Arts have unanimously recommended that the Arts classes shall be thrown open to women as soon as the ordinance has passed. The Senatus and the Court have approved, and the Heriot Trust are to offer entrance bursaries. Women will, therefore, enter upon Arts studies next session with full academic privileges. I am sure that I speak your mind as well as my own when I say that we give them a hearty welcome ... Several women were distinguished for humanistic culture during the early days of the revival of classical learning and from Hypatia to Madame Sophie Kovalevsky, who died only the other day, women have from time to time distinguished themselves as mathematicians ... I do not expect that any large number of women will enter my department, but if they work as enthusiastically as did pupils I used to have at Shandwick Place some years ago, they will be decided addition[s] to the elite of the university.

Bessie Craigmyle, who gained her L.L.A. in 1882.

In summary, there is no doubt that a process of Anglicization—or, less emotively, homogenization—took place in Scottish higher education during the Victorian period. The Scottish university curriculum gradually moved away from the traditional general M.A. to permit more specialization on the Oxbridge model (which had itself diversified). But lack of funding continued to hamper Scottish universities and inhibit growth. From the point of view of mathematics education, greater specialization was unavoidable if Scottish graduates were to compete on equal terms. As will be seen in later chapters, the body of mathematics itself changed dramatically during Victoria's reign, and although British contributions to research in pure mathematics were relatively modest at this time, there were stunning advances in the applications of mathematics to natural philosophy, with Scots such as William Thomson, Peter Guthrie Tait, and James Clerk Maxwell playing leading roles.

## Mathematicians in Victorian Scotland

In this section we present some of the personalities who influenced Scottish mathematical life in the 19th century. We look particularly at the Scottish context, rather than attempting to provide full overviews of the mathematical work of such major figures as Maxwell, Tait, and Thomson.

There was a strong and distinctive Scottish mathematical tradition in the universities, through James and David Gregory, Robert Simson, Colin Maclaurin, Matthew Stewart, John Playfair, John Leslie, and William Wallace. All were outstanding geometers who placed Euclidean geometry at the heart of the teaching syllabus. But James Gregory, Maclaurin, and James Stirling (who held no university post) were also able and original analysts, by the standards of their day, and Playfair was one of the first in Britain to draw attention to the superiority of continental calculus methods. Others, such as Sir William Stirling Hamilton, Edinburgh's influential professor of logic and metaphysics, resisted algebra and analysis as a threat to geometry. In 1838 he offered the following contrast of algebraic and geometrical methods:[48]

The process in the symbolical [algebraic] method is like running a rail-road through a tunnelled mountain; that in the ostensive [geometric] like crossing the mountain on foot. The former carries us, by a short and easy transit, to our destined point, but in miasma, darkness and torpidity, whereas the latter allows us to reach it only after time and trouble, but feasting us at each turn with glances of the earth and of the heavens, while we inhale health in the pleasant breeze, and gather new strength at every effort we put forth.

But, during the first three decades of the 19th century, Scottish mathematicians were among those who led the introduction of continental analysis into British mathematics. John Leslie, professor at Edinburgh, first of mathematics and then of natural philosophy, his rival and successor as professor of mathematics, the self-taught William Wallace, and the unfortunate James Ivory, whose mental health problems interrupted his academic career, were among the first British mathematicians to engage with continental analysis. John West in Jamaica and the solitary William Spence of Greenock were also able analysts, but too detached from the mainstream to receive recognition in their lifetimes. (But it was difficulties with analysis that led Thomas Carlyle, the translator of the 1824 English edition of Legendre's *Géométrie*, to abandon mathematics and concentrate on his literary career.)

As noted earlier, when Wallace retired because of poor health in 1838, Philip Kelland was chosen rather than Duncan Gregory. Gregory, who had studied under Wallace in Edinburgh before proceeding to Cambridge, was another able analyst. He explored the use of operator methods for solving differential equations and, as we will see in Chapter 7, was a co-founder, with fellow-Scot Archibald Smith, of *The Cambridge Mathematical Journal*.[49] But Gregory died at an early age, his full potential unfulfilled.

Philip Kelland had been senior wrangler and first Smith's prizeman at Cambridge in 1834. His early claim to fame was his book *Theory of Heat* (1837), praised by Sylvester and De Morgan, amongst others, in the testimonials that they provided for Kelland's application for the Edinburgh chair (although De Morgan noted that he had not seen the book until the day on which he wrote the testimonial).[50] In this book Kelland criticized some of the work of Fourier—although these criticisms were refuted in 1841 by the young William Thomson. Kelland's later researches on water waves were also flawed, and were soon superseded by those of George Stokes. Later, he edited a two-volume reissue of Thomas Young's *Lectures on Natural Philosophy*, and, late in life, wrote an *Introduction to Quaternions* together with P. G. Tait. Although he did not live up to the high expectations of him in research, Kelland was a worthy and effective teacher who modernized the Edinburgh mathematics curriculum in line with the educational reforms of J. D. Forbes, Blackie, and Lorimer.

In the sessions of 1854 and 1855, when controversy over university reform was at its height, Kelland used his opening lectures to set out his views on the Scottish university system, arguing against the German research-intensive model and identifying the need for greater opportunities and rewards for young scholars. His 1855 lecture concluded:[51]

> Our system is both excellent in itself, and admirably adapted to the people for whose benefit it exists. But whilst, on the one hand, it is without the substantial prizes of the English Universities, and can offer little beyond honour as a reward of high attainments, there is, and has long been needed, on the other hand, an efficient tutorial aid to the junior students... For this University, it is my earnest prayer that she may be permitted to continue her humble but deeply important labours. May she, as in time past, covet rather to be reverenced as a nursing mother to the remote isles of our land, than to be exalted as a proud beauty for the admiration of the few who are thought worthy of being admitted into her presence.

Philip Kelland (1808–79).

The elderly Kelland was later remembered fondly by Robert Louis Stevenson:[52]

> There were unutterable lessons in the mere sight of that frail old clerical gentleman, lively as a boy, kind like a fairy godfather, and keeping perfect order in his class by the spell of that very kindness...[But] for all his silver hair and worn face, he was not truly old; and he had too much of the unrest and petulant fire of youth, and too much invincible innocence of mind, to play the veteran well. The time to measure him best...was when he received his class at home. What a pretty simplicity would he then show, trying to amuse us like children with toys; and what an engaging nervousness of manner, as fearing that his efforts might not succeed...I still fancy I behold him frisking actively about the platform, pointer in hand...that which I seem to see most clearly is the way his glasses glittered with affection.

The next Englishman and Cambridge graduate to join the Scottish professoriate was Frederick Fuller, 4th wrangler in 1842. He had been a fellow of Peterhouse and tutored William Thomson in his first year there. Appointed in 1851 as professor of mathematics at King's College, Aberdeen, nine years later he was confirmed as professor at the joint university. He remained until 1878, expertly training a succession of mathematics students who subsequently did well in the Cambridge tripos. Along with his colleague Alexander Bain, he pressed for curricular reforms at Aberdeen and, after his retirement, was appointed as an additional member of the 1889 Royal Commission.[53]

Although an excellent teacher, Fuller did no research, but this was not unusual for the time. For example, successive St Andrews mathematics professors Nicolas Vilant, Robert Haldane, and Thomas Duncan likewise did none (though Vilant and Duncan wrote student texts), and later the Oxford-educated Peter R. Scott-Lang's long occupancy of the same chair was similarly undistinguished.[54]

William Thomson was born in Belfast in 1824, where his father, James, was professor of mathematics at the Belfast Academical Institution. James, who had studied at Glasgow, wrote many textbooks, including an edition of Euclid and a work on calculus. He was appointed professor of mathematics in Glasgow when William was 8, and after studying at Glasgow for seven years William took the Cambridge tripos, being 2nd wrangler in 1845. This result was something of a disappointment to him, but William compensated by winning the first Smith's prize.[55]

He then spent some time working in Paris in the physical laboratory of Victor Regnault and meeting leading Paris mathematicians such as Liouville, Sturm, and Chasles. Having enhanced his experimental credentials, in 1846 he was unanimously elected to the chair of natural philosophy in Glasgow, as his father had so earnestly desired. William's elder brother James was later to become professor of engineering at Glasgow from 1873 to 1889; but James Thomson senior had long since died during an outbreak of cholera in 1849. William tried to persuade his Cambridge friend, George Gabriel Stokes, a fellow-pupil of the tripos private tutor William Hopkins, to apply for the Glasgow mathematical chair in succession to James senior. While initially interested, Stokes, an Anglican, felt that he could not conscientiously claim to meet the religious test, despite Thomson's assurance that:[56]

the *amount* of conformity to the established Church which a conscientious observance by one in your position of the obligations imposed by the tests, would really be in no way inconvenient, or repugnant to your feelings.

Stokes having withdrawn, the chair went to another of Thomson's Cambridge friends, Hugh Blackburn, 5th wrangler of 1845, who had previously studied at Edinburgh Academy and Glasgow University.[57] Stokes was shortly afterwards appointed to the Lucasian professorship of mathematics at Cambridge.

Thomson's career was distinguished in every respect. His early work on heat and thermodynamics made his name, and he did important work on electricity, magnetism, and fluid mechanics. He also was responsible for what we now call 'Stokes's theorem'—after he communicated it to Stokes, the latter set it as a tripos exam question (see Introduction), with the result that *his* name was subsequently given to the theorem. In connection with his work on water waves, Thomson developed his well-known 'method of stationary phase' in asymptotics. He played a large part in the new telecommunications industry, making vital contributions to the transatlantic cable project. (When he was elevated to the peerage in 1892, there already being a peer with the name Thomson, he considered becoming 'Lord Cable' before settling on 'Kelvin', after the river that runs through the university area of Glasgow.) His telegraphic and other patents, and the compass he designed which became the Admiralty standard, were very profitable (see Chapter 8): as early as 1870 he was in a position to buy the 126-ton schooner *Lalla Rookh*.

When he visited the United States in 1897, the *Buffalo Express* wrote:[58]

Lord Kelvin is short and thin and gray and plain. He is very lame, but there is something in his appearance that does not belie his youthful record as an athlete...His small blue eyes are kindly and genial in their expression. His clothes fit badly, after the English fashion.

In retrospect, by the time of his death in December 1907, Kelvin had become somewhat out of date. His anti-evolutionary thermodynamic argument, that the rate of the earth's cooling meant that it had not been in existence long enough for Darwin's theory to be viable, turned out to be erroneous because of physical considerations he could not have known about. His famous dismissal of the possibility of heavier-than-air flying machines was disproved just before the end of his life. His ingenious vortex–aether hypothesis as the basis of a theory of matter, electricity, and magnetism, was shown to be incorrect (although it led to interesting theoretical studies of vortex motion). He was never totally convinced by Clerk Maxwell's revolutionary theory of electromagnetism, and he steadfastly opposed the use of vectors and quaternions in mathematical analysis. Nevertheless, he was the best-known scientist of his day and his contributions to science and engineering were immense.[59]

A close colleague of Thomson was William John Macquorn Rankine, Glasgow's professor of civil engineering and mechanics. He held the post from 1855 until his death, when he was succeeded by William's brother, James Thomson junior. Rankine taught the practical skills of ship design and conducted research on railway locomotion, but he was also concerned with fundamental science, including thermodynamics, aether theory, vortex motion, and water waves.

James Clerk Maxwell attended Edinburgh Academy, the University of Edinburgh, and finally the University of Cambridge. Tutored by Hopkins at Cambridge, Maxwell was 2nd wrangler in 1854, and tied for first Smith's prize. He gave what are now known as 'Maxwell's equations of electromagnetism' in a paper of 1861, and did fundamental work in electrodynamics and kinetic theory. His period as a professor in Scotland was short: he was appointed to the chair of natural philosophy at Marischal College, Aberdeen, in 1856. In 1859 he married Katherine Mary Dewar, the daughter of the principal of Marischal College, but as the junior professor he lost his post when the two Aberdeen Colleges merged in 1860. He also lost to Tait in the contest for the chair of natural philosophy at Edinburgh, and the rest of his academic career was spent south of the border, first at King's College, London, and subsequently back in Cambridge.

P. G. Tait had been a contemporary of Maxwell's at the Edinburgh Academy, where he won the 1846 mathematics prize, beating both Maxwell, who was third, and second-placed Lewis Campbell, subsequently professor of Greek at St Andrews. Tait and Maxwell entered Edinburgh University together, attending Kelland's class, but after just a year Tait moved to Cambridge, where (as another of Hopkins's students) he was senior wrangler in 1852. After some time as professor of mathematics at Queen's College, Belfast, he took up the Edinburgh chair of natural philosophy in 1860.

Peter Guthrie Tait (1831–1901).

Tait and Thomson collaborated on a famous textbook which J. M. Barrie, who took Tait's class, described as follows:

T. and T. was short for Thomson and Tait's *Elements of Natural Philosophy*[60] (Elements!), better known in my year as the *Student's First Glimpse of Hades*...I fear T. and T. is one of the Books Which Have Helped Me. This somewhat damps my ardour [for Tait].[61]

The book to which Barrie refers was distilled from Thomson and Tait's *Treatise on Natural Philosophy*, commonly known as *T and T'* (the *Elements* was *Little T and T'*). Thomson was *T* and Tait *T'*, while James Clerk Maxwell, writing to *T'*, signed himself *dp/dt*, because the *Treatise* contained the equation $dp/dt = jcm$.

Despite this unpleasant memory, Barrie does convey his appreciation of Tait:

Never, I think, can there have been a more superb demonstrator. I have his burly figure before me. The small twinkling eyes had a fascinating gleam in them; he could concentrate them until they held the object looked at; when they flashed round the room he seemed to have drawn a rapier. I have seen a man fall back in alarm under Tait's eyes, though there were a dozen benches between them. These eyes could be merry as a boy's, though, as when he turned a tube of water on students who would insist on crowding too near an experiment, for Tait's was the humour of high spirits. I could conceive him at marbles still, and feeling annoyed at defeat. He could not fancy anything much funnier than a man missing his chair.

From his time in Ireland, Tait was a committed advocate of the quaternions of his friend William Rowan Hamilton, which (before their importance today in computer game programming) were seen by many as an elegant mathematical idea with limited practical value. But Tait valued them as a convenient notation for the equations of natural philosophy. Thomson disagreed, recalling that:[62]

We have had a thirty-eight year war over quaternions. Times without number I offered to let quaternions into Thomson and Tait if he could only show that in any case our work would be helped by their use. You will see that from beginning to end they were never introduced.

Tait's pioneering work on the classification of knots, in collaboration with the English vicar Thomas Penyngton Kirkman, produced graphical tables of all projections of knots with up to ten crossings and correctly identifying equivalences.[63] Tait's other passion was golf, and he spent as much of his time as possible at St Andrews where he owned a second home. The German mathematician Hermann von Helmholtz, a friend of both Tait and Thomson, noted that:[64]

Mr. Tait knows nothing here besides golf. I had to go along with him. My first swings succeeded, but after that I hit only ground or air. Tait is a curious kind of savage—exists here, as he says, only for his muscles, and only today, Sunday, when he dared not play, though he didn't go to church either, would he be brought around to rational matters.

Tait worked on the mathematics of the motion of a golf ball, showing that the distances achieved by golfers were due to 'underspin' imparted to the ball. He was very proud of his third son Freddie, a brilliant golfer who twice won the Amateur

TREATISE

on

# NATURAL PHILOSOPHY

BY

SIR WILLIAM THOMSON, LL.D., D.C.L., F.R.S.,
PROFESSOR OF NATURAL PHILOSOPHY IN THE UNIVERSITY OF GLASGOW,
FELLOW OF ST PETER'S COLLEGE, CAMBRIDGE,

AND

PETER GUTHRIE TAIT, M.A.,
PROFESSOR OF NATURAL PHILOSOPHY IN THE UNIVERSITY OF EDINBURGH,
FORMERLY FELLOW OF ST PETER'S COLLEGE, CAMBRIDGE.

PART I.

*NEW EDITION.*

CAMBRIDGE:
AT THE UNIVERSITY PRESS.
1888

[*All Rights reserved.*]

The title page of Thomson and Tait's *Treatise on Natural Philosophy*.

Championship and was six times the leading amateur in the Open. Freddie Tait's death in the South African war was a terrible blow to his elderly father.[65]

Kelland's successor as professor of mathematics at Edinburgh was George Chrystal, who after taking first class honours at the now united Aberdeen University was joint 2nd wrangler at Cambridge in 1875 (bracketed with the group-theorist William Burnside) and was also second Smith's prizeman (after Burnside). He was appointed to the mathematics chair at St Andrews in 1877 in succession to Ludwig Fischer, and moved to the Edinburgh chair two years later.

J. M. Barrie was a student in Chrystal's first year. According to his account, the new professor came as something of a shock to the weaker students:[66]

When Chrystal came to Edinburgh he rooted up the humours of the classroom as a dentist draws teeth... Horse-play fled before the Differential Calculus in spectacles. I had Chrystal's first year, and recalled the gloomy student sitting before me who hacked 'All hope abandon ye who enter here' into a desk that may have confined Carlyle... Chrystal was a fine hare for the hounds who could keep up with him... The men who were on speaking acquaintance with its symbols revelled in him as students love an enthusiast who is eager to lead them into a world toward which they would journey. He was a rare guide for them. The bulk, however, lost him in labyrinths... Twenty per cent. was a good percentage in Chrystal's examinations; thirty sent you away whistling.

A later writer describes Chrystal at the end of his career:[67]

[O]ur clearest recollection of Professor Chrystal stands out from a mist of dingy lecture-rooms in the old buildings. It consists of a grey-headed but erect figure, dashing, with marvellous speed, cohorts and battalions of graphs, theorems, triangles, and symbols upon a silent, but secretly suffering blackboard. Then, when every available square inch of space had been filled, there was a triumphant swing round to his astounded audience, as the professor sped on to some other colossal piece of mathematical architecture.

George Chrystal (1851–1911).

Barrie also notes that Chrystal took to calling the register in the middle of the class, rather than at the beginning, to discourage students from slipping away. And Barrie also recorded Chrystal's one weakness: a propensity to make simple arithmetical mistakes. This was lampooned in a contemporary cartoon of 'Professor CHR–ST–L, who recently succeeded, amid a scene of unrivalled enthusiasm, in making a correct addition'—the Professor is shown with the calculation '$2 + 5 = 7$', having previously proposed answers of 8 and 13.[68]

Barrie's comments undoubtedly reflect the fact that Chrystal was a reformer who wished to raise the level of university mathematics:[69]

Ever since I became convinced that a majority of educated Scotsmen desired to break down the old curriculum of seven subjects, my watchword has been "Greater freedom and higher standards". It is obvious that in any subject which is generally compulsory the standards cannot be high. I was never very anxious that all Arts students should take either mathematics or natural philosophy; but I have all along striven to secure, so far as possible, that those who do take these subjects should be well prepared to receive them.

He was also involved in reforming school education, playing a large part in the introduction of the Scottish Higher Leaving Certificate examinations by the Scottish Education Department.[70]

Chrystal's most famous publication was his substantial and comprehensive book *Algebra: An Elementary Textbook for the Higher Classes of Secondary Schools and for Colleges*, which met a pressing need. But he remained interested in physical applications, having studied under Clerk Maxwell at Cambridge, where he did experimental work on electricity. This interest led to his later involvement with the Scottish Lake Survey, for which he calculated the periods of seiches in several lochs, using quite recent theoretical ideas on water waves.[71]

As mentioned earlier, the new University College in Dundee was founded in 1881 and soon became a part of the University of St Andrews. John Edward Aloysius Steggall was appointed Dundee's first professor of mathematics and natural philosophy (from 1895 he was professor of pure and applied mathematics).[72] Steggall was a Londoner, 2nd wrangler at Cambridge behind E. W. Hobson and first Smith's prizeman in 1878. He chose to marry and was therefore unable to take up his Cambridge fellowship. From 1880 until his move to Dundee he was Fielden lecturer at Owens College, Manchester, which was in many respects the model for the Dundee institution. Steggall taught at Dundee for over fifty years. From 1881 he was a life member of The Association for the Improvement of Geometrical Teaching (which later became the Mathematical Association), and was a member of the Dundee School Board from 1891 to 1900: he advocated oral rather than written examinations for mathematics at school level. He also wrote a guidebook, *Picturesque Perthshire*.[73]

His colleague at Dundee for a time was James Alfred Ewing, who was professor of engineering there from 1883 to 1890. A student of Tait and Fleeming Jenkin at Edinburgh, he was first a professor at Tokyo Imperial University from 1878 to 1883, where, with the Englishman John Milne, he did valuable work on seismology and magnetism. These researches were ably continued by Ewing's successor, Cargill Gilston Knott, and Japanese colleagues Sekiya, Omori, and others.[74] From Dundee, Ewing was recruited to Cambridge to succeed another Scot, James Stuart, as professor of mechanism and applied mechanics, and late in life he served as principal of Edinburgh University. James Stuart was educated at St Andrews and Cambridge, becoming 3rd wrangler in 1866. He was an associate of Clerk Maxwell: together with the brothers Charles and William D. Niven, two other Scottish fellows at Cambridge, he did much to promote Maxwell's electromagnetic theory after the latter's untimely death. Charles Niven returned to his first alma mater in Aberdeen in 1880, becoming professor of natural philosophy.[75]

Cargill Gilston Knott had been a student and assistant of P. G. Tait at Edinburgh University. In 1883, he was appointed professor of physics in Tokyo on the recommendation of William Thomson and the newly-returned James Ewing. He returned to Edinburgh University in 1891 as a lecturer and later reader

Cargill Gilston Knott
(1856–1922).

in applied mathematics. He was disappointed not to succeed Tait in 1901, and he never became a professor, despite his many and varied publications. Among the best known is his *Life and Scientific Work of Peter Guthrie Tait* (1911), still the standard biography. He also played a prominent role in the Royal Society of Edinburgh and he helped to found the Edinburgh Mathematical Society (see below).[76]

Not all mathematical work in 19th-century Scotland took place in universities. A notable example is the Kirkcaldy-born table-maker Edward Sang.[77] After entering Edinburgh University at the age of 13, where he impressed Leslie and Wallace, he worked in Edinburgh as a surveyor, civil engineer, mathematics teacher, and as an assistant lecturer. He was one of the unsuccessful applicants for the chair of mathematics in 1838 when Kelland was appointed, and after a period teaching in Manchester he went to Constantinople as an engineer, returning to Edinburgh after eleven years in 1854. He was an active member of the Royal Society of Edinburgh and the Royal Scottish Society of Arts, and was a founder-member of the Faculty of Actuaries in 1856, becoming its first official lecturer. He published extensively: idiosyncratic books on arithmetic, containing unusual but (at least for himself) effective algorithms for rapid computation; astronomical and actuarial tables; papers and encyclopedia articles on engineering, physics, and mathematics (including a pessimistic analysis of the structure of the Forth Rail Bridge, which fortunately has defied his prediction); and a table of five-figure logarithms. He also produced, over forty years and with significant help from his daughters Flora and Jane, forty-seven manuscript volumes of remarkably accurate 26-place and 15-place logarithmic, trigonometrical, and astronomical tables.[78]

Another unsuccessful candidate for the Edinburgh mathematical chair in 1838 was John Scott Russell. He subsequently had a career as an engineer and naval architect in London, but is now best remembered for his experimental discovery of the 'solitary wave', which has led to today's theory of solitons. As joint secretary of the Royal Society of Arts, he had a role in the planning of the Great Exhibition of 1851; as a ship designer he was responsible for the racing yacht *Titania* which unsuccessfully challenged the famous *America* in the inaugural America's Cup races, and the iron warship *Warrior*, and he designed the hull for Brunel's *Great Eastern*. He wrote *The Modern System of Naval Architecture for Commerce and War*, and advocated improved technical education. But in the later part of his life he was afflicted by business failures and financial problems.[79]

Mention should also be made of the astronomers who worked at the Royal Observatory in Edinburgh. Thomas Henderson had learned mathematics at Dundee High School from its rector, Thomas Duncan (later professor of mathematics at St Andrews). At 15 he entered a solicitor's office, but he continued his interest in astronomy and in the 1820s he devised methods of improving astronomical calculations. After a period as director of the observatory in South Africa, he was appointed the first Astronomer Royal for Scotland in 1834, at a salary of £300. According to his biographer, he died 'worn out by the workload and by the daily climb up Calton Hill'.[80]

Mary Somerville (1780–1872).

Henderson's successor was Charles Piazzi Smyth, who struggled with the university authorities over the chronic underfunding of the observatory and eventually resigned in protest in 1888. He was responsible for some influential, but now discredited, theories about the numerology of the designs of the Egyptian pyramids, which obscured his more deserving research in spectroscopy.[81]

Finally, we must mention the achievements of Mary Somerville (née Fairfax), who was born in Jedburgh and brought up in Burntisland, Fife. She was at first largely self-taught. But, following the death of her first husband Samuel Greig in 1807, she consulted John Playfair, who recommended William Wallace, who gave her mathematics instruction by correspondence; she was also tutored by Wallace's brother John. In 1812 she married William Somerville, an army doctor, and in 1816 the family moved to London, where she soon became well known in scientific circles. Though she was not an original researcher, her several books on astronomy and other sciences were best sellers that did much to educate the general population.[82]

## Mathematical societies in Scotland

The above-mentioned individuals were active not only in the universities, but also in the affairs of Scotland's main scientific and mathematical societies. Here we mention only the most influential of these; but there were many other local bodies that encouraged an interest in science, among them the mechanics institutes and the literary and philosophical societies in most of the larger towns.

The *Royal Society of Edinburgh* was created in 1783 by Royal Charter for 'the advancement of learning and useful knowledge'.[83] During the Victorian period its home was the Royal Institution building in Princes Street (now the Royal Scottish Academy), where Maxwell attended meetings as a schoolboy. It was under the secretaryship of George Chrystal that the Society was ejected from the Royal Institution and acquired its current home in George Street.[84]

Throughout this period, leading mathematicians in Scotland were members of the Society and published their work in its *Proceedings* and *Transactions*. Brewster (1864–68), Kelland (1878–79) and Thomson (three times, 1873–78, 1886–90, and 1895–1907) were all presidents, with Kelland dying shortly into his term. In addition Tait (1879–1901), Chrystal (1901–12) and Knott (1912–22) all served lengthy terms as General Secretary.

Cargill Knott gives an entertaining account of Thomson's style of presentation. Knott had to report on meetings of the Royal Society of Edinburgh for *Nature*, and on one occasion:[85]

Sir William, as he then was, had in his well-known discursive but infinitely suggestive manner so talked around the subject of the communication that I had some difficulty in quite understanding its real essence. Next morning I tried to get enlightenment from Tait. He laughed and said, "I had rather not risk it; but the great man is coming at twelve—better tackle him himself."

Transactions of the Royal Society of Edinburgh.

When in due time Sir William was "tackled", he fixed his gaze at infinity for a few moments and then, a happy thought striking him, he said, with a quick gesture betokening release from burden, "Oh, I'll tell you what you should do. Just wait till the *Nature* Report is published—that fellow always reports me well." Tait's merriment was immense as he unfolded the situation, and he chaffed Thomson as to his obvious inability to explain his own meaning. Not a few of both Kelvin's and Tait's communications to the Royal Society of Edinburgh were never written out by them; they appear as reports only in the columns of *Nature*.

In 1821 the *Royal Scottish Society of Arts* was founded by David Brewster and John Robison as 'The Society for the Encouragement of the Useful Arts in Scotland', and received its Royal Charter in 1841. It provided a forum for the presentation of scientific and technical papers: Maxwell also attended its meetings as a schoolboy. Sang contributed to its *Proceedings* for fifty years and served as secretary,[86] and Kelland served as president in 1852–53.[87]

The *Edinburgh Mathematical Society* was founded in 1883.[88] A circular dated 23 January 1883, signed by Cargill Gilston Knott, Andrew Jeffrey Gunion Barclay, and Alexander Yule Fraser, and sent 'to gentlemen in Edinburgh, in Cambridge and throughout Scotland generally whom they deemed likely to take an interest in such a Society' indicated that:

It is proposed to establish, primarily in connection with the [Edinburgh] University, a Society for the mutual improvement of its members in the Mathematical Sciences, pure and applied.

Membership would be formed from:

(1) present or former students in either of the Advanced Mathematics Classes of Edinburgh University,

(2) Honours Graduates of any of the British Universities, or

(3) recognised Teachers of Mathematics.[89]

Knott became the first secretary and treasurer just before leaving for Japan, and was later twice president, in 1893 and 1918. Fraser and Barclay were mathematics teachers at George Watson's College in Edinburgh and graduates (respectively) of Aberdeen and Edinburgh. The Society was set up with John Sturgeon Mackay as its first president. He was the chief mathematical master at the Edinburgh Academy and a noted scholar of ancient Greek geometry.[90] Tait and Chrystal were elected honorary members, and they were subsequently joined by William Thomson, Charles Niven, and (in 1894) Mackay.

Unlike the London Mathematical Society, some eighteen years older, the Edinburgh Mathematical Society was led from the beginning by school teachers, rather than university mathematicians. The first ordinary meeting admitted fifty-one ordinary members, and by the end of the session there were fifty-eight, of whom forty were teachers (five from Watson's) and fifteen had university connections. The first professor to be president was Steggall in 1891; the honorary members could not be officers.

The second president of the Society was Thomas Muir, whose presidential address is referred to above. He was then mathematics master at Glasgow High School and had previously been Blackburn's assistant at Glasgow University. In 1892 he was appointed Superintendent-General of Education in South Africa, where he spent the rest of his life.

The Society first met in Edinburgh University, and then from 1884 in the Edinburgh Institution in Queen Street. From 1900 onwards it held meetings in Glasgow, and later in the other Scottish universities. From the outset, the Society published its *Proceedings*,[91] the early volumes largely comprising the papers presented to the Society by its members. Although these contain little original mathematical work, they 'cover a wide range, but with a preponderance of Euclidean geometry. There are several articles on historical and pedagogical points, including collections of mnemonics'.[92]

## Conclusion

This brief survey of mathematics in Scotland in the reign of Queen Victoria has revealed a diverse panorama. Reform of the university curriculum was supported by some, but resisted by others who feared losing the traditional values of the 'democratic' Scottish system. Despite several Royal Commissions, it was only towards the end of the Victorian period that limited changes to the Master of Arts degree were finally achieved, allowing some specialization on the English model. (Even at the end of the 20th century, an English academic who worked at Glasgow University commented that it was only when he read *The Democratic Intellect* that he was able to make sense of the workings of the university and the attitudes of his colleagues.[93])

That the 'democratic' ethos persists, even today, can perhaps be seen from the different policies north and south of the border regarding university tuition fees. At the time of writing, this is enhancing the opportunities of Scottish students while causing some financial difficulties for the Scottish universities. The chronic underfunding of the 19th century, as well as arguments about access and standards, have their echoes today.

It was an inevitable consequence of the poverty of the Scottish universities that research was not uniformly strong during our period: Chrystal, Blackburn, Kelland, Duncan, and Fuller primarily devoted their energies to teaching and writing textbooks, and there were very limited opportunities for young graduates to devote themselves to original work. But outstanding work in applied mathematics and natural philosophy did come from the likes of William Thomson, Tait, and Maxwell—$T$, $T'$, and $dp/dt$—and several able young men received experimental training and theoretical stimulus in the physical laboratories in Glasgow, Edinburgh, and (later) Cambridge. Thomson's acquired wealth from his patents and engineering ventures provided an example for would-be mathematical entrepreneurs, and his international links with Europe and Japan widened the horizons of several former students.

The career possibilities for mathematicians were also expanding. While professorships were hotly contested when they became vacant, new professional opportunities were emerging for the mathematically-trained: for instance, Adam Anderson (engineer and school rector in Perth before his appointment at St Andrews), Edward Sang (engineer and actuary in Constantinople and Edinburgh), and John Scott Russell (naval architect). And geographical boundaries were enlarging, exemplified by the distant sojourns of Sang, Ewing, and Knott. But the commonest journey remained the 'high road to England', and especially to Cambridge, where rewards, as well as opportunities, were available to the brightest and best.

Nevertheless the biggest changes were still to come. The period of certainty and self-confidence of the Victorian age had been mirrored by the 'classical physics' of Thomson, Tait, Maxwell, and their co-workers. Soon, developments in early 20th-century physics, in relativity, atomic theory, and quantum mechanics, were to remodel the understanding of the universe as comprehensively as the coming First World War was to redefine the boundaries of Europe.

Hamilton's pocket-book note on quaternions, 1843.

CHAPTER 5

# Taking root

*Mathematics in Victorian Ireland*

RAYMOND FLOOD

I s there such a thing as Irish mathematics, as opposed to mathematics which happens to have been done in Ireland, or by those born in Ireland? Certainly, during the 19th century, Ireland was in one of its richest mathematical phases, with mathematicians such as William Rowan Hamilton, James MacCullagh, George Salmon, and George FitzGerald to inspire generations of students and researchers, as well as exporting to England such talented mathematicians as Lord Kelvin, George Gabriel Stokes, and Joseph Larmor. By examining the mathematics taught and researched, we explore whether the category of 'Irish mathematics' is one that makes sense for this period.

The 19th and early 20th centuries were an important and notable period in Irish mathematics and science. Ireland produced many fine mathematicians. Some, like William Rowan Hamilton and James MacCullagh, spent the whole of their working lives there, while others, such as George Stokes, William Thomson (Lord Kelvin), and Joseph Larmor, left Ireland for England or Scotland. Ireland also provided a base for English-born mathematicians, the most notable of whom was George Boole.

In this chapter we investigate whether there is any identifiable 'Irish mathematics' in the Victorian period. We therefore dispense relatively quickly with Stokes and Kelvin; while they were Irish-born, their Irish origin did not seem to influence their scientific work, although it did influence their political and religious views. (Joseph Larmor is different and will be part of the argument we develop.)

The Lincoln-born George Boole is best remembered for his enduring work on logic, as expounded in his 1854 book *An Investigation of the Laws of Thought, on Which are Founded the Mathematical Theories of Logic and Probability*. In 1849, he was appointed the first professor of mathematics at Queen's College, Cork, where he also worked on differential equations, an overlap of interest with Hamilton, although there was little communication between the two men. Indeed, although based in Ireland, Boole had more contact with Augustus De Morgan and other English mathematicians than he did with his Irish contemporaries. He did not set up any research group, was never elected as a fellow of the Royal Irish Academy, and remained geographically and intellectually isolated from the rest of the Irish mathematical and scientific community. Thus, despite his fine mathematical attainments, there is little more to say about George Boole with regard to his connection with Ireland and Irish mathematics; for some of his mathematical contributions, see Chapters 13 and 16.

Having dispensed so brusquely with Stokes, Kelvin, and Boole, one may wonder what else there is to say. Fortunately the vein of Irish mathematics in this period is rich and deep enough to sustain further discussion. We explore this vein mainly through the activities of William Rowan Hamilton, James MacCullagh, George Salmon, George Francis FitzGerald, and Joseph Larmor, and the principal themes of geometry and electrodynamics. There are, we suggest, characteristic Irish dimensions to these men and to these areas. But first, we survey some institutional developments.

## Institutional developments

Not surprisingly, the development of Irish university education during the 19th century was closely intertwined with the tensions between the Catholic Church and the state, as well as the strivings of Catholics and dissenters for equal access to education. At the beginning of the 19th century the only university-level institution in Ireland was Trinity College, Dublin. Established in 1592, it had worldwide renown in philosophy and literature but, up to the beginning of the 19th century, no significant mathematical reputation. Although Catholics were allowed to attend lectures at Trinity College, their rights were more limited than those of Protestants, and college offices were subject to religious requirements.

In 1845 the non-denominational Queen's Colleges, at Belfast, Cork, and Galway, were established; opened in 1849, they were linked together by royal charter as the Queen's University in 1850, and were modelled on University College, London, established over twenty years earlier (see Chapter 3). However, they were denounced by the Catholic Church in a papal rescript of 1847 as 'detrimental to religion'.[1] The Irish Catholic hierarchy was divided as to whether or not to support them, as indeed was the public. The issue formed one of the main topics of debate at the Synod of Thurles, held in 1850—the first synod held in Ireland since the middle ages. Supporters of the papal position eventually won, although some decrees—for example, two decrees excluding priests from the Colleges under pain of suspension—were approved only by the narrowest of margins. Catholics were prohibited from going to the Queen's Colleges, and consequently Cork and Galway had few students.

Another decision of the Synod was to establish the Catholic University of Ireland, modelled on Louvain, which opened in 1854 with John Henry Newman as its first rector. It was funded independently from the government and its degrees were not recognized by the state. Newman had many disagreements with the governing bishops and left four years later. One of the issues of contention between them was his staffing of the university with Oxford men who had lost their fellowships there because they had converted to Catholicism. John Casey, arguably the most distinguished Catholic mathematician at the time, was appointed to the chair of mathematics in 1873. He was a distinguished geometer who was elected to the Royal Society two years later in 1875, by which time the number of students at the Catholic University was extremely low.

A resolution of the problem of low numbers at the Queen's Colleges in Cork and Galway and the Catholic University, as well as problems of staffing, was attempted with the creation of the Royal University of Ireland in 1880 to replace

John Casey (1820–91).

## Box 5.1: Mathematicians at Belfast, Galway, and Cork, 1849–1901.

**BELFAST**

- 1849–54          William Wilson
- 1854–60          Peter Guthrie Tait
- 1860–62          George Slesser
- 1862–1901       John Purser

**GALWAY**

- 1849–53          John Mulcahy
- 1853–93          George Johnston Allman
- 1893–1901       Alfred Cardew Dixon

**CORK**

- 1849–64:        George Boole
- 1865–66:        Robert Romer
- 1867–80:        Charles Niven
- 1880–87:        John Christian Malet
- 1887–1913:     Arthur Henry Hallam Anglin

them. This also had a constituent Catholic college, University College, Dublin. This remained the position until 1908 when the Irish Universities Act dissolved the Royal University and established two new Universities in its place—the National University of Ireland and the Queen's University of Belfast.

Several good mathematicians were based in these Irish colleges during the Victorian period. Belfast had William Wilson from 1849 to 1854. A fellow of St John's College, Cambridge, Wilson had been senior wrangler in 1847. He left Belfast on being appointed the first professor of mathematics at the University of Melbourne (see Chapter 6). From 1854 to 1860 the chair was held by Peter Guthrie Tait, senior wrangler in 1852 and a fellow of Peterhouse, Cambridge; as we saw in Chapter 4, he moved to Edinburgh as professor of natural philosophy in 1860. George Slesser was another senior wrangler and a fellow of Queens' College, Cambridge, who held the chair from 1858 until his death in 1862. From then until 1901, the professor was John Purser, a Trinity College, Dublin, graduate and the first Irish graduate to hold the chair; one of his students was Joseph Larmor who eventually went on, via the chair of natural philosophy at Galway, to hold the Lucasian professorship of mathematics at Cambridge.

In Galway, John Mulcahy, a Trinity College Dublin graduate, was the first professor from 1849 to 1853. Another Trinity College graduate, George Johnston Allman, then held the chair for forty years until he was succeeded by Alfred Cardew Dixon who stayed until he moved to Belfast in 1901. As we have seen, in Cork, George Boole was the first professor of mathematics, a position he held until his death in 1864. He was followed by Robert Romer who stayed for a year before going off to pursue a career in law. Charles Niven and John Christian Malet then covered the next twenty years before Arthur Anglin, a County Cork man, was appointed.

But it was probably Trinity College, Dublin, where the situation for mathematics was the most fruitful. That said, at the beginning of the 19th century the syllabus for its undergraduate course was old-fashioned and undemanding, as was the syllabus for a fellowship, with a strong emphasis on Newton's *Optics* and *Principia Mathematica*, supplemented by Hugh Hamilton's textbook on conic sections, and Colin Maclaurin's on fluxions. But in 1813, Bartholomew Lloyd, a Trinity College graduate, was appointed to the chair of mathematics. With John Brinkley, who had been Trinity's professor of astronomy since 1792, Lloyd reformed the syllabus and introduced more current mathematics. Fundamental to this reform was the introduction of up-to-date French textbooks from the École Polytechnique, as well as the writing of new texts. As a result the undergraduate science course became more up to date with recent Continental developments.

In its reform of the syllabus, Dublin was slightly ahead of Cambridge: perhaps this was because Lloyd and Brinkley were already professors, while the reformers at Cambridge were initially undergraduates and took longer to become established. The range and availability of new and translated textbooks was also important and could draw on an existing tradition of mathematical textbooks published in Ireland.

Trinity College, Dublin.

So, by the start of the Victorian period mathematics in Ireland was emerging and becoming invigorated, but initially at least these reforms raised Ireland, as Ivor Grattan-Guinness put it, 'from the mediocre to the routine'.[2] However, a more prominent position would be achieved in particular through the work of four subsequent Irish mathematicians: Humphrey Lloyd, William Rowan Hamilton, James MacCullagh, and George Salmon. The first three worked in geometrical optics, with Hamilton also being productive in dynamics and algebra. Salmon did important work in algebraic geometry but was particularly influential through his textbooks. Indeed throughout the Victorian period a considerable number of mathematicians renowned on both sides of the Irish Sea helped to establish the Irish colleges and universities as centres of sound teaching and research.

Humphrey Lloyd (1800–81).

## Hamilton, MacCullagh, and Salmon: a geometrical triple

William Rowan Hamilton was undoubtedly Victorian Ireland's most important native resident mathematician. Born in Dublin in 1805, he had been a child prodigy, with legend having it that by the age of 13 he could speak as many languages. One of Hamilton's biographers, Sean O'Donnell, has analysed this claim and come up with a more modest assessment that he knew Latin, Greek, and Hebrew well enough to read them easily, with lesser facility to read French, German, and Italian. His knowledge of Persian, Syriac, Sanskrit, and Chaldee seems to have been more basic! Interestingly, O'Donnell also says that Hamilton seems to have claimed no knowledge of Gaelic, the language spoken by half the population of Ireland at the time.[3]

Hamilton was educated by his uncle James with an emphasis on classics—although he also showed remarkable calculating ability from an early age. But by the time he was preparing for entrance to Trinity his preferences were becoming clearer, as he indicated in a letter to his sister Eliza:[4]

One thing only have I to regret in the direction of my studies, that they should be diverted—or rather, rudely forced—by the College Course from their natural bent and favourite channel. That bent, you know is Science—Science in its most exalted heights, in its most secret recesses. It has so captivated me—so seized on … My affections—that my attention to Classical studies is an effort and an irksome one.

He began reading Laplace's *Mécanique Céleste* in 1822 and took first place in the entrance examinations for Trinity the following year, quickly scaling the academic ladder and becoming professor of astronomy and Astronomer Royal of Ireland in 1827 at the age of 22, just before he graduated. During this time and over the next five years he worked mainly on geometrical optics and dynamics, using the calculus of variations, introducing his characteristic function and building on the principle of least action.

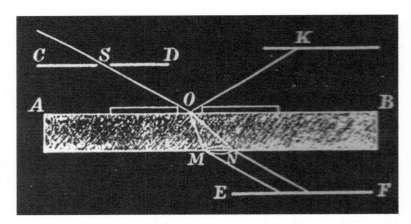

Conical refraction.

At the end of this period, in 1832, he obtained international recognition when he predicted that a ray of light incident on a biaxial crystal in a certain direction would be refracted into a cone of rays and emerge as a hollow cylinder; this unexpected theoretical prediction was experimentally verified by his colleague Humphrey Lloyd shortly afterwards. Hamilton's optical work was dominated by the study of the geometry of curves and surfaces in the tradition of Monge, and as Ivor Grattan-Guinness has commented:[5]

he welded together two great but vastly differing styles of French mathematics, mixing the algebraic and geometry styles.

This mixture of algebra with geometry is also a defining feature of Hamilton's work on quaternions—the area in which he spent most of the second half of his academic life.

The geometrical representation of complex numbers by the so-called 'Argand diagram' raised two questions in Hamilton's mind, and is expressed as follows by another of Hamilton's biographers, Thomas Hankins:[6]

- Is there any other algebraic representation of complex numbers that reveals all valid operations on them?

- Is it possible to find a hypercomplex number that is related to three-dimensional space, just as a regular complex number is related to two-dimensional space?

Hamilton answered the first question by proposing that a complex number is just an ordered pair of real numbers, with a specified way of adding and multiplying them.

So we add and multiply the complex numbers $(a, b)$ and $(c, d)$ as follows:

$$\text{addition: } (a, b) + (c, d) = (a + c, b + d);$$
$$\text{multiplication: } (a, b) \times (c, d) = (ac - bd, ad + bc).$$

If we write $(a, b)$ as $a + bi$, with the interpretation that $i^2 = -1$, then we have the usual representation of complex numbers. Note that $(a + bi)(a - bi) = a^2 + b^2$.

It is interesting to read P. G. Tait's explanation of complex numbers in his obituary of Hamilton in the *North British Review:*[7]

If an officer and a private be set upon by thieves, and both be plundered of all they have, this operation may be represented by negative unity. And the imaginary quantity of algebra, or the square root of negative unity, will then be represented by a process which would rob the private only, but at the same time exchange the ranks of the two soldiers. It is obvious that on a repetition of this process both would be robbed, while they would each be left with the same rank as the first. But what is most essential for remark here is that the operation corresponding to the so-called imaginary of algebra is throughout regarded as *perfectly real*.

Of course, this view of complex numbers had been implied by the geometrical representation, but Hamilton made it explicit in 1837. However, he had less success when he moved up one dimension to algebraic triples of numbers, $(a, b, c) = a + bi + cj$, where $i \neq j$ and $i^2 = j^2 = -1$. Adding them was easy—performed in a similar way as for pairs—but he could find no system that would obey the rules of multiplication. It was an impossible quest. Georg Frobenius showed in 1878 that, in modern terms, the only associative division algebras over the reals are the reals themselves, the complex numbers, and the quaternions. In other words, a triple system as envisaged by Hamilton does not exist.

A sketch of Hamilton writing on Brougham Bridge.

The solution came with Hamilton's discovery of a *quadruple* system of numbers, namely *quaternions*. We still have his own account of this discovery, written towards the end of his life in a letter to his son Archibald:[8]

On the 16th day of October (it was 1843), which happened to be a Monday and Council day of the Royal Irish Academy I was walking to attend and preside and your mother was walking with me along the Royal canal; and although she talked with me now and then yet an undercurrent of thought was going on in my mind which gave me at last a result, whereof it is not too much to say that I felt at once the importance. An electric circuit seemed to close; and a spark splashed forth, the herald as I foresaw immediately of many long years to come of definitely directed thought and work... Nor could I resist the impulse—unphilosophical as it might have been—to cut with a knife on a stone on Brougham Bridge, as we passed it, the fundamental formula with the symbols *i, j*, and *k* namely $i^2 = j^2 = k^2 = ijk = -1$ which contains the solution of the problem.

So what was this blinding flash of inspiration? Essentially it was to realize that he would have to sacrifice commutativity to preserve the law of the modulus, while keeping the property that the product of any two roots of unity gives a third.[9]

Quaternions satisfy all the usual arithmetical laws, apart from the commutative law, and were important in the development of abstract algebras (see Chapter 15). They could also be used to achieve the transformation of any directed line in three dimensions to any other directed line. Although they did not become the powerful and widespread tool that Hamilton hoped they would be, they were however important in the development of vector analysis. Hamilton used the notation of 'scalar part' for the term that does not involve *i*, *j*, or *k*, and 'vector part' for the terms that do, writing this decomposition as

$$Q = S.Q + V.Q.$$

If $Q$ and $Q'$ are two quaternions with no scalar terms, then $S.QQ'$ is the negative of the dot product of the vectors $Q$ and $Q'$, while $V.QQ'$ is the vector product of the vectors $Q$ and $Q'$.

As is mentioned in Chapters 4 and 8, Tait was an enthusiastic supporter of quaternions but not of vector analysis. Although the commutative law does not apply for quaternions, the dot product in vector analysis does not have the closure property and for the vector product both commutativity and associativity are lost. Perhaps for these reasons Tait viewed vector analysis as 'a hermaphrodite monster'.[10] But in spite of his opposition, vector analysis ultimately proved more useful and more popular than quaternions, and remains the tool of choice for most mathematical physicists today.

We have already mentioned Grattan-Guinness's comment on the algebra-with-geometry approach to Hamilton's optical investigations, which also seems to have influenced his discovery of, and work on, quaternions. But it also arose in his work on what he called the 'Icosian calculus'—an investigation of the dodecahedron and icosahedron—inventing (in 1857) a game based on his ideas; the aim is to traverse a graph so that each vertex is visited once and only once (see Chapter 17). There is a note by Hamilton in the *Philosophical Magazine* discussing it, which again illustrates the theme of algebra-with-geometry.[11]

Hamilton died in 1865 battling against alcoholism. In his obituary the following year, Augustus De Morgan wrote:[12]

Hamilton was not only an Irishman, but Irish: and this with curious oppositions of character. He was a non-combatant: there was too much kindness in his disposition to allow any fight to show itself. Impulsive and enthusiastic, with strong opinions and new views, he was never engaged in a scientific controversy. In this matter he was the Scotchman, and the Edinburgh Sir W. Hamilton—never quite out of hot water—was the Irishman. William Rowan Hamilton's preservative was his dread of wounding the feelings of others... He had a morbid fear of being a plagiarist; and the letters which he wrote to those who had treated like subjects with himself sometimes contained curious and far-fetched misgivings about his own priority. But, with all this, there was a touch of the national temperament in him. An Englishman who never strikes, can, nevertheless, clench his fists, which the most warlike Frenchman cannot do: an Irishman who never gets into a row may give quick but quiet symptoms of opposition of opinion, and of what, were it more than a rudiment, would be called pugnacity.

Throughout his life, Hamilton had been showered with honours and distinctions. But perhaps the highest recognition came shortly before his death when he was elected first in the list of the fifteen foreign associates of the newly formed American National Academy of Sciences in 1865. These associates were, in the view of the Academy, the most important scientists working outside the United States of America. According to Hankins, the work that won Hamilton recognition from the Academy was that on quaternions.[13] But his world-class research in geometrical optics had been no less groundbreaking. His approach, based on the principle of least action—which he also applied to dynamical systems—created a methodology that later became even more important in the 20th-century development of quantum mechanics.

A contemporary of Hamilton, someone who shared his passion for (and ability in) geometrical optics, and possibly also someone who fuelled his dread of plagiarism, was James MacCullagh. Four years Hamilton's junior, MacCullagh entered Trinity College, Dublin, in 1824 at the age of 15 and was elected to a fellowship in 1832—choosing one of the three lay fellowships. In 1835 he was appointed professor of mathematics at Trinity College, where he was to produce nearly forty mathematical publications, mainly in the area of geometry applied to optics. His first completed work was an elegant geometrical method for constructing the Fresnel wave surface, published in the *Transactions of the Royal Irish Academy* in 1830 with the title: 'On the double refraction of light in a crystallized medium according to the principle of Fresnel'.

This paper subsequently led to a dispute between Hamilton and MacCullagh. As we have already mentioned, Hamilton predicted conical refraction in a biaxial crystal late in 1832, which was then experimentally verified by Humphrey Lloyd. MacCullagh wrote in the August 1833 edition of the *Philosophical Magazine* that:[14]

When Professor Hamilton announced his discovery of Conical Refraction, he did not seem to have been aware that it is an obvious and immediate consequence of the theorems published by me, three years ago, in the *Transactions of the Royal Irish Academy*...The indeterminate cases of my own theorems, which, optically interpreted mean conical refraction, of course occurred to me at the time; but they had nothing to do with the subject of that Paper; and the full examination of them, along with the experiments they might suggest, was reserved for a subsequent essay, which I expressed my intention of writing. Business of a different nature, however, prevented me from following up the inquiry.

Again, at the 1842 meeting of the British Association in Manchester, Hamilton speculated on a possible model for the aether, drawing the response from MacCullagh that he too had:[15]

...indulged in speculations...involving this very conception of Sir W. Hamilton, and had even followed out some of its consequences...but he had abandoned it as mere speculation.

Naturally these and other controversies led to tensions in their relationship, as illustrated by this excerpt from a letter of Hamilton to his first biographer Robert Graves:[16]

The peculiarity of memory of the friend to whom you allude is one with which I do not well know whether to be amused or annoyed...In the person you refer to there seems to be a sort of Platonic reminiscence, by which any discovery or suggestion that strikes him in other people is recognised as having been known in some former state of existence.

But that there was also great regard is shown in a letter that Hamilton wrote to De Morgan in 1852, five year's after MacCullagh's premature death, referring to:[17]

...my intercourse with poor MacCullogh [sic], who was constantly fancying that people were plundering his stores, which certainly were worth the robbing. This was no doubt a sort of premonitory symptom of that insanity that produced his awful end. He could inspire love and yet it was difficult to live with him; and I am thankful that I escaped, so well as I did, from a quarrel, partly because I do not live in College, nor in Dublin. I fear that all this must seem a little unkind; but you will understand me. I was on excellent terms with MacCullogh...spoke of those early papers of his, in 1832, to the British Association, when it first met at Oxford; took pains to exhibit the merits of one of his papers on light...on the occasion of presenting him with a Gold Medal in 1838...followed his coffin on foot from the College through the streets of Dublin; cooperated in procuring a pension for his sisters; and subscribed to the MacCullogh Testimonial.

The 'Gold Medal' to which Hamilton refers is the Cunningham medal of the Royal Irish Academy, which was awarded to MacCullagh for his paper 'On the laws of crystalline reflexion and refraction', published in 1838. In this paper

MacCullagh developed a mechanical model of the aether, which he used to predict the reflection and refraction of light in a crystalline medium. Hamilton, presenting the gold medal to MacCullagh, said:[18]

The method which Mr. MacCullagh has adopted may be said to be in general the method of mathematical induction, as distinguished from mathematical deduction. He has not sought to deduce from any pre-supposed attractions or repulsions, and arrangements of molecules of the ether, any conclusions respecting the vibrations in the interior or at the boundaries of a medium, as necessary consequences of those dynamical principles or assumptions...He has sought to arrive at laws which might bear somewhat the same relation to the optical investigations already made, as the laws of Kepler did to the astronomical observations of his predecessor Tycho Brahe, without seeking yet to deduce these laws, as Newton did the laws of Kepler, from any higher or dynamical principle.

MacCullagh sought to answer the final point in the above portion of the address, about deduction from a higher or dynamical principle in the above paper, by constructing a potential function from which to deduce the oscillation of the aether and propagation of light waves. The potential function was a homogeneous function of the coordinates of the curl of the displacement vector, so giving the aether rotational elasticity but no compressional elasticity which would have had the consequence of introducing longitudinal waves. Although he did not give a mechanism for this rotational elasticity, his model of the aether later influenced other Irish mathematicians, as we shall see.

Although not a nationalist in politics, MacCullagh was very concerned with the state of the general Irish population. The mid-1840s was a terrible time for the

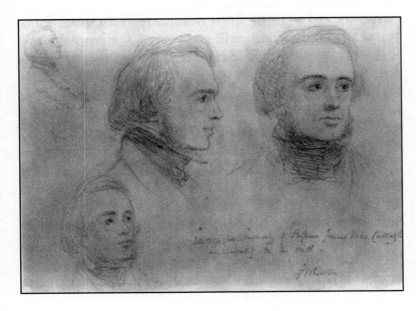

Sketches of James MacCullagh (1809–47).

country, with the potato blight destroying crops and causing widespread famine. MacCullagh was more concerned about this than many others and in the general election of 1847 he stood for one of the two University seats in Dublin. The previous MPs had been Oxford graduates, and in his election address MacCullagh said:[19]

…the University ought to be represented in parliament by her own sons… Other Universities, the Universities of England, have always acted on it. How was it that we had overlooked it? It was the principle of self-respect and therefore they had overlooked it; for that was not a principle usually acted upon in Ireland. The feeling of self-respect was, unfortunately, not one that exercised the influence that it ought over the conduct of Irishmen. It was to the want of this feeling that much of the evils and misfortunes of this unhappy country might be traced.

Of the four candidates, MacCullagh came last, but with a creditable share of nearly 17 per cent of the votes cast. Two months later he committed suicide by cutting his throat. T. D. Spearman has speculated that MacCullagh might have believed his election could have had an influence on the situation in Ireland, and concluded:[20]

Although he was not by any means an unsociable man, and he was regarded by affection by his friends, he could easily misunderstand people, and communication was frequently difficult. This difficulty of communication probably became acute when his need of friendship was greatest.

MacCullagh's *Collected Works* were eventually published in 1880. In their introduction, the editors wrote:[21]

A considerable part of his optical researches, more especially those of an earlier date, really belong to the domain of geometry.

This geometric approach, so characteristic of both MacCullagh and Hamilton's work, seems to have been more highly developed in Ireland than in England, if we are to believe the following interchange between De Morgan and Hamilton, early in 1852:[22]

There is another Irishman in whose writings I am interested, I mean Mr. Salmon. What age of man is he? How comes it that the geometrical extensions take such root in Ireland? Yourself, C. Graves, MacCullogh, Salmon, etc are all up in them. Hardly a soul in England cares about them. The pure differential calculus does not seem to interest you (I mean you plural) so much, except as something to be applied.

The reply from Hamilton to De Morgan said:[23]

I think there is a greater, or at least a more general, aptitude for pure geometry in Ireland than in England. The fellows of T.C.D. are nearly all geometers, and some of them are extremely good ones, although the public examinations for Fellowship do not turn much upon geometry, analytics having, I think, a larger share allocated to them.

George Salmon (1819–1904).

George Salmon, referred to by De Morgan above, graduated from Trinity College in 1838, and had been lecturing in mathematics (and divinity) there since the 1840s. His principal area of research was in algebraic geometry, in which his most celebrated discovery arose from a collaboration by correspondence with Arthur Cayley. In 1849, Cayley found that every cubic surface contained a finite number of lines; it was Salmon who actually discovered how many—27, to be precise.

Salmon was most famous in his day for a series of advanced mathematical textbooks which were very influential at the time. These textbooks, *A Treatise on Conic Sections* (1848), *A Treatise on the Higher Plane Curves* (1852), *Lessons Introductory to the Modern Higher Algebra* (1859), and *A Treatise on the Analytic Geometry of Three Dimensions* (1862), contained clear expositions of many recent research results, particularly in algebraic geometry and invariant theory. All four books were very well received, were translated into several languages, and went through multiple editions at home and abroad. Appointed Regius professor of divinity at Trinity College in 1866, Salmon gave up his mathematics lectureship, and soon after devoted himself entirely to theology.

A cubic surface.

# FitzGerald and Larmor: research in electromagnetism

Having surveyed a number of geometrically inclined Irish mathematicians, we now turn to the other connected theme of Irish mathematics that we left with James MacCullagh—mathematical physics. MacCullagh's potential and his rotationally elastic aether model eventually found favour in the 1880s, with the development of electromagnetism and the need for an aether theory to support electric and magnetic fields.

George FitzGerald (1851–1901).

George Francis FitzGerald, a fellow of Trinity College, Dublin, was appointed to the Erasmus Smith chair of natural philosophy in 1881 and showed that MacCullagh's aether could do exactly that—namely, support an aether theory of electromagnetic radiation in general. Although he is now best remembered for the FitzGerald–Lorentz contraction, FitzGerald's main scientific interest was on developing Maxwell's theory of electromagnetism. He also had an interest in education, being one of the Commissioners of national education in Ireland for 1898 with the task of revising primary education along more practical lines, and in the year preceding his death was appointed to the Irish Board of Intermediate Education. An active school of theoretical and experimental physics developed in Dublin under FitzGerald and continued the tradition of Humphrey Lloyd, Hamilton, and MacCullagh. This work was also continued in the 1890s by another Irishman, Joseph Larmor, who in 1903 was appointed George Stokes's successor to the Lucasian professorship at Cambridge. According to Ivor Grattan-Guinness in his *History of the Mathematical Sciences*, Larmor maintained an Irish tradition, drawing on Hamilton for mathematical methods, MacCullagh for physics, and working under the influence of his compatriot FitzGerald.[24]

Joseph Larmor (1857–1942).

The work for which Larmor is chiefly remembered took place between 1892 and 1901, a transitional period in physics. As A. S. Eddington argued at the start of his obituary notice of Larmor,[25] the main wave of advance in physics, after half-a-century of rapid progress, seemed to have run its course, and it seemed that the possibilities for fresh discoveries were limited. Before the end of that decade the discoveries of X-rays, electrons, and radioactivity were to rejuvenate experimental physics. However, at the time when Larmor started his main work in 1892, there was little to inspire new ideas. Of the researchers who tried to make progress at this difficult time, two names stand out—the Dutch physicist H. A. Lorentz and Larmor himself. In Eddington's view:[26]

Their work had much in common, so that it is sometimes difficult to assess their contributions separately. Larmor's reputation has perhaps been overshadowed by that of Lorentz. But on any estimate, Larmor's achievements rank high; and his place in science is secure as one who rekindled the dying embers of the old physics to prepare the advent of the new.

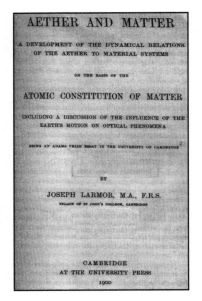

The title page of *Aether and Matter*.

In his major work, *Aether and Matter* (1900), Larmor developed MacCullagh's aether model by using his preferred technique of least action to explain electric charges. To him, the aether was not a material medium and he did not attempt to develop a model working on mechanical lines. Indeed, instead of thinking of the aether as matter, he viewed matter as an array of electric particles, as 'electrons' that moved about in the aether according to the laws of electromagnetism. But as he described it, the electrons were not material particles:[27]

An electric point charge is a nucleus of intrinsic strain in the aether if only we are willing to admit that it can move or slip freely through that medium much in the same way that a knot slips along a rope.

This model led Sir Horace Lamb, at a meeting of the British Association in 1904, to refer to Larmor's book as being better called *Aether and no Matter*.[28]

Larmor's success in becoming senior wrangler at Cambridge was celebrated by a torchlight procession at Queen's College, Belfast.

Larmor was also interested in the history of his subject, working on his own *Collected Works* as well as editing collections of papers by scientists like Henry Cavendish, James Thomson, George Stokes, and Lord Kelvin. An obituary notice of FitzGerald in March 1901 also reveals his interest in promoting Ireland (and Dublin in particular) as a centre of mathematical and scientific activity. Larmor writes:[29]

In those early years [i.e. the 1880s] there were three main centres of development of the new departure in electrical theory which has since revolutionised the whole domain of physical science. Maxwell's own presence as a professor had guided the trend of physical thought at Cambridge predominantly into that direction which it has since largely retained; in Berlin, Helmholtz was devoting his great powers and turning the attention of his pupils to the discussion and elucidation of the subject; while in Dublin its study and investigation became vital under FitzGerald's lead and influence.

And he finished the obituary by saying:[30]

His scientific place will be henceforth alongside Rowan Hamilton and MacCullagh and Humphrey Lloyd, and the other famous men who have secured for the Dublin school so prominent a position in the edifice of modern physical science.

It is noteworthy that Larmor includes Dublin as one of the three main centres for the development of electrodynamics, especially in view of the fact that he had strong historical interests, and that he placed it in a tradition stretching back to

Hamilton and MacCullagh. Indeed, according to Andrew Warwick, he also saw himself as continuing this Irish tradition:[31]

Larmor aligned himself closely with the distinguished school of mathematical physics associated with Trinity College, Dublin—including James MacCullagh, William Rowan Hamilton and George FitzGerald—and considered himself to be developing the tradition they had begun.

So by the beginning of the 20th century we see that a definite mathematical tradition had developed in Ireland, focusing on the two main areas of geometry and electromagnetism. Furthermore Larmor and FitzGerald viewed themselves as contributing to and developing this tradition which originated in the work of Humphrey Lloyd, William Rowan Hamilton, James MacCullagh, and George Salmon.

# Epilogue

Eamon de Valera (1882–1975).

This tradition influenced one further institutional development which, although it occurred well into the 20th century, will serve as a fitting conclusion to this chapter. It is that of the Dublin Institute for Advanced Study (DIAS), founded in 1940 with its two constituent schools, the School of Celtic Studies and the School of Theoretical Physics. A third School of Cosmic Physics was created in 1947.

The DIAS was established by Eamon de Valera who had a life-long interest in mathematics. De Valera was born in New York of an Irish mother and a Spanish father. At the age of 3, he was brought to Ireland by his uncle to live in his grandmother's house at Bruree, County Limerick. The family was poor, and he was able to continue to secondary and university education only by winning a number of prizes. He graduated in mathematics from the Royal University of Ireland in 1904, and two years later was appointed professor of mathematics at Carysfort Teacher Training College in Dublin. In 1916 he fought for the republicans in the Easter Rising, and as a result was sentenced to death, which was commuted to penal servitude. He was imprisoned in Dartmoor and later in Lewes Jail, where he continued his interest in mechanics, as we can see from this extract from a letter that he wrote on 21 February 1917:[32]

I am anxious then to lay my hands on some good work on celestial mechanics. I have my Appell here and he refers to *Mécanique céleste* de Tisserand. I wonder if it is possible to get books from Paris at present?...Since I got Appell my principal amusement has been to turn him into quaternions and if I get Tisserand I will try to do the same with him. I am never at a loss now for something to do.

After his release in 1924 he founded the Fianna Fail party and became head of the Irish Free State in 1932.

Throughout the 1930s, de Valera developed his ambition to found a research institute that would establish Dublin as a centre for both Irish studies and theoretical physics, linking his own two passions and placing Ireland on the academic world stage. He himself was conscious of the traditions which we have traced in this chapter. In introducing the bill which established DIAS on 6 July 1939, de Valera said:[33]

There is however a branch of science in which you require no elaborate equipment, in which all you want is an adequate library, the brains and the men, and just paper. We have already in the world an important place, or had in the past an important place, in mathematics and theoretical physics. The name of Hamilton is known wherever there is a mathematical physicist or theoretical physicist. This is the country of Hamilton, a country of great mathematicians. We have the opportunity now of establishing a school of theoretical physics which could be specialized as the school of Celtic studies can be specialized, and which I think will again enable us to achieve a reputation in that direction comparable to that which Dublin and Ireland had in the middle of the last century.

Modelling the Institute on that at Princeton, de Valera was successful in attracting mathematicians with worldwide reputations, some of whom were Irish-born such as John Lighton Synge, nephew of the playwright J. M. Synge, who during his tenure at DIAS developed one of the world's great centres in relativity theory with his focus being on geometrical methods in relativity. During this period, Synge's fellow researchers included the Hungarian Cornelius Lanczos who contributed to the development of relativity and quantum physics, Walter Heitler who continued his work on quantum electrodynamics and cosmic rays while at the Institute, and perhaps most renowned of all, Erwin Schrödinger, who became the first director of DIAS in 1939.

In this chapter we have identified a tradition of mathematical physics, as being a thread that runs through the interests and influences of many Irish mathematicians. Some were 'home-grown' and some came to Ireland to continue their research, but all contributed to this rich seam of mathematical development that can trace its origins to the Victorian era.

The University of Melbourne in the 1860s.

CHAPTER 6

# Wranglers in exile

*Mathematics in the British Empire*

JUNE BARROW-GREEN

After the defeat of Napoleon at Waterloo in 1815, Britain began a period of Imperial expansion that continued throughout the Victorian era. As the colonies grew, so too did the number of colonial institutions of higher education. For founders of mathematics departments looking to perpetuate the tradition of the mother country, Cambridge was the obvious model to follow, and many new colonial universities, especially those in Australia, provided a source of employment for Cambridge wranglers. Mathematics graduates from Scotland also found their way overseas, notably to New Zealand and South Africa. India provided employment not only on the sub-Continent, but in England too, the latter through the mathematics departments of the East India Company's training colleges of Addiscombe and Haileybury and the Royal Indian Engineering College at Cooper's Hill. Meanwhile, Canada, with its older tradition of higher education, did not so much provide employment for wranglers, as a place for them to visit. The British Association meetings held in Montreal (1884) and Toronto (1897) were the first to be held on foreign soil, and several mathematicians took the opportunity to make the trip across the Atlantic.

During the 19th century many British mathematicians spent part, sometimes almost all, of their careers working overseas in a British colony. Others remained in Britain but were employed by, or supported, an institution that was involved in some sort of Empire-related activity. Conversely, there were mathematicians who had begun their lives in one of the colonies or dominions but who came to Britain to study and/or to work. It was a period of both expansion and consolidation within the British Empire.[1] Several new universities were established and mathematics had a role to play in a number of ways. But what motivated mathematicians to travel? To what extent did their decision to travel affect their mathematical careers? What mathematical contributions were they expected to make in their new environment? What mathematical contributions did they make?

Our investigation of these questions focuses primarily on two complementary strands of activity:

• The setting up and maintenance of mathematical departments in the new colonial universities;

• The mathematical training in Britain for those sent out to promote British interests in India.

In our discussion we concentrate almost exclusively on university level and research mathematics. Thus we do not include either school mathematics or the use of mathematics as a strictly practical or commercial activity, as for example in the setting up of systems of weights and measures, the making of local maps and charts, and so on. Consequently we have not investigated the actual use of mathematicians in commercial enterprises such as the East India Company or in government departments. Finally, we make no claim for completeness in the topics we discuss—for example, not every colonial university comes under our gaze. Our goal is a broad overview rather than a definitive study.

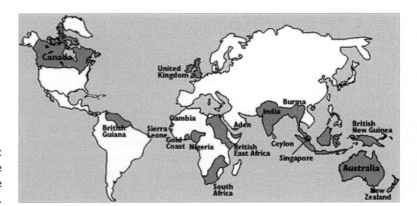

Map of the British Empire: the shaded areas show the countries belonging to the Empire in Victorian times.

## Australia

The University of Sydney, founded in 1852, is the oldest university in the Commonwealth and was the first colonial university to be established in the British Empire. Its first professor of mathematics, Morris Birkbeck Pell, was only 25 years old when he began his tenure in 1852, although as the senior wrangler and second Smith's prizeman of 1849, and a fellow of St John's College, Cambridge—with a published book, *Geometrical Illustrations of the Differential Calculus* (1850), already to his credit—he came with an impeccable mathematical pedigree for someone of his age.[2] Pell, who was born in the United States and arrived in England in late adolescence, came from a family with a tradition of education, being descended from George Birkbeck who founded the London Mechanics' Institute in 1823, the first Mechanics' Institute in England, that was later to become Birkbeck College.[3]

Certainly Pell's qualifications fulfilled the hopes of the founders of the university in their quest for a professor of mathematics. In their letter to the selection committee in England the first provost and vice-provost wrote:[4]

[W]e hope we shall not inconveniently fetter your choice by confining it to…the first ten wranglers in mathematics at the University of Cambridge… that the gentleman should be a Master of Arts of not more than six years' standing, and that there should have been no material interruption to the pursuit of their academical studies up to the time of their appointment by you.

The selectors included Sir John Herschel and Sir George Airy, the Astronomer Royal. Pell was chosen from among twenty-six candidates (the names of the others are not known). That the founders wished for a Cambridge man is not unexpected. As described in Chapter 1 of this book, the standard achieved by high-ranking mathematics graduates from Cambridge far exceeded that achieved by mathematics graduates from elsewhere, such as Oxford or London, during the Victorian period. Moreover, a Cambridge man would take with him first-hand knowledge of the Cambridge system of teaching and examining, as embodied in the mathematical tripos (with all of its connotations of status and success). That there was a desire for a relatively recent graduate probably had as much to do with physical fitness—the journey would be difficult and the conditions likely to be harsh—as it did to wanting someone who was (possibly) at the peak of their creative powers.

When Pell began teaching in Sydney, the university had twenty-four students, all of whom were required to take mathematics for three years. One of the students was W. C. Windemeyer[5] who, on 13 October 1852, noted in his diary 'Went to a lecture at 10 with Mr Pell who amused as well as instructed, think I shall like him…'.[6] When Pell retired twenty-five years later, in 1877, the number of students had grown to fifty-eight, although by this time students were required to study mathematics for only two years. Intermittently from 1860, Pell had the help of two assistants, both Cambridge men.[7] The first, William Scott, 3rd wrangler of 1848, was Government Astronomer at the Sydney Observatory from 1856 to 1862. He worked with Pell from 1860 to 1862 and again in the 1870s. The second, George Smalley, 28th wrangler of 1845, had spent several years in South Africa (see below), before ending up in Sydney where he took over from Scott as the Government Astronomer. He taught at the University in 1865 and examined with Pell from 1865 to 1870.

The students' mathematical background was very limited. The lack of suitable secondary schools meant that for matriculation the students had to know only the four rules of arithmetic, algebra to simple equations in one unknown, and the first book of Euclid. For the pass degree Pell aimed simply to get the students to the harder rules of arithmetic, quadratic equations, and the first four books of Euclid. The honours students were taught the more advanced subjects of differential and integral calculus and probability theory, although the lack of suitable textbooks meant that Pell had to prepare his own lecture notes; and he devised his own somewhat idiosyncratic method of examining:[8]

To candidates for honours at the B.A. degree he was wont to give on each of two successive days a paper with which they began at half past nine in the morning, and which they were at liberty to struggle with till the shades of evening compelled them to retire—a cold collation being allowed to any whose spirit might be willing, but whose flesh was weak.

It was hardly comparable to the eight-day ordeal of the Cambridge tripos examination that Pell himself had endured. Pell continued to develop the course, and by the end of his tenure in 1877 the subjects required for an honours degree had advanced to include conic sections, differential and integral calculus, dynamics, higher algebra, and trigonometry.

The less than challenging nature of the teaching gave Pell time to pursue other academic interests. He embarked on actuarial studies, publishing several papers, and for several years acted as a consultant to a leading assurance company. He also became very involved with the development of secondary education and largely due to his efforts the position of mathematics in secondary schools was greatly improved.

When Pell retired in 1877, another St John's man, Theodore Gurney, was selected from a large number of applicants to succeed him. Gurney, who was 3rd wrangler in 1873, came from a family with a Cambridge tradition; his father, who graduated as 23rd wrangler in 1837, had been an exact contemporary of George Green and James Joseph Sylvester. He was recommended for the position by George Stokes, the Lucasian professor of mathematics. As a matter of fact, Stokes ignored an application for the position from another promising Cambridge man who was already in Australia, Horace Lamb.[9] However, since Lamb had recently taken up the chair of mathematics in Adelaide (see below), it is likely that Stokes thought it easier to find someone new for Sydney, rather than a replacement for Lamb in Adelaide.

Gurney gained a good reputation as a teacher, but he did not live up to his early promise as an original mathematician and produced nothing in the way of research. The local authorities were disappointed and in 1902, just prior to Gurney's retirement, they wrote to Joseph Larmor, a member of the selection committee, and who succeeded Stokes as Lucasian professor in the following year, making their feelings known:[10]

There is very considerable activity in all other branches of Science here (i.e. at Sydney), but research in mathematical matters is absolutely non-existent. The present professor, Gurney of your College, has held the chair for 25 years.

Mentally equipped with every gift except ambition, he has, as you know, never published a line…

Since the university barely grew during Gurney's tenure, it is unlikely that pressure of work was the cause for his lack of output: perhaps it was simply that he did not find his environment with its mathematical isolation conducive to research. But Gurney did not neglect the teaching of his subject, and under his stewardship there was a marked increase in the breadth and difficulty of the mathematics taught. By 1902 the third-year students were expected to study Isaac Todhunter's *Algebra*, *Conics*, *Calculus*, and *Spherical Trigonometry*, Charles Smith's *Algebra* and *Solid Geometry*, Joseph Edwards' *Differential* and *Integral Calculus*, Daniel Murray's *Differential Equations*, William Besant's *Dynamics*, Edward Routh's *Analytical Statics*, and Hugh Godfray's *Astronomy*.

The small but steady increase in students attending the university—the number rose from eighty-one to two hundred and forty-four in the twenty years between 1881 and 1901—made it necessary both to raise the number of staff and to expand the number and type of courses on offer. The first lecturer of mathematics was appointed in 1880, and by 1887 a second appointment had been made to accommodate the demand for evening classes.

When Gurney retired in 1903 the authorities were careful to employ someone who already had a good track record in creative mathematics. Horatio Carslaw had graduated from the University of Glasgow in 1891, before entering Cambridge and becoming 4th wrangler in 1894. He returned to Glasgow in 1896 as an assistant in the mathematics department, before embarking the following year for Göttingen, Rome, and Palermo. In Göttingen he had been one of Arnold Sommerfeld's first students, and it was there that the foundations were laid for the rest of his mathematical career. He returned to Glasgow, staying there until he was appointed to the Sydney chair. He held the chair from 1903 until his retirement in 1935 and, renowned for his own work in Fourier series, led the department to worldwide recognition. He is perhaps best remembered for his book on *The Introduction to the Theory of Fourier Series and Integrals and the Mathematical Conduction of Heat*, first published in 1906 and considered by many to be a good introduction to the theory of functions of a real variable (see Chapter 13). However, despite Carslaw's success in the research field, the level of mathematics studied by undergraduates at Sydney still remained far behind that of Cambridge, and Carslaw, in line with his own career, encouraged the best of his students on graduation to go to Cambridge to take the mathematical tripos.

The University of Melbourne, which was founded in 1853, also began its mathematics department with a senior wrangler, William Wilson, a first Smith's prizeman who graduated from St John's and became a fellow of his college in 1847. As we saw in Chapter 5, in 1849 he was appointed founding professor of mathematics at Queen's College, Belfast, where he was instrumental in the establishment of the astronomical observatory.[11] He came to Melbourne in 1854 with good credentials, having already published *A Treatise on Dynamics* (1850). As one member of the four-man selection committee, John Herschel, wrote to another, George Airy: 'For my own part, I cannot see how it is possible to find a better man or [one] more completely uniting every qualification than Professor Wilson'.[12]

Although Wilson came highly recommended to Melbourne, his name appears rarely in the university records and little is known about him apart from the fact that he was apparently 'a temperamental little bachelor' who entertained liberally.[13] As a teacher he had a reputation for being clear and concise, but exacting. As time went on, he became more interested in physics, experimenting in electricity, magnetism, and light. He also maintained his interest in astronomy, making valuable observations, and in 1871 journeyed to the northern-most tip of the continent to observe the eclipse of the sun. He died suddenly in 1874, having only recently established an observatory at Mornington, south of Melbourne.

It is conspicuous that Wilson, in making the move to Melbourne (in contrast to his compatriots in Sydney), had relinquished a good position in Britain. Wilson had a professorship in Belfast and he had invested time and energy into establishing an observatory there. Furthermore, the journey to Melbourne was long and not without danger, the conditions in Melbourne were unknown, and the opportunity for academic discourse limited. So what persuaded him to make the move? In a word: money. In Belfast the annual salary was in the region of £200 to £250, plus lecture fees from students. In Melbourne the annual salary was £1000 for life, plus a free house and (possible) lecture fees depending on the number of students.[14] It was a temptation too alluring to resist.

On Wilson's death the University Council sent a telegram to John Couch Adams, the Lowndean professor of astronomy and geometry at Cambridge, with the following request: 'Professor Wilson dead. Please select best man procurable any university, not in orders. Duties Melbourne University Calendar'.[15] The position was duly advertised and a notice was seen by Sylvester who, on 15 February 1875, wrote to Arthur Cayley:[16]

I see from the newspapers that Adams is to appoint to the vacant Mathematical professorship at Melbourne. I am more than half induced to go out to the Antipodes rather than remain unemployed and *living upon Capital* in England if he should think me a suitable man for the place—Perhaps you will not mind putting the question to him.

But any enthusiasm Sylvester might have had for the position was soon dampened by Cayley, and a few days later Sylvester wrote to him again to thank him for his 'desire to save me from the Antipodes'.[17] (Sylvester did in fact leave England the following year to take up his appointment as the first professor of mathematics at the newly founded Johns Hopkins University in Baltimore.)

Wilson's death also prompted a glowing report of his activities from the 'Victoria correspondent' of *The Times*:[18]

The Professor was everywhere respected. He was at the head of all scientific movements, devoting himself energetically to anything which promised to promote the intellectual progress of the Colony, and his practical abilities were recognized by men of business who sought and quoted his opinion as the highest authority procurable upon any difficulty which could be solved by figures.

Unlike in Sydney, the candidates for Melbourne were not restricted to Cambridge graduates, although the man selected was such a graduate, and a very recent one. Edward Nanson had been 2nd wrangler and second Smith's prizeman in 1873, so had been one place above Gurney in the order of merit. On leaving Cambridge, he had spent a year as professor of applied mathematics at the Royal Indian Engineering College at Cooper's Hill (see below) before accepting the position in Melbourne. Fortunately for Melbourne, not all the class of 1873 were tainted with inactivity. From a research point of view, Nanson proved an admirable choice. Described as 'a disciple of Cayley, Sylvester and Salmon...of the dominant English 'aesthetic' school of pure mathematics',[19] he published on a range of subjects, including hydrodynamics, theory of differential equations, methods of election, geometry, and determinants, as well as contributing numerous short notes of a didactical nature to journals such as the *Messenger of Mathematics*, the *Mathematical Gazette*, and the *Educational Times*. In the public arena he was best known for his work on electoral reform, a subject on which he spoke and wrote prodigiously.

Nanson had a number of successful students including John Henry Michell, who graduated in 1884, and, on Nanson's recommendation, went to Cambridge where he was bracketed senior wrangler in 1887 and joint first Smith's prizeman in 1889.[20] He returned to Melbourne and in 1891 became Nanson's teaching assistant, shortly afterwards being appointed as an 'Independent Lecturer in Mixed Mathematics'.[21] He went on to gain renown in the fields of hydrodynamics and elasticity, being elected a Fellow of the Royal Society in 1902. When Nanson retired in 1922, having been in post for forty-eight years (a record in the history of the university), he was succeeded by Michell who took up the appointment in 1923.

Edward Nanson (1850–1936).

Although the mathematics departments of Sydney and Melbourne had their moments, the strongest mathematics department in Australia during the latter part of the 19th century was undoubtedly that of the University of Adelaide. The University was founded in 1874 and its first professor of mathematics was Horace Lamb. Lamb, who had spent a year at Owen's College, Manchester, before going up to Trinity College, Cambridge, was 2nd wrangler and second Smith's prizeman in 1872. In the same year he was elected a fellow and lecturer of his college. He gave advanced courses on rigid dynamics, sound, and hydrodynamics, the latter providing the foundations for his celebrated textbook on the *Motion of Fluids* (1879), which was later republished as *Hydrodynamics* (1895). *Hydrodynamics* ran into several further editions and is considered to be one of the finest textbooks in applied mathematics of the period.

In 1875 Lamb's attention was drawn to the position at Adelaide by the Reverend Stanley Poole, then an Adelaide resident, but who had been a schoolmaster at Stockport Grammar School when Lamb had been a pupil. Although the

Horace Lamb (1849–1934).

opportunity to become the head of a new department must have been attractive in itself, there was another (and perhaps more compelling) reason which drew Lamb to Adelaide. In 1875 the rule of celibacy still pertained for fellows of Trinity and Lamb had recently become engaged to be married.

Lamb commenced work in 1876 when the university had a total of eight matriculated students, with another fifty-two students attending lectures but not reading for a degree. Initially conditions were difficult as the new university moved from one set of inadequate rented quarters to another, and it was in September of his second year that Lamb wrote to Stokes offering himself for the Sydney chair.

Sir Thomas Hudson Beare, one of Lamb's former students and later professor of engineering at Edinburgh, recalled:[22]

Lamb was a wonderful teacher; he was carrying out at that time a good deal of his original work in hydrodynamics. He was an excellent lecturer, very clear, very lucid, and, as he had to deal with somewhat raw material it was a difficult task for him.

As Beare correctly recalled, Lamb was indeed continuing with his research. During his tenure at Adelaide he published several papers in distinguished journals, including the *Proceedings of the Royal Society* and the *Proceedings of the London Mathematical Society*, on statics and electricity as well as on hydrodynamics.

At the end of 1883, Lamb requested a year's leave of absence, pointing out that he had been in the service of the university for nine years, during which time he had undertaken duties that did not fall strictly within the scope of his professorship (a reference to his physics teaching). His request resulted in a protracted debate with the University Council which was finally resolved in Lamb's favour in the middle of 1885, by which time there was little doubt as to his underlying motive. He had made an application for the chair of pure mathematics at Owen's College, Manchester, and the application had been successful, subject to a satisfactory interview and testimonials from Adelaide. In 1885 he left Australia for Manchester and did not return.

Lamb's successor at Adelaide was William Bragg who, like Lamb, was a Trinity man and had been 3rd wrangler in 1884. Although physics was the subject in which Bragg would later become one of the most eminent men in his field—in 1915 he won a Nobel prize for his work on X-rays—at the time of his Adelaide appointment his training had been largely in mathematics. He had, however, spent some of his final year at Cambridge in the Cavendish Laboratory attending the lectures of J. J. Thomson, and it was as the result of a chance remark by Thomson (one of the electors for the Adelaide chair) that Bragg applied for the position. Many years later, when Bragg wrote to Thomson on the occasion of the latter's 80th birthday, he reminded him of the appointment process:[23]

Perhaps you were the one who asked a certain Adelaide man—then visiting London—whether the Council of the University of Adelaide was likely to prefer a senior wrangler who occasionally disappeared under the table after dinner to a young man who had so far shown no signs of indulging in the same way. The Adelaide man was Sir Charles Todd, whose daughter I married a few years afterwards.

At all events, the move to Adelaide as professor of mathematics and physics proved decisive for Bragg as his interests moved gradually into the field in which he made his name, although for several years his energies were almost exclusively focused on teaching, university affairs, and the Australasian Association for the Advancement of Science, which left him with very little time for actual research. In the first eighteen years of his twenty-two-year tenure, from 1886 to 1904, he published only three minor papers on electrostatics and the energy of the electromagnetic field.[24]

Bragg's teaching commitment was particularly heavy, and in his second year he recorded the details of a typical week (Box 6.1).[25]

---

## Box 6.1: Bragg's teaching week

|  |  | hours per week |
|---|---|:---:|
| **B.A.** | | |
| 1st year | *Mathematics* | 2 |
| | *Physics* | 2 |
| 2nd year | *Mathematics* | 2 |
| 3rd year | *Mathematics* | 2 |
| **B.Sc.** | | |
| 1st year | same as 1st year B.A. | 1 |
| 2nd year | *Mathematics* (extra) | 2 |
| | *Physics* | 2 |
| | *Practical physics* | 2 |
| 3rd year | *Physics* | 2 |
| | *Practical physics* | 2 |
| **Mus.B.** | | |
| | *Acoustics* | 1 |
| **Evening classes** | | |
| | *Mathematics* | 2 |
| | *Physics* | 2 |
| | *Practical physics* | 2 |
| | Total | 26 |
| **Additional in 1888** | | |
| | Honours in 3rd year *mathematics* | 2 |

---

William Henry Bragg
(1862–1942).

At this time he had no academic colleague to assist with student difficulties and setting examination papers (he had twenty-one to write) and he had only one part-time laboratory assistant to help build, prepare, and supervise lecture demonstrations and laboratory experiments. And that was not all: he also set and marked five mathematics, one natural philosophy, and one English history matriculation examination, and a further five mathematics and two physics public examinations!

Having made the burden of his position clear to the Education Committee, an assistant was appointed in 1888, selected by Bragg:[26]

I have chosen [Robert] Chapman as assistant lecturer: he knew a good deal more than the other man, was energetic and strong in appearance, whilst the other was a scholastic weak-eyed type. I think Chapman will do very well. By the way he is an oarsman...

The workload and conditions being what they were, a strong physique was clearly a necessity. Bragg himself was an enthusiastic sportsman and Chapman's reputation as an oarsman would have added to his appeal. It was a good choice. Chapman turned out to be an excellent teacher and went on to have a distinguished career in applied science. Bragg remained in post until 1908 when he left to take the Cavendish chair of physics at Leeds and his chair in Adelaide was awarded to Chapman.

The mathematics department at Adelaide could hardly have had a better start. Both Lamb and Bragg were excellent teachers and both became well known through work either begun or continued at the University. The Reverend Poole put it very well when he said: '... the University of Adelaide lies under a great debt for the good lead which the first two professors of mathematics gave it'.[27]

# New Zealand

New Zealand, unlike Australia, based its university system on the London model. The individual universities where the teaching was done were each affiliated to The University of New Zealand, a single examining and regulating body founded in 1870. The University of Otago, the first university to be established in New Zealand, was actually founded the year before the University of New Zealand, but it did not open its doors until 1871, and it was only merged into the University of New Zealand in 1874. University College, Auckland, was founded in 1872, and Canterbury College, Christchurch, in the following year; Victoria University followed some twenty years later in 1894. One noticeable but, given the historical ties, not unexpected, difference between the mathematicians in New Zealand and those in Australia was the strength of the former's Scottish connection. To put it in general terms, 'The idea of creating a University in the new land was born in Scotland, and came to the colony with the first settlers'.[28]

The first professor of mathematics in Otago was John Shand, who had graduated in 1854 from King's College, Aberdeen, where he was reputed to have been one of the best students of his year 'notable for his energy, persistence and, as one of his teachers put it, moral tone'.[29] After graduation Shand had spent the next fifteen years as a schoolmaster, first at Ayr Academy and then at Edinburgh Academy where, for three years, he worked alongside the geometer John Mackay. He arrived to take up his position as professor of mathematics and natural philosophy at Otago in 1871. He was clearly a talented teacher: one student reported 'He made everything so clear that we began to think there were no difficulties in Mathematics'.[30]

Although he did not engage in research, Shand devoted considerable energy to the wider business of education in general and university administration in particular. With regard to his teaching, he presumably followed the syllabus laid down by the University of New Zealand, as was done at other New Zealand universities (see Auckland below). In fact, his courses were probably not dissimilar to those he had previously taught at the Edinburgh Academy, a school that had a very good reputation in mathematics. In 1886, when the chair was divided into mathematics and natural philosophy, he chose the latter.

The chair of mathematics was awarded to Frederick Gibbons. Gibbons had been 2nd wrangler in 1877 and had gone on to study law. Having been admitted to the Bar in England in 1881, he had travelled to New Zealand in 1884. It was a time of financial restraint for the university and the authorities were keen to appoint a 'local' in order to minimize travel costs. Having a 2nd wrangler on their doorstep, they looked no further. It turned out to be an unfortunate choice. That Gibbons was a very poor lecturer was evident from the start: his inaugural lecture turned into a 'bear garden' with students singing 'Hang old Gibbons on a sour apple tree'.[31] One can only suppose the difficulties were compounded when in 1895, in addition to his other duties, he was appointed lecturer in political economy. He retired due to ill health in 1906 leaving no mark of significance on the mathematical world.

Auckland College was affiliated to the University of New Zealand in 1872, but for the first ten years the teaching of mathematics was almost non-existent. When the mathematics classes were inspected by a member of the University senate in 1874, the standard was described as frequently lower than that of the corresponding classes at the grammar school. Nine-tenths of the students in the class were labelled incompetent, and most were persuaded to leave. From then on, the students requiring university level instruction were taught in classrooms with schoolchildren. The students not only had poor instruction, but they also had to study in a room that had a floor not safe to tread on and a roof open to the sky. As a residence it would have been deemed uninhabitable.

Things took a turn for the better in 1882 with the founding of Auckland University College and the allocation of government funds for the appointment of the first two professors, one for classics and English, the other for mathematics and mathematical physics, and shortly afterwards further finance was made available for a chair in chemistry and physics, and one in natural sciences. The selection for the mathematics chair was put in the hands of Peter Guthrie Tait, professor of natural philosophy at Edinburgh, and John Jellett, provost of Trinity College, Dublin. The first appointee was George Walker, who had been 2nd wrangler in 1879. Walker arrived in Auckland in April 1883 but, due to a tragic drowning accident, he did not survive to deliver his first lecture.[32]

The vacant chair was filled by William Steadman Aldis, professor of mathematics at the Durham College of Science, Newcastle. As it happened, Aldis had been the selectors' original choice for the chair, but the information had not reached him in time and he was prevented from taking up the position by his duties at Newcastle, and so he had recommended Walker. Prior to his appointment to Newcastle, Aldis, who had been the senior wrangler of 1861, had spent ten years coaching for the mathematical tripos in Cambridge, being unable to take a fellowship as a Non-conformist. He also authored a number of well-received textbooks.

Since the majority of the students in Auckland were employed in business or teaching (some of the female students were unemployed), most of the lectures were given outside working hours. The syllabuses[33] were set by the University of New Zealand, but they were examined externally by British examiners, the latter including the Cambridge coach Edward Routh, 'by far the most influential mathematics teacher in mid and late Victorian Britain'.[34] Routh was later followed by Horace Lamb and the geometer Henry Frederick Baker. It was a system that seemed to work well, provided that the number of students was small, and in the 19th century the number was small—in 1893 only thirteen students graduated from the college in total. But from this small number Aldis produced some good mathematicians, the most notable of whom was R. C. Maclaurin who culminated his career as president of the Massachusetts Institute of Technology.

Aldis produced only one more textbook after his arrival in Auckland, *A Textbook of Algebra*, published in 1887. However, it was clearly not aimed at his current crop of students. It was based on earlier lectures given in Newcastle and

William Steadman Aldis
(1839–1928).

was published in Oxford, with the proofs being revised in Cambridge by his brother Thomas,[35] due to 'the remoteness of the Author's present residence'.[36] With an almost non-existent market and the difficulty of getting accurate mathematical type-setting, there was little incentive to publish locally.

Aldis was a talented and appreciated teacher and it was therefore much to his students' surprise when, in 1893, he was dismissed on the grounds that he had not given some of his advertised lectures. But, needless to say, there was a subtext:[37]

It involved a personality clash between Aldis and Sir Maurice O'Rourke, Chairman of the A. U. C. [Auckland University Council] and a power struggle between the Council and the Professorial Board... Aldis and his wife had made themselves unpopular in some quarters with their outspoken views and uncompromising statements on moral and social issues. (For example, on the death of the Duke of Clarence they protested against the military salute "on the score of noise and expense". A [New Zealand] Herald correspondent wrote that he would sooner be a champion shot than a Senior Wrangler, for he would be more likely to be of use to society.)

The original accusation was based on almost negligible evidence: there had been no complaints against Aldis, he had kept sufficient hours, and there had been no criticism of his teaching. The news of the dismissal engendered heated debate and efforts were made to get Council to reverse the decision. E. A. Abbott, a school friend of Aldis and well known as the author of *Flatland* (1884), initiated a press campaign in England and appealed to university men to boycott Aldis' chair. Unfortunately, this served only to arouse antagonism, and Council voted against the motion to rescind the decision. Further appeals and petitions to Parliament were all to no avail. Aldis remained in Auckland for a further three years before returning to England. He did some occasional examining, but never again held a teaching position.

There is a further aspect to Aldis's career that is of interest, and that is his concern for the education of women. Aldis held strong views on the topic, and the absurd situation created by the success of Philippa Fawcett in the mathematical tripos of 1890—her examination marks were the highest but, as a woman, she was debarred by the university from assuming the title of senior wrangler (given instead to the man who achieved the next highest marks)—prompted a public statement from Aldis, which in turn prompted the following comments in the *New Zealand Herald* on 30 August 1890:

We sincerely trust that the suggestion thrown out by Professor Aldis at the meeting in the Choral Hall when the University degrees were conferred will be acted upon without delay. Professor Aldis suggested that the statutes of the University should be so altered as to allow the degrees of BA and MA to be conferred on women who have passed such examinations as would have entitled them, had they been men, to receive these degrees. It appears that the conservatism of Oxford and Cambridge does not yet allow women to graduate, although they are graciously permitted to undergo the same examinations as men.

As everybody knows, Miss Fawcett, the daughter of the late Professor Fawcett, has passed in the mathematical tripos in Cambridge with such marks as would have entitled her to rank above the Senior Wrangler. Now, Professor Aldis, himself a Senior Wrangler, says that he cannot imagine the existence of a human being above Senior Wrangler so that Miss Fawcett is *de jure* Senior Wrangler of this year. What is proposed is that she should be offered the degree of MA in the University of New Zealand, the first University in the British Empire which conferred its degrees on women.

This would be a graceful and appropriate compliment and there can be hardly a doubt that Miss Fawcett would accept it. It would also be a hint, and a pretty broad hint, to the Cambridge men, that we think them rather behind the times. A few such hints might produce some effect on the British intellect in the course of a quarter century or thereabouts, which is the usual time required for a new idea to penetrate the British skull and reach the British brain.

New Zealand may not have been ahead of Britain in its mathematics, but it was certainly far more enlightened in its attitude towards the education of women.

After Aldis's dismissal, the chair was offered in 1894 to the 6th wrangler of 1882, Alfred Johnson, but he declined it having heard about the Aldis affair from Abbott. However, on his recommendation it was offered to the 2nd wrangler of 1890, Hugh Segar, who did accept it, although at a salary of £500 (£200 less than Aldis had received). Abbott's appeal seems to have had little deterrent effect, since it was reported that there were thirty-seven applicants for the chair. In making their selection, the College Council had made it clear to the Commissioners

how desirable it is that the Professor of Mathematics should not confine his duties to theoretical mathematics, but should assist in popularising in applied Mathematics, Astronomy and Optics.[38]

Prior to coming to Auckland, Segar, who had had a short spell as a lecturer at University College of Wales, Aberystwyth, had published a number of short papers on the properties of determinants and related topics in algebra. But on reaching the southern hemisphere his research came to an abrupt halt. He had an enormously heavy teaching load: for almost the entire period of his appointment (he retired in 1934) he taught the subjects of both pure and applied mathematics up to honours standards, to ever increasing numbers. He also spent a lot of time on university committees. He started to publish again in 1900, and over the course of the next eight years produced twelve papers, mainly on statistical and economic topics, after which his output dried up once more.

Nevertheless, Segar enjoyed significant success as a teacher. He taught a number of outstanding students, although his most notable, such as L. J. Comrie, the calculating machine pioneer, fall outside the Victorian period—and this in spite of teaching rather old-fashioned mathematics (clever mathematical manipulation at the expense of modern mathematical analysis) based on the Cambridge of his youth. Throughout his forty-year tenure he would buy multiple copies of old textbooks as they were going out of print to lend to his students.[39]

# India

Prior to 1857, when the examining universities of Bombay (Mumbai), Calcutta (Kolkata), and Madras (Chennai) were founded, based on the model of the University of London—they only became teaching institutions in their own right towards the end of the century—there were twenty-seven colleges situated in different cities. With the founding of the new universities, a university system was superimposed on the colleges, the objective being to disseminate European culture in the form of both arts and sciences. The universities set standards for courses, prescribed the texts, and had the power to confer degrees in arts, law, medicine, and civil engineering. While these activities seem in line with those of their British counterparts, underlying their foundation was a rather different agenda. The British in India were not looking to the Indian universities for the promotion of intellectual activity, but rather the production of 'a sizeable group of Indian 'civil servants'.[40] To take an example, the B.Sc. degree was established at Bombay in 1879, but in the sixteen years to 1895 it was awarded to only forty-three students, although one of them, Raghunath Paranjpye, did go on to Cambridge and become senior wrangler in 1899.

Asutosh Mookerjee (1864–1924).

The most talented Indian mathematician of this era to learn all his mathematics in India was Asutosh Mookerjee, who studied at the Presidency College in Calcutta,[41] obtaining an M.A. in mathematics in 1885, and went on to found the Calcutta Mathematical Society in 1908. So unrivalled were Mookerjee's talents that the Indian historian of mathematics, Ganesh Prasad, considered him 'After Bhaskara, to be the first Indian to enter the field of mathematical researches'.[42]

However, Mookerjee was not altogether impressed by his teachers, writing in his diary that he found it 'really shocking' when one of them confessed to being unable to teach him Maxwell. He believed that these 'vaunting Anglo-Indians', as he described them—they included H. W. McCann (6th wrangler of 1876), J. H. Fisher (an Oxford graduate), and W. Booth (a Dublin graduate)—envied him for his election (in 1884) to the London Mathematical Society.[43] To supplement his studies, Mookerjee had many books shipped from Europe, including works by Laplace, Fourier, Poisson, Cayley, Lamb, and Forsyth.[44] Of these, he was particularly enthusiastic about Forsyth's *Treatise on Differential Equations* (1885), echoing Sylvester's remark that he considered it 'in my opinion, the best mathematical book extant in the English language', even reading it on the night before his marriage.[45] Mookerjee published a number of mathematical papers and a textbook, *An Elementary Treatise on the Geometry of Conics* (1893), but finding it difficult to carve a career in mathematics, eventually settled on university administration.

Several Cambridge men went out to India to take up positions as 'professors of mathematics', but the designation should not be misunderstood. Most of these men were very young, often fresh from the tripos; they may have been ready and qualified to take on a professorial role in India, but they would not have been ready or qualified to take one on in Cambridge (or indeed Oxford or London). George Kuchler, 9th wrangler in 1883, was professor of mathematics at

Presidency College, Calcutta, from 1884 to 1906, while Paranjpye, after a three-year fellowship at Cambridge, was appointed principal and professor of mathematics at Fergusson College, Poona, positions he held from 1902 to 1926. Francis Candy, the 23rd wrangler of 1854 who also took a first class in the natural science tripos, and James Hathornthwaite, 7th wrangler in 1870, were both professors of mathematics at Elphinstone College, Bombay, the latter being in post from 1872 to 1900. Towards the end of his time at the college, Hathornthwaite wrote *A Manual of Elementary Algebra* (1894) specifically 'For the entrance examinations of the Indian universities'. Although by the time of the book's publication, the universities were beginning to change from being examining bodies to teaching institutions, Hathornthwaite's book gives an idea of the elementary level of mathematics required for university entrance. Typical of the problems it contains is the following, in which Hathornthwaite demonstrates his readiness to adapt to his surroundings:[46]

83. A man starts from Madras on his bicycle for Bombay and travels at a uniform rate nine hours a day; half way between Madras and Bombay an accident happens to his machine and he has to reduce his speed by one quarter during the remainder of his journey and arrives in Bombay three days later than he had intended; his speed originally being six miles an hour, how many days did his journey occupy and how far is Madras from Bombay by road.

Textbooks for the Indian market were also produced by English authors working in England. The best known of these were the books of the Cambridge mathematician and prolific textbook writer Isaac Todhunter.[47] Editions of Todhunter's books sold around the globe and in several different languages, and he was particularly successful in the Indian market with editions of his texts appearing in both Urdu and English.[48] Urdu editions of his *Euclid* and his *Algebra* were published in 1871, while in 1876 Macmillan published four of his texts especially designed for use in Indian schools: *An Abridged Mensuration with Numerous Examples for Indian Students*; *Algebra for Indian Students*; *The Elements of Euclid for the Use of Indian Schools*; and *Mensuration and Surveying for Beginners, adapted for Indian Schools*. But Todhunter's association with India was not restricted to textbook writing, as he also examined for the East India Company's military college at Addiscombe and the Indian Civil Service. The extent to which such institutions relied on his textbooks can be gauged from a letter that Todhunter wrote to his wife while examining for the Indian Civil Service in 1878:[49]

There is a library of mathematical books provided by the Civil Service Commissioners for the use of the Examiners. It consists of fourteen volumes, ten of which are by myself. Thus you see I am able to do much of that labour which Matthew Arnold thinks distasteful, namely, that of perusing your own books.

Not all of the mathematics texts produced by Englishmen in India were written by men employed in academic mathematics. John Henry Pratt, 3rd wrangler in 1833 and Archdeacon of Calcutta from 1850 to 1871, published extensively over the course of his career. Most of his writings were on mechanics and applied topics, such as the shape of the earth; one paper—on the attraction of the Himalayan mountains upon the plumb line in India—was singled out for special attention by George Airy,[50] resulting in its subject becoming known as Airy and Pratt isostasy (see Chapter 8). Pratt's book *The Mathematical Principles of Mechanical Philosophy*, first published in 1836 while he was still in Cambridge, was expanded and republished in 1860 as *A Treatise on Attractions, Laplace's Functions, and the Figure of the Earth*. In between the two publications a large part of the book had been openly used by Isaac Todhunter in his *Treatise on Analytical Statics* of 1853. Although produced legally—Pratt had disposed of the copyright—Pratt was not at all happy about Todhunter's book since, as he made only too plain in the preface to his *Treatise on Attractions*, it interfered with his own publication plans. But being in India, he had been unable to keep abreast of activities in the burgeoning textbook industry back home.

# South Africa

The first college to be founded in South Africa was the South African College in Cape Town which began its life in 1829 as a high school for boys, although in many respects it was little better than a primary school. The professors soon discovered that before they could initiate their charges in the mysteries of metaphysics, trigonometry, and astronomy promised in the prospectus, they first had to teach them to read! Colleges were later established in Grahamstown, Graaff-Reinet, and Stellenbosch. When the University of the Cape of Good Hope, modelled on the University of London, was founded in 1873, only two colleges were given government recognition as teaching colleges: South African College and Graaff-Reinet College; these two were then joined by the College at Stellenbosch (although by 1885 Graaff-Reinet was no longer considered worthy of inclusion). After 1880, assisted by the earlier discovery of gold and diamonds which had created a demand for mining expertise and also provided an important financial boost, higher education began to develop seriously, although it was not until the turn of the 20th century that these colleges became true university institutions.

## South African College

In 1851 the South African College appointed George Smalley (previously mentioned in connection with Sydney University), who was an assistant at the Cape Observatory, as professor of mathematics, although he resigned after less than a year due to the inadequate salary (£100 p.a. plus fees). Not a high wrangler, he was nevertheless more than competent. On returning to England he taught mathematics at King's College, London, from 1854 to 1862. It was there that he came to Airy's attention, and on the latter's recommendation he was appointed government astronomer at the Sydney Observatory, a post he took up in 1864.

George F. Childe (d. 1897).

Smalley was succeeded in 1852 by George Childe, who held the position until ill health forced him to resign in 1877. Childe had graduated from Christ Church, Oxford, with first-class mathematical honours in 1837 and, like Smalley, had begun his career in South Africa as an assistant at the observatory, arriving there in 1845. He published on topics in geometry and his work was not simply didactic. In the introduction to his book *Investigations in the Theory of Reflected Ray-Surfaces* (1857), Childe drew attention to an obvious, but none the less very real, problem with regard to such publications:[51]

The printing of a work of abstract science in an isolated colony cannot but be subject to inconveniences, arising from various causes; among which are to be considered, not less the difficulty and delay of obtaining appropriate symbols; even when procured, as in this instance, for the special purpose; than that of insuring correct typography.

Childe went on to thank the printer who, in his opinion, had secured for the book 'a degree of accuracy in the execution of which it is believed will render it no discreditable specimen of colonial printing'. Childe was also the author of a book on the *Singular Properties of the Ellipsoid* (1861) which, patriotically, he dedicated to 'His Royal Highness Prince Alfred, in remembrance of his visit to the Colony of the Cape of Good Hope'[52] in 1860. Rather fancifully, one writer imagined 'the Prince drawing inspirations from its figures and formulae during his progress through South Africa'.[53]

1860 was also the year in which another British mathematics graduate, Charles Abercrombie Smith, 2nd wrangler and second Smith's prize winner in 1858, arrived in Cape Town. Although Smith practised as a surveyor, he held several public offices, including that of vice-chancellor of the university.[54] He made important contributions as a member of Council to the advancement of scientific studies and was for many years a moderator and examiner in mathematics.

When Childe retired in 1878, the new professor of mathematics was Francis Guthrie, a graduate of London University and erstwhile student of Augustus De Morgan. Guthrie had begun his career in South Africa in 1861 as a lecturer in mathematics at Graaff-Reinet College, and although he remained in his position at South African College for twenty years, he made a greater impression in botany than in mathematics and is remembered in South Africa for having plants rather than theorems named after him.

Nevertheless, Guthrie is not an entirely forgotten figure in the history of mathematics, although the reason for which he is remembered dates back to just after he left De Morgan's class, almost a decade before he arrived in South Africa. In 1852, while colouring a map of the counties of England, he postulated what became known as the 'four-colour conjecture' (see Chapter 17). This celebrated conjecture—which states that any map in the plane can be coloured, using no more than four colours, so that adjacent areas are coloured differently—was proved only in 1976, and then only with the assistance of a computer. Although Guthrie never published anything on the subject, and it was his brother Frederick (who was in De Morgan's class at the time) who described it to De Morgan, the credit for posing the problem is certainly his.

Francis Guthrie (1831–99).

But while Childe, Abercrombie Smith, and Guthrie left little behind them in the way of mathematical research, the same is not true of a fourth mathematical emigrant to Cape Town. Thomas Muir (whom we met in Chapter 4) graduated from the University of Glasgow in 1868 and studied in Berlin and Göttingen before returning to Glasgow, first as assistant to Hugh Blackburn, the mathematics professor at the university, and then as chief mathematical master at Glasgow High School. During these years Muir played an important role in the encouragement and criticism of mathematics as it was then practised,[55] and he began in earnest the work on determinants for which he is primarily remembered. In 1892 he was persuaded to go to the Cape as superintendent general of education, a job he took in preference to the chair of mathematics at the newly opened Stanford University in California which he had just been offered.

Muir's activities in South Africa were essentially of a broad educational nature and his work was much more in the promotion of science and education than mathematics in particular, and the duties associated with his new post left him little time for research. It was only when he retired in 1915 that he took up the reins of research again and devoted himself to his history of determinants. The final result, which runs to five volumes, is a classic and monumental work. When Muir died in 1934 he bequeathed his mathematical library of periodicals and books to the South Africa Public Library (Cape Town), and at the time it was reputed that only the mathematical collections of the British Library and of Cambridge were better.

However, Muir's arrival in 1892 marked a high spot for 'colonial' mathematics in South Africa in the last half of the 19th century. Altogether there was very little by way of higher-level mathematics education or mathematical research being undertaken in this period, and certainly nothing to compare with the progress being made in the Antipodes. Despite the stimulation provided by mining, the situation was still rather more akin to that of India, where the colonists generally had more pressing local problems to deal with than the setting up of academic departments.

But by the turn of the century, the situation had begun to change for the better, not least due to a number of mathematical emigrants from Scotland. Apart from Thomas Muir, others included John Carruthers Beattie, Lawrence Crawford, and Alexander Brown, all of whom were appointed to chairs at the South African College and all of whom made significant contributions to the administration and development of their departments, often at the expense of their own research.[56] Beattie was a graduate of Edinburgh University who had studied physics at Munich, Vienna, and Berlin. On the recommendation of George Chrystal, he was appointed professor of applied mathematics and experimental physics in 1897. Crawford, who been a student at Glasgow High School during Muir's tenure there, had taken a first degree at Glasgow University before going to Cambridge and graduating as 5th wrangler in 1890. He was appointed professor of pure mathematics in 1899, having spent the early part of his career as a lecturer at Mason College, Birmingham. Brown had taken a first degree at Edinburgh before going up to Cambridge and graduating as senior wrangler in

1901. He took over the chair of applied mathematics in 1903 when Beattie's position was split.

## The Cape Observatory

Founded in 1820, the Royal Observatory at the Cape of Good Hope was the first organized scientific institution to be established in South Africa.[57] Until the 1880s, the astronomical work at the Observatory was of a fairly pedestrian nature and lacking in innovation. But by the turn of the century the Observatory had become established as the leader in the southern hemisphere and arguably equal to the best in the northern.

During the Victorian period the Observatory provided employment for a number of British migrants from mixed educational backgrounds. Although preference for the leadership of the Observatory was for a Cambridge graduate, suitable candidates were not always forthcoming, leading to the employment of the rather more home-spun variety of astronomer.

The Royal Observatory at the Cape of Good Hope, in a drawing by John Herschel, 1837.

Among those who held the position of Her Majesty's Astronomer at the Cape was Edward Stone, 5th wrangler of 1859. At the time of his appointment in 1870, Stone was the chief assistant to Airy at Greenwich and a Fellow of the Royal Society. He had an impressive research record, having devoted considerable time to what Airy called the 'noblest problem of astronomy', that of deriving the distance of the earth from the sun; he eventually calculated it to be 91,700,000 miles, an achievement for which he was awarded the Gold Medal of the Royal Astronomical Society in 1869. But when Stone arrived at the Observatory he did not find life at all easy. His fellow workers, who were oblivious of his reputation, knew only that he had been an assistant at Greenwich, and not, as he complained in a letter to Airy, that he was 'a university man or had done anything or obtained any position at all in England'.[58]

Stone's period at the Cape was remarkable for his industry in preparing the definitive catalogue of over twelve thousand southern stars and for removing some of the backlog of observations. The work was extremely arduous and there were frequent complaints from the junior staff, particularly as the pay was less than adequate. William Finlay, the first assistant, felt the lack of remuneration to the extent that he felt compelled to take in private pupils to the detriment of his work at the Observatory. But Stone was lucky to get Finlay, as he reported to Airy in October 1876:[59]

There is however a good deal to say on Finlay's side…Finlay was about 33 wrangler and although of course not much of a mathematician yet he is a quick sharp fellow and there are only 5 wranglers of any kind in South Africa, Bishop Colenso, C A Smith, myself, Bard[60] and Finlay.

Stone remained at the Cape until 1879, when he returned to England to take up the position as Radcliffe Observer in Oxford.

He was succeeded by David Gill, a graduate of Aberdeen who had had the good fortune to be taught by Clerk Maxwell and by whose teaching he was greatly influenced.[61] Although agreed to be of senior wrangler material, he was required by his father to enter the family clock-making business, where he remained from 1860 to 1871. Retaining his interest in science, he gradually became enthralled with practical astronomy and, eventually, on the strength of an exceptionally high-quality photograph of the moon, he was offered the directorship of Lord Horatio Crawford's private observatory. He accepted, placing the family business in the hands of a manager.

Having lost out to Stone in his bid for the position of Radcliffe Observer, Gill was one of only two candidates for the job in Cape Town. To the chagrin of the other, William Christie, Airy's chief assistant at Greenwich, Gill got the job. Two years after his appointment he was elected a Fellow of the Royal Society, with several leading mathematicians, including Cayley and J. W. L. Glaisher, signing his election certificate. Gill's arrival at the Cape heralded a new lease of life for the Observatory. Old-fashioned methods were abandoned and there was a development of new techniques and expansion of instrumentation and personnel. Under Gill's guidance the Observatory was transformed 'from a shabbily maintained institution…to the splendid observatory, perfect in every refinement, which he left to his successor'.[62] Ill health forced his eventual retirement to England in 1906.

# Canada

The first college to be founded on Canadian soil was the Queen's College of New Brunswick, Fredericton, which was originally established in 1785 and eventually metamorphosed into the University of New Brunswick in 1859.[63] This was followed in 1821 by McGill University in Montreal, which, in 1884, played host to the first colonial meeting of the British Association for the Advancement of Science. King's College, Toronto, which was chartered in 1827, became the University of Toronto in 1850—and hosted the second colonial meeting of the British Association in 1897—while Queen's University, Kingston, was established in 1841. Unlike the other countries surveyed so far, Canada based its universities on a variety of different models; for example, McGill was based on the Oxford model while Toronto was based on that of Edinburgh. There were also universities, such as Laval in Quebec, founded along French lines.

The first known course of mathematics to be studied was at King's College, Windsor (Nova Scotia) in 1814, where in the third year 'Euclid and Wood's Algebra' was in the syllabus. At Queen's College in Fredericton in 1824, where all the instruction was provided by the principal, there appears to have been a similar amount of mathematics. By 1860 an increased emphasis on mathematics is evident. At the University of New Brunswick, for example, the mathematical requirement for a B.A. included Euclid, algebra, trigonometry, and logarithms in the first year; plane and spherical trigonometry, conics, and calculus in the second year; and Books 1–3 of *Principia Mathematica* and hydrostatics in the third year. At the same time mathematics was required in each of the four years at the University of Toronto and in three of the four at McGill University. Concomitant with this was an increase in general scientific awareness: the Canadian Institution (later the Royal Canadian Institute) was founded in 1849, and the Royal Society of Canada was founded in 1882.

Three men in particular made an important contribution to the progress of Canadian mathematics during the Victorian period, both by their involvement in university teaching and in their production of texts.[64] John Bradford Cherriman was professor of mathematics at Toronto from 1853 to 1875. Born in England, he was 6th wrangler in 1840 and spent some time as a master at Sedburgh School before going to Toronto in 1848. He published several papers in Canadian journals, and although there was nothing of startling originality—they were mostly either recreational or didactic—these papers do mark the beginning of research at the university. Cherriman's successor, James Loudon, who was also professor of physics, was a home-grown product who graduated from the University of Toronto in 1862. He eventually became the university's first professor of mathematics and physics in 1875. A strong promoter of research, although he had little time for research of his own, he was president of the university from 1892 to 1906. Like Cherriman, he published articles mostly on teaching concerns.

The third man who contributed greatly to the progress of Canadian mathematics was Nathan Dupuis, professor of mathematics at Queen's College from 1880 to 1911. He had graduated from the college in 1866 at the age of 30, and the following year had been appointed as professor of chemistry and the natural sciences. Although it is not certain why he changed to mathematics in 1880, he had always maintained a strong interest in the subject, his *Geometrical Optics* being published in 1868. It seems most likely, therefore, that he was brought in to raise the standard of mathematics teaching for, as he himself later admitted, 'mathematics from the modern point of view was at a very low ebb in the college...Hard and unceasing work offered the only hope for a better state of things'.[65] It was no empty rhetoric. Dupuis worked hard both in teaching and in administration and, as is evident from the textbooks he wrote himself, he did not shy from presenting his material in a modern fashion. He was avowedly against the teaching of Euclid. His *Elementary Synthetic Geometry of the Point, Line and Circle in the Plane* (1889), which provided 'a complete break from the traditional geometry of Euclid', was 'the first venture of its kind in Canada'[66] and was very successful. By 1914 it had run to at least five editions, and it was a significant factor in the demise of Euclid as a standard text in Ontario schools. He also produced textbooks on algebra, solid geometry, trigonometry, and astronomy.

By the 1890s most of the established Canadian institutions had a professor of mathematics, and in some cases, such as at McGill and Toronto, a mathematics assistant was also employed. Although the teaching of mathematics gained considerable ground over the century, there was little opportunity for research. The mathematical community was small, mathematicians had little exposure to international literature, and there were few rewards for engaging in research beyond personal satisfaction. The professors were teaching a wide variety of courses and, even though many were at elementary level, this left little time for much else. Furthermore, the universities, being based on the English and Scottish models of the early 19th century, did not inherit a great tradition of research—at least not of theoretical research—and practical results were much more the order of the day.

At the beginning of this section, reference was made to the meetings of the British Association in Montreal in 1884 and Toronto in 1897. While this is not the place to go into detail about these meetings,[67] it is not without interest to note the participation of British mathematicians. It is impossible to say exactly how many of the seven hundred and fifty or so delegates from Britain at the Montreal meeting were mathematicians, but among those who attended were Lord Rayleigh (president of the meeting), John Couch Adams, William Grylls Adams, Robert Stawell Ball, George Darwin, J. W. L. Glaisher, George Greenhill, Olaus Henrici, and Sir William Thomson (president of the mathematics and physics section). Lord Rayleigh's presidential address on 'Recent progress in physics' was widely reported, with *Science* reproducing it in full.[68] *Science* also featured a biographical sketch of Rayleigh, which closed with the following remarks:[69]

Lord Rayleigh's countenance will soon become familiar to every American man of science; and we hope that even the uneducated Americans will learn to see in him, not the lord of the manor of Terling and the patron of two livings, but a peer of the distinguished school of mathematicians in Cambridge, Eng., the pre-eminence of which, in mathematical science, American centres of learning can honor, but not dispute.

Nevertheless, despite the relatively strong showing of British mathematicians, and the appreciation of their merits articulated in *Science*, they either chose not to show off their wares or were presented with little opportunity to do so. The only British contributors to the mathematics subsection, out of a total of twelve, were John Couch Adams and R. S. Ball.

Nathan Fellowes Dupuis (1836–1917).

In 1897 the Toronto meeting had the competing attraction of the first International Congress of Mathematicians, which had been held in Zürich only two weeks before. Indeed, the Toronto meeting was cited as the reason for the small number of British mathematicians in Zürich—only E. W. Hobson, Joseph Larmor, and John Mackay attended—although the number of British mathematicians in Toronto was not large either. Greenhill and Lord Kelvin made the journey across the Atlantic again, and were joined by Andrew Forsyth (president of the mathematics and physics section), William H. H. Hudson, Percy MacMahon, and E. T. Whittaker.[70] The number of mathematical talks was few—nine in total, of which three were contributed by British mathematicians (Henrici, MacMahon, and Larmor), but only MacMahon's was given in person. Despite this, Forsyth's sectional address on the importance of pure mathematics made an impact, and mathematics achieved greater press coverage than it had in 1884. As the physicist Hugh Callendar reported, Forsyth's address was:[71]

…an eloquent and convincing vindication of the importance of studying mathematics for its own sake with the single aim of increasing knowledge, and not, as some would have it, from a utilitarian point of view, as an instrument for the use of the engineer, the physicist or the astronomer. The path of practical utility, as he justly remarked, is too narrow and irregular, not often leading far. It is evident from the demeanor of the audience that they were all thoroughly interested and engrossed in the subject of the address, which, it is to be hoped, may do something to moderate the utilitarian and technical spirit, and to check its inroads upon the sanctuaries of university education.

At the end of Forsyth's address, James Loudon, who was by this time president of Toronto University, remarked that 'he was glad to have the address for the Canadian public to read as Toronto University had been charged with paying too much attention to mathematics'.[72] How many members of the Canadian public actually read the address is, of course, unknown. Nevertheless, Forsyth obviously hit the right note and being the holder of a Cambridge chair—he had succeeded Cayley as Sadleirian professor in 1895—he had the credentials to be listened to.

The showing of several high-calibre mathematicians at each of the British Association meetings in Canada reveals, at the very least, British interest in the promotion of colonial mathematical communication (albeit combined with an interest in colonial travel). While their visits to Canada may not have had a major effect on the academic lives of the travellers, the presence of William Thomson, Lord Rayleigh, and Forsyth—household names in mathematical or scientific terms—and the other British mathematicians would certainly have provided a welcome stimulus for the local scientific communities.

## Inside going out: the case of India

As well as the mathematics taught and practised in the colonies, there was also mathematics taught for practice in the colonies. Both the British government and commercial organizations, such as the East India Company, required large numbers of scientifically educated personnel, such as engineers and surveyors, who had to be trained before embarking on their travels. Mathematics was an important part of their curriculum and they needed mathematicians to teach it. Of all the colonies, nowhere was the need for mathematical expertise greater than in India, and the demand was met by three institutions: Haileybury College, Addiscombe Military Seminary, and the Royal Indian Engineering College.

## Haileybury College and Addiscombe Military Seminary

In 1806 the East India Company set up twin colleges at Haileybury in Hertfordshire and Addiscombe in Surrey, the former for training administrators and the latter for training military cadets for the Company's private army. Mathematics was a significant part of the curriculum at each establishment although, since the entrance qualifications were minimal and the age of entry was about 17, the level of mathematics taught was not high. Neither college was long lived: with the cessation of East India Company rule in India and the transfer of military power to the British Army (both consequences of the Indian Mutiny of 1857), Haileybury closed in 1857 and Addiscombe closed in 1860. Nevertheless, for several decades the two colleges provided a welcome source of employment for wranglers (and some others) wanting to stay within the realms of academic mathematics.

In 1816 the young Charles Babbage, armed with 'strong recommendations from Ivory and Playfair',[73] applied for the professorship of mathematics at Haileybury.[74] That Babbage was unsuccessful in his application, however, probably says more about his attitude towards teaching than it does about the

Old Haileybury College.

Charles Webb Le Bas
(1779–1861).

standard of mathematical ability expected from the staff. The job was given to Henry Walter, 2nd wrangler of 1806, who three years later became a Fellow of the Royal Society, but was otherwise undistinguished. Others who taught mathematics at Haileybury included Charles Le Bas, 4th wrangler of 1800, who held the combined appointment of dean and professor of mathematics and natural philosophy from 1813 to 1837, when he became principal of the college. He was succeeded by J. W. L. Heaviside, 2nd wrangler of 1830 and a tutor of Sidney Sussex College, who held the position until the college closed in 1857. As a student at Cambridge, Heaviside had been close friends with Charles Darwin and a member of the latter's 'Glutton' club. An accomplished teacher, his lectures 'proved him to be a rare example of a great mathematical scholar largely endowed with practical common sense, and able to make intricate problems intelligible to ordinary minds'.[75] He seems to have taken dedication to teaching to extremes, however, since the story is told of him lecturing—enunciating mathematical formulas, to be precise—while at the same time extinguishing a blaze from the bottom of his gown.[76]

At Addiscombe, mathematics always played a prominent part in the curriculum, although the scale by which the order of merit was settled—that is, the comparative value of each subject in the end-of-course examinations—was frequently revised. In 1836, out of seven subjects listed, mathematics counted for 29 per cent of the total marks awarded, equalled only by fortification, while in 1843 the total for mathematics had increased to 32 per cent, with fortification at 21 per cent being the nearest rival. In 1854, out of fifty-four hours of study, twenty-two hours were allocated to the different branches of mathematics, the subjects being divided according to the cadets' needs. For example, the infantry course required geometry, arithmetic and algebra, logarithms, plane trigonometry, and mensuration, while the engineers' course required the theory of equations and expansion of series, differential and integral calculus, spherical trigonometry, and astronomy.[77]

Of the staff at Addiscombe, the most successful in mathematical terms was Jonathan Cape, who had been 5th wrangler in 1816. Cape was professor of mathematics and classics at the college for almost forty years, from 1822 until it closed in 1860. He was elected to the Royal Society in 1852, with George Peacock and William Whewell amongst his proposers, and was the author of *A Course of Mathematics Principally Designed for the Use of Students in the East India Company's Military College at Addiscombe* (1839, 1844), a two-volume work that ran into several editions. It was based on the well-known *Course of Mathematics* by Charles Hutton (composed for students at the Royal Military Academy in Woolwich, and first published in 1798), which had been in use for many years at Addiscombe.

Cape recognized the need for an updated text which would take into account 'improvements from the Continent' and thus reflect 'the state of Mathematics in the present day'.[78] Cape, who was well read in French mathematics, did not shy from revealing his sources, some of which he drew on extensively. He favoured 'Lagrange's principle' over 'the doctrine of fluxions or method of limits', and in geometry he 'borrowed so largely from Legendre's work' that 'it required a distinct acknowledgement'. He also endeavoured to keep his text up to date by incorporating recent developments; for example, in the second edition (1844) he included 'the beautiful theorem of Sturm' by means of which the number of real roots on a given interval can be determined. Cape had a reputation as an excellent teacher and, mindful of the career paths of his students, he emphasized the importance of 'rigorous reasoning':[79]

I consider it quite as important that the engineer, or the practical man, should have his mind well disciplined with strict and rigorous reasoning, as the mere theorist or the philosopher in his study. The one, from the want of it, will merely delude himself; the other may do injury to millions.

Cape also compiled a book of *Mathematical Tables* (1838) which, as well as standard tables of logs, sines, etc., also included astronomical tables 'requisite to be used with the Nautical Almanac'. His aim, as stated in the preface, was to produce a collection which included 'everything which a Military officer might require in his professional duties, in a compendious form' and which omitted nothing 'which might be useful, either in the closet, in the practice of surveying, or in finding the latitude and longitude from celestial mechanics'. As with his *Course*, the *Tables* were popular and ran to at least three editions.

Almost immediately after graduating as 17th wrangler, Alfred Wrigley went to Addiscombe as professor of mathematics and classics, and remained in position until the college closed. He subsequently became headmaster of Clapham Grammar School, where he taught G. H. Darwin and other sons of Charles Darwin.[80] Described as 'a good drill master rather than an inspiring teacher',[81] Wrigley was the author, jointly with W. H. Johnstone (13th wrangler of 1842 and chaplain of the college from 1842 to 1860) of *A Collection of Examples in Pure and Mixed Mathematics, with Hints and Answers* (1845). It was a popular text that ran into several editions.

Teaching mathematics at Addiscombe was not always an easy task, and sometimes young wranglers found trying to manage a class of unruly cadets simply too demanding, as in the case of John Witt (7th wrangler of 1854), who resigned after a year. But some, despite having discipline issues, managed a longer stint. These included Arthur Dusautoy, the 4th wrangler of 1848 and a man with talents of 'high order', who went to Addiscombe straight from Cambridge and stayed there until the college closed. Although acknowledged as an excellent teacher, Dusautoy was of nervous disposition and stories abound of his inability to keep control of the cadets.[82]

## The Royal Indian Engineering College

The Royal Indian Engineering College at Cooper's Hill, Surrey[83] was founded in 1871 to cope with the shortfall of good candidates for the increasing number of appointments becoming available in the Public Works Department of the Government of India, and in particular to satisfy the expanding need for well-trained civil engineers. Each year the college admitted fifty students for a three-year course; the students were aged between 17 and 21 and were selected by competitive examination. Initially, those who successfully completed the course went on to work in some form of engineering capacity for the Public Works Department. From 1878 the college also began to admit candidates for the India Telegraph Department, and in 1881 it also accommodated students who were not necessarily going to work in India. Between 1883 and 1906 it provided over one hundred trained men for the Royal Engineers, the Royal Artillery, the Admiralty, the Egyptian Government, the Uganda Railway, and some other non-Indian services.

But the college was not long lived. The growth of facilities for training engineering specialists elsewhere, especially at the new civic universities in England, was a strong argument against continuing such a specialized institution which was expensive to run and maintain, and it was closed in 1906 with its work being transferred to India. Nevertheless, in its thirty-five years of existence at Cooper's Hill, the college employed several mathematicians and provided a useful stepping stone in a number of careers.

Both pure and applied mathematics formed part of the basic obligatory curriculum. Other compulsory subjects included mechanical drawing and descriptive geometry, as well as subjects catering for the obvious needs of engineering such as the theory and practice of construction, Hindustani, and the history and geography of India. Higher mathematics was offered as an optional subject. From 1872 a number of scholarships were awarded, including one for mathematics. The students were drawn from both public schools and grammar schools, with a number coming from schools in India and abroad.[84]

On arrival at the College the students were expected to have had only the most elementary mathematical training; consequently, the mathematics they were destined to study was not very advanced. There was thus little need for the mathematical staff to be top-flight mathematicians, and yet in its short history the college counted one or two rather good mathematicians amongst its number.

The first professor of mathematics was Joseph Wolstenholme, who had been 3rd wrangler in 1850 (the year when the well-known coach William Besant had been senior wrangler) and had remained in Cambridge until he moved to Cooper's Hill in 1871. He wrote several papers, mostly on analytical geometry, described as 'marked by a peculiar analytical skill and ingenuity'.[85] Wolstenholme was also the co-author, together with his former coach Percival Frost, of a book on solid geometry (1863), but he is best known for his book of mathematical problems on subjects included in the tripos course. When the latter book was first published in 1867 it contained some sixteen hundred problems, but when the second edition appeared in 1878 it had been enlarged to contain almost three thousand, no doubt many of them tried out on his students at Cooper's Hill. As Forsyth was to say, it was 'a curious and unique monument to ability and industry, active within a restricted range of investigation'.[86] Wolstenholme also wrote an elementary calculus textbook, aimed specifically for his students at Cooper's Hill. The book, published in 1874, included a short contribution by A. G. Greenhill on the theory of couples.

Once at Cooper's Hill Wolstenholme settled into a routine of teaching which, as the years passed, he was reluctant to adapt to accommodate the changing needs of the college. Things came to a head in 1889 when the president of the college, General Sir Alexander Taylor, wrote to the Under Secretary at the India Office:[87]

Wolstenholme has been a leading member of the college staff for 18 years and is a brilliant and most distinguished mathematician but his teaching has not that practical bearing which is required for a young man studying for the practical profession of an engineer under the conditions which now obtain at Coopers Hill—This want of adjustment to our requirements has latterly come more and more under my notice. It cannot be allowed to continue and I have no hope that the radical change which is required can be made so long as the teaching remains in Dr Wolstenholme's hands.

In fact, the college president had already written to Wolstenholme to ask him to resign, resting his case on the observation that Wolstenholme had reached the age of 60 and was hence eligible for retirement. On the face of it, asking Wolstenholme to resign without giving him a chance to mend his ways would seem rather a drastic measure. However, an extract from a letter by Leslie Stephen, the father of Virginia Woolf, who had come to know Wolstenholme while he was at Cambridge, sheds a bit more light on the matter. The letter, which is autobiographical in nature, is from Stephen to his children:[88]

I think especially of poor old Wolstenholme, called 'wooly' by you irreverent children, a man whom I had first known as a brilliant mathematician at Cambridge, whose Bohemian tastes and heterodox opinions had made a Cambridge career unadvisable, who had tried to become a hermit at Wastdale. He had emerged, married an uncongenial and rather vulgar Swiss girl, and obtained a professorship at Cooper's Hill. His four sons were badly brought up; he was despondent and dissatisfied and consoled himself with mathematics and opium. I liked him or was rather fond of him, partly from old association and partly because feeble and faulty as he was, he was thoroughly amiable and clung to my friendship pathetically. His friends were few and his home life wretched.

Wolstenholme spent several vacations (without his wife) in the Stephen household. It appears that he made a lasting impression on Virginia Woolf, since it is generally agreed that the character of Mr Augustus Carmichael in *To the Lighthouse* is based on him.

Wolstenholme was replaced at Cooper's Hill by Alfred Lodge, the younger brother of the celebrated physicist Oliver Lodge. Lodge, unusually, was an Oxford graduate who had obtained first-class mathematical honours in 1876. Having spent some time in the family business, he returned to Oxford to continue with mathematics before he was appointed to Cooper's Hill in 1889. It appears that Lodge immersed himself in his teaching. Apart from a few minor papers and some 'laborious calculations on the values of Bessel's Functions' for the British Association,[89] his published output was confined to elementary textbooks. When in 1904 the forthcoming closure of the college was announced, Lodge resigned and moved to Charterhouse school.

The task of finding Lodge's replacement for the two years that the college still had left to run fell to the Gabbitas & Thring agency, but with the stipulation:[90]

It is absolutely essential that the lecturer should be a gentleman and that he should be a good teacher rather than a man of very high mathematical attainments.

There is no ambiguity here about what was required. The subjects to be taught were listed as practical work in logarithms and some mensuration, plane analytical geometry, differential and integral calculus sufficient to give facility in practical applications, differential equations to the same extent, and spherical trigonometry.

Many of the men proposed by the agency were rejected by the college because they were too young: they were recent Cambridge graduates and younger than the men they were to teach. One of the latter was the future astrophysicist James Jeans. The eventual appointee was Richard Haygarth who, although a Cambridge man, was notably not a wrangler: he was 14th optime in the tripos of 1890. Apart from his two years at Cooper's Hill, Haygarth spent the rest of his career as a schoolmaster.

With respect to applied mathematics, the first appointment was a man of high calibre, Alfred George Greenhill, 2nd wrangler in 1870 who in the same year was elected to a fellowship at his college, St John's. He was the most able of the mathematicians employed at Cooper's Hill, but his tenure at the college was short—he was there for only two years—and, apart from his collaboration with Wolstenholme, he left little evidence of his presence. Finding the martial regime uncongenial to his independent spirit, he moved on to teach the advanced class of artillery officers at Woolwich. In later life, however, he had rather rosy memories of the college, speaking about his time there with pleasure and affection and lamenting the college's loss as a disaster to India and the Empire. He remained at Woolwich for the rest of his working life, publishing on a wide range of subjects in applied mathematics, including the dynamics of mechanical flight, hydrodynamics, and elasticity, and he had a particular interest in the application of elliptic function theory.

The second incumbent of the applied mathematics chair was Edward Nanson, who also did not stay long, spending only a year at the college before taking up the professorship at the University of Melbourne (as described above). The third and final holder of the applied mathematics chair was George Minchin, a Dublin graduate, who was appointed in 1875 and stayed until the college closed in 1906, when he retired to Oxford. Minchin did his most original scientific work on photo-electricity and his interest in experimental physics, which dated from his student days, was revived after his appointment to Cooper's Hill where he had access to a good physical laboratory. He was made a Fellow of the Royal Society in 1895. Considered a fine expositor, he published research papers and textbooks in mathematics and physics, on a variety of subjects: from statics and kinematics to geometry and mathematical drawing. He also had an active interest in mathematical pedagogy, being president of the Association for the Improvement of Geometrical Teaching in 1889 and 1890.

In his memoirs Oliver Lodge, who was a regular visitor to Cooper's Hill, wrote warmly of Minchin, praising his academic ability and empathy with students. Minchin and the Lodge brothers were good friends, and in one of his books Minchin acknowledged their help. Minchin had an excellent reputation as a teacher, as Sir George Chesney, the president of the college, testified in 1888:[91]

There is one point on which I may usefully testify from an experience of ten years as head of Coopers Hill, and that is to his [Minchin's] remarkable gifts as a lecturer. He combines in a peculiar degree the power of lucid explanation, and of interesting his pupils in subjects which, under ordinary handling, may easily be made dry and repulsive. Over and over again when visiting his lecture room with the intention of staying only a few minutes, I have found myself sitting out the whole lecture, so interesting, and indeed charming, did I find his demonstrations on the black-board of the processes of both pure and applied mathematics.

And Minchin's students appreciated him too, as the following tribute from one of them reveals:[92]

… there must be many in this country who would acknowledge that they came out to India stronger in mind and spirit for having come under the influence of his [Minchin's] fine teaching and unobtrusive example.

The level of mathematics taught at Addiscombe, Haileybury, and Cooper's Hill was not high, but it did not need to be. They were colleges for training engineers and administrators, not mathematicians. Nevertheless, many of the professors employed were more than qualified for the job they had to do. Several of them did original research—the elementary and rather restrictive nature of the teaching allowing them time to pursue their own interests—and although the colleges in themselves were mathematically isolated, their proximity to London helped to ameliorate this disadvantage.

## Conclusion

If judged by their performance in the tripos examination, the calibre of mathematical emigrants to the colonies was generally rather high. Indeed of the wranglers appointed to positions at universities in Australia or New Zealand, the 'weakest' was ranked 5th (and that was Carslaw, who was arguably the best mathematician in any case). Of course, not all mathematical emigrants had studied at Cambridge—Oxford, London, and the Scottish universities also figure—but wranglers constituted the majority. Since Cambridge was the premier training ground for mathematicians during the Victorian period, the fact that it produced the greatest number of mathematical emigrants comes as no surprise. However, that so many of that number were high-ranked, at a time when a position in the tripos examination was a passport to a job in Britain in almost any field, is less expected. Why did these men make a perilous journey to an unknown, potentially difficult, and (generally) academically poor environment, when there were so many comfortable opportunities at home?

Putting aside the excitement of foreign travel—which was certainly a factor for many of those attending the British Association meetings in Canada—two main motivations emerge. The first was the difficulty in Britain of engaging in a career in academic mathematics. Not only were there few institutions of higher education, but outside the University of Cambridge, with the exception of Oxford and to a lesser extent London, mathematics departments were extremely small, almost invariably consisting of a professor and an assistant. Job opportunities were few and far between, as Sylvester made palpably clear. That Cooper's Hill attracted mathematicians of the calibre of Wolstenholme, Greenhill, Nanson, and Minchin, helps to underline the paucity of teaching positions elsewhere. The second motivation, as exemplified by Wilson's move from Belfast to Melbourne, was financial. Thus, for those with an academic inclination, a (possibly temporary) move to the colonies, especially if it involved attractive financial remuneration, was not altogether unappealing.

That said, it was not only high-ranking wranglers who took advantage of the openings provided by the expanding British Empire; many of those lower down the list, or graduates from other institutions, went into the colonial service or to ministries overseas. The observatories too provided openings, as our discussion of the Cape Observatory attests. Teaching mathematics in a colonial school or college was also an option. The novelist Walter Besant (younger brother of William), who was 18th wrangler in 1859, spent six years teaching mathematics at the Royal College, Mauritius, leaving only when his health failed in 1867.

Nevertheless, it cannot be denied that the atmosphere in most of the colonial outposts was not conducive to either advanced level teaching or research. The entry standard of students was low and the facilities meagre. It was not an attractive prospect for a talented mathematician with research ambition. But sometimes the reason a mathematician went out to the colonies rather than stay in England was not directly related to mathematics at all. Lamb provides a case in point: an extremely promising mathematician who as a result of marital plans ended up having to spend time battling with colonial administration. Not that in his case it stopped him from doing research; others, however, never got into the habit—like many of their colleagues back home in Victorian Britain.

The title page of *The Ladies' Diary* for 1723.

CHAPTER 7

# A voice for mathematics

*Victorian mathematical journals and societies*

SLOAN EVANS DESPEAUX

This chapter explores the establishment and growth of two conduits for mathematical communication in Victorian Britain: mathematical societies and journals. Although commercial mathematical journalism was well established by 1837, this period saw a rise in the number of journals devoted to the subject, as well as dedicated learned societies. The successful foundation of the country's first national mathematical society in 1865 was a clear indication of the emergence of mathematics as a profession in Britain. By the end of the Victorian era, the number of major British mathematical societies specifically catering to mathematicians' needs at both the research and educational levels had risen to three, as a growing group of scholars increasingly identified themselves with the discipline of mathematics.

At the beginning of the Victorian era, how and where could a British mathematician communicate mathematics? Learned scientific societies certainly provided an option for communication, and both general societies (such as the Royal Society of London) and specialized ones (such as the Royal Astronomical Society) counted many mathematicians among their members. However, these mathematicians would have to wait almost three decades for a major British society devoted solely to mathematics.

Beyond the face-to-face communication of a society meeting, publication represented another option for mathematical communication and, although they authored monographs, mathematicians increasingly used the periodical format as a means for communicating mathematical ideas. The journals of all the major general scientific societies accepted mathematical contributions; in fact, soon after its foundation in 1821, the *Transactions of the Cambridge Philosophical Society* was dominated by Cambridge mathematicians. Mathematical articles were also welcomed into the journals of scientific societies devoted to related disciplines, such as the *Memoirs of the Royal Astronomical Society* (founded in 1822), the *Proceedings of the Royal Statistical Society* (founded in 1834), and later, the *Journal of the Institute of Actuaries* (founded in 1850). In addition, general science journals that were operated as commercial ventures also represented a publication option for British mathematicians, a notable example being the *Philosophical Magazine*. But while all these journals offered an array of options for mathematical publication, none was devoted exclusively to mathematics.

Unlike their situation with societies, however, British mathematicians did not have to endure a long wait for journals specifically devoted to mathematics. In fact, a tradition of mathematical journalism existed well before the dawn of the Victorian era. In an 1880 article for the journal *Nature*, J. W. L. Glaisher characterized the evolution of mathematical journals:[1]

The history of mathematical journalism in all countries seems very similar: first, there is the Annual or other periodical, containing at the end puzzles, problems for solution, &c., the best solutions and the names of those who sent in correct solutions being given in the following number; at length these are supplemented by short articles on particular subjects—frequently suggested by the problems—by the leading contributors. The next step is the mathematical journal, consisting of two parts, the one containing original papers, and the other—quite distinct—containing a limited number of problems and solutions. Finally we have the strictly scientific journal, differing in no essential respect from the *Transactions* of a society.

Mathematical journalism in Britain certainly followed this pattern. In this chapter we trace the development of this publication medium for mathematics in Britain during the Victorian era. We also discuss the establishment of Victorian professional societies for mathematicians and the journals they published, and we explore the cumulative publication infrastructure that made this process possible.

## Part 1: Commercial mathematical journalism in Victorian Britain

By the 1830s, mathematical journalism in Britain already had an established tradition of puzzles and problems for solution, and it was well on its way to the next stage in Glaisher's description—a journal with a mixture of original papers and problems. These journals operated as commercial ventures, and could sometimes also devote pages to pursuits such as education, literature, or the entertainment of British ladies and gentlemen. In 1929, Raymond Archibald, professor of mathematics at Brown University, characterized these journals as 'minor mathematical serials'[2] (see Box 7.1).

## Box 7.1: Victorian minor mathematical serials

| | |
|---|---|
| The Ladies' Diary | 1704–1840 |
| The Gentlemen's Diary | 1741–1840 |
| The Lady's and Gentleman's Diary | 1841–71 |
| The York Courant — mathematical column | 1829–46 |
| The Northumbrian Mirror, or, Young Student's Literary Mathematical Companion | 1837–41 |
| The Mathematician | 1843–50 |
| The Educational Times Mathematical Column | 1847–1915 |
| The Western Miscellany | 1849–50 |
| The Alnwick Journal | 1862–63 |
| Mathematical Questions . . . from the "Educational Times" | 1864–1918 |

## The pre-Victorian legacy: the *Diary* dynasty

Possibly the first British commercial journal containing mathematics, *The Ladies' Diary*, was initially established in 1704 as an almanac with articles for homemakers, but by 1707 it had replaced many of its domestic features with mathematical ones.[3] While a typical issue would have information fitting an almanac, over half of its content would be devoted to enigmas, puzzles, mathematical questions, and their solutions.[4] The popularity of the *Diary*'s format sustained it for 136 years and inspired scores of similar journals throughout the 18th and 19th centuries.

During the last half of the *Diary*'s life, from 1773 onwards, the editors of, and many of the contributors to, the journal were centred at the Royal Military Academy in Woolwich. Charles Hutton and Olinthus Gregory served on the mathematics staff at the Academy and as successive editors of the journal. Gregory had also edited *The Gentlemen's Diary* since 1804. After editing both ventures for two years, he passed the *Gentlemen's Diary* editorship to Thomas Leybourne, a teacher of mathematics at the Royal Military College in Marlow, later Sandhurst. Leybourne's editorship ended with his death in 1840; during the next year, the journal merged with *The Ladies' Diary* to form *The Lady's and Gentleman's Diary*, which continued until 1871.

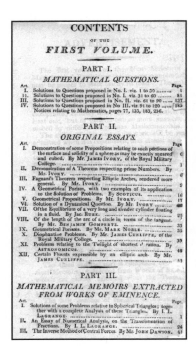

The table of contents of Leybourne's *Mathematical Repository*, 1804.

Leybourne had established a more exclusively mathematical journal in 1795 with his *Mathematical and Philosophical Repository*. Like the *Diaries*, this journal was composed of problems-for-answer. In addition, it also included original papers, translations, and abstracts. Leybourne's *Repository* ran for forty years, ending just before the beginning of the Victorian era, while the *Diaries* enjoyed substantially longer runs. However, these journals formed the exception rather than the rule among minor mathematical serials.

A commercial mathematical journal was constantly in danger of economic demise. In his 1848 review of mathematical periodicals for the *Mechanics Magazine*, an exasperated Thomas Turner Wilkinson evaluated such mathematical journals as 'unproductive speculations'. Of those that did survive, Wilkinson explained that:[5]

probably some under-current of *self-sacrifice* enabled these vessels of science to ride so long in safety. Nor is the *lack* of support a thing unknown even in the present dearth of such publications. *Already* the "*Cambridge and Dublin Mathematical Journal*" has hoisted its signal of distress, and if such is the condition of a journal published at "the first University in the world," who can venture to predict the long continuance of its worthy contemporary, the *Mathematician*.

## The Mathematician

Started in 1843 by T. S. Davies, William Rutherford, and Stephen Fenwick, all mathematical teachers at the Royal Military Academy in Woolwich, *The Mathematician* attempted to fill a void left by the cessation of Leybourne's *Mathematical Repository*. *The Mathematician*, like its predecessors, contained problems-for-answer, but the editors restricted the problems to those that

involve some new principle, or require for their solution some new modes of investigation ... we hope to render this department free from the reproach so often applied to works of this class—that of 'creating a race of mere problem-solvers'.[6]

At least seventy-three contributors either proposed questions (one hundred and eighty-nine in total) or submitted papers to the journal's departments of algebra (twenty-one papers), calculus and differential equations (eight), mechanics (six), plane and spherical trigonometry (eleven), plane geometry (forty-five), probability (two), solid geometry (nineteen), or miscellaneous (twenty-three), as well as several small mathematical notes. Fenwick, Rutherford, and Davies were among the top eight contributors of problems or papers, and these eight submitted almost half of the items. Including the problem solvers, a little over one hundred contributors provided all of the material for this periodical during its eight-year existence.

One contributor, Hugh Godfray, contributed a short paper on the 'Approximate rectification of the circle' to the first volume of *The Mathematician*. Godfray wrote that 'The following approximation to the rectification of the circle is closer, I believe, than any one yet given', and cited earlier constructions given by Davies in Leybourne's *Repository*. He also claimed that his construction 'is about one-fourth shorter than his, as I have tested on several occasions, by comparing the times in which the same person, equally acquainted with both methods, could perform them'. This construction yielded an approximation of 3.14159261546 for the circumference of a circle with diameter 1. In a note at the end of this article, the editors wrote that:[7]

Mr. Godfray also favoured us with two other constructions, the operations of which are very brief and simple: but, as might be expected, their degree of approximation is much less than some known constructions furnish.

At the time of this publication, the 23-year-old Godfray resided in his birthplace, the island of Jersey. Four years later he would commence studies at St John's College, Cambridge, where he would graduate as 3rd wrangler.

The following mathematical problem-for-solution, submitted by a Dr Burns of Rochester, gives an idea of the flavour of this department of *The Mathematician*; but it is the places of residence listed for its solvers that give a sense of the readership of the journal:

$$\text{Given } \tan^{-1} x^2 + \tan^{-1} x = \tan^{-1} \tfrac{1}{3}, \text{ to find } x.$$

Two solutions to this problem were published, and the solvers hailed from London, Manchester, Castleside, Burnley, Lamesley, Cuminestone, and Newcastle.[8]

Although *The Mathematician* enjoyed support from a devoted band of contributors, and was even reprinted in part in Germany, the journal ceased publication after the first three volumes. Rutherford and Fenwick, who had edited the last two volumes without Davies, reported to their readers that they had hoped that the journal could be continued 'under other auspices', but 'That hope has not been realized'. They had made great efforts to continue the enterprise, and had even corrected the proof sheets themselves to save money. 'This labour', they explained, 'became so continuous, that...[it has] occupied almost the whole of...[our] disposable time'.[9]

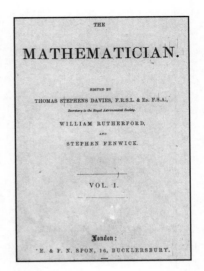

THE

# MATHEMATICIAN.

EDITED BY

THOMAS STEPHENS DAVIES, F.R.S.L. & ED. F.S.A.,
*Secretary to the Royal Astronomical Society.*

WILLIAM RUTHERFORD,
AND
STEPHEN FENWICK.

VOL. I.

London:
E. & F. N. SPON, 16, BUCKLERSBURY.

A title page from
*The Mathematician*.

## The mathematical column of *The Educational Times*

Much longer lived than *The Mathematician*, the mathematical column of *The Educational Times* became what has been called a 'veritable thesaurus for questions of all sorts'.[10] This publication began in 1847 in association with the newly founded College of Preceptors, an organization that sought to maintain professional and academic standards among teachers. Accordingly, *The Educational Times* focused on pedagogical themes, including methods for teaching mathematics. Mathematical questions soon infiltrated the journal, and by 1849 the department of 'Mathematics Questions and Solutions' had been established.[11] Initially edited by two members of the College of Preceptors, Richard Wilson and James Wharton, the editorship of the mathematical department passed to William Miller after Wharton's death in 1862. Miller continued to serve in this capacity until 1897.[12]

Under his editorship, the nature of the 'Mathematical Questions' changed considerably, as Miller encouraged more original problems and targeted established mathematicians. Besides changing its character, Miller wanted to publish a separate reprint of the mathematical department. This twice-yearly reprint, entitled *Mathematical Questions with their Solutions taken from The Educational Times*, began in 1864. It contained all of the mathematical material from *The Educational Times* and provided extra space for new solutions or papers. After one year, the journal could boast of contributions from British mathematicians of the calibre of J. J. Sylvester, Arthur Cayley, William Clifford, Augustus De Morgan, Isaac Todhunter, and Thomas Archer Hirst.[13]

## Box 7.2: Some mathematical questions taken from *The Educational Times*

1507 (By the late Professor Clifford, F.R.S.)—Consider six planes A, B, C, D, E, F, and join the point ABC to the point DEF, and so on; we have thus ten finite lines, and their middle points lie in a plane.

1591 (By Professor Hirst, F.R.S.)—Find the polar equation of a curve whose radii vectores [sic] are each divided into segments having a constant ratio, when, upon the same, the respective centres of curvature are projected orthogonally.

1831 (By Professor Paul Serret)—Une ellipse et l'un de ses cercles directeurs étant tracés, il existe une infinité de triangles simultanément inscrites au cercle et circonscrits à l'ellipse; le point de rencontre des hauteurs est le meme pour tous ces triangles.

1927 (By Professor [William] Burnside, M.A.)—Find the conic of least eccentricity which can be drawn through four given points.

7254 (By Professor Matz, M.A. [of King's Mountain, North Carolina])—Given the axes $CA = 2a$ and $CB = 2b$ of an elliptic quadrant $AP_1P_2P_3B$; also $\angle ACP_1 = \omega = 30°$, $\angle P_1CP_2 = \phi = 15°$, $\angle P_2CP_3 = \theta = 30°$: find (1) $D_1P_2$, $D_2P_3$, $CD_1$, $CD_2$, where $P_2D_1$, $P_3D_2$ are perpendicular to $CP_1$; also (2) these values for $a = b = 1$, $\omega = 0$.

7436 (By Satish Chandra Basu [of Calcutta])—Find the general value of x from

$$a + b + c = a^2 + b^2 + c^2 = a^3 + b^3 + c^3 = a^{2x} + b^{2x} + c^{2x} = 0.$$

7337 (By H. L. Orchard, M.A. [of Burnham])—P is a particle moving with uniform angular velocity, $\omega$, in the circumference of a circle of radius a and centre C. If O be any point in the plane of the circle such that $CO = a \sin 45°$, find the maximum angular velocity of P with regard to O.

The commitment by several of these mathematicians extended long after the debut of *Mathematical Questions*. Sylvester's questions appeared in each of the first seventy volumes; he considered that some of those that remained unanswered 'really contain the germs of theories'.[14] Clifford, like Sylvester, actively contributed to *Mathematical Questions*, and he believed that the mathematical department of *The Educational Times* 'has done more to suggest and encourage original research than any other European periodical'.[15]

When thumbing through the pages of *Mathematical Questions*, one is struck by the great variety of both the problems and their contributors. Some conundrums from the 1885 volume of the journal appear in Box 7.2.[16]

## Part 2: University-related or affiliated mathematics journals

As we have seen, many of the minor mathematical serials were affiliated with the military colleges, and in one case, with the College of Preceptors. We now turn to commercial mathematical journals affiliated with universities—in particular, Cambridge, Oxford, and Trinity College, Dublin (see Box 7.3). These journals represent the ultimate phase in Glaisher's history of mathematical journalism: a movement away from problems-for-answer, and towards what Glaisher described as 'the strictly scientific journal, differing in no essential respect from the *Transactions* of a society'.[17]

---

## Box 7.3: Victorian mathematical journals affiliated with British universities

| | |
|---|---|
| *The Cambridge Mathematical Journal* | 1837–45 (4 vols.) |
| *The Cambridge and Dublin Mathematical Journal* | 1846–54 (9 vols.) |
| *The Oxford, Cambridge, and Dublin Messenger of Mathematics* | 1862–71 (5 vols.) |

---

## The Cambridge Mathematical Journal

In December 1836, Archibald Smith introduced the idea of founding a mathematical journal to Duncan Gregory. Smith was at that time a 24-year-old fellow of Trinity College, Cambridge.[18] Gregory, also of Trinity, was preparing to sit the mathematical tripos examination in the following month, and he agreed to act on Smith's publication idea after the completion of his exams. As we saw in Chapter 1, he was placed as 5th wrangler, a very respectable ranking considering that the highest-ranked wranglers for 1837 included Sylvester (2nd) and George Green (4th).[19]

In his preface to the first volume of *The Cambridge Mathematical Journal*, Gregory pointed out that many recognized the lack of a 'proper channel...for the publication of papers on mathematical subjects, which did not appear to be of sufficient importance to be inserted in the Transactions of any of the Scientific Societies'—that is, research by unknown mathematicians. Gregory felt confident that Cambridge contained many 'who are both able and willing to communicate much valuable matter to a Mathematical periodical'.[20] In fact, twenty-one of the twenty-six identified contributors to the journal had a Cambridge affiliation, and this group provided at least two-thirds of the contributions.[21]

Through his journal, Gregory encouraged this Cambridge cadre to embrace (in particular) the calculus of operations. As will be seen in Chapter 13, the calculus of operations involved separating symbols of operation, such as differentiation, from symbols of quantity. Using general properties of algebra, one then simplifies the separated symbols of operation in order to reach a simple solution to an analytical problem. Gregory's contributors quickly accepted and extended the calculus of operations, and produced an abundance of *Journal* articles on the differential and integral calculus that featured the method. This research led them and other British mathematicians 'to the notion that it was not the nature of the objects under consideration which was most significant, but rather the laws of combination of their symbols'.[22] The prevalence of this method in the *Journal* is evident in De Morgan's remark to John Herschel in 1845 that 'It is done by the younger men...It is full of very original communications. It is, as is natural in the doings of young mathematicians, very full of symbols'.[23]

Besides the calculus of operations, *The Cambridge Mathematical Journal* provided the first publication venue for the ideas that came to characterize invariant theory. In his 1841 article, 'Exposition of a general theory of linear transformations', George Boole showed, among other results, that the discriminant represented a simple example of what would later be called an 'invariant'— that is, a function in the coefficients of a homogeneous polynomial that remains unaltered (up to a power of the determinant) after linear transformation.[24] Boole was a self-taught teacher in Lincoln at the time he wrote his 'Exposition', acquiring university ties only later in life. As an outsider, Boole used the *Journal* as a means to get his work noticed in Cambridge and elsewhere. Tony Crilly has asserted that 'Without the [*Journal*]...Boole would have almost certainly languished in Lincoln, there to remain, an erudite school proprietor, instead of becoming a leading mathematician of the Age'.[25] Boole's article blossomed into a new field of research that an international collection of mathematicians actively pursued throughout the rest of the 19th century (see Chapter 15).

By 1843, Gregory, who had encouraged Boole and many other mathematicians at the beginning of their research endeavours, was in the midst of a recurring illness. The November 1843 number of the journal was the last issue that he would edit, and he finally succumbed to his illness in 1844 at the age of 30. After a short

Preface from *The Cambridge Mathematical Journal*.

interim period, the young William Thomson (later to become Lord Kelvin) began a new series of the journal.

## The Cambridge and Dublin Mathematical Journal

Thomson had been involved with *The Cambridge Mathematical Journal* since the age of 16, when he published papers under the pseudonym 'PQR'. He had just returned from a semester-long sojourn in Paris after graduating as 2nd wrangler. During this trip, he began a productive friendship with Joseph Liouville and introduced French mathematicians to the work of George Green.[26] With international experiences fresh in his memory, Thomson tried to widen the contributorship of his newly acquired journal. To this end, he added 'Dublin' to the title, and his decision was soon rewarded by a devoted group of Dublin-based contributors.[27] While only two Dublin mathematicians had contributed to the original Cambridge journal, eighteen members of this group contributed almost one-quarter of the articles to the Cambridge and Dublin one. This Dublin group was second only to the twenty-five Cambridge mathematicians who contributed half of the articles.[28]

One of these Dublin contributors was George Salmon. As we saw in Chapter 5, Salmon, unlike so many of his colleagues at Trinity College, Dublin, who researched synthetic geometry, was mainly attracted to the analytic approach to the subject, a technique also preferred by Arthur Cayley. The two mathematicians first met in 1848 when Cayley was visiting Dublin in order to attend William Rowan Hamilton's lectures on quaternions; this visit marked the beginning of a life-long friendship between Cayley and Salmon. As Cayley's biographer, Tony Crilly, points out, 'The two schools of mathematics, Dublin and Cambridge, were drawn together by the recently created *Cambridge and Dublin Mathematical Journal* and Cayley's visit can be seen as an effort to cement the union'.[29]

The first research that they conducted together was probably also the most surprising: through their correspondence, they found that twenty-seven lines lie on a cubic surface. No lines lie entirely on a surface of degree $n$ in three-dimensional complex projective space if $n > 3$, and for $n < 3$ there are infinitely many lines.[30] But Cayley found that for non-singular cubic surfaces, there are finitely many lines, and Salmon found the exact number. Cayley published this result in *The Cambridge and Dublin Mathematical Journal* in 1849, giving full credit to his collaborator.[31] This result, in the estimation of Rod Gow, 'remains one of the deepest and most intriguing subjects in algebraic geometry'.[32]

Besides wanting to encourage this more inclusive and unified contributorship, Thomson also wanted his publication to have a more professional character. He abolished the convention of printing articles anonymously or with pseudonyms, a standard practice used in many of the minor mathematical serials discussed

THE CAMBRIDGE AND DUBLIN
MATHEMATICAL JOURNAL.

EDITED BY

WILLIAM THOMSON, M.A.,
PROFESSOR OF NATURAL PHILOSOPHY IN THE UNIVERSITY OF GLASGOW; AND LATE FELLOW OF ST. PETER'S COLLEGE, CAMBRIDGE, F.R.S., F.R.S.E., FOREIGN MEMBER OF THE ROYAL SWEDISH ACADEMY OF SCIENCES, HONORARY MEMBER OF THE MANCHESTER LITERARY AND PHILOSOPHICAL SOCIETY;
AND
N. M. FERRERS, B.A.,
FELLOW OF GONVILLE AND CAIUS COLLEGE, CAMBRIDGE.

VOL. VIII.
(BEING VOL. XII. OF THE CAMBRIDGE MATHEMATICAL JOURNAL.)

Δυῶν ὀνομάτων μορφὴ μία.

CAMBRIDGE:
MACMILLAN AND Co.;
GEORGE BELL, LONDON;
HODGES AND SMITH, DUBLIN.
1853.

A title page from *The Cambridge and Dublin Mathematical Journal*.

above, and one that had also been common, and indeed had been his own practice, in papers in *The Cambridge Mathematical Journal*.

Thomson's *Journal*, unlike its predecessor, was not *for* students *by* students, nor was it completely *for* established researchers *by* researchers. It was, instead, a mixture of articles at differing levels by mathematicians with differing abilities. However, in trying to please everyone, it seems that Thomson ended up pleasing no one; at least, not enough people to keep *The Cambridge and Dublin Mathematical Journal* afloat financially.

## The Quarterly Journal of Pure and Applied Mathematics

In 1854, *The Cambridge and Dublin Mathematical Journal* ceased publication, but was soon reincarnated as *The Quarterly Journal of Pure and Applied Mathematics*. With Norman Ferrers—a co-editor of the later volumes of *The Cambridge and Dublin Mathematical Journal*—and J. J. Sylvester at the helm of an editorial team that included Cayley, George Gabriel Stokes, and Charles Hermite, the *Quarterly Journal* left its predecessors' university focus for wider, international goals.

These goals were celebrated in the inaugural volume by many short articles and notes from international mathematicians; these included Hermite, the Italian mathematicians Enrico Betti and Francesco Brioschi, and Reinhold Hoppe, a privatdozent at the University of Berlin. This level of international participation was not matched in subsequent 19th-century volumes of the *Journal*; however, later in the century, Leonard Dickson would provide a stream of contributions from the United States.

With an 1896 University of Chicago Ph.D. degree in hand, Dickson pursued postdoctoral studies in Leipzig and Paris. The research that Dickson accomplished while on the faculties of the Universities of California, Texas, and (ultimately) Chicago, strengthened the foundation of algebraic excellence in the United States.[33] To *The Quarterly Journal of Pure and Applied Mathematics*, Dickson contributed a series of papers that generalized group-theoretic results from Camille Jordan's *Traité des Substitutions*; he also investigated linear groups in a *Quarterly Journal* article of 1900.[34] Dickson soon codified his linear group research in the book, *Linear Groups with an Exposition of the Galois Field Theory*, which appeared in 1901 and inspired American interest in finite group theory.[35]

The accomplishments of Dickson and the more than two hundred other 19th-century contributors to the *Quarterly Journal* illustrate the high calibre of authors that the editorial team was able to attract. Through its two reincarnations, *The Cambridge Mathematical Journal* had matured towards more original research and a more international group of contributors. However, the result of this maturation, the *Quarterly Journal*, left room for a mathematical journal that catered exclusively to the needs of British undergraduates.

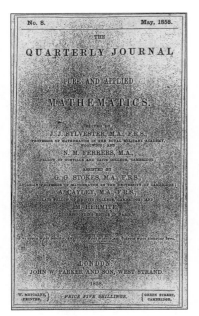

A title page from *The Quarterly Journal of Pure and Applied Mathematics*.

# The Messenger of Mathematics

Enter *The Oxford, Cambridge, and Dublin Messenger of Mathematics*, 'a journal supported by junior mathematical Students'. The seven editors, representing the three universities at the journal's launch in 1862, were either current students or recent graduates. In their lengthy introduction to the journal's first volume, the editors wrote that the mathematicians of these three universities 'are more and more widely separating in style and selection of subjects ... Let us have an English school of mathematics by all means, but sub-divisions in that school are simply an evil'. In order to improve this situation, the editors founded their journal to 'induce such students to attempt original investigation in their favourite branches of mathematics'. However, as recent graduates, they recognized that their target audience needed an incentive to contribute to the journal, and they promised that 'the distinctness of conception, and the exercise of imagination required for such work will be found to react on themselves with profit in University examinations'.[36]

This ecumenical effort was not to last. By the fifth volume, only three of the original editors remained, and they were all from Cambridge. At the end of this volume, in 1871, a new initiative cut the university ties of the journal. By naming the new series of their journal *The Messenger of Mathematics*, the editors hoped to 'appeal directly to the mathematical world at large, and to remove from their title-page any words which might be supposed to limit the sphere of usefulness of the Messenger'.[37]

James Whitbread Lee Glaisher, the last addition to this editorial team, graduated from Cambridge as 2nd wrangler in 1871 and obtained a fellowship at Trinity College, where he resided for the rest of his life. By the end of the 1870s, he alone was left to lead this journal into the 20th century. Although it had formally cut its university ties, with Glaisher as the only editorial gatekeeper, the *Messenger* formed a conducive publication venue for young mathematicians.

In 1879 Glaisher had also joined the editorial team of the *Quarterly Journal*, which by 1887 had lost Sylvester, Hermite, and Stokes. By its 28th volume (1895–1896), Glaisher was again the only one left to direct the enterprise, and he continued to edit both the *Quarterly Journal* and *The Messenger of Mathematics* until his death in 1928. To Glaisher therefore fell the inheritance of two journals founded by university students for university students for the encouragement of mathematical research.[38]

In a 1929 article memorializing Glaisher, G. H. Hardy characterized him as the 'last of the old school of mathematical editors, the men, who, like Liouville, contrived to run mathematical journals practically unaided'.[39] The level of commitment of the editors was only one of the many similarities between mathematical journalism in Victorian Britain and that of Europe and the United States. Journals devoted to mathematics proliferated during this period across the Channel and the Atlantic, and a small but motivated group of British mathematicians contributed to these foreign journals as well as their domestic ones (see Box 7.4).[40]

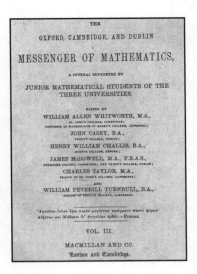

A title page from *The Oxford, Cambridge, and Dublin Messenger of Mathematics*.

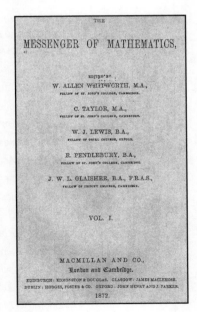

A title page from *The Messenger of Mathematics*.

# Box 7.4: Mathematical journals in Europe and the United States[41]

| Country | Journal | Years of existence |
|---|---|---|
| Belgium | Correspondance Mathématique et Physique | 1825–39 |
| | Nouvelle Correspondence Mathématique | 1876–80 |
| | Mathesis: Recueil Mathématique | 1881– |
| Bohemia | Časopis pro Pěstování Matematiky a Fysiky | 1872– |
| Denmark | Mathematisk Tidsskrift (later Tidsskrift for Mathematik, then Nyt Tidsskrift for Matematik) | 1859– |
| France | Journal de Mathématiques Pures et Appliquées | 1836– |
| | Nouvelles Annals de Mathématiques | 1842–1927 |
| | Bulletin de la Société Mathématique de France | 1872– |
| | Revue de Mathématiques Spéciales | 1890– |
| | Enseignement Mathématique | 1899– |
| German states | Journal für die Reine und Angewandte Mathematik | 1826– |
| | Archiv der Mathematik und Physik | 1841–1920 |
| | Mittheilungen der Mathematische Gesellschaft in Hamburg | 1873– |
| | Jahresbericht der Deutsche Mathematiker-Vereinigung | 1890– |
| | Zeitschrift für Mathematik und Physik | 1856–1917 |
| | Matematische Annalen | 1869– |
| | Zeitschrift für Mathematischen und Naturwissenschaftlichen Unterricht | 1870– |
| | Monatshefte für Mathematik und Physik | 1890– |
| | Unterrichtsblätter für Mathematik und Naturwissenschaften | 1895– |
| Hungary | Mathematikai és Physikai Lapok | 1892–1943 |
| Italy | Annali di Scienze Matematiche e Fisiche | 1850–57 |
| | Annali di Matematica Pura ed Applicata | 1858– |
| | Giornale di Matematiche | 1863– |
| | Rivista di Matematica | 1891–1906 |
| | Il Pitagora. Giornale di Matematica | 1895– |
| | Rendiconti del Circolo Matematico di Palermo | 1884– |
| | Periodico di Scienze Matematiche (e Naturale) per l'Insegnamento Secondario | 1873– |
| Netherlands | Niew Archief voor Wiskunde | 1875– |
| Poland | Prace Matematyczno-Fizyczne | 1888– |
| | Wiadomości Matematyczne | 1897– |
| Russian Empire | Rapports of the Харьковское математическое общество | 1879– |
| | Mathematischeskii Sbornik | 1866– |

continued

| Sweden | *Tidskrift för Matematik och Fysik* | 1868–85 |
| | *Acta Mathematica* | 1882– |
| United States | *Bulletin of the New York (later American) Mathematical Society* | 1888– |
| | *The Analyst: A Journal of Pure and Applied Mathematics* | 1874–83 |
| | *Annals of Mathematics* | 1884– |
| | *American Journal of Mathematics, Pure and Applied* | 1878– |
| | *American Mathematical Monthly* | 1894– |

Like those in Britain, these journals represented both commercial ventures and proceedings of mathematical societies. In the latter type of journal, Britain held a pioneering role, as the French geometer Michel Chasles reported to his compatriots in 1870:[42]

a mathematical society was founded in London in 1865 with a membership of one hundred, and this number is increasing; a society whose *Proceedings*, like those of the Royal Society of London, … publishes abstracts, more or less extended, of many papers. Is not [the existence of the *Proceedings of the London Mathematical Society*], which we applaud, an ingredient of future superiority in mathematical culture that should worry us?

## Part 3: Victorian mathematical societies and their journals

Although no major mathematical society existed in Britain until 1865, such learned bodies had not been totally absent before this time. There were, in fact, a number of amateur mathematical societies across the country (see Box 7.5).

### Box 7.5: Mathematical societies in Britain before 1865

| | |
|---|---|
| Society of Ingenious Mathematicians | 1710–24 |
| Spitalfields Mathematical Society | 1717–1845 |
| Manchester Mathematical Society | fl. 1718 |
| Lewes Mathematical Society | 1730s |
| York Mathematical Society | mid-18th century |
| Mathematical Society of Wapping | c.1750 |
| Oldham Mathematical Society | fl. 1794 |

The best known of this list, the Spitalfields Mathematical Society, was founded in 1717 and originally consisted of men seeking an intellectual recreational break from a hard day's work, each with 'his pipe, his pot, and his problem'.[43] In time, the members of the society became more bourgeois and began offering public lectures on a variety of subjects. However, by the 1840s, its membership was flagging, and in 1845 it was absorbed into the Royal Astronomical Society.

Other early mathematical societies were much less tenacious than the one in Spitalfields. For example, the well-known Analytical Society was established in Cambridge in 1812, producing a volume of *Memoirs*, but survived only two years.[44] Despite the dissolution of these earlier societies, British mathematicians enthusiastically received the London Mathematical Society (LMS) when it was founded in 1865.

## The London Mathematical Society and its *Proceedings*

The original idea for this society came from Arthur Cowper Ranyard and George Campbell De Morgan, two recent graduates of University College, London. Augustus De Morgan, George's father, served as the first president of the new body, initially called the University College Mathematical Society. As its original title suggests, the organization had both the name and flavour of a student club. Of its twenty-seven initial members, all but one had University College ties. However, after five months it had doubled its membership, and by 1866, it numbered ninety-four members, with over half unconnected with University College.[45]

In their account of the foundation of the LMS, Rice, Wilson, and Gardner state that:[46]

The sharp rise in membership also illustrates the very real need which existed for a mathematical society at this time; such a scheme was clearly long overdue.

Compared to similar mathematical societies abroad, however, the London Mathematical Society was not a latecomer; in fact, it represented one of the first national mathematical societies in America or Europe (see Box 7.6).[47]

Soon after its inception, the London Mathematical Society began publishing a journal. The *Proceedings of the London Mathematical Society* adopted a careful and rigorous editorial process, as Glaisher, an early and sustained member of the LMS council, later recalled:[48]

every paper was invariably considered by two referees, who sent in written reports which were read to the Council; and when the reports differed the paper was sent to a third referee. Every paper was balloted for, to decide whether it should be printed ... At the Astronomical Society, on the contrary, it was rarely that a paper was refereed, and a verbal report from a single referee was generally accepted.

Thomas Archer Hirst, Henry Smith, and Arthur Cayley were just three of many conscientious members who accepted what Glaisher described as the 'rather irksome duty' of refereeing papers.[49] Cayley especially supplied papers as well; in fact,

## Box 7.6: Some 19th-century mathematical societies outside Britain

| Country | Society | Year of foundation |
|---|---|---|
| Bohemia | Spolek pro volné přednášky z matematiky a fyziky | 1862 |
| Bulgaria | Сыюз наьблгарските математици | 1898 |
| Denmark | Mathematisk Forening | 1873 |
| Finland | Suomen matemaattinen Yhdistys | 1868 |
| France | Société Mathématique de France | 1872 |
| German states | Göttinger Mathematiker Gesellschaft | 1873 |
|  | Deutsche Mathematiker Gesellschaft | 1890 |
| Hungary | Mathematikai és Physikai Társulat | 1885 |
| Italy | Circolo Matematico di Palermo | 1884 |
|  | Mathesis | 1895 |
| Netherlands | Wiskundig Genootschap: Onvermoeide Arbeid Komt Alles te Boven | 1778 |
| Russian Empire | Московское математическое общество | 1867 |
|  | Харьков'ское математическое общество | 1879 |
|  | Санкт петербургское математическое общество | 1890 |
|  | Каеан'ское математическое общество | 1890 |
|  | Киевское математическое общество | 1890 |
| Sweden | 'Lunds Mathematical Society' | 1862 |
| United States | American Mathematical Society | 1888 |

at least seventy-eight of his contributions are printed in the *Proceedings*, almost nine per cent of all papers published in the *LMS Proceedings* during the 19th century.[50]

New members in the later part of the century maintained the number and quality of contributions. For example, William Burnside published several times in the *Proceedings* on his innovative work on groups of finite order. Besides this very pure topic, Burnside also used the journal to publish research on various areas of applied mathematics.[51] In fact, while pure mathematical papers consistently outnumbered those in applied mathematics, contributions in the latter area were far from absent and counted among their authors James Clerk Maxwell, Lord Rayleigh, J. J. Thomson, Joseph Larmor, and Horace Lamb.[52]

As a mathematician involved in a variety of scientific and learned bodies, Thomas Hirst described the London Mathematical Society as 'my favourite Society,—the one in which I have taken the greatest interest'.[53] But while the LMS was devoted to the progress of mathematical research, it failed to be active in mathematical pedagogy. Hirst, however, was interested in both the research and educational realms of mathematics and, in 1871, a year before he was elected the fifth president of the LMS, also served as the first president of the Association for the Improvement of Geometrical Teaching (AIGT).

## The Association for the Improvement of Geometrical Teaching

AIGT notice in *Nature*, 29 December 1870.

As we will see in Chapter 14, the Victorian period witnessed a lively debate on the teaching of geometry—and in particular the use of Euclid's *Elements*—in schools and colleges. This new association, whose goals were declared in its name, took as its first task the creation of a school syllabus of plane geometry that departed from the traditional course of Euclid.[54]

In 1868, James Maurice Wilson, senior mathematics master at Rugby School, complained that 'there are scores of schools where boys learn and say their Euclid like declensions'.[55] In order to improve this educational situation, Wilson wrote a textbook on geometry which diverged from the restrictive structure of Euclid's propositions. Wilson discussed his textbook reform and views on Euclid at a meeting of the London Mathematical Society in 1868 but, in his words, 'was well "heckled"'.[56] After this meeting, the LMS disengaged itself from issues in teaching; this event, in the estimation of AIGT historian Michael Price, 'seems to have set an important precedent for the LMS of non-involvement in mathematics education and of exclusive concern for the advancement of the subject'.[57]

The Association for the Improvement of Geometrical Teaching began in 1871, twenty years prior to the creation of any other association in Britain devoted to the teaching of a secondary school subject. At its beginning, the Association was mainly composed of schoolmasters. Their first goal, the creation of a geometry syllabus, proved difficult to attain; after much effort, their plan was finally published in two parts in 1873 and 1875, but it was a compromise and did not prove to be as effective as its creators had hoped.[58] Considering the formidable foes of these geometrical reforms, the limited success of the AIGT in this arena is not surprising. Influential Cambridge mathematicians, including Cayley and Todhunter, took a conservative stance on geometrical education.[59] The Oxford mathematician Charles Dodgson (better known as Lewis Carroll) mocked the AIGT by dubbing it the 'Association for the Improvement of Things in General'.[60]

While not immediately successful in its primary goal, the AIGT did succeed in providing a publication venue for mathematics education through its *Report*. In 1883, the Association began accepting and publishing papers on a variety of mathematical subjects in order 'to cater for both mathematical and educational interests, and thereby to widen the AIGT's Appeal'.[61] In that year, the Cambridge lecturer William Henry Besant presented a paper on 'The teaching of elementary mechanics'; George Minchin, a Royal Indian Engineering College professor, gave 'Notes on the teaching of elementary dynamics'; and Horace Lamb, from the University of Adelaide, contributed 'The basis of statics'. In the following nine years, the *Report* printed, among others, papers by Charles Taylor on 'The discovery and the geometrical treatment of conic sections', Robert Baldwin Hayward (Harrow School) on 'The correlation of the various branches of elementary mathematics'; Robert William Genese (professor at University College, Aberdeen) on 'Elementary mechanics', and Alfred Lodge (another Royal Indian Engineering College professor) on 'The multiplication and division of concrete quantities'.[62]

In 1894, to reflect the Association's widened interests in mathematics and pedagogy, the AIGT renamed itself the Mathematical Association and its *Report* developed into the *Mathematical Gazette*—a journal devoted to mathematics and mathematics teaching that continues to this day. Edward Mann Langley, the *Gazette*'s first editor and a schoolmaster at Bedford Moderate School, wrote in the first volume of 1894 that:[63]

We hope to extract from desk and pigeon-hole many MSS, which have remained unpublished for want of a suitable organ for making them known… But we intend to keep strictly to 'Elementary Mathematics'.

The *Gazette* served dual functions as a minor mathematical serial and an educational jo2urnal. Along with mathematical articles on the subjects described above by Langley, it contained valuable book reviews, historical notes, and questions for solution. The *Gazette*'s educational content, as calculated by Michael Price, varied from three to sixteen per cent during the 19th century.[64]

# The Edinburgh Mathematical Society and its *Proceedings*

Besides the Mathematical Association, another British mathematical society emerged to further the agenda of mathematical educators. According to an 1883 circular announcing its creation, the Edinburgh Mathematical Society began with the goal of

the mutual improvement of its members in the Mathematical Sciences… [through] Reviews of works both British and Foreign, historical notes, discussion of new problems or new solutions, and comparison of the various systems

The *Proceedings of the Edinburgh Mathematical Society.*

of teaching in different countries, or any other means tending to the promotion of Mathematical Education.[65]

Two of its founders, Alexander Fraser and Andrew Barclay, whom we met briefly in Chapter 4, worked as mathematical masters in an Edinburgh school, and their profession was shared by forty of the Society's fifty-eight founder members. Even by 1926, the proportion of university members of the Society was only about one-third.[66] In accordance with the Society's initial objectives and the profession of most of its membership, the *Proceedings of the Edinburgh Mathematical Society* contained pedagogical, historical, and many geometrical articles. Frequent contributors included the schoolmasters John Mackay and Arthur Pressland, both of the Edinburgh Academy, and George Crawford of Harrow School. The *Proceedings* also published articles by those well within the university sphere, such as Peter Guthrie Tait and George Chrystal, professors of natural philosophy and mathematics (respectively) at the University of Edinburgh, and John Steggall, mathematics professor at University College, Dundee.

The establishment of the Edinburgh Mathematical Society and its *Proceedings* helped to enrich a Scottish mathematical environment that, as we saw in Chapter 4, had long been overshadowed by Cambridge, while also enabling many Scottish mathematicians to become active in a British mathematical journal.

## Conclusion

Although they were founded for different specific reasons, the Mathematical Association and the mathematical societies of Edinburgh and London all emerged in response to a growing group of scholars who increasingly identified themselves with the discipline of mathematics. While British mathematicians had no major mathematical society of their own before 1865, twenty years later they had three from which to choose. These new societies gave mathematicians the freedom to present research that scientists in other disciplines found increasingly difficult to follow. The meetings of the Edinburgh Mathematical Society and the AIGT also allowed its members to discuss issues specific to the teaching of mathematics.

Besides publishing in the proceedings of its societies, Victorian mathematicians had access to a broad spectrum of periodical outlets. Throughout this period, there existed commercial outlets for research, pedagogical, and recreational mathematics. With commercial mathematical journals they could enjoy an open reception and swift publication of their articles. Moreover, budding mathematicians could enter the publication community through the junior departments of any number of minor mathematical serials. Mathematical journals with university roots encouraged students to explore and communicate mathematics that lay outside the confines of the classroom.

When writing his 1880 description of the three-step evolution process for mathematical journalism cited in the introduction of this chapter, Glaisher considered Britain to have passed through all of these phases, ending with the 1865 foundation of the *Proceedings* of the London Mathematical Society.[67] However, as the above discussion of Victorian mathematical journals illustrates, the establishment, activity, and distribution of these journals was not simply a progressive and evolutionary process. As British mathematicians embraced new journal formats, they did not spurn the previous ones. In fact, at the end of the Victorian era, mathematical problems for solution could still be found in the *Mathematical Questions from the Educational Times*, although the general level of these questions was higher than those found in most of the earlier minor mathematical serials. New formats emerged to satisfy the needs of a developing mathematical publication community. This development occurred in more than one direction, with some community members gravitating towards original research, some towards pedagogy, others towards recreational mathematics, and some exploring more than one of these.

The vagaries of business ensured that commercial mathematical ventures would never enjoy the stability of society publications. Yet as soon as one journal failed, motivated editors rallied to start a new periodical. This devotion ensured the continued existence, even if under constantly changing titles, of the great variety of mathematical journalism in Victorian Britain.

While the survival of the fittest certainly occurred at the level of individual journals, at the genre level the development of British journals for mathematics during the 19th century was a cumulative process, rather than one of natural selection; *The Quarterly Journal of Pure and Applied Mathematics* provides an instructive example of this point. As a national mathematical journal evolved from a student publication, the *Quarterly Journal* represented a new genre for mathematical articles, but left a vacuum in university-centred student mathematical journals. Instead of signalling the death of the student genre, however, this evolution encouraged the subsequent foundation of *The Oxford, Cambridge, and Dublin Messenger of Mathematics*, a journal that later enlarged its audience while continuing to cater to young researchers throughout the 19th century.

One reason that different genres of mathematical journals could exist simultaneously in Victorian Britain was the flexibility of their contributorship. Unlike those of the society journals, the contributors to minor mathematical serials drew heavily from a recreational problem-solving tradition outside the university sector. However, the university and recreational mathematical spheres were certainly not disjoint; in particular, several university-trained mathematicians developed their early tastes for mathematics as young problem-solvers. The university and recreational spheres also met in several instances when these serials were produced by professors at England's military colleges. These military-college-centred serials contained original research and mathematical innovations alongside the more routine questions and answers, and research mathematicians continued to publish in each genre.

This cumulative publication network for mathematics cultivated support for future mathematical enterprises in Victorian Britain. Minor mathematical serials could pique the imagination of a young student; with the mathematical tools honed at a British university (most often Cambridge or Dublin), the recent graduate could begin publishing original mathematical articles and receive the constructive criticism of peers through a student journal; by submitting later mathematical articles to the *Cambridge and Dublin Mathematical Journal*, or the *Quarterly Journal*, the developing researcher could become imbued with the mathematical standards being developed by an ambitious and increasingly internationally-aware group of mathematicians. With this publication infrastructure it is no surprise that, by 1865, there was sufficient support for the foundation of the London Mathematical Society and its *Proceedings*, a society and a journal practically born as full-grown enterprises.[68] The British mathematical publication community had finally come of age.

Lord Kelvin with his compass.

# CHAPTER 8

# Victorian 'applied mathematics'

A. D. D. CRAIK

Aremarkable flowering of applied mathematics took place during Queen Victoria's reign, establishing Britain as the major research centre in many areas. After sketching the educational milieu, we outline major achievements in research under the broad headings: celestial and rigid-body mechanics; fluid mechanics and elasticity; light and the 'aether'; heat and thermodynamics; and electricity and magnetism.

During Queen Victoria's long reign, British applied mathematics underwent a remarkable flowering from small beginnings to lead the world in several areas. At this time were established many theories and results now regarded as 'classical'. The volume of writings on applied mathematics during this period is enormous, and no-one can know it all. Accordingly, to attempt a comprehensive history of 'Victorian Applied Mathematics' within a small compass invites comparison with the 19th-century publisher and lecturer Sir Richard Phillips, memorably described by Augustus De Morgan as having 'four valuable qualities; honesty, zeal, ability, and courage. He applied them all to teaching matters about which he knew nothing'.[1]

But what do we mean by 'Victorian Applied Mathematics'? Even today, the term 'applied mathematics' means different things in different countries. In Britain, the earlier expression 'mixt' (or mixed) mathematics was used throughout the 18th and early-19th centuries to designate mathematics employed to investigate the physical world, as distinct from mathematics *per se* which came to be called 'pure mathematics'. For instance, Charles Hutton's *Mathematical and Philosophical Dictionary* of 1795 distinguished the two as follows:

*Pure Mathematics* is that branch which considers quantity abstractedly, and without any relation to matter or bodies.

*Mixed Mathematics* considers quantity as subsisting in material being; for instance, length in a pole, depth in a river, height in a tower &c.
*Pure Mathematics*, again, either considers quantity as discrete, and so computable, as arithmetic; or as concrete, and so measurable, as geometry.
*Mixed Mathematics* are very extensive, and are distinguished by various names, according to the different subjects it considers...such as Astronomy, Geography, Optics, Hydrostatics, Navigation, &c., &c.

According to Hutton:

Pure Mathematics has one particular advantage, that it occasions no contests among wrangling disputants...because the definitions of the terms are premised, and every person...has the same idea of every part of it...[But] in Mixed Mathematics...such just definitions cannot be given as in geometry: we must therefore be content with descriptions.[2]

But even the term 'pure mathematics' had not settled down in the 19th century. For instance, the Cambridge-based historian of mathematics W. W. Rouse Ball wrote in 1888 that:

The mathematicians of the nineteenth century have mostly specialized their work in one or more departments. They may roughly be divided into those who have specially studied pure mathematics (in which I should include theoretical dynamics and astronomy) and those who have specially studied physics: the latter subject requiring as it develops a fair knowledge of mathematics even on the part of those who treat it from the experimental side alone.[3]

That Rouse Ball includes 'theoretical dynamics and astronomy' as branches of *pure* mathematics seems surprising today. But his rationale was that, since these are subjects with well-established governing equations, their further investigation is purely mathematical. In contrast, other subjects that still involved speculation about their underlying principles and equations should properly be regarded as 'physics', or 'natural philosophy' as it was then usually called. But all who studied physics, even those who were primarily experimentalists, by then needed a considerable knowledge of mathematics.

Throughout this chapter, we interpret the expression 'applied mathematics' widely, to include dynamics, physics (natural philosophy), and technology: that is, we include most of the main areas where mathematics was a major investigative tool. But we largely exclude those areas in which probability and statistics were used, as these are treated elsewhere. The topics we investigate here fall mainly into five categories, although these are not entirely distinct:

- celestial and rigid-body mechanics;
- fluid mechanics and elasticity;
- light and the 'aether';
- heat and thermodynamics;
- electricity and magnetism.

# The mathematical environment in early 19th-century Britain

In the first two decades of the 19th century, British universities were in a poor state. For most of the preceding century, the two English universities of Oxford and Cambridge had been largely moribund (though with a few notable scholars), while by 1820 the Scottish Enlightenment had lost its impetus, the Scottish universities being ill-organized and under-funded. As John Playfair of Edinburgh lamented in 1808, few in Britain were able to begin to read Laplace's *Mécanique Céleste*, and many were even 'stopped at the very first page of the works of Euler and d'Alembert'.[4] The focus of the new applied mathematics was definitely in France, where Laplace, Lagrange, Legendre, Poisson, Cauchy, and others made many important advances.[5]

Only two British applied mathematicians of the immediate pre-Victorian period deserve mention: James Ivory (who studied at St Andrews, and then with Playfair in Edinburgh), and the largely self-taught George Green of Nottingham (who became a mature student at Cambridge *after* publishing his major work). By their private study of French works, Ivory and Green absorbed the new analytical methods and went on to apply them fruitfully in physical contexts. Ivory made significant contributions to the much studied problem of the 'figure of the Earth' under gravity and rotation, to atmospheric refraction of light, to capillary attraction, and to the calculation of the orbits of planets and comets. Green's privately printed *Essay on the Application of Mathematical Analysis to the Theories of Electricity and Magnetism* (1828), at first ignored, was later recognized for its brilliant early insights, and his few later papers made useful contributions to hydrodynamics and to optics.[6]

The growth of applied mathematics in Britain owed much to the resurgence of the University of Cambridge. The mathematical tripos provided intense competition, with fellowships and probable career enhancement as rewards. Due to the efforts of George Peacock and Richard Gwatkin as moderators or examiners during 1817, 1819, and 1821, the syllabus was modernized to embrace the Continental form of the differential and integral calculus. William Whewell also supported this reform, but became more reactionary as time passed, favouring geometrical over analytical demonstrations. Before 1830, the lectures and writings of professors Robert Woodhouse and George Biddell Airy fostered interest in astronomy, mechanics, and optics. Most importantly, the best private tutors were able to provide sound instruction in advanced tripos subjects, helping their pupils to gain top places in the list of 'wranglers'.

The most successful and famous of these tutors were William Hopkins, from 1828 to 1860, and Edward John Routh, from 1856 to 1888. Among the pupils of Hopkins and Routh were many who became outstanding scientists and university professors, as listed in Chapter 1. It is noteworthy that a substantial number of these individuals were Scots or Irish, and that many had studied at colleges or universities elsewhere before going to Cambridge. Later, many staffed the universities and colleges of England, Scotland, and Ireland.[7]

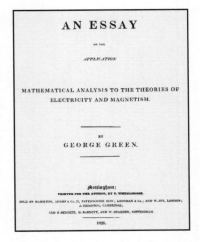

AN ESSAY

ON THE

APPLICATION

OF

MATHEMATICAL ANALYSIS TO THE THEORIES OF
ELECTRICITY AND MAGNETISM.

BY

GEORGE GREEN.

Nottingham:
PRINTED FOR THE AUTHOR, BY T. WHEELHOUSE.

SOLD BY HAMILTON, ADAMS & Co., PATERNOSTER ROW; LONGMAN & Co.; AND W. JOY, LONDON;
J. DEIGHTON, CAMBRIDGE;
AND S BENNETT, H. BARNETT, AND W. DEARDEN, NOTTINGHAM.

1828.

The title page of Green's *Essay*.

As well as those who received at least part of their mathematical education at Cambridge, some others made notable contributions to applied mathematics. As Chapter 5 makes clear, Trinity College in Dublin fostered an illustrious Irish tradition, with luminaries that included Bartholomew Lloyd, William Rowan Hamilton, James MacCullagh, and later George Francis FitzGerald. In Glasgow, the engineers William J. M. Rankine and James Thomson Jr. (brother of William) were also capable mathematicians, and former students of Peter Guthrie Tait in Edinburgh, such as Cargill Knott and James Ewing, did similarly good work.

In contrast, as we saw in Chapter 2, mathematics and its applications at the University of Oxford did not then have a high reputation, although successive Savilian professors of geometry, Baden Powell and Henry Smith, brought improvements. Baden Powell conducted research in optics about which he corresponded with Stokes and Airy, while Smith's interests were mainly in pure mathematics. Oxford's professor of experimental philosophy from 1866 to 1921 was the Cambridge-educated Robert Bellamy Clifton, who had previously been a professor at Owen's College, Manchester; although he published next to nothing, he was instrumental in establishing Oxford's Clarendon Laboratory. Bartholomew Price, Sedleian professor of natural philosophy from 1853 to 1898, was a sound teacher and administrator, but no researcher.

Oxford's Savilian professor of astronomy from 1842 to 1869 was William Fishburn Donkin, whose few but varied scientific papers were of some significance: his subjects included the method of least squares, Hamiltonian mechanics, solutions of Laplace's equation, and secular acceleration of the moon's motion. A treatise on acoustics remained unfinished at his death: though this was published, it was soon superseded by Lord Rayleigh's *Theory of Sound*, which appeared in 1877–78.

Donkin was succeeded by the 62-year-old Charles Pritchard, who held the chair for twenty-three years until 1893. A Cambridge graduate and former headmaster of grammar schools, Pritchard supervised the building of a new observatory and pioneered the use of stellar photography, the observations being made by assistants. Robert Main, another Cambridge graduate and former assistant to Airy at the Royal Greenwich Observatory, held the post of Radcliffe Observer from 1860 until his death in 1878. There he undertook major observational and computational work that was published in the first two volumes of the *Radcliffe Star Catalogue* and sixteen volumes of *Radcliffe Observations*. His successor was yet another Cambridge wrangler, Edward James Stone, who returned to England after nine years at the Cape Observatory in South Africa (see Chapter 6).

Another whose name is associated with Oxford—though being a woman she never studied there—was Mary Somerville; Somerville Hall (later College) was established and named after her in 1879. She was a capable though unoriginal mathematician, a translator, a popularizer of Laplace's *Mécanique Céleste*, and a much-read writer of broad scientific surveys encompassing astronomy, physics, geology, and geography.[8]

The development of institutions across the country—colleges, universities, learned societies, local literary and philosophical societies, and mechanics' institutes—did much to meet the growing demand for scientific knowledge.

Specialist scientific journals were established, 'self-help' textbooks were published by the Society for Diffusion of Useful Knowledge and by its imitators, such as Dionysius Lardner's *Cabinet Cyclopaedia*, and large-scale encyclopedia projects recruited distinguished contributors. The increasing demand for instruction in science and engineering provided more opportunities for employment, and many who obtained teaching posts also found time to pursue research.

## Celestial and rigid-body mechanics

Early Victorian mathematicians inherited a strong, if ossified, Newtonian tradition in which geometry and the fluxional calculus were applied to physical problems, and particularly to those of planetary astronomy. Indeed, several parts of Newton's *Principia Mathematica* remained canonical texts at Cambridge University well into the 19th century. But by the early 1830s, more up-to-date textbooks in English had been published on differential and integral calculus, on plane and spherical trigonometry, and on their applications to astronomy. Appreciation of the analytical and mechanical works of Continental scientists, particularly Euler, d'Alembert, Lagrange, Laplace, Poisson, and Cauchy, had been partial and belated, but by the 1830s translations of the first few books of Laplace's *Mécanique Céleste* finally brought Continental analytical methods to the fore. Although the basic laws of the mechanics of a particle under applied forces and the law of gravitational attraction were undoubtedly those formulated by Newton, the methods now applied became primarily analytical, rather than geometrical or fluxionary.[9]

The famous 'three-body problem' for the motion of three mutually-attracting bodies (for example, the sun, earth, and moon) presented a major challenge. It could be solved only approximately by perturbation techniques pioneered by

A French cartoonist disparages British claims to have discovered the position of Neptune.

Laplace, who gave an early theoretical description of the observed irregularities of the moon's motion. Unexplained small deviations of the orbit of the planet Uranus led to the conjecture that there might exist an undiscovered planet. By careful analysis and intelligent guesswork, independent predictions of the position of the new planet were given almost simultaneously during 1845–46 by John Couch Adams, recently graduated from Cambridge, and Urbain Leverrier of the École Polytechnique in Paris. Urged on by Leverrier, Johann Gottfried Galle of the Berlin Observatory searched for the planet, and on 23 September 1846 he found it. Later named Neptune, its discovery was a stunning demonstration of the predictive power of mathematics.

Unfortunately, a bitter priority dispute developed, in which British (and especially Cambridge-educated) scientists were accused of trying to rob Leverrier of his

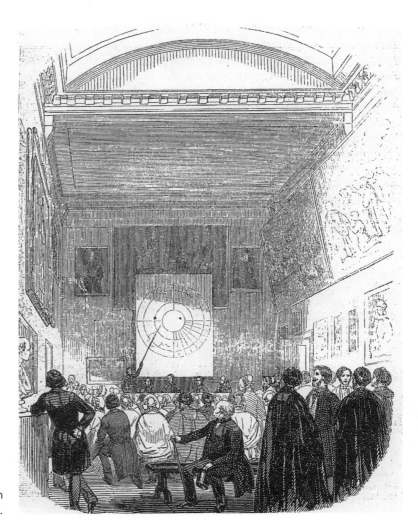

A British Association lecture on the discovery of Neptune, 1847.

rightful claim as predictor of the new planet. Although Adams had indeed performed similar calculations, he had not made them public by the time of Neptune's discovery, and he had been ill served by the Cambridge astronomer James Challis who had only belatedly and half-heartedly searched for the planet.[10] (For more on the amateur nature of astronomy in Victorian Britain, see Chapter 9.)

Later, and by then a Cambridge professor himself, Adams made a painstaking theoretical analysis of the irregularities of the moon's motion which surpassed all previous efforts. As well as a new planet, more than two hundred comets were discovered during the 19th century as a whole. One who made an extensive study of the orbits of comets and meteors between 1860 and 1880 was Alexander Herschel, the third generation of astronomical Herschels.

Important for astronomical observations were the small periodic variations in the rotational motion of the earth known as 'precession' and 'nutation', under the influence of perturbations of the net gravitational force exerted by the combined action of the sun and moon. Originated by Leonhard Euler in the 18th century, this theory was one of the most advanced topics in the Cambridge tripos examinations of the 1830s, and was expounded in the *Mathematical Tracts* of George Biddell Airy, Astronomer Royal at Greenwich from 1835 to 1881.[11] Improved knowledge of these 'wobbles' of the earth's motion, and of the distorting effects of atmospheric refraction, were required to establish the true position of a star or a planetary orbit from the apparent one observed by telescope through the variable atmosphere of an orbiting, rotating, and slightly wobbling earth. Better star catalogues, and documentation of the orbits of planets and comets, were produced by various European observatories: the processing (or 'reduction') of each observation entailed considerable computational effort.

Also important for both astronomy and geodesy was the study of the shape of the earth itself, the so-called 'figure of the earth', which remained a major mathematical preoccupation for much of the 19th century. At its simplest, the theory proposed an earth composed of a self-gravitating fluid of constant density, rotating with uniform angular velocity about its axis. Newton had correctly predicted that the shape of its surface must be an oblate ellipsoid of revolution, flattened at the poles. Improved derivations were given by Clairaut and Maclaurin, and some subsequent advances were made by Laplace and Ivory. In 1834, Jacobi and Liouville made the unexpected mathematical discovery that such a rotating self-gravitating mass of fluid can assume *non-symmetrical* shapes of equilibrium. Though still ellipsoidal, these could resemble a slightly squashed rugby ball, a not-quite-symmetric discus, or a long thin shape like a squashed cigar.

In the 1830s and 1840s, studies of the figure of the earth were extended by Airy, John Henry Pratt, and William Hopkins to include variable density and an earth assumed to be part-solid and part-liquid. Stokes, Airy, and Pratt also examined variations in the force of gravity due to mountain ranges, and concluded that the earth's density must be less beneath mountain ranges than that beneath lower ground (a phenomenon now known as Airy and Pratt isostasy).[12] The precession and nutation of a part-solid and part-liquid earth were also studied by Hopkins, in the hope of discovering more about the earth's internal composition. The latter topic was later taken up by Hopkins's former student William Thomson (Lord

Kelvin). In the 1880s, Thomson also studied the asymmetric Jacobi–Liouville ellipsoids, showing that under some circumstances these were unstable and went pear-shaped. Thomson, and later Henri Poincaré and George Darwin, speculated that moons of planets and double stars might have so resulted.[13]

An important advance in mechanics was the variational formulation by the Astronomer Royal of Ireland, William Rowan Hamilton. In the 1830s, he established a general principle of least action with wide applications in mechanics, optics, and electricity. This is enshrined in a compact and elegant set of first-order partial differential equations satisfied by a Hamiltonian function, now known as *Hamilton's equations*. They apply in all physical contexts where a Hamiltonian function is required to take a maximum or minimum value, and many of the laws of physics can be recast in this form. The Hamiltonian formulation was later to play a major role in quantum mechanics.[14]

The textbook *A Treatise on the Dynamics of a Particle* (1856) by recent Cambridge graduates P. G. Tait and W. J. Steele (both Scots and both pupils of Hopkins) was a considerable improvement on its precursors. Later, Tait collaborated with William Thomson on their groundbreaking *Treatise on Natural Philosophy* (1867). This collaboration was at times frustrating: though the treatise was envisaged as just the beginning of a larger project, no further volumes appeared. Nevertheless, it marked a milestone in mechanics, placing energy methods centre-stage and emphasizing the interconnection of heat, motion, and energy dissipation, as formulated in the recent theory of thermodynamics.[15]

Valuable publications on the stability of spinning bodies were written by Routh, and Francis Bashforth made a major study of ballistics. As the latter topic was still not amenable to exact mathematical study, empirical estimates of air resistance of projectiles were used to compile improved artillery tables for range-finding. Bashforth's *Mathematical Treatise on the Motion of Projectiles* (1873) became a standard work, and he developed a valuable experimental device called the 'Bashforth chronograph'. He also collaborated with Adams in a study of the shapes of liquid drops under surface tension, when the still-used 'Adams–Bashforth method' for numerical solution of differential equations was first developed. Another who examined the effect of air resistance on rapidly-moving projectiles was Tait, by then Edinburgh's professor of natural philosophy, who studied the flight of spinning golf balls. A keen golfer, Tait realized that 'underspin' imparted lift and so prolonged the flight of the ball, making it differ from that of Bashforth's ballistic tables.

The compiler of tables *par excellence* was Edward Sang, whose massive unpublished computations of 26-place and 15-place logarithmic, trigonometric, and astronomical tables fill forty-seven manuscript volumes. Sang graduated from Edinburgh University in 1824, and had a varied career as surveyor, civil engineer, astronomer, actuary, and mathematical lecturer. He typifies the Victorian mathematical Jack-of-all-trades, with publications on many mechanical, optical, and actuarial topics. These range from novel works on arithmetic (he was a prodigious mental calculator) to a treatise entitled *A New General Theory of the Teeth of Wheels* (1852), works on life assurance, and an unduly pessimistic 1887 paper on the strains on the Forth Bridge, then under construction.[16]

## Fluid mechanics and elasticity

The development of theoretical fluid mechanics lagged far behind the mechanics of point masses and rigid bodies. Studies of the figure of the earth and of capillary attraction employed hydrostatics, but fluid in motion was an altogether different matter. The governing equations of non-viscous fluid motion had been derived by Euler in the 1750s, after pioneering work by Johann and Daniel Bernoulli and Jean d'Alembert, but even by 1840 only a few particular solutions of no great practical interest were known. The famous 'd'Alembert paradox' showed that inviscid steady flow past a rigid body could not exert any drag force upon the body, but drag was in fact experienced by every body moving through a fluid; for this reason, empirical methods, such as those used by Bashforth for projectiles, long remained essential in practical applications. As a result, theoretical fluid mechanics and practical engineering hydraulics long developed independently of one other.

The sole important exception was the theory of water waves and tides. Following on from work by Laplace, Lagrange, Poisson, and Cauchy, the theory of linear and weakly non-linear water waves was developed in Britain during the 1840s by Airy, Green, Stokes, and Philip Kelland. Whereas Poisson and Cauchy, in their quest for generality, had created mathematics of great complexity that repelled their readers, the British workers confined themselves to more manageable yet practical situations, such as plane waves of constant frequency propagating in channels of constant or slowly varying depth and cross-section.

The tidal theory of Laplace described the response of a shallow sea covering the whole earth to the periodically varying net force of gravitation due to the sun and moon, while Airy's theory was restricted to tides propagating within a channel of uniform width. Obviously, neither was appropriate for the tides actually observed at coasts and harbours, and it was these that were of greatest importance to shipping. With support from the British Association for the Advancement of Science, John Lubbock and William Whewell made extensive observations in the 1830s of high and low tides around British coasts, with their seasonal variations; this data enabled them to devise empirical methods of tide prediction. A later practical development from the 1870s, still independent of the equations of fluid mechanics, was William Thomson's 'harmonic analyser': this, the first automatic 'Fourier analyser', enabled tide prediction from a far shorter time-record than that previously needed.[17]

But the hydrodynamic theories of water waves developed by Airy, Green, Kelland, and Stokes were successful, within their limits. Most notably, in 1847 Stokes gave a masterly analysis of weakly non-linear waves of finite amplitude: not only did he find significant corrections to the linear theory for small-

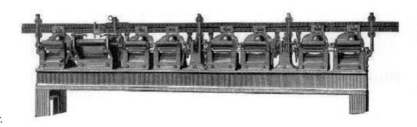

Thomson's harmonic analyser.

amplitude waves, but he provided a powerful analytical method based on series expansion in ascending powers of a small parameter (in his case, a measure of the wave steepness), that has been followed in many later studies up to the present day.

Unfortunately, none of these theories could describe the solitary waves observed by John Scott Russell in experimental and field studies supported by the British Association (in 1837 and 1844). As a result, Russell's work was misunderstood and criticized for alleged errors, but he was later vindicated when Boussinesq (1871) and Rayleigh (1876) found solutions corresponding to Russell's 'great wave of elevation'. In recent times, Russell's solitary waves have been found to play a central part in the 'theory of solitons' that describes non-linear waves in many contexts, from water waves to lasers.[18]

The *Navier–Stokes equations* of viscous fluid dynamics were first derived by Navier, Cauchy, and Poisson in the 1820s, and then under improved hypotheses by Saint-Venant (1837) and Stokes (1845). These equations contain more terms than do Euler's non-viscous equations, and even the latter were then regarded by many as intractable. Nevertheless, Stokes found some useful exact solutions of the equations for viscous incompressible flow, and several approximate ones—most famously, the 'Stokes flow' around a sphere falling slowly through a viscous liquid—thereby laying the foundations of one of the most important and challenging areas of applied mathematics.[19]

Later advances in fluid dynamics included:

George Gabriel Stokes
(1819–1903).

- Lord Rayleigh's studies of sound waves in compressible gases and of instability of jets;

- papers of William Thomson, Hermann Helmholtz, and William J. M. Rankine on vortex flows;

- Thomson's asymptotic analysis of ship waves, using his 'method of stationary phase', now a standard tool of asymptotic theory;

- the work of Thomson, Osborne Reynolds, and William McF. Orr on hydrodynamic instability of shear flows;

- Reynolds's averaged formulation of the equations of motion, introducing the concept of turbulent 'Reynolds stresses'.[20]

British mathematicians of the next generation who made significant contributions to fluid mechanics included Andrew Russell Forsyth, George

Greenhill, W. M. Hicks, M. J. M. Hill, Horace Lamb, A. E. H. Love, and J. H. Michell. Suffice it to say that, between 1830 and 1900 the subject had advanced to mathematical maturity, though still with limited practical applications. In the 1830s, the only student texts on hydrodynamics at Cambridge were poor offerings by W. H. Miller and T. Webster, barely deriving Euler's equations of motion. By the century's end, the treatises by Alfred B. Basset, Lamb, and Rayleigh were truly authoritative, and are still consulted by researchers today.

The mathematical theory of elasticity developed alongside fluid mechanics and optics, with which it has close connections. In viscous fluid mechanics, the key postulate (used by Stokes) is the linear relationship between the components of stress and those of rates-of-strain, while in classical elasticity there is a similar linear relationship between stress and strain.[21] Early French work on the governing equations by Cauchy, Poisson, and Fresnel usually had light propagation in mind. This was taken up and extended by Green (1837) and Saint-Venant (1855).

Augustus Edward Hough Love (1863–1940).

Stokes, Airy, Thomson, Lamb, Love, Rayleigh, and Greenhill all contributed to the advance of the subject. Solutions were found to many statical problems involving the bending, torsion, and distortion of elastic solid bodies (typically, plates, beams, and cylinders of specified cross-section) under external forces and couples. Dynamical problems concerned the various kinds of elastic waves that can propagate though elastic media and along their boundaries. The classic text on linear elasticity theory, still valuable today, is that of Augustus Love, first published in 1892–93.[22]

Elasticity theory was soon employed in the new subject of seismology: this originated in Japan, where opportunities to study earthquakes are disconcertingly frequent. Between 1870 and 1885 several British scientists were recruited to establish the new Imperial College of Engineering and the University of Tokyo.[23] There, Liverpool-born John Milne, James Alfred Ewing from Dundee, and Thomas Gray from Glasgow, built and operated a sophisticated seismological laboratory, while Cargill Gilston Knott, another Scot from Midlothian, developed the mathematical theory, including the Fourier analysis of earthquake records and the formulation of the correct boundary conditions for elastic waves at an interface between different media. Ewing, Gray, and Knott were protégés of Tait, William Thomson, and Edinburgh's professor of engineering, Fleeming Jenkin. As we saw in Chapter 4, Ewing returned to Britain as professor of engineering, first at Dundee University College and then at Cambridge, while Knott returned to Edinburgh University, where he played leading roles in the Royal Society of Edinburgh and the Edinburgh Mathematical Society, writing prolifically on a wide range of scientific topics and publishing several textbooks and mathematical tables.[24]

Engineering practice still relied largely on empirical rules. Wind stresses on proposed bridges had to be estimated, together with the strength of beams, girders, and brickwork under extreme loading. Though much had been achieved, such calculations remained beyond the reach of the exact sciences. The hydraulics of water pipes, rivers, canals, and sewers still owed comparatively little to the fundamental equations of fluid dynamics. Similarly, ship design relied more on art, experience, and trial-and-error than on precise calculations, though William Froude and Rankine applied some mathematical expertise. Inevitably there were mistakes in design and instances of ill-managed construction, sometimes with disastrous consequences.[25] Nevertheless, Victorian engineers accomplished marvels of innovation, many still effectively serving today's cities, roads, and railways.

## Light and the 'aether'

By the early 19th century, geometrical optics based on ray theory had reached a state of sophistication. Refraction and reflection of light by lenses and mirrors were well understood, and the troublesome defects of spherical and chromatic aberration of telescope lenses had been conquered. But there was still a major dispute over whether light consisted of particles or 'corpuscles' of the various colours, as Newton had suggested, or whether it was composed of waves of different lengths. Robert Hooke had been an early supporter of a wave theory, and around 1800 Thomas Young demonstrated that such a theory was consistent with observations of interference and diffraction. Some twenty years later, French experimental and theoretical work on polarization—notably that of Fresnel—provided strong evidence that light consisted of *transverse* waves having no motion along the direction of propagation. Fresnel's wave theory was broadly accepted by the influential Cambridge-educated scientists John Herschel, Airy, and Stokes, but some others, most notably David Brewster in Edinburgh, remained loyal to Newton's corpuscular theory.

Although some connection had been made between light and radiant heat, the identification of light as just the visible part of a wide spectrum of electromagnetic waves was still not appreciated. This had to await the development of electromagnetic field theory by James Clerk Maxwell between 1865 and 1873, and the experimental discoveries of Heinrich Hertz in 1887, to be mentioned later.

The corpuscular theory envisaged tiny particles streaming from bright sources, and so explained how light could travel through empty space. In contrast, it was unclear how waves could propagate in this way, since waves normally had to be supported by some medium that permitted bodily deformations. For instance, water waves and vibrating elastic membranes involve displacements of a bounding surface, and sound waves are longitudinal compressions and rarefactions of air. To resolve this dilemma, light waves were envisaged as supported by an invisible and otherwise undetectable 'lumeniferous aether', filling all space but not resisting planetary motion. This aether was supposed to possess elastic, or jelly-like, properties that allowed it to vibrate.

The elastic aether assumption led to a close connection between the mathematical theories of elasticity, fluid mechanics, and optics. For example, Cauchy, Green, Stokes, and Archibald Smith were among those who made contributions in all three fields. But Stokes found it hard to accept that an elastic aether could pass freely around, or perhaps even *through*, solid bodies like the planets. Thus, his 1850 theoretical study of the motion of viscous air close to an oscillating 'ball pendulum' was undertaken, in part, as an analogue of the motion of the earth through the surrounding aether. To explain how the aether flowed around planets, Stokes proposed elsewhere that it must resemble 'glue water' (like wallpaper paste) that could undergo rapid elastic vibrations but also allow the passage of slow-moving solid bodies. Many other aether hypotheses were to occupy the minds of the leading scientists of the time.

Stokes's mathematical analyses gave clear explanations of many optical phenomena: these included supernumerary rainbows, Newton's rings, and related interference phenomena, diffraction, stellar aberration, and fluorescence. His simple experiments, carried out at home, served well to stimulate his theoretical imagination, but they were hardly state-of-the-art. From the 1870s, Stokes corresponded with the experimenter William Crookes on cathode rays and X-rays. As Stokes and Crookes interpreted cathode rays as due to particles, rather than waves, they reignited the wave-versus-particle debate that had seemed decided in favour of waves.

Stokes's work on supernumerary rainbows was a notable contribution to the theory of caustics: in its course, he developed an asymptotic method to locate the zeros of the Airy function, as it is now known. This led him to the discovery of the universal *Stokes phenomenon* in asymptotics, whereby asymptotic representations of smooth functions exhibit discontinuities across so-called 'Stokes rays' in the complex plane.

In 1835 and 1839, James MacCullagh of Dublin had proposed an alternative model of the aether, in which vibrations of light were associated with local *rotations* of particles of aetherial matter (see also Chapter 5). Although this model produced equations that resembled those of Maxwell's later electromagnetic theory, MacCullagh's hypothesis was considered too fanciful to find favour.[26] In 1867, William Thomson adapted MacCullagh's notion, combining it with ideas of Faraday, Hermann Helmholtz, and Rankine. The result was Thomson's ingenious *vortex–aether* hypothesis: a bold attempt at a 'theory of everything'.

In this, the aether was regarded as a fluid full of microscopic vortex motions. Since vortices can sustain oscillations, resist deformation, and even bounce off one other, they mimic elasticity and so in theory could transmit light just like the supposed elastic or 'glue–water' aether. But Thomson (and Rankine before him) went further in suggesting that *matter itself* was composed of such vortices. Each 'atom' of a particular substance was envisaged as a closed vortex with a particular 'knotted' form, with a different kind of knot to represent each substance. The recently observed spectral lines of incandescent matter—such as the double line of sodium—might correspond to the characteristic vibrations of the particular vortex atoms of the material. Thomson even hoped to incorporate electrical and magnetic effects into this theory.

Maxwell was also interested in vortex–aether theories, and had proposed his own elaborate variant during 1861–62. This envisaged arrays of hexagonal vortex cells, separated by thin layers containing 'idle-wheel' particles. The rotation

A postcard from Maxwell to Tait.

within the vortices was supposed to represent the magnetic field, and the movement of 'idle-wheel' particles modelled the flow of electric current.

Thomson's speculations led to a considerable body of work on the supposed vortex atoms and on vortex motions generally. Thomson's original 'vortex atom' paper was devoid of mathematics, but he followed it with several important mathematical papers on vortex motion in fluids, clearly motivated by his grand idea. Though much interested in Thomson's work on vortices, Stokes remained sceptical, rightly believing that such arrays of vortices would be highly unstable. But J. J. Thomson (appointed as Cambridge's Cavendish professor of experimental physics in 1884) was sufficiently intrigued to write a treatise on *The Theory of Vortex Rings* (1883), before his own discovery of the electron in 1897 dealt a final death-blow to the vortex–atom conjecture.[27]

The doomed 'vortex atom' idea had another benefit for mathematics: Tait, in collaboration with an elderly vicar named Thomas Penyngton Kirkman, produced a definitive classification of all knots with up to ten crossings. Their work duly became part of the mathematical theory of topology, then in its infancy.[28]

## Heat and thermodynamics

Early theories of heat, beginning with 'Newton's law of cooling', had some partial success. In Glasgow during the 1760s, their interest in the steam engine led James Watt and Joseph Black to discover the latent heat of steam, and worthwhile experiments on heat and cold were subsequently conducted by John Leslie of Edinburgh and others. A correct mathematical description of heat conduction was first achieved by Joseph Fourier in his *Théorie Analytique de la Chaleur* (1822), a work that described for the first time the powerful method of Fourier series.

Philip Kelland's attempt at a *Theory of Heat* (1837) was rightly criticized by William Thomson. At the time, Kelland was a Fellow of Queens' College, Cambridge, and recently appointed as professor of mathematics in Edinburgh, while Thomson was a precocious 16-year-old just graduated from Glasgow University.[29] Thomson went on to take the mathematical tripos at Cambridge, where his tutor was Hopkins. Thomson's subsequent researches were wide-ranging, with heat a major preoccupation. It was probably through Hopkins that he first came to study the cooling of the earth from a supposed initial molten state. Hopkins, Thomson, and Helmholtz all tried to estimate the age of the earth and the thickness of its solid crust by calculating how it had cooled.

Between 1851 and 1857, Hopkins collaborated with James Joule and the engineer William Fairbairn of Manchester to investigate changes in the melting points and conductivities of substances under large pressures. Thomson took a keen interest in the Joule–Fairbairn–Hopkins work, which showed that the melting point increases with pressure, at least in some cases. By implication, this supported Hopkins's view that very hot rock under sufficient pressure may be solid rather than liquid. In a related study, J. D. Forbes of Edinburgh measured the thermal conductivity of volcanic 'trap rock', finding that this increases with depth and so with pressure.

The idea that the earth was gradually losing its original 'primitive' heat accorded with the occurrence of fossil tropical vegetation in now temperate climes, but it seemed inconsistent with the existence of a previous, and not-too-distant, ice age. But Hopkins argued that such climatic changes were superficial phenomena largely independent of the earth's internal heat, influenced more by changes in the distribution of land and sea, and particularly by ocean currents such as the Gulf Stream of the Atlantic.

During 1854–56, both Thomson and Helmholtz addressed the question of the cooling of the earth. Helmholtz estimated that it would have taken 350 million years for the earth's core to cool from 2000 to 200 degrees Celsius. He concluded that the heating of the earth's surface from below now gave an increase of only one-thirtieth of a degree above that due to heating by the sun alone. Thomson's calculations, based largely on Fourier's theory of heat conduction within *solids*, led to upper and lower estimates of four hundred million and twenty million years for the time since the earth was a uniformly hot molten body. He thought that plants could have existed on the surface only for a few million years.

Having estimated the time it would take for all the sun's energy to be consumed due to depletion by radiation of heat and light, Thomson became convinced that the sun was not simply losing its residual 'primitive heat', and he examined various mechanisms for heat generation. He quickly dismissed frictional heating by internal motions, and chemical heating by burning of its matter, as insufficient to maintain the nearly steady state that had persisted for the past six thousand years. He considered replenishment of the sun's mass by

Philip Kelland (1808–79).

a regular supply of falling meteorites: these would not only provide chemical fuel but, more importantly, would convert kinetic energy into heat on impact. In the early 1850s, he believed that this 'meteoric matter theory' provided the answer, despite sceptical objections by his friend Stokes. Thomson estimated that the sun had illumined the earth only for about 32,000 years, which even then was considered an implausibly short time.

In 1854, Helmholtz realized that a cooling sun, contracting in size, must convert some of its own gravitational potential energy into new heat. As a result, cooling would take place far more slowly than predicted by radiation alone. By 1861, Thomson was convinced that this effect was greater than that of his supposed meteorites, and he adopted Helmholtz's model to increase his estimate of the age of the sun to between 20 and 100 million years, but this estimate was still far too small for geologists to accept. Thomson was not to know that his estimates were invalidated by the energy released by thermonuclear reactions within the sun: the actual age of the earth is now estimated as 4500 million years.

Through his work on the age of the earth and the sun, Thomson was drawn into the controversies raging among scientists and theologians. He met much opposition from geologists, especially Sir Charles Lyell, who supported the Uniformitarian non-progressive theory of the earth: but this was unsustainable if the energy of the solar system was continually being depleted, as both Hopkins and Thomson rightly claimed. Thomson's estimate of the age of the earth was also too short for the Progressionists: it contradicted a growing weight of geological evidence that supported a far greater age, and it was similarly far too short a time for Darwin's proposed biological evolution to have taken place. In retrospect, Thomson's estimates of the age of the earth and sun were taken more seriously than they deserved, because of his high standing as a natural philosopher. His many speculative hypotheses, required to reach even order-of-magnitude estimates, have not withstood later scientific advances.[30]

Of far more lasting importance were Thomson's general ideas on energy conservation and dissipation. The experiments of Joule convinced him that *total energy is conserved* when mechanical energy is lost to heat generated by friction, and also when heat is converted into mechanical energy. Though energy tended to dissipate into less useful forms, it never disappeared. The temperature, density, and pressure of gases were connected by known laws. These were part of the new theory of *thermodynamics*, pioneered in France in the 1820s by Sadi Carnot's studies of steam engines, and taken up by Joule, Thomson, Rankine, and a few others in Britain in the 1840s and 1850s. Carnot had proposed that heat is conserved as a substance known as 'caloric', but this had been disproved by Joule. Versions of the first and second laws of thermodynamics were enunciated by Thomson in a paper of 1851, where he gave due credit to Joule, Carnot, and Rudolf Clausius. It was for this work that the absolute temperature scale, in degrees Kelvin, was later named in honour of Thomson (or Lord Kelvin, as he became in 1892). The new subject of thermodynamics developed alongside revolutionary technological improvements in steam engines and other machinery. As mentioned above, energy was the major theme of Thomson and Tait's two-volume *Treatise on Natural Philosophy* (1867), and Tait's popular *Thermodynamics* (1868) was the first textbook so named.[31]

It was recognized that the temperature of gases is related to the degree of agitation of its particles (in fact, its molecules), and that pressure is the average force per unit area exerted by the gas particles on colliding with a surface. In the 1860s, this idea was taken to new heights by Rudolf Clausius and James Clerk Maxwell, and later by Ludwig Boltzmann: their kinetic theories of gases showed how statistical distributions of randomly moving particles give rise, as averages, to the observable properties of density, temperature, and pressure; and they explain both heat conduction and viscosity in terms of collisions of particles. An even earlier meritorious attempt at a kinetic theory of gases, by the Edinburgh-educated John James Waterston, had been denied publication by the Royal Society in 1845. Waterston had spent time in India and had few scientific contacts in Britain. Such innovatory work by an unknown outsider was met with scepticism and lack of understanding. The rediscovery of his manuscript by Lord Rayleigh, and its belated publication in 1892, gave him posthumous recognition.

Maxwell's kinetic theory of gases started with the assumption that the particles of a gas behave like perfectly elastic spheres. They have velocities that are randomly distributed, and they are free to collide with, and bounce off, each other. Maxwell showed that the gas is in a state of equilibrium, in an averaged sense, when the particle velocities are distributed according to the normal law, now also known as the 'Maxwellian distribution'. The averaged kinetic energy of the particles, per unit volume of space occupied by the gas, is a measure of its temperature. Thermal diffusion takes place when a hotter gas is placed next to a colder one because, *on average*, hotter particles enter the space occupied by the colder gas, and colder particles enter the space of the hotter gas.

Rather similarly, diffusion of average momentum takes place when layers of gas are in mean relative motion, and this transfer of momentum determines the viscosity of the gas. Maxwell's 'elastic sphere' model of the particles predicted that viscosity should increase with the square root of the temperature of the gas, but his experiment of 1865 contradicted this, showing instead that viscosity increases linearly with temperature. He therefore developed another model, postulating that the particles repel each other with a force that varies as some inverse power of the distances between them. On carrying through his analysis, he found that an inverse fifth power of the distance gave the required behaviour. Maxwell's theory was later usefully expounded in H. W. Watson's *A Treatise on the Kinetic Theory of Gases* (1876).[32]

James Clerk Maxwell and his wife Katherine.

## Electricity and magnetism

Following on from the work of Lagrange, Laplace, and Poisson on 'potential theory' applied to electrostatics and magnetostatics, important mathematical advances were made, in unlikely circumstances, by George Green. His long-neglected *Essay on the Application of Mathematical Analysis to the Theories of Electricity and Magnetism* (1828) predates the Victorian era, but it was not until its discovery and republication by William Thomson in 1850–54 that its value was widely recognized. In it are to be found a mature account of magnetostatics and electrostatics, including the now well-known *Green's theorem* connecting surface and volume integrals, and the method of 'Green's functions' for solving differential equations.

During the 1830s, Whewell and others tried to introduce electricity and magnetism into the Cambridge tripos syllabus. But this proved unsuccessful as the questions were routinely ignored, and the subjects were deleted in 1849, just before Maxwell became a student. Related to this attempt, Whewell prepared a 'Report on the recent progress and present condition of the mathematical theories of electricity, magnetism, and heat' for the 1835 meeting of the British Association, and Robert Murphy wrote his *Elementary Principles of the Theories of Electricity, Heat and Molecular Actions* (1833), intended for student use, with Whewell's encouragement. Much later, William Thomson commended Murphy's work, but its direct influence was small. Murphy also published twenty-two research papers on electricity and other topics, characterized by a mathematical elegance at odds with his chaotic and ultimately destructive lifestyle.

A major impetus to the fundamental theoretical advances of William Thomson and Maxwell was provided by the experiments on electricity and magnetism conducted by Michael Faraday at the Royal Institution in London. His 1831 discovery of electromagnetic induction inextricably linked the two fields. Faraday used no mathematics in his descriptions, being opposed to all symbolic representations because of his Sandemanian religious beliefs, but in the hands of Thomson and Maxwell, Faraday's clear verbal accounts were transformed into a coherent, if sometimes ill-organized, group of mathematical theories. Thomson's papers of the 1840s sought to unify electricity and magnetism, and in the 1860s Maxwell built on these to develop his own ideas, culminating in the publication of his *Treatise on Electricity and Magnetism* (1873). Although Maxwell's work displaced the rival Continental theories of Oersted, Ampère, and Wilhelm Weber, Thomson himself was never totally convinced.[33]

Throughout his career, Maxwell used geometrical and physical analogies to sharpen his insight. For instance, his study of Faraday's 'lines of force' in electricity and magnetism exploited similarities with fluid flow, where streamlines took the place of magnetic or electrical lines of force. His analogues and theoretical models of electromagnetic phenomena were not necessarily compatible with each other, though each served its immediate purpose. For this reason, and also his mixing of experimental and theoretical chapters, Maxwell's *Treatise* is difficult and sometimes confusing to read. Although he included clear statements of what are now known as *Maxwell's equations of electromagnetic field theory*, Maxwell did not himself seem to regard these equations as the cornerstone of the subject that they are now recognized to be. Only later was the theory of electromagnetism definitively established, following the experiments of Hertz on electromagnetic waves, and the initially controversial theoretical reformulation by the English outsider Oliver Heaviside.

William Whewell (1794–1866), Master of Trinity College.

Like his contemporaries, Maxwell believed in an all-pervading aether, but from the 1880s the electric and magnetic fields themselves were placed at the heart of the theory. Maxwell's equations now represented the coupled behaviour of electric and magnetic fields, charges, and currents, at last divorced from any supposed aether that had previously been thought necessary to support them. Reworkings of Maxwell's theory by Heaviside and the American Josiah Willard Gibbs gave it its modern form, while at Cambridge a new generation led by J. H. Poynting, Joseph Larmor, and J. J. Thomson vigorously advanced the Maxwellian tradition.

The interplay between theory and experiment was a dominant feature of the researches of both Thomson and Maxwell. One of Maxwell's preoccupations, both before and after moving to Cambridge, was the establishment of standardized units of electricity and magnetism. For the British Association, he and his co-workers determined the standard unit of resistance, the 'ohm', that was subsequently used by all electricians and physicists.

Maxwell's determination of the units of electrostatics (based on the force between charges) and magnetism (based on the force between magnetic poles) led to a fundamental discovery. By his electromagnetic theory, he knew that the ratio of these units had the dimensions of a velocity, and that this quantity was the velocity of propagation of plane electromagnetic waves. But it was only when he discovered that this ratio was nearly the same as the known velocity of light that he realized that light waves themselves were almost certainly electromagnetic, as later confirmed by Hertz.[34]

At the instigation of Tait, Maxwell had given a version of his electromagnetic equations in the compact notation of quaternions: only later did Gibbs and others express these in the now-customary notation of vector calculus. As we saw in Chapter 5, the theory of quaternions had been created by William Rowan Hamilton in 1843. A novel algebraic system, quaternions promised to provide a natural language for much of mathematical physics. Tait was an early enthusiast, devoting several papers and expository treatises to the topic. His far-from-elementary *Elementary Treatise on Quaternions* first appeared in 1867 and was translated into both German and French. A more elementary work, *Introduction to Quaternions* (1873), was a joint publication by Tait and the then elderly Philip Kelland. Tait, the most enthusiastic user and popularizer of quaternions, was also a noted polemicist, vigorously debating with those who doubted their importance. But William Thomson was not a convert, either to quaternions or to the later vector calculus that displaced them.[35]

Electricity and magnetism quickly played a leading role in technology, and the electromagnetic properties of materials were subject to intense experimental study. Electricity was soon employed for lighting, at first as a novelty, but eventually replacing gas as the major source; electric motors were built, and the electric telegraph revolutionized communications.

For a time, the laboratory of the Imperial Engineering College in Tokyo was one of the best. There, William Edward Ayrton and John Perry conducted experiments on electromagnetism and telegraphy to such effect that Maxwell opined that they threatened 'to displace the centre of electrical development...

quite out of Europe and America to a point much nearer to Japan'. Cargill Knott was also active, conducting a major magnetic survey of all Japan with the help of Nagaoka Hantaro and Tanakadate Aikitsu; these two later became leaders of Japanese physics. Ayrton and Perry, and later Tanakadate, all spent some time working in the Glasgow laboratory of William Thomson, and all spoke highly of the experience.

On their return to England, Ayrton and Perry worked at Finsbury Technical College in north London, as professors of applied physics and of mechanical engineering, respectively, continuing their electrical researches and inventing many new electrical devices. Later, from 1896 to 1913, Perry was professor of mathematics and mechanics at the Royal College of Science in London (which, as mentioned in Chapter 3, became part of Imperial College in 1907). He campaigned for the reform of mathematics teaching, arguing against a too-rigid adherence to Euclid's *Elements* in geometry (see Chapter 14), and maintained that non-specialists should instead be taught in a way that emphasized the *usefulness* of mathematics. He had already promoted these ideas in Japan, where he is still remembered as a pioneer in mathematics education, and this approach was successfully followed in his popular textbook *Calculus for Engineers* (1896).[36]

As the construction of iron-reinforced ships rendered traditional magnetic compass needles unreliable, correcting pieces of metal had to be situated around the ship's compass in order to compensate for the attraction of the hull. Archibald Smith, a lawyer and keen amateur sailor, made a major study of this important problem between 1842 and 1862. He published practical tables, formulas, and graphical methods, and in 1862 helped to compile *An Admiralty Manual for Applying the Deviations of the Compass Caused by Iron in a Ship*. Smith had employed methods developed by Poisson and by Fourier, improving on earlier work by Airy which made approximations that were potentially dangerous to shipping. William Thomson corresponded with Smith about this work and, after Smith's untimely death in 1872, Thomson designed an effective ship's compass, manufactured in a Glasgow factory that he jointly owned. From 1883, this compass began to be fitted in ships of the Royal Navy, and by 1890 it had totally displaced the Admiralty's own. For another ten years, it remained the state-of-the-art choice for both warships and commercial vessels.[37]

Thomson's involvement with the Atlantic telegraph cable project made him a household name and earned him a fortune, as well as a knighthood. In one of the most ambitious and useful technological projects of his generation, he brilliantly—and very unusually—combined fundamental physical research with its implementation in a spectacular engineering venture.

An earlier expensive failure had operated for only a few days: though Thomson had provided some instruments for this, his advice often went unheeded. Together with Fleeming Jenkin, an electrical engineer who later became professor of engineering at Edinburgh, Thomson then played a leading role in designing its replacement. They concluded that the previous failed cable had been too thin, that the copper wire and the reinforced gutta percha insulation had been of insufficient quality, and that the cable had been subjected to too large currents.

MAULL & POLYBLANK                    LONDON

Archibald Smith (1813–72).

*The Great Eastern*, the largest ship of its time, was designed by Isambard Kingdom Brunel, built by J. Scott Russell and Co., and launched in 1858.

Thomson devised new sensitive measuring instruments to record very small currents, and he advised on the construction of thicker and better-insulated cables. The only ship large enough to lay the new cable was the *Great Eastern*, designed and built by Brunel and Scott Russell: following its conversion to a cable-layer, the mission was triumphantly accomplished at the second attempt in 1866.[38]

Thomson's other engineering ventures also met with commercial success. He held many patents for electrical instruments, a siphon telegraph recorder, his ship's compass, and a sounding apparatus for measuring depths at sea; most were manufactured in his own factory, Kelvin and James White, Ltd.[39]

## Conclusion

As any area of research reaches maturity, its sophistication naturally increases and the sheer number of published papers poses a daunting challenge for the historian. Thus the above account is necessarily biased towards the first two-thirds of Victoria's reign. While some mention has been made of later work, to attempt a thoroughgoing account of all that was achieved in the later period of Victorian applied mathematics within the constraints of the present volume is next to impossible. When Victoria came to the throne, William Whewell's 1824 Cambridge textbook *An Elementary Treatise on Mechanics* was still popular: this treated mechanics as an illustration of the simpler parts of Newton's *Principia*, carried out in Newtonian fashion. By the end of her reign, new subjects, barely taught in the universities of the 1830s, had become cornerstones of physics and applied mathematics. The advances during the Victorian period created much of what is now regarded as 'classical' applied mathematics and physics.

Perhaps inevitably, the above account has been dominated by the four major figures of Stokes, Thomson, Tait, and Maxwell, whose successes in applying mathematics to the physical world far surpass those of their British, and most foreign, contemporaries. Only William Rowan Hamilton and James MacCullagh in Ireland did work of comparable importance at that time.[40] But, as the 19th century advanced, many others built on the foundations laid by these pioneers. Applied mathematics in Britain developed from small beginnings into a mature group of interrelated subjects, pursued by a thriving research community whose achievements led the world.

Most, though not all, of this scientific community had studied the mathematical tripos in Cambridge. This intensive and competitive training, mainly under the direction of private tutors, did much to define, and was later redefined by, the major research areas of the day. But technical excellence in mathematics is rarely enough. A. E. H. Love observed how it was often emphasized that 'most great advances in Natural Philosophy have been made by men who had first-hand acquaintance with practical needs and experimental methods'.[41]

Only a few years after Victoria's death in 1901, 'modern physics' was born. Ernest Rutherford first proposed his nuclear theory of the atom in 1906, Albert Einstein's special theory of relativity dates from 1905, and Max Planck's theory of quantum mechanics began with his 1901 study of radiation. Although the mathematical study of these new fields came to be labelled 'theoretical physics', they may still be subsumed within 'applied mathematics', if broadly interpreted. These advances extend mathematics to regions where the 'classical' models break down, with very fast speeds and very small length scales. They delimit the range of validity of the Victorian 'classical' theories, but in no way invalidate them in the majority of contexts. Indeed, modern developments of 'classical' subjects, into such areas as aerodynamics, acoustics, and microchip and laser technology, continue to influence our daily lives.

The 10-inch heliometer by Adolf Repsold for the Radcliffe Observatory, Oxford, 1848.

# CHAPTER 9

# Victorian astronomy

*The age of the 'Grand Amateur'*

ALLAN CHAPMAN

Victorian Britain saw a burgeoning of astronomical activity, which was inspired by earlier Georgian achievements. In its social composition, however, British astronomy was very different from that of continental Europe, being run not by paid professionals in universities, but by well-educated self-funded 'Grand Amateurs': wealthy independent men with a passion for science, such as Sir John Herschel, Lord Rosse, and William Lassell. Using their own resources, these people did crucial work in celestial mechanics and nebula studies, pioneered the new sciences of solar- and astrophysics, and developed the big reflecting telescope as an instrument of both optical and engineering precision. And in addition to this cutting-edge research and innovation, astronomy flourished on a popular amateur level, and became part of the public consciousness.

During the sixty-four-year reign of Queen Victoria, astronomy went through many fundamental developments. In 1837 the science was still largely dominated, both technically and intellectually, by Newtonian celestial mechanics, but by 1901 cutting-edge astronomical research had become synonymous with astrophysics. The physical composition of space and the chemical characteristics of the bodies which occupied it had caused astronomy to move along lines of inquiry that could scarcely have been imagined when the young Queen ascended her throne, let alone when she was born eighteen years previously in 1819. Yet running through all of this burgeoning research was a clear mathematical unity, as astrophysicists and chemists found laws applying to the spectrum, to radiant heat, and to thermodynamics, which were every bit as elegant and all-pervading as those that governed the motions of the planets around the sun.[1]

Yet while no scientist in 1837 could have remotely guessed at the level of knowledge about the universe that would have been attained by 1901, we can see, in hindsight, how many of the necessary ingredients were present when Victoria became Queen. For astronomy was not only the oldest of the sciences, with a coherent ancestry going back to classical antiquity; it was also the first *technically based* science. For while physicians, botanists, zoologists, and early chemists, in let us say AD 150, were still at the stage of collecting, collating, and comparing their respective branches of nature, astronomers already possessed the mathematics of the 360° circle, a set of unfalsifiable geometrical axioms, and a series of graduated wooden and bronze instruments with which to make precise angular observations of objects in the sky.

These observations were related to already established constants, such as the stellar constellations encompassing the ecliptic, the celestial equator, and the first point of Aries, and the resulting measurements were expressed in degrees and minutes of arc, which provided the basic data from which mathematical models of the heavens could be built. The rate of the precession of the equinoxes, the prediction of eclipses, and the expression of terrestrial and lunar sizes and proportions in terms of Greek *stadia* were all part of established knowledge by the time of Claudius Ptolemy in AD 150. And as Ptolemy's *Magna Syntaxis* (*Almagest*) made clear, the enduring axioms of astronomy depended on some kind of instrumental measurement as a prelude to mathematical analysis.[2]

This relationship between astronomical technology, computation, and theory was to develop over the centuries, as instrumentation became more and more sophisticated, especially after the invention of the telescope around 1608 and the precision engineering revolution of the 18th century. And by the 1830s, one became aware of glaring disparities in explanatory capability across the range of the sciences. For, when cholera struck England in 1832, physicians still tried to explain the epidemic in classical Greek Hippocratic terms, as being caused by bad airs and foul smells, and for several decades to come they would remain powerless in the face of all manner of infections.

Yet in that same decade, astronomers possessed a routine knowledge of the laws that bound the solar system together, knew by the 1830s that Newton's laws applied to remote binary star systems, measured the distances of three 'fixed' stars as part of a growing knowledge of what might be called the geometrical demography of stellar space, and in 1846 would go on to use data provided by precision instrumental observations to form the foundation for a Newtonian gravitational analysis that would lead to the discovery of the planet Neptune.

## The social composition of the British astronomical community

Astronomy, like the rest of the classical inheritance, had been a part of a common European culture since at least early medieval times. Yet while on the European continent academic astronomy was by 1830 firmly established in great universities and state-funded research institutes, in Great Britain that creative initiative which lay at the heart of new discovery resided in the hands of private individuals: wealthy independent gentlemen who may be referred to collectively as 'Grand Amateurs'. They were indeed self-confessed 'amateurs' in the noble Latinate sense of the word, in so far as they *loved* astronomy (for the word 'amateur' derives from the Latin *amat*, 'he loves'), and while addressing themselves to the great intellectual issues and cutting-edge researches in contemporary astronomy, they were truly 'grand' in their aspiration.

Yet how was it that, in 1840 and down to the early 20th century, research work of a quality that would have been undertaken by university professors or directors of imperial institutes in Paris, Berlin, or St Petersburg, was undertaken by a private gentleman in Slough (Sir John Herschel), a well-beneficed clergyman in Leicestershire (the Revd. Dr William Pearson), a brewer in Liverpool (William Lassell), an aristocrat in central Ireland (Lord Rosse), a retired Royal Navy officer in Bedford (Admiral W. H. Smyth), the heir of a City of London mercery business (Sir William Huggins), and a sanitary engineer in Ealing (Andrew A. Common), along with many other individuals in similar circumstances across the country?

The difference lay not in the science itself, which was of a recognized uniform standard across Great Britain, Europe, and America—as witnessed by the honorary degrees, academic distinctions, and corresponding memberships of the great continental Academies conferred on astronomers of all nationalities—but in different national styles of patronage. For in continental Europe, high culture in general tended to receive liberal state patronage, with universities, opera houses, musical conservatories, fine art academies, and science paid for out of taxation, especially as nations ravaged by Bonaparte's wars tried to rebuild their identities around excellence in high culture. Yet Great Britain, the victor of these wars, prided herself on low taxation, minimalist state action, and letting business and industry generate private entrepreneurial wealth, so that it was up to the burgeoning commercial and industrial classes to take the cultural initiatives, rather than the Crown or the government, and grand civic buildings, art galleries, concert halls, and new schools and colleges sprang up as a result of their munificence, especially in the major industrial cities.

'Grand Amateur' astronomy was but one of the products of that wider national tendency, as scientifically-minded rich men either did the scientific research themselves—such as the lunar observer and Scottish–Manchester iron master James Nasmyth—or else paid professional astronomers to 'assist' them, as was the way with the London wine merchant George Bishop at his Regent's Park observatory. It was, moreover, these 'Grand Amateurs' who had founded and run the Royal Astronomical Society after 1820, who pretty well controlled the Royal Society (at least up to the 1870s), and who paid out of their own pockets to advance new astronomical technologies. And, as we indicate in a later section, spectroscopy, photography, big-mirror telescopes, and large-aperture achromatic refractors, along with new engineering structures for large precision equatorial mounts and observatory domes, all came into being because private individuals were putting up the venture capital to try out new designs, in exactly the way that

George Biddell Airy (1801–92).

these same men risked their money in the opening-up of new coal seams or experimental designs for novel railway locomotives.[3]

In contrast to the 'Grand Amateur' tradition, British public astronomy was poorly provided for. It is true that George Biddell Airy, Astronomer Royal from 1835 to 1881, was a veritable dynamo in his reform and re-equipping of the Royal Observatory in Greenwich, but he saw his work as essentially supplying a useful service to the nation. He indeed provided an electric telegraphic time service across Great Britain after 1852, checked and warranted admiralty chronometers, systematized the correction of navigational compasses in iron ships, recorded meteorological and geomagnetic changes, and much more besides, but he saw all of this as giving good value for money to the taxpayers of a proudly expanding mercantile country, as well as the world's foremost maritime imperial power. And even Airy, thoroughgoing Victorian that he was, considered that it was *not* the Royal Observatory's job to undertake time- and money-consuming 'speculative' researches, such as spectroscopy or cosmology. No, these should be left to the self-funded Grand Amateurs![4]

Indeed, one of several reasons why Airy was not willing to search for the as-yet-undiscovered planet Neptune in 1845 was his view that the Royal Observatory did not exist to assist private individuals in pursuing their own researches, but rather to observe, record, analyse, and publish the positions of *known* bodies to the highest level of accuracy. And while, with hindsight, we may consider this a blinkered view, especially when John Couch Adams of Cambridge had computed a relatively precise position for the unknown planet by the late summer of 1845, we must not forget that it was an approach entirely in keeping with official attitudes to the use of resources provided by the taxpayer.

Between 1837 and 1901, the Royal Observatory's established, or warranted staff, including the Astronomer Royal himself, grew from around eight to about a dozen, although their staple and standard work remained routine meridian astronomy that would be useful for time-keeping and the generation of data for nautical tables. But in addition to this permanent establishment, the Royal Observatory had a long tradition of employing at any one time a dozen or more short-term 'computers', or calculating assistants, to undertake the routine mathematical reduction of astronomical, magnetic, and meteorological observations. For such a system was seen as cheaper than maintaining a large permanent staff of mathematical assistants.[5]

Around ten academic observatories were in operation in Great Britain in Queen Victoria's time: the Radcliffe Observatory in Oxford, Cambridge University Observatory, Calton Hill and the Royal Observatory in Edinburgh (Lord Lindsay's private observatory, bequeathed to the Scottish nation in 1892), a small teaching observatory in Glasgow, the Liverpool Dock Board Observatory at Bidston Hill, the Dunsink Observatory of Trinity College, Dublin, the Archbishop's observatory at Armagh, and the Cork Collegiate Observatory, being like the College itself a benefaction from Horatio Crawford. There were also a small number of publicly accessible semi-civic observatories, such as those at Leeds, Preston, Dundee, Liverpool, and Airdrie, given to their respective towns by local benefactors for local educative or maritime safety purposes.

The Dunsink Observatory, Ireland.

The Armagh Observatory buildings, as seen from the south in 1882.

Yet every single one of these academic foundations, from the grand Dunsink and Oxford to small ones such as the Airdrie Observatory in Scotland, was run entirely on private money and received not a penny from the government. Sir Robert Stawell Ball, for instance, might hold the title 'Royal Astronomer of Ireland' through his Trinity College (Dublin) Andrews Professorship of Astronomy and Directorship of Dunsink, and Charles Piazzi Smyth that of 'Astronomer Royal for Scotland' because of his Regius chair of astronomy at Edinburgh. Yet both men received their livelihoods from their universities—which were themselves independent of government money—and not from state research grants, as their counterparts in France and Germany routinely did.

This mode of operating the academic observatories had two consequences. First, it gave their directors total intellectual freedom as far as their chosen research projects were concerned, and second, it often meant that state-of-the-art research equipment could not be afforded, so that research projects were usually confined to more conservative or routine branches of astronomy, such as measuring binary star motions or the taxonomy of star populations; these were cutting-edge researches in 1830, it is true, but were distinctly routine by 1880. Indeed, this was the very position that Sir Robert Stawell Ball found himself in when he was looking for a research programme that he could undertake upon assuming the Directorship of Dunsink Observatory after 1874. Having a $11\frac{3}{4}$-inch-aperture refractor with a fifty-year-old object glass as his main instrument (albeit re-mounted in the 1860s by Thomas Grubb of Dublin), and no significant photographic or spectroscopic instruments at his disposal, he embarked on a painstaking project to measure the parallaxes of four hundred and nine stars manually.[6]

So whether the patron of an institution was the government or a university, British 'public' astronomy remained conservative throughout the 19th century. For, if one wanted to study the distribution of nebulae in space, establish whether such nebulae were made up of individual stars or of glowing gas, conduct spectroscopic studies into stellar chemistry, use high-resolution photography in conjunction with large-aperture telescopes to ascertain whether nebulous bodies changed over time, commission a large private solar observatory, or travel around the world photographing eclipses to lay the foundations of solar physics, one needed to have the time and resources to pay for the development of the necessary technology oneself—and for this, one had to be a Grand Amateur.

## Royal patronage and British astronomy

It may appear rather idiosyncratic that while British astronomy was almost entirely privately funded, the two premier learned societies to which committed research astronomers belonged both enjoyed royal patronage, at least in the titular sense. Yet both the Royal Astronomical Society (founded in 1820, chartered in 1830) and the Royal Society (1660) were established and operated on the accepted constitutional principle that while the monarch graced a title and bestowed a royal approval upon these (and other) societies, no Crown or government officers played any role whatsoever in the running of their affairs. Beyond drinking the monarch's health in the Loyal Toast at their monthly dinners, the Royal Societies did as they themselves saw fit. Very importantly, they regulated their own membership by self-election, unlike the practice in most European academies, where ministers of state would generally hand out patronage from on high; and it is not for nothing that *all* the British learned societies did not have academicians, but *fellows*. This mode of self-government, indeed, was deliberately modelled upon the self-regulating fellowships of Oxford and Cambridge colleges, and before them on medieval monastic communities.[7]

But this way of doing science was very much in keeping with wider constitutional practice. For it had long been the recognized procedure that the monarch granted *titles* and *graces* to institutions of education, the arts, the sciences, the professions, and the Church, and then held back from any further interference.

Of course, some monarchs had a strong personal interest in the sciences, as was the case with Charles II who founded the Royal Society, with George III, and with Victoria's consort, Prince Albert. King George's scientific, and especially astronomical, enthusiasm was so great that he actively collected scientific instruments, had his own private observatory at Kew, and personally encouraged the work of astronomers such as Sir William Herschel, Dr Stephen Demainbray, and the chronometer inventor John Harrison, whose last chronometer H5 was tested under direct royal supervision at Kew. And when the highly intellectual Prince Albert became consort in 1840, he made it clear from the start that he enjoyed the company of scientists. When Victoria was making a royal visit to Cambridge in 1847, for instance, he told Professor James Challis that he wished to come up to the Observatory after the day's junketings were over, so that he could look at the heavens through the relatively new Great Northumberland refractor; Mrs Sarah Challis left a letter describing the event.[8] Yet Albert often felt thwarted that the British constitution kept royal well-wishers at arm's length and discouraged direct intervention in the actual running of science. This was particularly frustrating for a German prince, who back at home would have been openly solicited as a fount of royal patronage. Perhaps Prince Albert's greatest single achievement on the technical front was the role he played as an inspirer and driving force behind the Great Exhibition of 1851.

Indeed, government involvement in science, beyond that connected with naval and military institutions, was effectively limited to the bestowal of Civil List pensions or knighthoods as rewards for scientific achievement, such as those given to Sir George Airy, Charles Babbage (who turned down a baronetcy), and a handful of others. Sir Robert Peel had been active in his official encouragement of scientists and scholars, and in February 1835 had offered a Civil List pension to the 34-year-old Airy. Airy declined it for himself, yet arranged to have the £300 per annum bestowed upon his wife Richarda, to give her an independent income. Then in December 1835, and now in office as Astronomer Royal, Airy was offered a knighthood, but refused it on the grounds that his income was insufficient to maintain the lifestyle appropriate to that dignity.[9] Airy turned down further offers of a knighthood in 1847 and 1863, before finally accepting the Order of the Bath in 1872.

Yet after the receipt of such favours, scientists were left entirely to themselves to devise, fund, and pursue whatever research projects they deemed worthwhile, for a knighthood or a Civil List pension was seen as a grace granted to acknowledge past achievement, rather than as a contract of employment for the future. Even so, in the 1830s and 1840s some scientists resented any suggestions of a government honour on the grounds that it might indicate that their allegiance had been somehow 'bought' and their intellectual integrity compromised.

# The intellectual concerns of Victorian astronomy

While the Victorian age certainly saw many branches of modern astronomy coming into being, and major technological innovations being made, it is essential, if we want to make sense of science in Queen Victoria's reign, to look at how astronomy was moving before 1837—and in this respect, it is impossible to overestimate the fundamental role played by the Herschel family (Sir William, Miss Caroline, and Sir John) in setting the agenda for what would become the 'new astronomy' of the 19th century. Indeed, their primacy in this respect was fully acknowledged by that pioneer historian of astronomy, Agnes Clerke who, in her books *History of Astronomy during the Nineteenth Century* (1885) and *The Herschels and Modern Astronomy* (1895), saw the Herschels as setting the scene for what would take place in the Victorian age.[10] But before discussing the wider cosmological researches of the Herschels and their intellectual heirs, we must look at how astronomy as a science was perceived across Europe and in America in 1837—which, in a nutshell, meant astrometry and gravitational mechanics.

## Celestial mechanics and the astronomy of angular measurement

At the start of the Victorian era, almost every major observatory, on both sides of the Atlantic, was equipped to do two things: to measure the positions of the stars, sun, moon, and planets in their declination (vertical angles) as they came to the meridian, and to measure the right ascension (east-to-west angles) of the same bodies. Towards this end, a large transit circle would be set up in the plane of the meridian: this would resemble a great cartwheel balanced upon massive trunnions like a cannon, and supported by a pair of massive stone piers with foundations buried deep in the earth, so that the transit could delineate the meridional plane through a possible 180° of arc. The graduations upon the circle's limb might produce a critically accurate value for the position of a star or planet, by means of multiple microscopes reading down to 0.1 seconds of arc or less, and through a process of error analysis and elimination. In this way, an astronomer, whose own 'personal equation' or physical reaction time had already been quantified, could read off the north polar distances or declinations of objects.[11]

A few firms specialized in building these icons of precision heavy engineering: British firms such as Troughton & Simms in London, Thomas Cooke of York, and Thomas and Sir Howard Grubb of Dublin, along with companies such as Merz & Mahler, Reichenbach, Repsold, Eichens, Secrétan, and Gautier on the Continent. The great meridian transit, designed by Airy and built by Ransome & May of Ipswich and William Simms of London, was to define the Greenwich meridian from 1851 onwards. It was in continuous daily use until 1954, after which time the Royal Observatory moved to Herstmonceux, although Airy's 1851 instrument on its original mountings still defines the Greenwich meridian, and

The Greenwich meridian transit.

also the prime meridian of the world, at what is now the Royal Observatory in Greenwich.[12]

The right ascension was also obtained by these same instruments, when used in conjunction with a precision regulator clock. This right ascension angle would be measured simultaneously with the declination angle, so that the exact position of a star or planet could be obtained in two coordinates at the same time as it came to the meridian. To obtain a body's declination, an observer at the eyepiece would record the exact seconds and fractions of seconds of time as it crossed the meridian wire of the transit circle, so that after months or years of such timings the right ascension angles of every object in the sky had been precisely measured as a fractional part of the sidereal or star-rotation day. A simple formula enabled the star time differences to be converted to lateral angular differences, so that when the results were finally published in tables of star positions, the declination angles were given in accordance with the 360° circle, while the right ascension angles were in 'horae' or hours, minutes, seconds, and decimal fractions of a second of time.

Yet, by 1870, why were astronomers in Pulkowa (St Petersburg), Greenwich, Palermo, Madrid, Berlin, Madras, Capetown, Harvard, Washington, and over a dozen other places besides, all making observations of seemingly identical objects, with similar instruments and mathematical reduction techniques? Let us not forget that they were not always observing the same stars, even in the northern hemisphere. Pulkowa, for instance, was better placed to observe the circumpolar stars than Palermo, Madrid, or Madras, whereas these south European observatories could see deeper into the southern skies than could Cambridge or Königsberg. The Royal Observatory at the Cape of Good Hope, which was established in 1820 under the original directorship of the Revd. Fearon Fallows, and which grew greatly under his successor, Sir William Maclear (a former Bedford surgeon and 'Grand Amateur' who took up this prestigious appointment and title 'Royal Astronomer at the Cape' in 1833), was intended to do for the stars of the southern hemisphere what Greenwich and the other observatories were doing for the northern.[13] But quite apart from geographical extremes, observatories often developed their own specialist lines of inquiry, such as ecliptic stars or minor planet positions. Between them, they were building up a rich harvest of primary data first for the northern hemisphere and then for the southern, on which all manner of analyses might be based.

What manner of analyses were based on this minute 'sweeping' of the meridian? The oldest of these concerned stellar cartography, and by 1840 the publication of annual volumes of tables recording the exact coordinates of thousands of stars was routine for many observatories. These coordinates, either in tabular or visual cartographic form, provided the mathematical astronomer with a grid, or series of fixed reference points, against which to observe the positions of the sun, moon, planets, and comets. From these reference points one could determine the exact length of the year, monitor the precession of the equinoxes, quantify planetary and cometary periods, and maintain an exact calendar. And from the 1760s, and extending right through the Victorian era, the Greenwich data had been of such a high quality that the tables and formulas in the *Nautical Almanac*, first constructed by the Astronomer Royal Nevil Maskelyne in 1767, had been able to use the observed orbit of the moon to enable navigators to find their longitude at sea.[14]

As the body of data on right ascension and declination grew, mathematicians found that they could use it as a mine from which to extract all kinds of unexpected facts about the solar system. Since the brilliant 'earth-grazing' comet of 1770, for instance, was found to have had no measurable effect upon the length of the terrestrial day or other aspects of the earth's orbit, this enabled physicists to conclude that the feeble gravitational attraction of comets indicated that they were *not* massive planet-like bodies (as previously believed), but extremely flimsy ones.[15] It had been from a far less extensive and less exact body of such data that Edmond Halley had been able to discover the 'proper motions' of three stars in 1718:[16] such proper motions, the tiny independent movements of stars against each other as they move in three-dimensional space over time, first taught astronomers that the stars were not 'fixed' in a static geometry, but displayed a line-of-sight effect as the earth, sun, and solar system moved amongst them through space. By 1840, several scores of these proper motions had been detected and quantified, and from the ways in which the stars moved, it had been possible to propose a three-dimensional geometrical model for stellar distribution.

Yet why did certain stars display very large proper motions, and others hardly any at all? And why did dim stars, such as the sixth magnitude 'flying' star discovered in Ursa Major by Stephen Groombridge in 1830, show enormous proper motions (in this case, seven arc seconds per year), whereas many bright stars displayed much smaller ones? The brilliant first magnitude Arcturus, for instance, has an annual proper motion of only 2′ 29 arc seconds. Could it mean that not all stars were equally bright, and that visual dimness was not simply a function of distance from the earth? For the bigger the line-of-sight proper motions, the closer to the earth the star now seemed to be.[17]

By 1830, therefore, astronomers had realized that the size of proper motions gave a criterion by which they might select suitable stars for stellar parallax determination. On the other hand, proper motion was not an infallible guide to distance, for if a relatively close star was independently moving in a direction that was radially the same as the earth and sun, then it would display little or no proper motion, and thereby appear very remote.

But in addition to straightforward meridian work, there was another type of mathematical astronomy that had been growing since the pioneering observations of Sir William Herschel in the 1780s: *double star astronomy*. From his high magnification 'sweeps' of the heavens made with powerful reflecting telescopes, Herschel had come to recognize that certain stars seemed to form close pairs, an arc second or so apart. At first he hoped that such stars might provide convenient trigonometrical pointers whereby their parallaxes, and hence distances, could be measured, although by 1802 he had realized that several of them had moved over the previous couple of decades' worth of observation. It seemed therefore that these stars were in some kind of gravitational rotation around each other. Astronomers across Europe realized that if sufficient data for these mutual orbital motions could be accumulated, then these could be used to extract the gravitational masses of the respective stars—and if this could be done, then it would furnish mathematical proof that Newton's laws of gravitation were truly universal and did not just apply to known solar-system bodies. In this way, observational astronomy would provide a stunning physical substantiation for a body of mathematical data, and demonstrate the true unity of scientific knowledge.

By the 1820s, observatories across Europe were engaged in the business of discovering, identifying, and measuring these binary (and sometimes triple) star systems. By the time that the Grand Amateur Admiral William H. Smyth published his authoritative *A Cycle of Celestial Objects* in 1844, no fewer than thirty-four of these star systems were known.[18] Some had already been observed to pass through almost complete cycles of rotation, such as gamma Virginis which was a favourite object of study for Smyth. Meanwhile, Félix Savery in France and Sir John Herschel in England had developed mathematical techniques with which to analyse this growing body of data, and could confidently announce to the world in 1830 that Newton's inverse square law of attraction had indeed been demonstrated to act in remote stellar space, and was clearly universal.

Admiral William Henry Smyth and his wife Annarella.

But generally speaking, these double star data were not best gathered by great meridian instruments. Large-aperture refracting telescopes, set upon meticulously engineered equatorial mounts and guided by clockwork, were better. For, in this way, astronomers could set their telescopes upon the chosen star, whose motion the equatorial mount would track automatically throughout the night, leaving the astronomers free to work unhindered at the eyepiece. There they would use a variety of precision micrometers to measure the positions of the

Part of the table of contents of
*A Cycle of Celestial Objects.*

stellar components in a binary system, both from each other and from adjacent stars in the field. This task, performed for a couple of hundred selected binary systems over the space of twenty years or so, could provide a corpus of exact data upon which the mathematician could set to work in his search for gravitational action in stellar space.

The design and manufacture of the highly specialized lenses, telescopes, and equatorial mounts needed for such work, gave rise to a whole new industry in itself. Fraunhofer in Germany led the way around 1820 with his large-aperture achromatic refractors on their 'German mounts', followed by firms such as Dollond, Merz, Alvan Clark, Cooke, Grubb, Gambey, and Secrétan, to name the leaders in the field. We discuss this more fully later.

## Cosmology, or 'the length, breadth, depth, and profundity of space'

The large-mirrored reflecting telescopes developed by Sir William Herschel in the early 1780s began a transformation in mankind's understanding of deep space, as they listed those misty patches of light traditionally called 'nebulae', along with densely-packed clusters of stars. By the early 1780s, a total of just over one hundred of each group had been listed by Charles Messier. By 1789, this number had risen to two thousand,[19] following Sir William's discoveries, and by 1838, when Sir John Herschel returned from his four-year southern hemisphere sky survey at the Cape of Good Hope, their number had escalated to an incredible five thousand and sixty-three nebulae and five thousand four hundred and forty-nine double stars.[20]

What were these objects? Were they star clusters, such as the spectacular M13 in Hercules, intense gravitational fields that drew more stars into their vast combined mass of light and matter? Was the Milky Way itself disintegrating as a uniform stellar structure, as gravitationally powerful star clusters sucked in more and more of its stars? And then there was the question of the nebulae, or what we now call *galaxies*. Could they be dense clouds of stars that were so remote that individual stellar members of the cloud were not visible, or were they made up of what Sir William Herschel had called 'luminous chevalures', or some kind of glowing tenuosity, perhaps gas? Considering what was known even in 1860, how could a gas be self-glowing in the vacuum of space? Why did stars differ in colour, such as the brilliant blue-white of Vega, or the red of Aldebaran? And were some stars older than others?

Indeed, the large telescopes developed by the Herschel family, with their 18-inch and 48-inch-aperture mirrors, had an unprecedented 'space-penetrating power', or capacity to gather up faint light, to see deeper into space than anyone had seen before. They suggested a universe that was infinitely vast and populated by a strange variety of stellar and nebulous objects, in which gravitational forces brought matter together to form glowing stars, and then dense star clusters which probably broke up, only to re-distribute their matter, perhaps to form the stuff of the nebulae, in a universal process of steady-state recycling.

Generally speaking, this is where deep-space cosmological understanding stood in 1850. Yet this was also astronomy at its grandest and most visionary—indeed, a cosmology for the Romantic Age. It inspired poets, philosophers, and theologians, all the more so because none of it would have been possible without giant telescopes to provide, or to suggest, the data. Then in the late 1830s, William Parsons, Third Earl of Rosse, an intellectual protégé of Sir William Herschel and a friend of his son Sir John, became the next individual to take up the cosmological quest, to see whether he could decide once and for all what the Herschels' nebulae were really made of. And to do this, he used the considerable private resources of his Irish earldom and the proceeds of his marriage into Yorkshire landed and woollen money—combined with his own scientific training obtained at Trinity College, Dublin, and at Oxford and his natural flair as an engineer—to build telescopes bigger than those constructed by the Herschels, so that he might hopefully see even deeper into space than they had.

By the early months of 1845, Lord Rosse was making the first observations with his recently-completed 'Leviathan' telescope, mounted in the grounds of his seat at Birr Castle in central Ireland. It had a mirror of heavy 'speculum' metal with a clear aperture of 6 feet in diameter, set in a factory-chimney-like wooden tube 52 feet long, and operated by heavy industrial-standard machinery. This Leviathan had cost the Earl £12,000 to build, and would remain the world's biggest telescope down to the 20th century.[21] And all of this ingenuity and expenditure—perhaps the supreme single achievement of 'Grand Amateur' astronomy—was aimed at one primary goal: to see whether the nebulae could be resolved into individual stars. For if they could, then it would suggest that they were relatively local, in cosmological terms, and may even be outflanking parts of the Milky Way. But if they could not be so resolved, then several questions arose. Were the nebulae made up of some sort of glowing gas, and still therefore fairly 'local'? Or did their inability to be resolved into stars mean that they were at such unbelievable distances as to be entirely outside the Milky Way: 'island universes', in fact? And if this were the case, could our Milky Way be merely one of thousands or more of such star systems extending for ever into deep space? However, this seemed too incredible to be true, and it is hardly surprising that writers as diverse as Lord Tennyson and Thomas Hardy found horror, awe, and majesty in this vastness.

Lord Rosse's Leviathan of Parsonstown.

Within a few weeks of its coming into operation in the spring of 1845, Lord Rosse had used his Leviathan to detect stars and structure in just over a dozen nebulae. His most spectacular find was what came to be called the 'whirlpool' nebula M51 in Canis Venetica, with its clear spiral structure and what appeared to be a satellite nebula appended to it by some sort of curving filament of glowing material. For very clearly, from its spiral shape, the whirlpool was rotating slowly, over vast periods of cosmological time, begging the inevitable question: 'do Newton's laws act within it just as they do in our own solar system?' Yet if such massive optical technology as Lord Rosse possessed could reveal structure only in a handful of presumably fairly nearby nebulae, then what was going on in those thousands of already known nebulae that were so far away as to appear as the merest specks of light in the great 72-inch mirror?[22]

Indeed, with ideas like these coming into circulation, through newspapers, books, and popular magazines, it is hardly surprising that astronomy came to hold such a grip on the contemporary imagination. John Pringle Nichol, professor of astronomy at Glasgow University, in particular, was highly influential in getting these cosmological ideas out into the wider intellectual market-place, in books such as the successive editions of *The Architecture of the Heavens* (first published in 1843), which carried not only detailed accounts of nebulae and their possible natures and locations, but also high-quality engravings, reproducing those published in academic journals by the Herschels and Lord Rosse. Yet it seemed by 1850 that cosmology had come up against a brick wall, for if Lord Rosse's vast telescope could reveal structural detail in a mere handful of the five thousand and sixty-three known nebulous objects, then what more could be done? Bigger telescopes seemed unimaginable. The way forward, however, came from astronomy's borrowing and co-opting techniques being developed in other branches of science, and finding that they could yield spectacular results when applied to distant cosmological and stellar bodies. For the process of discovery is rarely ever linear, and breakthroughs in research are invariably the products of lateral thinking.

## Solar physics

Nichol's *The Architecture of the Heavens*.

In 1835 Auguste Comte, the French philosopher and founder of 'positivism', had argued that a knowledge of the internal physics and chemistry of astronomical bodies constituted a good example of something that mankind could never know.[23] Yet, by the late 1850s, the identification of incandescing metallic elements by means of the laboratory spectroscope was already under way, and the sciences of solar physics and astrophysics were about to be born.

But solar physics had really begun sixty years earlier, in Observatory House, Slough, when Sir William Herschel was conducting a series of experiments into the passage of solar heat through glasses of different colours. On 11 February 1800, he noticed that not only did the temperature on his thermometer continue to rise as he moved it into the red end of the spectrum, but it rose even more when he moved it into the darkness beyond the red. Was the sun emitting 'black light' that was thermally, but not optically, discernible? This black light we now call 'infrared', and it was the first indication that the sun gives out more than simple light and heat. A year later, the German chemist Johann Wilhelm Ritter was conducting experiments into the darkening effects of sunlight on certain silver salts, when he found that not only did the silver salts rapidly go black in the blue light of the spectrum (though hardly at all in the red), but that the blackening continued in the darkness beyond the blue light. We now call this area of the spectrum 'ultra-violet'.[24]

In 1802, the English physician and experimentalist William Wollaston modified Newton's classic pinhole beam of light and prisms that Newton had demonstrated to the Royal Society in 1672. Wollaston admitted his sunlight through a fine slit, rather than a pinhole, and found, in addition to the Newtonian spectrum, a series of thin black lines that ran horizontally across the colours. Just over a decade later, the Munich optician Josef von Fraunhofer revisited Wollaston's black lines, but with more sophisticated apparatus, attaching his prism and slit before the telescope object-glass of a good theodolite. Fraunhofer saw a spectacular array of lines and mapped five hundred and seventy-six of them in 1814, using the precision scales and micrometer of his theodolite. He grouped and classified them under the letters A to I, extending from visible red to blue light. Then, at letter D, he noticed a very dark and conspicuous pair of lines close together—lines that fifty years later would have great significance for the sciences of solar and stellar physics. We now name these black bands *Fraunhofer lines*, rather than after Wollaston, their original discoverer. And there solar physics rested for another quarter-century.[25]

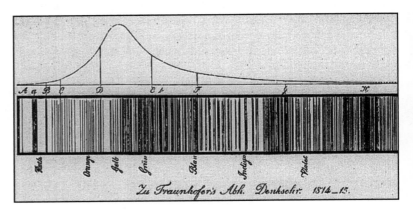

Fraunhofer's 1814–15 spectrum map of the sun.

In 1842, astronomers from across Europe flocked to northern Italy to observe a total eclipse of the sun. The area around Turin was a favoured location, and there at totality they noticed 'pink protuberances' or 'pink flames' appearing at certain places around the silhouette of the moon. There had been reports of these pink tints during earlier eclipses, such as tints that Edmond Halley thought he saw during the total eclipse visible in London in 1715, but generally they had been explained away as optical or instrumental aberrations. But those seen at Turin on 7 July 1842 were observed with excellent instruments, and were too well defined and too well reported to be ignored.

One of the many observers who saw the pink flames in Italy (and Austrian astronomers also saw them as the eclipse track later moved across Vienna) was George Airy, the Astronomer Royal, who had travelled to Turin with his wife Richarda to see the eclipse. In a long and hitherto unrecognized letter sent to his maternal uncle Arthur Biddell, a country gentleman of Playford in Suffolk with an interest in science, Airy speculated about the nature of the 'flames':[26]

But what can these wonderful things be? Are they belchings of flame from the interior of the Sun? Or as most people who have looked well at the spots since Herschel's time, believe that the spots are holes in the sun's skin, can it be that some red light is shining from the sun's interior through some holes in the external atmosphere of the Sun? It is all a puzzle.

What is more, as Airy emphasized to his uncle:

nobody knew it when the eclipse came: and the consequence was that everybody was taken by surprise with these an[d] no two persons have given the same positions.

There was also speculation as to whether the pink colours were even produced by a vestigial lunar atmosphere.

At the next total eclipse that was conveniently accessible, one that swept over southern Scandinavia in 1851, well-equipped astronomers were out in force to look for the 'pink flames', possible solar atmospheric phenomena, and anything else that might be learned during the brief period of totality. One of these was the Liverpool brewer and Grand Amateur astronomer William Lassell, who travelled to Scandinavia with his fine portable refractor by Merz of Munich. Observing by direct vision, as was often the Victorian custom, Lassell nonchalantly mentioned that the heat of the sun prior to the eclipse cracked several of his eyepiece dark glass filters, obliging him to replace them with others.[27]

By the time of the eclipse that swept across Portugal and Spain in 1860, there were so many astronomers, British and foreign, professional and Grand Amateur, who wished to see it that the Admiralty provided *H.M.S. Himalaya* to convey them, under the leadership of George Airy. Included in their company was Warren de la Rue, the paper manufacturer Grand Amateur who, perhaps more than anyone else, came to pioneer photography as a technique for investigating the sun.[28]

Indeed, the 1850s had seen several key advancements in our understanding of the sun. In the late 1840s the Frenchman Hervé Faye had developed a 'cyclonic' theory of sunspots, based on an idea of swirling vortices within the upper regions of the sun, and Sir John Herschel had been impressed with this. For, under very high magnifications, sunspots seemed to resemble dark holes in the solar atmosphere, leading down almost volcano-like to turbulent regions below: these were the 'holes in the sun's skin' referred to by Airy above.[29]

Then, in 1851, the researches of the amateur Heinrich Schwabe were published. Schwabe was an apothecary of Dessau in East Germany, who had made an almost daily visual record of sunspots with a small telescope over several decades, and from recordings made between 1826 and 1843 in particular, it appeared that the solar spots displayed a periodicity: from maximum to minimum and back to maximum in around ten years. What is sometimes called 'Schwabe's law' of periodicity was perhaps the most important new thing about sunspots since Thomas Harriot first discovered them in December 1610, although Galileo's account of his own independent parallel discovery, published in 1613, is generally given priority. Later observers, working on larger cycles of spots, came to readjust Schwabe's periodicity to eleven years, although the 'law' itself still held.[30]

Richard Christopher Carrington was very differently circumstanced from Schwabe, being a Grand Amateur who, in 1853, built the most advanced solar observatory in the world on his property at Redhill, Surrey, on the strength of his inheritance of a brewing fortune. Here he began to monitor the appearances and behaviour of sunspots on a regular and detailed basis, and amongst many other things made several significant discoveries. In particular, he came to the conclusion that the sunspots were not part of the solid body of the sun, but were in its uppermost atmosphere.

Then Carrington noted from his long run of observations that the body of the sun did not rotate at a uniform rate, but that its equator rotated faster than its 'temperate' polar regions. He then came to realize, statistically, that when a new eleven-year sunspot cycle began, the spots tended to appear at the higher solar latitudes, at about 35°, but as the cycle progressed, new spot groups formed closer and closer to the equator, finally forming at about 6°.[31] In short, the sun displayed some of the rotation characteristics of what might be called a semi-fluid body, and at least those parts of it that we could see were not solid in the way that the moon and the earth were.

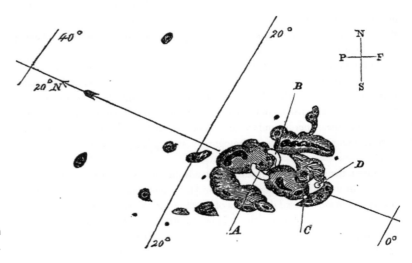

The flare of 1859, as drawn by Richard Carrington.

It was also Carrington who confirmed that, in addition to heat, visible light, infra-red, and ultra-violet, the sun gives out magnetic energy. This came about by one of those chance occurrences that could only have happened once electrical technology had reached a certain level of sophistication and a solar observatory of the kind operated by Carrington was in daily operation. For between 28 August and 4 September 1859, the rapidly growing commercial electric telegraph system that was spreading across Great Britain, Europe, and the east coast of America, was subject to a strange malfunction. Telegraphic apparatus began to chatter and to be activated when not in use, and even when batteries had been disconnected. Magnetic compasses also went haywire, and there were spectacular auroral displays, not only in both hemispheres but even in the tropics. A grand magnetic storm was clearly sweeping the globe, and although such storms were by no means unknown and were an obvious hazard to navigation, their connection with the sun had not previously been demonstrated.

At the height of this invisible storm, at 11.18 a.m. on 1 September, Carrington was studying a group of sunspots in his observatory when he saw a brilliant white elongated shape emerge amongst them, near the middle of the sun. From its fraction of the known solar diameter, he estimated that it travelled 35,000 miles into space in five minutes. Simultaneously observed by another astronomer, Mr Hodgson of Highgate, London, this was what we now call a *facula*, part of a violent magnetic storm raging on the sun, whose energy, when bombarding the earth eight minutes later induced electrical currents in the miles of telegraph wires and related apparatus, not to mention exciting the spectacular auroral displays?[32]

In June 1860, nine months after this magnetic storm, James Nasmyth was observing the solar surface under high magnification at his private observatory at Penshurst, Kent. During periods of perfect atmospheric transparency and still-ness, he made what his friend Sir John Herschel later styled 'a most wonderful discovery'. He saw that, apart from the sunspots, the solar surface was not a bland glowing entity, but was rather made up of countless elongated and interlocking granular particles resembling willow leaves. Nasmyth estimated that each one was around ten thousand miles across, and that 'our world might be dropped into [any one of] them'.[33]

By the late 1850s, therefore, a great deal had been discovered about the physical, optical, and magnetic properties of the sun, although it was to be within a Heidelberg chemistry laboratory that a series of experiments, observations, and inspired guesses would take place which would, within a decade, fundamentally transform mankind's knowledge of the universe, and disprove Comte's 1835 statement that the chemistry of the sun and stars must remain a closed book.

After 1825 or so, Germany came to be recognized as Europe's centre of excellence in chemistry, especially in Giessen, Berlin, and Heidelberg. Robert Wilhelm Bunsen (of bunsen burner fame) was professor of chemistry at Heidel-berg in the late 1850s, and was conducting a series of experiments to find out whether the coloured light emitted by pure metallic substances when they were incandesced at high temperatures could be used as a tool of precision chemical analysis. Bunsen was joined in some of his experiments by his friend, the Heidelberg physics professor Gustav Kirchhoff, and they began to use a prism to disperse the incandesced light as a more exact way of testing a chemical colour. Kirchhoff and Bunsen found that, when an element was incandesced and its light passed through a prism, it produced its own signature of colours. This heralded a wonderful way of analysing chemical substances in the laboratory: sodium, for instance, when incandesced before the early spectroscope prism always produced two brilliant yellow lines that in all respects (other than that of colour itself) resembled the bold black D-lines that Fraunhofer had found in the sun.

Before one could reliably perform such analyses, however, it was essential to prepare pure test substances or reagents in the laboratory, to form benchmarks against which compounds could be identified. And this is why an alliance between a chemist and a physicist was essential for the establishment of spectroscopy.

Next followed one of those examples of brilliant lateral thinking that was destined to enter the canon of 'great moments in science'. Working together one night in their laboratory, Kirchhoff and Bunsen noticed a fire burning brightly in the town of Mannheim, about ten miles from Heidelberg. Focusing its light through the spectroscope, they found clear signatures of barium and strontium in the Mannheim blaze, a fact confirmed by subsequent investigation into what had been burning. Soon after, when the two professors were taking a stroll together, the question arose: if we can detect chemical elements in a fire in

Spectroscopic apparatus.

Mannheim, why can we not detect them in the sun and stars? The real puzzle, however, lay in trying to fathom the relationship between the coloured lines displayed by burning substances in the laboratory and the black 'Fraunhofer' lines seen in sunlight.

Experiments conducted at Heidelberg in the autumn of 1859 offered a solution. For if they simply burned salt in a bunsen burner flame in the optical train of the spectroscope, they saw the two familiar yellow lines of sodium. But if they then used a mirror to direct a focused ray of sunlight through the spectroscope, simultaneously with the burning salt, then the yellow lines instantly became intensely black—and exactly the same happened when they used the brilliant spark of a carbon arc lamp within the laboratory instead of the sun. It seemed that, when the higher heat and energy of the sun or electric arc encountered the lower cooler energy of a bunsen flame, the colours were 'reversed' to blackness.

On 15 November 1859, Bunsen wrote to his former pupil and friend, Henry Roscoe of Manchester, who was also interested in chemical spectroscopic researches, to say that Kirchhoff seemed to have found a physical cause for the 'reversal' of colours to blackness. This came to be known as *Kirchhoff's law*, and built upon his earlier researches into the nature of radiation and energy absorption. For, as Kirchhoff demonstrated, the reversal of colour to blackness took place only when a high-energy source of incandescence encountered a lower-energy one, which in turn posed all manner of questions about the sun's radiant energy and the transition between its hotter and cooler parts. Why did the Fraunhofer lines appear in a straightforward solar spectrum, such as those that Fraunhofer had first mapped in 1814? Was it because the intense heat and energy of the sun's interior was being modified and partly absorbed by the cooler solar atmosphere through which that energy must pass prior to commencing its journey across space, in the same way that the yellow sodium flames in a relatively cool bunsen flame were 'reversed' to blackness when mixed with the higher energy of a carbon arc lamp?[34]

After this initial series of discoveries had been made and their consequences assessed, between 1859 and 1860, the chemical investigation of the sun and stars suddenly became possible, just as Bunsen had suggested to Roscoe. First one needed to burn a succession of pure chemical elements in the laboratory to get their 'continuous' or coloured spectra; then one correlated their exact positions with those of the black Fraunhofer lines on an absorption spectrum. All of a sudden one could identify sodium, nitrogen, iron, and all manner of elements in the sun—a wonderful demonstration of the unified chemical nature of the solar system, giving substantiation to the suggestions of Laplace and others that the earth had formed out of cooling solar ejecta. The application of Kirchhoff's and Bunsen's work to astronomy, and the rapid development of the science of astrophysics, will be dealt with in the next section.

The application of the spectroscope to the sun did produce two spectacular, and unexpected, discoveries within the context of what has already been discussed. On 18 August 1868, many astronomers who had travelled to India observed the predicted total solar eclipse, and the Frenchman Dr Jules Janssen was successful in observing 'pink flames' or solar prominences with a fine spectroscope attached to his refracting telescope, thereby becoming the first astronomer to analyse them spectroscopically. But solar eclipse totality is very brief, inevitably restricting what one can have time to see. Janssen later related that, even while viewing the prominences through his spectroscope slit, it occurred to him that he might perhaps see the prominences *without* an eclipse by simply placing the fine slit of the spectroscope tangential to the solar limb. For the slit and prism train of his high-dispersion spectroscope would have prevented the glare and scattered light of the continuous solar spectrum from obscuring the solar limb, and would have given him a beam of monochromatic light coming tangentially from whichever part of the limb on which he chose to adjust his spectroscope. At 10 a.m. on the day after the eclipse, he tried out the experiment and saw prominences. At a stroke, solar physics was liberated from the bondage of eclipses, for by this simple yet inspired usage of existing equipment one could see prominences whenever one chose.

Surprisingly, Janssen seemed to be in no hurry to communicate this momentous discovery to European scientists, and was unaware of the fact that the same idea had occurred to the English Grand Amateur solar astronomer Joseph Norman Lockyer in London, where no total eclipse was visible. Using a newly-commissioned high-dispersion spectroscope at his private observatory, Lockyer also saw the prominences of an uneclipsed sun. Ironically, when both men got around to communicating their wholly independent yet identical discoveries in the autumn of 1868, their respective notifications arrived at the French Académie des Sciences within minutes of each other. Rising nobly above any priority disputes, the Académie recognized the genius of each man's work, regarded it as a joint discovery, and struck a celebratory medal upon which both men's names and portraits appeared. Lockyer's own medal is now preserved in the Museum collections of the Norman Lockyer Observatory at Sidmouth, Devon.

Joseph Norman Lockyer (1836–1920).

This double discovery was almost one of a tradition of Anglo-French scientific simultaneous 'firsts' in the 19th century, and brings to mind the announcement of the invention of photography by Daguerre and Fox-Talbot in 1839, and the more acrimonious discovery of Neptune by Couch Adams and Leverrier in 1846. But the Janssen–Lockyer 'double' was entirely amicable and the men became good friends.

Janssen's observation of prominences, on 18 August 1868 and then on the uneclipsed sun the next morning, instantly advanced solar physics, intellectually as well as technically. For during the eclipse, Janssen had seen and photographed an enormous prominence leaping an estimated 89,000 miles up from the sun, yet next morning it had almost gone. Individual prominences were clearly very transitory, as later astronomers would confirm, but to discover this fact more time was required than was available over the brief few minutes of eclipse totality.[35]

Seeing prominences without an eclipse in 1868 was one of the great discoveries in solar physics. The next took place at the total eclipse of 22 December 1870, and was observed by Professor Charles Augustus Young of Dartmouth College, USA. Young was carefully observing the narrowing crescent of the sun with his spectroscope in the last few seconds before the dark edge of the moon snuffed it out; as expected, he obtained the black Fraunhofer lines of what we now call an 'absorption spectrum'. Then suddenly, during the couple of seconds between the moon's edge blotting out the solar disc proper and beginning gradually to eat into the solar atmosphere, Young found that his black Fraunhofer spectrum instantaneously blazed into full colour, turning into a continuous spectrum such as one gets when incandescing chemical elements in a bunsen flame in the laboratory.

This brilliantly coloured spectrum lasted no more than a second or two, but it was a physical reality, and was also a wonderful confirmation of Kirchhoff's explanation of why, in the laboratory, a carbon arc spark could darken the continuous spectrum of a substance burning in a bunsen flame. Just above the sun's surface, and yet below the solar chromosphere where the prominences were seen, there was clearly a physical layer that 'reversed' the intense energy flowing from within the depths of the sun, absorbed some of its energy, and sent out into space the monochromatic light of a Fraunhofer line spectrum. This was an absorption spectrum whose light had passed through a 'reversing layer' which could only be seen on the tangent edge of the sun at a time when no direct light from the general solar mass could get through. A physical process, discovered and explained in a Heidelberg laboratory in 1859, had been confirmed as taking place in the sun in 1870.[36]

## The Victorian physics revolution

The rapid development of solar physics and astrophysics after 1850 must be seen within the wider context of major strides taking place in contemporary physics and chemistry. Kirchhoff's researches into the physics of radiant heat and other forms of what was coming to be seen as 'energy' were not unique, brilliant as they were. Earlier researchers, such as Sadi Carnot and Joseph Fourier, had made significant progress, although it was William Thomson (later to become Lord Kelvin) who after 1850 would lay the foundations for an integrated understanding of the physics of heat and energy in his laws of thermodynamics (see Chapter 8).

It was after 1861, moreover, that James Clark Maxwell began to publish those epochal equations that demonstrated that light, magnetism, and electro-magnetism were all physically and mathematically related natural phenomena, while in 1869 Dmitri Mendeleev of St Petersburg announced the elegant mathematical unity by which chemical substances were built up, in what came to be immortalized as the 'Periodic table of the elements', based on a clearer understanding of existing atomic weights. By such discoveries, chemistry and physics were shown to be intimately related and capable of precise mathematical expression, dealing as they did with different aspects or effects of what researchers across the physical sciences were coming to think of as 'energy'.

## Astrophysics

By November 1859, Bunsen, Kirchhoff, Roscoe, and others had recognized that if the spectroscope could reveal the chemical and possibly physical secrets of the sun, then there was no reason why it should not do so for the stars. For by this time, and for the preceding couple of centuries, astronomers had realized that the sun is a star that simply happens to be very close at hand. Yet Bunsen and his colleagues were not astronomers, and in the same way that Carrington, Janssen, Lockyer, and others had developed the science of solar physics, so it was astronomers who first pointed spectroscopes at the stars. And while everyone agreed that the sun was really a mere star, it was nonetheless a conveniently close star and a bright object to work with, whereas even the most brilliant stars of the night sky produced only feeble traces of light when viewed with the spectroscope. Right from the start, therefore, the infant science of astrophysics stretched the existing optical technology to the limit and, within a few years, telescopes of new design and spectroscopes more sensitive to starlight were being built.

Astrophysics came as 'a spring of water to a dry and thirsty land' to William Huggins when he learned of the Heidelberg and related discoveries, following their publication by Kirchhoff in 1861.[37] Huggins had inherited, and sold, the family silk mercery business in the City of London, before 'retiring' at the age of 30 to the then rolling meadows of Tulse Hill in south London. For astronomy, rather than business, was Huggins's real passion, and to this end he had purchased an excellent 8-inch-aperture equatorially-mounted and clock-driven refracting telescope, with optics by Alvan Clark of Boston, USA. Working from the garden behind the spacious house which (as a bachelor) Huggins then shared with his parents, he had done the rounds of the solar system, double-star measurements and such, by the early 1860s. Then he heard of the German spectroscopic discoveries, and felt inspired by the fresh 'spring of water' which they promised to the science of astronomy. For here was a potential programme

William Huggins (1824–1910).

of research that could help to transform mankind's understanding of the universe.

Working with his friend and near neighbour, Professor William Allen Miller of King's College, London, Huggins had built a special high-dispersion spectroscope that he could attach to his 8-inch refractor. There followed, after 1863, a stream of papers by Huggins to the *Philosophical Transactions* of the Royal Society and other journals, which either solved or cast light upon one old cosmological problem after another. For yes, the sun *is* a star, but it was clear from the arrangement of the stellar Fraunhofer lines that the presence or absence of specific elements suggested that the stars had a variety of chemical mixes in their composition, and that while sharing a basic chemistry with the sun they were by no means identical to it. Moreover, the very presence of black Fraunhofer-line spectra indicated that all the stars must also share a basic physics with the sun, as they too must possess 'reversing-layers' in their gaseous atmospheres.

Huggins began, as Bunsen and Kirchhoff had done in Heidelberg a couple of years previously, by preparing a set of laboratory spectra for the principal elements. He prepared twenty-four of them, and in 1864 published an account of his work with the Royal Society.[38] Indeed, 1864 was to be something of an *annus mirabilis* for Huggins and, in many ways, was the true foundation year for astrophysics. For in that year, Huggins produced not only the paper on the spectra of the twenty-four elements, but also (with William Miller) a paper on the spectra of the fixed stars, followed by a supplement 'On the spectra of some of the nebulae'.

The nebulae so far examined, it seemed, revealed Fraunhofer line spectra, indicating stellar composition, and in this way confirmed the idea of Sir John Herschel and Lord Rosse that they were really great star clusters that were so far away that no individual stars could be discerned, only their collective glow. Then, on the night of 29 August 1864, Huggins turned his telescope and spectroscope to the 'cat's eye' nebula in the constellation of Draco the dragon. Here he discovered, instead of the now familiar Fraunhofer lines, no more than a few simple, pale, luminous lines. Clearly this was a gas spectrum, or what Sir William Herschel, from his 48-inch telescope in 1790, would have called a 'luminous chevalure' or glowing hairiness. So nebulae really did contain two types of material after all: stellar and gaseous.

Yet the constituent matter of the cat's eye nebula did not seem to correspond to any known terrestrial element; so, wondering whether it was a material unknown on earth, he provisionally named it 'nebulium'. Later scientists would discover, as spectroscopy advanced, that Huggins's nebulium was no more than ionized nitrogen and oxygen. Indeed, Huggins announced the gaseous composition of no fewer than six nebulae in 1864, and followed them up in 1865 by an analysis of the great nebula in Orion. Huggins also saw how Christiaan Doppler's 1842 discovery of the change of sound pitch with relation to motion could also be applied to waves of starlight, and in 1868 he computed, from the mathematical proportions implicit within its spectral shift, that Sirius was receding from the sun at 29.4 miles per second, although he subsequently amended this figure. Even so, by 1868 it was clear that the spectroscope, when harnessed to a powerful telescope, could reveal hitherto unimagined wonders about deep space.

Huggins's spectroscopic work made him famous. The 8-inch Alvan Clark refractor was, after all, relatively modest for major research purposes in the 1860s, and in 1870, upon receiving the revenues of the Oliveira bequest, administered by the Royal Society, he was able to commission what would turn out to be a prototypical configuration of telescopes from the great Dublin telescope engineer, Sir Howard Grubb. This instrument would mount a 15-inch refractor and an 18-inch Cassegrain reflector on a common equatorial axis, so that stellar spectra and deep-space photographs could both be taken, if necessary, at the same time. Huggins was to use this instrument for the next thirty years.

In a curious way, his fame as a spectroscopist was also to lead the bachelor Huggins to romance. For Margaret Lindsay Murray, the child of a Dublin banking family with a passion for science, had been captivated by Huggins's discoveries since her girlhood, and their scientific correspondence led to marriage in 1875, when she was 27 and he was 51 years old.[39] The marriage rapidly became a scientific partnership, as Margaret proved to be an expert spectroscopist, both in the observatory and in the laboratory, and also a skilled photographer.

Margaret Lindsay Murray (1848–1916).

Margaret also instigated an important new procedure at Tulse Hill. Instead of spending hours, or even successive nights, at the telescope in painstakingly recording the spectra of an individual star or nebula by hand and eye, why not *photograph* the positions of the black Fraunhofer lines on one of the new and relatively fast emulsion plates? And once all the lines had been recorded, they could go on to photograph several more in the same night and then, during daylight hours, could use a special micrometer to measure off the lines from the glass plates at leisure.

In this way, they were able to use precious observing time more productively, especially in the capricious English climate, and to increase greatly both the collection and analysis of spectral data. A similar technique was to be used by the contemporary American Grand Amateur spectroscopist, Dr Henry Draper of New York, who also worked in partnership with his wife, the heiress Anna Mary Palmer, at their observatory residences on Madison Ave in New York City, and in the country at Dobbs Ferry.[40]

The other country, outside Britain and America, where astrophysics took off was Italy. In some ways this was unexpected, for the mid-19th century was a deeply disturbed and unstable period in Italian history that contrasted markedly with the peace and advancing industrial prosperity of England and post-Civil War America. Italy had witnessed violent revolution in 1848, there had been civil war in the south, with French invasion and Austrian occupation in the north and the turbulent events leading to the proclamation of Victor Emmanuel as first King of a united Italy in 1871, not to mention trouble regarding the autonomy of the Papal States around Rome. Yet human ingenuity seems irrepressible.

Father Angelo Secchi was a Jesuit priest who had been driven out of Italy by the 1848 Revolution, but after a time in America and England, returned to take up the directorship of the Pope's own observatory at the Speciola Astronomica del Collegio Romano in the Vatican City. Learning of the Heidelberg discoveries,

Secchi obtained a spectroscope for the observatory, and between 1863 and 1867 made a study of the light of over four thousand stars.

Secchi came to the conclusion that, spectroscopically speaking, these stars could be classified under four basic 'types'; indeed, in spite of obvious modifications and improvements and expansion over time, these four types still lie at the heart of the basic star categories used by modern astronomers and (although this was not fully realized at the time) were to provide the key to understanding star populations. For, as astronomers following on from Secchi were to discover, given features in a star's spectrum supply vital information about its age, temperature, and internal chemistry.

It is apposite to think of the Papal Collegio Romano as having some parallels with an English Grand Amateur observatory, in so far as it was a small private institution within the Roman Catholic Church, and independent of many of the usual management and funding aspects of the big continental university observatories. It is also notable that Secchi announced his 'mature' stellar classification system, including his most recent work on the carbon stars, at the 1868 meeting of the British Association for the Advancement of Science. Staunchly protestant as Victorian England might have been, it was always willing to give safe haven to, and help publish the work of, Roman Catholic scientists.[41]

Yet Father Secchi was not the only Italian spectroscopist. For on 5 August in 1864, that *annus mirabilis* of astrophysics, Giovanni Baptista Donati discovered at the Milan Observatory that, while the sunlight reflected from the tail of a second-magnitude comet exhibited the predictable Fraunhofer lines of the solar spectrum, it also contained some additional lines that indicated that there were as-yet-unidentified gases present in the comet's tail.

Like so many new or novel branches of astronomy, solar physics and astrophysics began life as Grand Amateur subjects, at the hands of enthusiasts like Huggins, Carrington, Lockyer, Draper, and perhaps the Papal astronomer Secchi; such people had quick and easy access to private resources and did not need to persuade government or university committees that the new line of research was worth the investment of time or resources. By the late 1860s, however, and once solar physics and astrophysics were 'up and running' as exciting new lines of scientific research, official public and academic institutions were gradually established to study them.

The first of these institutions was Germany's Potsdam Astrophysical Observatory in 1874, famously directed by Hermann Karl Vogel after 1882, followed by the Meudon Observatory in France (largely at Jules Janssen's instigation). Moreover, reflecting the Grand Amateur origins of American astronomy, an astrophysical observatory was established at Harvard College in 1882 on the strength of a large private bequest. For after Draper's sudden death from pleurisy, his widow Anna Mary used part of their fortune to found the Henry Draper Memorial and Catalogue at Harvard. Here, luminaries of astrophysics such as Edward Pickering, Annie Jump Cannon, and Henrietta Leavitt were later to make their enduring contributions to astronomy.[42]

In Great Britain, however, solar physics and astrophysics remained very much in private hands. Sir William and Lady Margaret Huggins remained firmly in harness at Tulse Hill until their deaths in 1910 and 1916, respectively. Carrington died tragically, after taking a mysterious overdose of chloral hydrate in 1875. Norman Lockyer, it is true, became professorial Director of the Solar Physics Observatory in South Kensington (now Imperial College, London) in 1879. But in 1911, at the age of 74, he left university solar physics and 'reverted to type' as a Grand Amateur, establishing a state-of-the-art private observatory at Sidmouth in Devon, largely on the strength of his second wife's fortune, thereby providing a science career base for his son William. The Norman Lockyer Observatory still thrives, although it now operates largely with the amateur astronomical community and with Exeter University.

Sadly, Britain had no major public or academic solar or astrophysical observatories that could rival those of Potsdam, Meudon, or America; and while spectroscopic researches were undertaken in Greenwich, Oxford, Cambridge, Edinburgh, Armagh, and elsewhere, they could not really compare with those conducted overseas, especially those performed under the pristine skies of the American desert and mountain observatories, which continued to enjoy a peace that vanished from mainland Europe after August 1914.

## Big telescope technologies

It had been Josef Fraunhofer in Munich who revolutionized the science of achromatic lens making (in the wake of John Dollond's original invention of 1758) and first mapped the dark lines in the sun. He also produced the enduring prototypical mounting for large achromatic refracting telescopes: this was the famous 'German mount', which was kinematically elegant, beautifully engineered, stable in its motions, and capable of being driven around the sky by a precision clockwork motor in order to track a star.

In the wake of Fraunhofer, European master opticians, such as Georg Merz, Robert Aglaé Cauchoix, and Thomas Cooke of York, had developed a succession of ever larger object glasses, each of which posed new problems in the engineering of bigger stable mounts. By the 1840s, industrial revolution heavy engineering and precision optics had joined hands. Thomas Grubb, a Dublin manufacturer of iron-bed billiard tables who also possessed a flair for engineering design, became the first British person to construct a precision mounting for large object glasses. In 1831, Edward Joshua Cooper, a wealthy Irish astronomer, had purchased a magnificent 13.3-inch-diameter Cauchoix object glass of 25-feet focal length, and asked Grubb to mount it in the equatorial plane at his observatory home at Markree Castle, County Sligo. The design was a brilliant success, and before long Grubb was abandoning billiard tables to mount, and then manufacture, fine-quality large telescope object glasses. His son, knighted as Sir Howard Grubb, carried on the business, and for around one hundred years the firm of Grubb of Dublin was an international benchmark of excellence in the making and mounting of fine telescope optics.[43]

At around the same time the former York schoolmaster, Thomas Cooke, was discovering his own largely untutored genius for the making of fine object glasses

Howard Grubb (1844–1931).

and mounting them on beautifully engineered stands. Indeed, by 1860 the two British 'amateurs', Grubb and Cooke, who had taken up telescope-making by chance, now headed businesses of the highest world repute, rivalling the more professionally based continental firms of Merz of Munich and Marc, August, and Georges Secrétan of Paris.

Curiously enough, telescope manufacturing in America developed along very similar lines to that of England, as Alvan Clark, a successful portrait painter of Boston, Massachusetts, became the founder of an American optical dynasty on the strength of early experiments in lens making. By the 1860s he was also competing with the English, French, and Germans in the already global market of big telescope engineers.[44]

Yet whether, by 1870, one chose to commission a 24-inch-diameter object glass from Merz, Cooke, Grubb, Clark, or some other company, and to set it on a stand that combined the heavy engineering of a railway locomotive with the precision of a marine chronometer, one was dealing only with *refracting* telescopes. And while refractors were ideal for spectroscopy and planetary photography, the *reflecting* telescope had also been transformed, optically and mechanically, between 1820 and 1870. For although the mirror telescopes, with which the Herschel family had conducted their ground-breaking cosmological researches between 1780 and 1840, had possessed excellent optics for the period, their mountings had been fairly rudimentary. A 4-foot-diameter slab of 'speculum' metal, weighing over a ton and carrying a good optical figure and polish, had been set at the bottom of an iron tube that was manhandled around the sky by means of blocks, tackles, capstans, and ropes in pretty much the same way as *H.M.S. Victory* would have brought her great sails into the wind—by a combination of coordinated brute force, conjoined with the delicate final touch of the observer–designer.[45]

But by the 1840s, a greater all-round precision was being demanded, and the new generation of large-aperture reflecting telescopes of William Lassell, James Nasmyth, and Lord Rosse were using industrial revolution precision technology, both for the manufacture of their optics and for mounting them. All three of these Grand Amateur pioneers of the post-Herschelian big reflecting telescope, for instance, used steam-driven machines to figure and polish their big mirrors.

And it was not for nothing that Nasmyth, whose great engineering factory at Patricroft in north Manchester made him a fortune from the manufacture of precision giant steam hammers and railway and marine steam engines, should build what was to be a prototypical large mirror and lens polishing machine. Nasmyth's machines produced the exquisite speculum optics not just for his own 20-inch mirrors, but for even larger mirrors for his friends. With William Lassell he cast, figured, and polished 24-inch and 48-inch mirrors which in their day were the finest in the world. He also undertook mirror-making for Warren de la Rue. Yet exactly who *designed* this mirror-making machine is unclear, for the man who wrote the detailed published account of it, with the implication that it was his own design, was Nasmyth's good friend, William Lassell. But there is no evidence that Nasmyth made any attempt to claim *design* priority.[46]

Lord Rosse, who knew Nasmyth and Lassell well, used the latest metal-casting techniques and a steam-driven machine of his own devising to produce a pair of 72-inch mirrors for his giant 'Leviathan of Parsonstown' telescope in 1845. Yet Lord Rosse's telescope was mounted in the altazimuth, just like Herschel's, and was capable only of a full vertical movement from pole star to southern horizon, combined with a lateral movement of just 15 arc degrees across the meridian. And although his giant telescope was the first to resolve a handful of spiral nebulae in 1845, to use it he had to wait patiently until the object he wished to observe came around to the meridian. The great 52-foot wooden tube was slung between a pair of stone walls and moved horizontally and laterally by a system of cables and cast-iron winches.[47]

But it was Lassell, the Liverpool brewer, and his friend Nasmyth, the Manchester engine manufacturer, who first successfully addressed themselves to the mounting of large-aperture speculum metal mirrors in the equatorial plane. In 1845, at his private 'Starfields' observatory near Liverpool, Lassell brought into operation a telescope of revolutionary construction. It had been designed and built by himself with unspecified contributions from Nasmyth, and had a steam-figured speculum mirror of 24-inches diameter which gave the superlative planetary images that Lassell sought. But it was positioned in a cast-iron tube, 20 feet long, and was pivoted in a great cast-iron fork mount capable of tracking objects in the equatorial plane. There was no wood, no rope, and no winches: only exquisitely exact cast-iron parts that were so finely balanced as to enable the whole mass to respond to a lever mechanism. What is more, its mirror was supported by an astatic array of pressure plates so that supports on the back of the mirror came into, and went out of, action automatically as the position of the tube varied, thereby preventing any hairsbreadth sagging that could cause image distortion of very fine detail when seen at the eyepiece. And while Lassell had not invented the astatic mirror support—Thomas Grubb and Lord Rosse had already experimented with, and used, balanced mirror supports—Lassell's telescope greatly improved on earlier support systems.[48]

Lassell's 24-inch-aperture telescope, based on his earlier 9-inch equatorial reflector, was a milestone in telescope engineering. In Liverpool, on 10 October 1846 (only three weeks after the Berlin discovery of Neptune itself), it enabled Lassell to discover Triton, the satellite rotating around the newly-discovered Neptune, as well as showing new detail on Saturn. Then, between 1852 and 1853, Lassell took it to Malta—an engineering feat in its own right—where, from the latitude of 35° north, he could see in the skies what no big telescope had ever revealed before. At last, the reflecting telescope had become as exact and as versatile to use as a Fraunhofer refractor, but with the added advantage of a much bigger optical surface, for in the 1840s it was possible to make mirrors that were three or four times bigger than one could make lenses. And what these 'light buckets' did was to enable an astronomer to see ever deeper into space.

When we consider the ease with which Lassell's 24-inch reflector seems to have picked up Neptune during the night of 2–3 October 1846 (from details in *The Times* of its original discovery in Berlin on 23 September), and the ease with which he discovered its satellite Triton a week later, we might wonder why Lassell was never approached about searching for Neptune in the first place. Indeed, with such formidable optical power at his disposal and (unlike Lord Rosse) being able to direct it to any part of the sky within minutes, Lassell would not have needed Challis in Cambridge or Heinrich D'Arrest in Berlin to conduct a careful astrometric survey in the hope of tracking down a slowly moving object within a star field. For the 24-inch reflector made Neptune stand out as a distinct disc against the starry background—so distinct, in fact, that it was also seen with its satellite Triton.

In the autumn of 1846, moreover, Lassell was widely regarded as a pillar of the British astronomical establishment—an already distinguished Fellow of the Royal Astronomical Society, and soon-to-be-elected Fellow of the Royal Society. Yet, for some reason that lacks any logical explanation, the British search for Neptune was confined entirely to the Cambridge University Observatory, with encouragement but no instrumental help from Greenwich. None of the superbly equipped and leisurely Grand Amateurs was asked to join in, or was even properly informed that a search for a new planet was under way; but as soon as Lassell was alerted, his great 24-inch reflector showed at a glance what a big reflecting telescope could do.

Then, having seen all that he could with his 24-inch telescope, Lassell set about building another one, twice the size. His new 48-inch mirror and 57-foot tube fork mount was the size of a railway locomotive, with a base circle the size of a large ship's paddle wheel. When it was set up, originally just outside Liverpool in 1859, it became an object of wonder. Yet because the local Merseyside skies did not enable its optics to be used to optimum effect, he dismantled it and shipped it out to his old observing site in Malta, where he worked with it between 1861 and 1865.

All of this innovation, and the planetary discoveries that flowed from it, stood four-square upon Lassell's brewing fortune. But the 48-inch telescope, with its revolutionary use of a skeleton tube to facilitate ventilation and hence prevent heat distortion of the optics, was to inspire the design of the government-funded 'Great Melbourne' Australian telescope of 1867 which replicated many of the details of Lassell's instrument. Sadly, this first large telescope to be set up in the southern hemisphere was not a success, due to a number of factors that had nothing to do with Lassell's prototype.[49]

Yet the transformation of the big reflecting telescope into a reliable astronomical research tool required something besides a revolution in engineering design: it also demanded a parallel revolution in optical manufacture. For all reflecting telescopes, from Newton's original experimental model of 1671–72 down to the late 1850s, had used speculum metal mirrors. But speculum (a complex alloy of tin and copper) cracked easily, was dense and heavy, and even when exquisitely polished often tarnished over a matter of weeks or months, demanding regular and time-wasting refiguring. As everyone realized, glass would be a much better base material for mirrors, being much lighter (inch for inch) than speculum, easier to figure, and less thermally troublesome.

The only problem lay in finding a technique for depositing a thin, uniform coat of silver on to the optical surface of the glass. Although silvered glass artefacts were displayed at the Great Exhibition of 1851, it had been the German chemist, Justus von Liebig, who first developed a viable process whereby a chemical reduction of silver nitrate could be used to bond a thin coat of metallic silver to a glass surface, and in 1856 Carl August von Steinheil of Munich and Léon Foucault in Paris produced the first surface-silvered mirrors for reflecting telescopes. In 1864 a telescope with an 80-cm mirror by Foucault was being used at the Marseilles Observatory.

Silver-on-glass mirrors had a profound impact upon the further development of the reflecting telescope. For not only was glass much lighter than speculum metal, which in turn presented fewer engineering problems when balancing big mirrors in equatorial mounts, but it also had a much higher reflectivity than speculum. Good-quality silvered-glass mirrors reflect almost 96 per cent of the light that hits them, whereas even freshly polished speculum rarely reflected more than 70 per cent. Consequently, there was a 25 per cent reflectivity gain for any given mirror aperture, while some early experimenters believed the reflectivity gain to be as much as 50 per cent. Silver on glass, moreover, could keep its brilliance for years under good conditions, whereas all the great speculum astronomers, the Herschels, Lord Rosse, and others, always built a couple of identical mirrors for their telescopes (although Lassell and Nasmyth developed purer and less easily tarnished specula): one mirror was currently in use in the tube while the other was being re-polished, so that as little observing time as possible would be lost when a switch due to tarnishing was necessary.[50]

Although a Franco-German invention, it was really in Great Britain, and then in America, that silver-on-glass reflecting telescopes took off. Glass mirrors, moreover, were much better than speculum for the burgeoning new technology of astronomical photography, although both de la Rue and Huggins continued to use speculum.

By the late 1870s, when Andrew Ainslie Common in Ealing (west London) and Henry Draper in New York were in friendly neck-and-neck rivalry to photograph nebulous and other objects from their private observatories, it was taken for granted that silvered glass formed the reflecting surfaces for their large privately designed and built reflectors. For although 3 feet and 1 foot smaller (respectively) in aperture than Lord Rosse's 1845 telescope, Common's 36-inch and 60-inch silvered reflectors were much more optically sensitive than Rosse's, and, on their finely engineered equatorial mounts, were specifically designed as giant cameras. Draper likewise worked with splendid $15\frac{1}{2}$-inch and 28-inch telescopes of his own construction.

Common, who was a civil engineer by profession, not only conceived and built what was *de facto* the biggest telescope in the world in what was then a rural London suburb, but engineered it with a sophisticated easily correcting clock-drive mechanism that made long exposures possible. For the revolutionary potential of photography, where photographic plates were made to work in harmony with precision optics and excellent engineering, made it possible to make out things in space that the human eye at the eyepiece could never see: a photographic plate can store photons of light and build up discernible traces of nebulous structure or dim stars in astronomical bodies which, to a living eye using the same optics, would appear as no more than a fuzzy trace at most.[51]

By the 1880s, therefore, cutting-edge research astronomy was less about visual observing than about photography. For, as everyone agreed, was it not by means of the giant 'astronomical camera' telescope that one was going to see and record structure in the nebulae, and attempt to discern changes in deep-space objects over time (as later became invaluable with the 20th-century studies of the 'supernova remnant' crab nebula), as well as to record spectra? None of this would have been feasible without a revolution in precision heavy engineering, both optical and mechanical, much of which was pioneered in Britain, both by Grand Amateurs such as Lassell and Common and by the new firms of optical engineers like Grubb and Cooke.

## Popular astronomy in Victorian Britain

By the 1850s, in spite of the appalling slum conditions in which many people lived and worked, Great Britain had the biggest and richest middle class in the world. And quite apart from the wealthy men who pioneered Grand Amateur research, there were tens of thousands who had varying degrees of leisure and modest incomes with sufficient spare cash to spend on subjects that interested them. In their ranks were the better-off office clerks, teachers in the new 'Board' and provincial grammar schools, parish clergy, local solicitors, general practitioner doctors, shopkeepers, skilled craftsmen, and so on. These people were also avid readers of books, magazines, and newspapers, attended one-shilling-a-head lectures on a variety of subjects, went to concerts, were involved with churches, chapels, and Sunday School teaching, took holidays in Eastbourne or Southport, wore good clothes, and filled their houses with knick-knacks—and many of them were fascinated by contemporary scientific discoveries.

In 1859 the Revd. Thomas William Webb, M.A., Vicar of Harwicke (Herefordshire) and a man with almost forty years' experience as a keen amateur astronomer, wrote what would become an enduring scientific bestseller—Webb's *Celestial Objects for Common Telescopes* led the aspiring amateur through the whole of practical astronomy in one volume. How should a person with £25 to spend go about choosing a 'common telescope' of 3 or 4 inches aperture? And how might one spot a second-hand bargain? The Revd. Mr Webb took his reader on a tour of the solar system, supplied a detailed map of the moon, and discussed the latest ideas about the stars and nebulae. And Thomas Webb knew his stuff: he was an Oxford graduate and a Fellow of the Royal Astronomical Society, as well as a devout hard-working country clergyman whose love of God and love of science worked effortlessly together and never ceased to inspire him. They also inspired countless others, and we should be cautious about reading too much into the so-called 'Victorian crisis of faith' that supposedly followed the publication of Darwin's *Origin of Species* in 1859, for time and time again Victorian astronomical enthusiasts drew attention to the religious inspiration that they found in their hobby.[52]

Thomas William Webb (1807–85).

Moreover, as if by a providential synchronicity, Webb's book coincided with the first British popular astronomical magazine, the *Astronomical Register* (1863), and with Captain William Noble's regular astronomical section in *The English Mechanic* magazine after 1865; all of these came on the scene when the silver-on-glass reflecting telescope had just become a viable technology. England by the 1860s (as was abundantly substantiated in the weekly pages of *The English Mechanic*) was becoming a land of 'do-it-yourself' addicts and technological hobbyists.

For, while it was enormously troublesome and potentially dangerous for people in modest circumstances to build the metallurgical foundry that was the essential preliminary to making speculum mirrors, the construction of glass-mirror telescopes could be easily within their reach. They could order a ready-cast glass blank from Chance Brothers of Birmingham, Pilkington's of St. Helens, or some other glass manufacturer, and receive it through the post—and then, with no more than commercially available abrasives, home-made tools, and hours of patience, they could figure and polish it into a good parabolic curve. Next, with easily available nitric acid and silver nitrate (albeit with a risk to their skin and lungs) aspiring enthusiasts could give the glass blank a lovely silver coating. All that then remained was to exercise their ingenuity in devising a tube and stand, and after buying a couple of eyepieces, view the heavens! *The English Mechanic* articles especially were full of advice about what to do, and letters from novice astronomers invariably produced replies and helpful tips from more experienced hands. Amateur astronomy was taking off as a serious hobby across middle England.

Everywhere one looked, there were clergymen: Anglicans, Methodists, and some Roman Catholic priests. The Revd. Edward Lyon Berthon, Rector of Romsey Abbey (Hampshire), built a series of spectacular large reflectors for his own observing purposes, and published a set of detailed plans and instructions in the 13 October 1871 issue of *The English Mechanic* whereby amateurs could build, or have built for them by a local carpenter, an elegant 'telescope house' or observatory that would keep their instrument safe and dry and act as a charming adornment to their gardens!

By the 1880s, if you lacked the manual dexterity or the inclination to build your own telescope, there were several manufacturers who could supply you by post with everything you needed, from an eyepiece sun filter costing a couple of shillings to an 'off the peg' full-scale high-quality observatory, complete with equatorial refractor, all the accessories, and dome, for several hundred pounds or more. Charles Grover, a former brush-maker who rose in the world through serious amateur astronomy, worked in the 1870s for Messrs Spencer, Browning, and Co., optical manufacturers in London. His diary records how much he enjoyed advising clients in the shop, and the pleasure he experienced in travelling around Britain to visit the homes of his firm's richer clients, 'running in' their new observatories for them and giving instruction as to how they might get maximum satisfaction from their purchases.[53]

By the 1880s, indeed, British astronomy was not only a major research science, but had also become a popular hobby, a do-it-yourself industry, and a lucrative

Robert Stawell Ball (1840–1913).

business. Furthermore, it was being fed by a publishing industry and generating 'celebrity astronomers', such as the earnest Richard Anthony Proctor and the genial Irishman Sir Robert Ball, who in addition to his Dublin professorship lectured to paying audiences in town halls and mechanics' institutes across the whole of Great Britain, and across the English-speaking world as well. Both Ball and Proctor lectured in America and Canada, while Proctor's lecture tours even extended across Australia and New Zealand. Both men, moreover, were gifted writers, and produced a succession of astronomical bestsellers, such as Proctor's *Other Worlds than Ours* (1870) and Ball's *The Story of the Heavens* (1886), which enhanced their wider fame and earnings. But Proctor and Ball were only the most outstanding of a small army of popular writers and lecturers on astronomy in Victorian Britain.

While much has been said about the comfortable middle classes and their astronomical passion, we must not forget that this passion embraced the working classes as well, as they too became enthusiasts, telescope makers, and attenders of lectures. Roger Langdon was a self-educated employee of the Great Western Railway in rural Devon, who built an excellent $8\frac{1}{4}$-inch silver-on-glass reflector in the late 1870s, gave informal teaching to interested local people, and was even invited by Webb to give a paper to the Royal Astronomical Society on his Venus observations. John Glass was a Highland Scottish miller, who built and used fine reflecting telescopes, while John Leech, a Frodsham (Cheshire) shoemaker–astronomer, even had a correspondence with Sir George Airy, the Astronomer Royal, and received a modest government gift of instruments in the hope that he might teach his fellow-artisans.[54] There was also the self-taught Welsh linguist–theologian–astronomer John Jones, who built a substantial silver-on-glass reflecting telescope from bits and pieces, and actually figured the thick dense glass from a discarded ship's porthole into prisms in order to make himself a spectroscope. When Jones was not reading his Greek Bible or observing the heavens, he earned a modest wage helping to pack roofing slates into ships on Bangor docks.[55]

The Victorians, further, had a passion for organizing things, and it was they who created the social prototype for that network of amateur astronomical societies which covers present-day Britain and Ireland, the USA, Japan, Australasia, and many European countries. Although academic or research science, even when self-funded, had operated through learned societies stretching back to the Royal Society in 1660 and the Royal Astronomical Society in 1820, it was Leeds in Yorkshire that became the first city to found a society explicitly for enthusiastic amateur astronomers. The Leeds Astronomical Society, founded in 1859, even acquired a good brass refracting telescope that was mounted in a corrugated iron observatory, and was a very lively organization, with lectures and observing sessions in the early 1860s. Those patriarchs of British astronomy, Sir George Airy (who became its first President) and Sir John Herschel, headed a list of distinguished patrons, although Herschel, residing in rural Kent and having indifferent health, never attended any of its meetings.

The ball really got rolling for amateur astronomical societies after 1881, with the founding of the Liverpool Astronomical Society. Indeed, things now suddenly took off, as people living all across Britain and well beyond north-west England began to take out membership of the Liverpool Society. An affiliated branch of the Society was established in the Isle of Man, and there was even one in Pernambuco, South America, which had an expatriate British community. The Liverpool Astronomical Society founded specialized research and observing branches to cater for membership interests, including a solar section (directed by Miss Elizabeth Brown, a wealthy amateur of Cirencester) and sections devoted to spectroscopy and astronomical photography. It also started a journal, from which we can learn a great deal about the social demography, instrumental capacity, and intellectual interests of serious amateur astronomers in Britain and overseas in the 1880s. In fact, we might argue that British amateur astronomy had reached a 'critical mass' by the late 1880s, and had perhaps extended beyond the resources of a local society, for it was Elizabeth Brown who, through the columns of the universally read *English Mechanic*, began to urge the setting up of a national amateur astronomical society based in London—an all-too-familiar pattern, where a provincial success feels obliged to relocate to London.

The British Astronomical Association was indeed established in London in October 1890. Miss Brown now became its solar director, and was one of four women on the Society's governing council. This association has remained the central organizing force in serious British amateur astronomy since 1890, although in the latter half of the 20th century other societies, dealing with children's, elementary, and 'popular' interest needs, also came into being.

But late Victorian and early Edwardian Britain also saw a boom in the establishment of other regional astronomical societies. These included Manchester as a B.A.A. branch in 1892 and as an independent Society in 1903, the re-founded Leeds Astronomical Society in 1892 (following an inspiring public lecture by Sir Robert Ball), the Cambrian Natural History Society's astronomical branch in Cardiff in 1894, the Ulster Astronomical Society in Belfast in the early 1890s, and Newcastle in 1904. Many of these societies came to attract the scientifically 'great and good' of their localities and to run social events and soirées of a scientific character, and they did much to popularize scientific knowledge generally. Significantly, they opened their membership to women on equal terms with men, with many societies having a 10–15 per cent female membership (from surviving membership lists), with women sitting on their governing councils as in the B.A.A.

Nor should one forget that, by the 1870s, astronomy was even beginning to creep into schools, or at least into some of the better-resourced fee-paying public ones. We know, for instance, that the Revd. Frederic Hall, M.A., was providing extra-curricular astronomy teaching and telescope observation sessions at Haileybury School by 1874 (when he was telling his class, including the future clergyman astronomer Thomas Espin, about the current Coggia's Comet), that Henry Maden was similarly teaching the science at Eton (where in 1878 he proposed the names Phobos and Deimos for the recently discovered satellites of Mars), and that Marlborough, Rugby, and the Jesuit school Stonyhurst in Lancashire (to name but a few) had well-equipped astronomical observatories on their premises. It is all the more sad, therefore, that over a century later, most children leave school without any properly taught knowledge of astronomy.[56]

# Conclusion

Between 1837 and 1901, astronomy advanced almost beyond recognition, moving away from the traditional concern with gravitational dynamics of the 1830s and on to solar physics and astrophysics, with a fundamental revolution in its instrumentation and technical capacity. Yet what ran through it all was an unchanged capacity for mathematical expression, for the laws of gravitation which governed the behaviour of the solar system and binary stars had immediate mathematical cognates in the spectroscopic discoveries of Kirchhoff and Huggins, and these in turn related to Clark Maxwell's equations of electrical and optical energy, Lord Kelvin's thermodynamics, and Mendeleev's periodic table of the chemical elements. Everything in the cosmos, it seemed, could be understood and expressed by means of a vast web of mathematical linkages.

And while all of this cutting-edge science was perhaps fully accessible only to a relatively small number of people who possessed the requisite physics and mathematical skills, the whole of astronomy had so captivated the populace at large as to generate a growing army of enthusiastic practical observers, telescope-makers, book and magazine readers, and popular lecture attenders, which extended from civic dignitaries at one end of the scale to shoemakers and slate packers at the other.

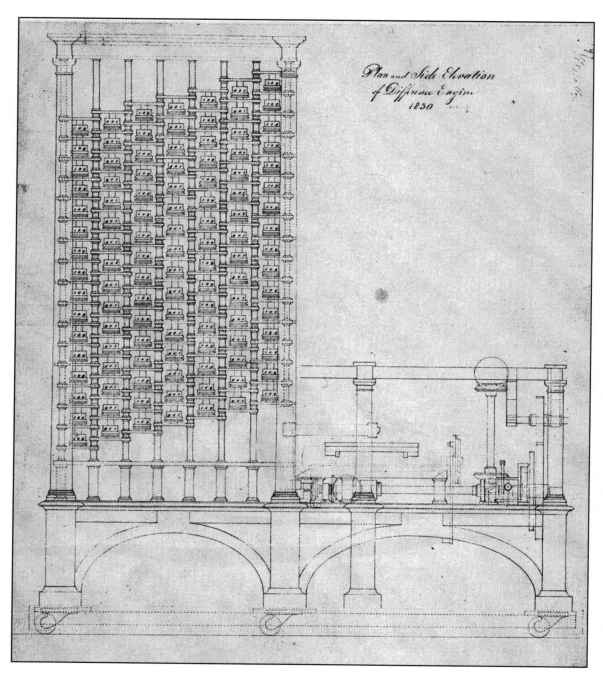

Plan and Side Elevation
of Difference Engine
1830

Babbage's Difference Engine No. 1, design drawing, 1830.

# CHAPTER 10

# Calculating engines

## *Machines, mathematics, and misconceptions*

### DORON D. SWADE

Industrialization and the mechanization of physical labour were hallmarks of the 19th century. This chapter describes attempts to extend the metaphor of industrial production from physical to mental labour, specifically to numerical calculation. In concept, design, and promise, the most remarkable automatic calculating engines of the period went beyond the mechanization of arithmetic to the mechanization of mathematics, and with it the notion of computation as a new mathematical method.

The early and middle decades of the 19th century were times of unprecedented engineering ambition. Machines, contrivances, devices, and contraptions were the obsession of the times. 'It was the Age of Machinery' wrote Thomas Carlyle, 'in every outward and inward sense of that word'.[1]

It was not just the role of machines in the conquest of the external world. Mechanism had become coterminous with understanding and had begun to usurp the terms in which the world was understood. 'There is no end to machinery...', Carlyle wrote in implied lament, 'what cannot be investigated and understood mechanically, cannot be investigated and understood at all'. The model was that of the emerging sciences which embraced material agency as first cause and its promised deterministic accounts of all phenomena. Carlyle extended the reach of mechanism to the conduct of science itself—to the social organization of its academies, societies, and committees. Intellectual life was not alone in succumbing to mechanism; religious and moral life fell under its spell as well: 'faith in Mechanism has now struck its roots down into man's most intimate, primary sources of conviction...our true Deity is Mechanism'. He refers not just to machine-worship in a metaphorical sense, but to the expression of spiritual life in the social machinery of organized religion and the overall embrace of efficiency. Mechanism was the sign of the times.

There seemed no limit to the variety and invention of the industrial arts in the exploitation of new processes and materials. Catalogues featured new household products that poured from the new manufactures and filled shopkeepers' shelves. Science journals reported daily on some new wonder—life-saving devices to reduce loss of life at sea after shipwrecks, home exercisers, a stethoscopic device to help locate the position of fog horns sounding warnings, flying machines using compressed air or pedal power, a miniature railway to transport food platters from the kitchen to the dining table, a printing press driven by solar energy, a vacuum-operated milking apparatus, a spring mattress, a moustache guard, a device for electroplating the dead…The lists are endless and the variety is thoroughly baroque and not infrequently bizarre.[2]

Amongst the most poignant illustrations advertising a new device is that of a mother rocking a sleeping infant in a cradle. The image depicts the mother or female carer in a rocking chair, with both hands engaged in needlework. The motion of the rocking chair is transmitted via cord and pulleys to operate a milk churn as well as to rock the cradle. The image is not presented as an idyll of motherhood or of domestic bliss, but as an announcement for a 'new domestic motor'. The caption explains that through this invention 'the hands of the fair operator are left free for darning socks, sewing or other light work while the *entire individual* is completely utilised'.[3] The device 'afforded an

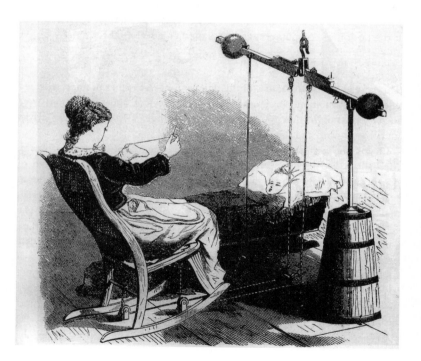

A 'new domestic motor', 1873.

effective method of diverting latent feminine energy, usually manifested in the pursuit of novels, beaux, embroidery, opera boxes, and bonnets into channels of useful and profitable labour'. Here the 'entire individual' is equated with the ability to provide motive power, clearly prized above all. In the language favoured in explanations of scientific apparatus, the inventor described the operation of the device by reference to 'lever A' and 'adjustable arms, B'. The announcement was evidently serious; were it not, it might easily be mistaken for a Swiftian satire of scientific writing.

The purpose of mechanical invention was for the most part directed at relieving or replacing physical labour, and the values that endorsed such enterprise were inseparably part of the 'industrial revolution'.[4] But there was another revolution in the making, one that has received less attention by historians, largely because it failed—the mechanization of mental labour, specifically the mechanization of calculation and mathematics, and with it the extension of the model of industrial production from things to thought.

## Manual calculating aids

The drudgery of numerical calculation had been an abiding lament for centuries. In the 17th century Gottfried Wilhelm Leibniz, philosopher and mathematician, wrote:[5]

it is unworthy of excellent men to lose hours like slaves in the labour of calculation which could be done by any peasant with the aid of the machine.

Numerical calculation, at least of a routine kind, was clearly regarded as a blight—menial, and therefore inferior to the abstract analytical practices of mathematics, especially when the calculations consisted of 'mechanical' tasks— that is, repetitive and low level, and unworthy of the creative intellects of the elite.[6]

In the 19th century, during the fierce debates on the utility of automatic calculating engines, the refrain bemoaning the numbing grind of routine calculation was taken up anew, this time by the protagonists of automatic calculating engines. Charles Babbage, Francis Baily, Dionysius Lardner, and James Jerwood wrote variously about 'the intolerable labour and fatiguing monotony of a continued repetition of similar arithmetical calculations'; 'that wearisomeness and disgust, which always attend to monotonous repetition of arithmetical operations'; 'the dull and tedious repetition of many thousand consecutive additions and subtractions'; and the 'mental drudgery' of constructing mathematical tables using repeated addition.[7]

In 1842 Luigi Menabrea, then an engineer and later prime minister of Italy, echoed Leibniz when he wrote of the stultifying effect of calculation on higher thought:[8]

And what discouragement does the perspective of a long and arid computation cast in the mind of a man of genius, who demands time exclusively for meditation, and who beholds it snatched from him by the material routine of operations!

Relief from the burdens of calculation was sought through a variety of mechanical aids. Pebbles (calculi), tally sticks, and tokens were used to aid counting and numerical record-keeping in practices that date back thousands of years. We find incipient mechanism in the abacus which can be seen as an early artefactual transition from counting to calculation. Beads or counters, constrained by wires in a frame, belong to a positional system of value—a representational device, rather than one in which the rules of calculation are embodied in its physical structure; the 'algorithm', the sequence of operational rules through which a calculation is executed, is dependent on human action.

The shift from counting to mechanical calculation received significant impetus in the 17th century when several leading savants in continental Europe sought to produce devices for simple arithmetic: Wilhelm Schickard built his 'calculating clock' (1623), Blaise Pascal, his 'Pascaline' (1642), and Leibniz his 'reckoner' (1674).[9] In these devices decimal numbers were entered on circular dials or sliders, sometimes using a stylus, and results were displayed on engraved or annotated discs. The reckoner is celebrated in the chronicles of history more for its ambition than for any practical accomplishment. Through a combination of design and manufacturing deficiencies the 'carriage of tens' failed to work as intended, and only a largely unsuccessful prototype appears to have been made. The Pascaline stimulated philosophical debate about the mechanization of mental process: it was paraded before royalty, and demonstrated in the drawing rooms of merchants, government officials, and university professors. But it was expensive and insufficiently robust for daily use, and not many were made.

For all the ingenuity of their makers and the seriousness of their purpose, mechanical calculators prior to the 19th century were largely *objets de salon*, many exquisite and delicate, sumptuous testaments to the instrument maker's art, but unsuited to daily use in trade, finance, commerce, science, or engineering. While makers struggled to produce viable mechanical calculators, anyone

Johan Müller's 'Universal calculator' of 1794, a testament to the instrument-maker's art.

needing to perform more than trivial calculations relied for the most part on printed mathematical tables or on the slide rule.

The mechanical calculators so far mentioned were essentially digital devices, in that each digit value was represented by a distinct position of a mechanical part. Only discrete positions of moving parts were valid representations of numerical values, and transitional positions were logically indeterminate. In contrast analogue devices, in which numerical values are represented on a continuous scale, represent a distinct class of calculating aid.

Throughout the 19th century, slide rules, a prime example of an analogue calculating device, were in widespread use for both general and specialized calculation.[10] They offered the convenience of portability and the assurance of robustness. Such were their charms that Richard Delamain, who first published a description of a circular slide rule in 1630, wrote that his device was 'as fit for use as well on horseback as on foot'.[11] In addition to 'universal' slide rules for general mathematical calculation (multiplication, division, logarithms, and trigonometric functions), many variants were produced with scales and divisions customized for special purposes, some exotically specific: rules for estimating excise duties (conversion scales for cubic inches to bushels, and finding the mean diameter of a cask), calculating the volume of timber or the weight of cattle, estimating varieties of interest rates, and scales for a host of specialized engineering applications.[12]

Standard slide rules served well for quick and convenient calculation, but accuracy was limited. The graduated scales and divisions were read by eye, and there was an element of subjective judgement in reading the last decimal places. Precision was variable and depended in part on the separation of the divisions, which tend to be compressed at the extremities of the ranges. Accuracy was typically limited to between two and four figures, with increasing uncertainty in the last one or two significant digits. This was adequate for many applications, but not all: interest payments, insurance premiums, and financial accounting often required exactness to the level of pennies in thousands, or tens of thousands, of pounds; land surveys, especially for cadastral tables used for property taxation, required calculations with the precision of metres in distances of kilometres and tens of kilometres, and this level of precision was unattainable using analogue scaled devices.

The precision of slide rules was stretched to the limit by extending the effective length of the scales. Fuller's 'spiral rule', designed in 1878, consisted of a cylinder with the graduated scales arranged helically, as in a screw thread, on the surface. With scales wrapped around the cylinder in this way, a Fuller's rule with a cylinder six inches long had an effective working length of over forty-one feet, and could be read reliably to four places.[13] The Stanley Company of Glasgow exhibited a Fuller's rule with logarithmic scales that was effectively eighty-three feet long. The readable precision was cited as '4 and sometimes 5 figures', representing the limits of precision achievable by a scaled rule that remained conveniently portable.[14] Yet when it came to a comparison between slide rules and tables, there was no contest. Leslie Comrie observed in 1948 that 'today schools are equipped with four-figure tables, which are ten times as accurate as the common ten-inch slide rule with which the great majority of engineering calculations are done'.[15]

The mechanical calculator that made a serious bid for widespread take-up was the arithmometer, patented and made public by Thomas de Colmar in 1820.[16] This was a desk-top device with sliders for entering numbers, numbered dials to display results, a moveable carriage for shifting decades, and a rotary crank handle. While often described as the first successful commercial calculator, the arithmometer was far from being an instant success. It took over fifty years of modification and improvement before it commanded even a small market.

The use of arithmometers at the General Register Office (GRO) illustrates the time lag between the announcement of the device and its adoption for routine use. The first recorded request for the purchase of an arithmometer, at a cost of £20, dated from January 1870 when George Graham, the Registrar General, requested its purchase following trials of the device at the GRO—this was some fifty years after de Colmar's first announcement.[17] By July 1873 there were still only three arithmometers in use at the GRO and annual sales of arithmometers were reckoned at no more than one hundred.[18]

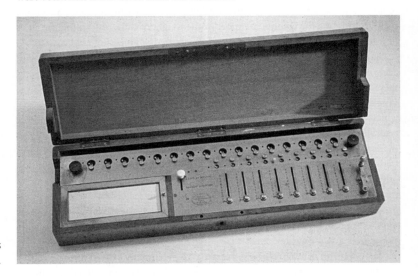

Thomas de Colmar's arithmometer, c.1880.

While Graham was consistent in his support for the benefits of arithmometers, it is clear that, even in the 1870s, they were still troublesome, noisy, subject to derangement, imprecisely made, and in frequent need of repair. In a tilt at the supposed inferiority of French manufacture, he suggested that 'the liability of "arithmometers" to get out of order and their noise would be greatly diminished if they were made by better workmen—perhaps Englishmen'.[19] But problems persisted and the market was limited. In March 1877 Graham wrote that his repeated requests and suggestions for improvements were rebuffed by the manufacturers on the grounds that there was insufficient demand.[20] Private use appears to have been equally patchy. In 1872 an arithmometer costing £12 was acquired as a novelty by Henry Brunel (Isambard Kingdom's son), who, though clearly taken with the device, paid it an ambiguous tribute when he described it as 'really a very useful article worth its weight in brass'.[21]

By the 1890s the arithmometer still remained unproved to the Treasury's satisfaction, and this persisted into the early 1900s. Arithmometers went on to sell in the tens of thousands, but it took the better part of a century for them to mature as a product.[22]

## Automatic calculating engines: calculating by steam

The calculators described so far are part of an unbroken thread of desk-top manual calculators, starting with Schickard's calculating clock in the 17th century, moving through the heyday of mechanical devices in the late 19th century, and most recently evidenced by electronic pocket calculators. In a little under four hundred years, four-function calculators evolved from expensive experimental devices that challenged ingenuity, design, and fabrication, to electronic giveaways or on-screen pop-ups bundled as incidental software add-ons to personal computers, mobile phones, and electronic personal organizers.

Against this background of unsurprising evolution, the calculating engines of Charles Babbage represent a startling intervention. While his earliest ideas on mechanized calculation date from the 1820s and early 1830s, and predate the period of interest here, his pre-Victorian thinking informed and laid the foundation for decades of sustained development that followed. As we shall see, the aspirational template and underlying motivation for Babbage's work was seeded in Georgian England, but pursuit, development, and the ultimate *dénouement* played itself out well into the Victorian age.

All the devices in the previous section were manual, in the sense that useful operation relied on the continuous informed intervention of the operator— entering numbers, operating carriages or sliders, cranking handles, or reading and writing down results. The 'algorithmic' element of the computation was provided by a human operator. The designs for Babbage's massive engines, on the other hand, represent a quantum leap in logical conception and physical size, and marked the start of a new thread—automatic machine computation in which operational and mathematical rule was transferred from the human user and embodied in mechanism. By exerting physical effort, one could achieve results that had been attainable up to that point in time only by mental effort.

The significance of this transference of intelligence from mind to machine was not lost on Babbage or his contemporaries. 'The marvellous pulp and fibre of the brain had been substituted by brass and iron, he [Babbage] had taught wheelwork *to think* or at least to do the office of thought' wrote Harry Wilmot Buxton, a junior colleague of Babbage and his posthumous biographer. Dionysius Lardner, an energetic and prolific popularizer of science, commented: 'the proposition ... to throw the powers of thought into wheel-work could not fail to awaken the

The last known portrait of Charles Babbage, 1860.

attention of the world', while Lady Byron (mother of Ada Lovelace) wrote: 'We both went to see the *thinking* machine (for such it seems)'.[23]

The story of Babbage and his engines is increasingly well known, and the saga of his failed attempts to construct an engine remains an object of study.[24] Less well known is the critical role that mathematics played in Babbage's motives and aspirations for his invention. Printed mathematical tables were thought to be riddled with mistakes, and the received perception is that the principal purpose of Babbage's engines was to eliminate the risk of error from the calculation and production of such tables. Almost every account of Babbage and his engines argues the central role of errors in Babbage's motivational landscape, and this is a view that has consolidated over time. It is worthwhile unpicking how history came to favour this account.

There are two main sources responsible for the legacy of this narrative. The first is the event captured in the well-known vignette of Babbage and his friend John Herschel who met in 1821 to check the newly calculated astronomical tables commissioned by the recently formed Astronomical Society.[25] The results had been calculated manually by two separate human computers without collaboration. In the course of checking the independently computed results, Babbage, dismayed by the discrepancies in the two sets of calculations, proclaimed 'I wish to God these calculations had been executed by steam'—an utterance described by the novelist and historian Peter Ackroyd as 'one of the most wonderful sentences of the 19th century'.[26]

'Steam' here serves as a metaphor for the infallibility of machinery, as well as for automatic production in response to the problem of 'supply'. The engines would be a 'manufactory of numbers', and in Babbage's invocation of 'steam' we have the essential extension of the industrial metaphor from goods to information, from the physical to the mental. Not only would infallible machinery calculate unerringly, but the engine would automatically print results and produce stereotype moulds from which printing plates could be made. At a stroke the engines would eliminate the risk of human error to which each of the manual processes involved was prone—calculation, transcription, typesetting in loose type, printing, and proof-checking. The episode with Herschel in 1821 was Babbage's mechanical epiphany and, starting immediately on the design of Difference Engine No. 1, he spent the better part of the next half-century in pursuit of the ideal of mechanized computation.

That tabular errors provided the jumping-off point is well evidenced: Babbage documented the episode at least three times, appropriating more credit with each retelling.[27] He may well have dramatized the episode and aggrandized his role, but that the episode happened, and that it was the starting point of his efforts to build automatic calculating engines, is supported by published accounts and was never contradicted by Herschel in whose presence the event took place. Historians, endlessly charmed by the episode, appear to have translated this initial stimulus into an enduring motive. The fact that Babbage continued to work on the design of his machines on and off for the rest of his life, and never publicly disabused his contemporaries of any misconception, allows us to sustain the notion that his motives remained unchanged. But it is not only the vignette that steers us towards subscribing to the principal role of errors as the driving purpose of the engines.

The second source that gives tabular errors a central role is an article by Dionysius Lardner, published in 1834. Lardner, a prolific popularizer of science, has been the butt of ridicule. Anthony Hyman portrayed him as a clown—'a scientific Falstaff...even now...occasionally mistaken for a serious figure'; 'Lardner was the comedy act of the show: he ballooned across the engineering landscape of the time sustained by an inexhaustible supply of hot air'.[28] For all this, Lardner was a gifted, energetic, and often melodramatic lecturer, famed for his semi-technical lectures on industrial topics, with live demonstrations and working models a signature feature. He was a consummate showman, played to packed houses, and commanded high fees.[29] His writing and lecturing brought him wealth and celebrity, as well as ambivalence from the scientific establishment who were partly gratified by his success in raising the public profile of science, and at the same time wary and at times alarmed by his sometimes erratic pronouncements and predictions.

Lardner's article, the most substantial contemporary account of Babbage's difference engine, has had a defining influence on historical perceptions of the purpose of Babbage's machines.[30] Based on the large number of published errata, Lardner argued that printed mathematical tables were generically flawed. He asserted that each discovered error represented countless yet undiscovered errors, and that tables were therefore immune to improvement by correction, evidently taking no account of the rolling discovery of errors and progressive 'purification' through use. He stated, with the full force of grandiloquent rhetoric, that it was only Babbage's engines and the 'unerring certainty of mechanical agency' that could remedy the problem: 'It is only by the *mechanical fabrication of tables* that such errors can be rendered impossible'.[31]

The motivational landscape is clear: errors were the problem, machines the solution—a classic problem–solution pairing, not unlike the clear-cut story of Harrison's clocks a century earlier in the quest to solve the problem of determining longitude at sea. Framing the utility of the engines in this way has the appeal of simplicity. It also provided Lardner with a rousing cause to be championed—a thwarted genius (Babbage), a wrong to be righted (dilatory Treasury funding by unimaginative government bureaucrats), high stakes and public good (maritime safety assured by accurate navigational tables).

Because Babbage collaborated with Lardner on the article and provided most of the technical material, and because the article championed the engines with almost unseemly vigour, it has always been assumed that Lardner was Babbage's mouthpiece and that the views that Lardner expressed were Babbage's. More than anything else, it is Lardner's article that so explicitly identified tabular errors as the 'problem' and the engines as the 'solution', and from which we have inherited the primacy of errors in our account of Babbage's efforts. However, there is evidence that Lardner's emphasis on errors was a deliberate simplification that has misled historians ever since.

# Errors and misconceptions

Lardner's interest in the engines dates from 1830 when he first wrote to Babbage to explore the possibility of an article.[32] The article was published in 1834, thirteen years after Babbage's first conception of the difference engine. To uncouple the historical conflation of Lardner's views and Babbage's, we need to excavate Babbage's earliest writing on the engines before Lardner framed utility the way he did.

In the six months between June and December 1822, Babbage wrote five papers on his expectations for his calculating engine. This suite of papers was written shortly after Babbage had conducted the first trials using a small working model completed in the spring of 1822, but before he showed any serious ambition to build a full-sized machine. The audience for these papers was the scientific community, mathematicians and astronomers, many of whom were Babbage's colleagues and friends. The ideas that he shared give revealing motivational clues to his subsequent efforts.[33]

The focus of these early papers was overwhelmingly mathematical and it is clear that, suggestive and undeveloped as his ideas were, it was the mathematical potential of the machines that intrigued and excited him. The elimination of errors begins to feature only as a device of persuasion for financial support, for which benefits needed to be framed in utilitarian terms. It is evident from these papers that for Babbage the engines represented a new technology for mathematics, in relation to which the elimination of errors was only incidentally important and comparatively mundane.

While Babbage's ideas on the difference engine immediately predate the Victorian period, they were a constant and defining influence on his later work. The designs for his analytical engine describe a 'universal' calculating machine, through which the 'whole of arithmetic now appeared within the grasp of mechanism'.[34] The designs reached a defining stage in 1840, and their development occupied him until his death in 1871. Between 1847 and 1849 he broke off his work on the analytical engine to design an advanced and elegantly simple machine, Difference Engine No. 2. So while Babbage's earliest ideas were seeded in the immediate pre-Victorian period, they provided the basis for much of what followed.

# Computation as systematic method

The first mention of new mathematical implications appeared in Babbage's open letter to Humphry Davy, president of the Royal Society. Here Babbage advertised the ability of his difference engine to solve equations with no known analytical solution. The prospects for machine solutions to unsolved equations signalled something fundamentally new.[35]

The roots of an equation are the values of the independent variable which reduce the function to zero. The standard analytical technique for solving equations is to equate the expression to zero and to solve for the unknown. There is no systematic process for doing this, and the success of the process depends on ingenuity, creativity, and often an ability to manipulate the problem into a recognizable form that has a known class of solution. Not only is there no guarantee of solution using such techniques, but there is no way of determining whether the equation in question can be solved in principle. If analytical methods fail, then trial-and-error substitution can be tried. This involves substituting trial values of the independent variable, and repeating this process to see whether a value of the argument can be found that reduces the function to zero. But the technique is hit and miss: it is regarded as 'inelegant' by mathematicians, and does not guarantee success.

What was new in Babbage's description of solving equations with machines was the use of computation as a systematic method of solution. Starting with an initial value of the independent variable, each cycle of the engine generated a new tabular value, and the machine found a 'solution' when the figure wheels giving the tabular result were all at zero. Finding a solution reduced to detecting the 'all-zero' state, and the number of machine cycles taken to achieve this represented the value of the independent variable, which was the solution sought.

Rather than rely on visual detection of a particular tabular value, Babbage incorporated a bell in his second machine that rang to alert the operator to the occurrence of specific conditions in the column of tabular values. If the bell mechanism could be set to detect the all-zero condition, or a sign change as the value passed through zero, then 'extracting the roots of equations' involved the operator in setting the initial values and cranking the handle to cycle the machine until the bell rang. The operator would then halt the machine and read off the number of cycles that the machine had run on a counter automatically incremented after each machine cycle. This number was the first root of the equation; if there were multiple roots, the operator kept cranking until the bell rang again.

So as not to rely on an operator responding to a bell to halt the machine, Babbage incorporated devices that allowed his machines to halt automatically. In 1842, Luigi Menabrea published a description of the unbuilt analytical engine and made specific reference to the all-zero state automatically halting the machine.[36]

Difference Engine No. 2, designed between 1847 and 1849 and completed in 2002, also features automatic halting. Here a device halts the machine to prevent the recorded results overrunning the limits of the printed page. In the course of tabulation each new result is printed automatically in inked hard copy as it is generated. The same result is simultaneously impressed into soft material, wet plaster, or papier mâché, for example. If the printing apparatus ran out of paper, or the stereotype trays containing the soft material overran their borders, the lost results could not be recovered without resetting the initial values and re-running the sequence of calculations from scratch. The halting mechanism prevents this.

The output apparatus for Difference Engine No. 2, showing the printing and stereotyping apparatus.

When the end-of-page condition is reached, the drive to the apparatus is automatically broken by the release of a clutch, the drive handle suddenly runs free, and the machine stops dead. After the trays are replenished and the machine restarted, the output resumes in an unbroken sequence of results. Automatic halting spares the operator from having to keep track of how many cycles had been run before stopping the engine, and also from having to halt the engine at exactly the right point in the cycle to allow the recording medium to be renewed. So automatic halting mechanisms were a feature of the designs for stopping the engines when a particular state was detected.[37]

In the event that there were no solutions, the machine would continue ad infinitum without the zero state or sign change occurring. The relationship between halting and the existence of solutions pre-echoes aspects of Alan Turing's work in the 1930s on 'computability' and the role of the halting criterion in the formulation of a solution. His notions of 'definite method' and 'mechanical process' are seminal concepts in modern computer science, formulated and realized in physical form by Babbage a century earlier.

Babbage's engines represented a new technology of mathematics which rendered practical methods that would otherwise be prohibitively labour intensive, tedious, and prone to error. With the aid of machines the repetition of computational operations would incur physical rather than mental cost. The feature of the engine that would allow this was its automation—that is, it embodied mathematical or computational rule in mechanism.

Difference Engine No. 2 (2004), the first complete Babbage engine built to original 19th-century designs: the engine weighs 5 tonnes, consists of eight thousand parts, and measures 11 feet long and 7 feet high.

## Heuristics, stigma, and class

A second mathematical feature, flagged by Babbage in his letter to Davy, refers to the ability of the machine to calculate a series for which there is no known analytical law. In a letter to David Brewster, Babbage expands:[38]

I can by setting an engine, produce, at the end of a given time, any distant term which may be required; or if a succession of terms are sought, commencing at a distant point, these shall be produced. Thus, although I do not determine the analytical law, I can produce the numerical result which it is the object of that law to give.

The power and appeal of analytical formulation derives from its generality—that is, the ability to represent, in a single statement through symbols, any and all specific instances of the relations expressed. The unspoken values of analytical science elevate generality and universality above example and instantiation. A silent premise of contemporary mathematics and philosophy was that example was inferior to generalization, induction inferior to deduction, empirical truths inferior to analytical truths, the synthetic inferior to the analytic, and the *a priori* inferior to the contingent. Stretching the analogy to social class, we find parallels with the social inferiority of trades and manual activity compared with philosophical and intellectual occupations. Journeyman and gentleman separated by social rank. Calculation, which involves a specific numerical example, was, in the prevailing culture, implicitly inferior to formal analysis. The existence of a series that could be produced by computational rule for which no formal law was known fell outside the comfort zone of analytical tradition.

This was new territory and Babbage was clearly intrigued by the general question of how to find analytical laws for series suggested or produced by the engine. In his letter to Brewster he tabulated the first set of values and differences of a new series suggested by the engine, and proceeded to derive a general expression for the general term.[39] The process was essentially one of induction, and Babbage acknowledged that the process was at odds with the traditions of mainstream mathematics. He wrote that the unusual route he took in arriving at a general form for a new series was 'much more conducive to the progress of analysis, although not so much in unison with the taste which at present prevails in that science'.[40] While this statement predates our period of interest, it demonstrates his awareness that his work on engines of mathematics was 'off piste,' and even disdained by the custodians of mathematical convention.

The example of the new series he described to Brewster was prompted by the layout of the engine, and represents one of several suggestive and actual instances of the heuristic value of the engine in stimulating new enquiries in mathematical analysis. In his second notice to the Astronomical Society, Babbage traced the process that led to his speculations on the new series:[41]

I will now advert to another circumstance, which, although not immediately connected with astronomical tables, resulted from an examination of the engine by which they can be formed. On considering the arrangements of its parts, I observed that a different mode of connecting them would produce tables of a new species altogether different from any with which I was acquainted. I therefore computed with my pen a small table such as would have been informed by the engine had it existed in this new shape and I was much surprised at discovering that no analytical method was yet known for determining its $n$th term.

He then listed the series in which each second difference was given by only the units value of the current tabular value, with the larger-value digits ignored. The gist of the issue was that the new series was formed by using only some of the digits in the tabular value as the basis for the calculation of each next value.

The passage quoted indicates that the idea for new series of this kind arose from the spatial representation of numbers in the machine. In all Babbage's calculating engines, numbers were represented by figure wheels engraved with the numbers 0 to 9, arranged in vertical stacks or columns, with one wheel for each digit of the number. The columns stood alongside each other and were coupled with internal gearing that orchestrated the motion of the figure wheels to perform the repeated additions required for tabulation by the method of differences.

The broadside view of the engine presented a rectangular matrix of figure wheels representing the value of the tabular value in one column, the first difference in the next column, the second difference in the next column, and so on. The matrix of wheels suggests that, apart from the internal gearing for repeated addition, individual figure wheels could influence others by *external* connection. Specifically, by externally gearing wheels together, he could add the value of any given figure wheel (representing units, tens, hundreds, and so on, of a given number) to the value of a wheel in another column. The technique allows feedback, feed-forward, or cross-feeding of individual digits in a way that influences the step-wise generation of successive results. For example, the tens wheel on the second difference column could be coupled to the hundreds wheel on the tabular value column, leaving intact the machine's internal gear train linking the two columns. Cycling the machine would produce a new series for which there was a clear computational rule by which to generate each next value, but for which there was no analytical formula known or available.

Babbage's speculations on finding general analytical laws for empirically generated series were undeveloped. They were inconsequential to the development of mathematics and represent one of the topics outside mainstream analysis that he focused on but did not pursue. However, in the history of the development of mathematical ideas, his line of enquiry represents the earliest realization that there was a theoretical dimension to computational method that was important and unexplored, and that computation involved more than the contingent specifics of numerical example.

The prospect of calculating machines elevated computation to systematic method, and in these early papers Babbage appears to be struggling to communicate its new status. His writing was often unclear and suggestive, and tended to slide between specifics and generalities. However, he appears clearly to have sensed that there was something new and fundamental in step-wise mechanical process as a realization of algorithmic procedure, and that the prospect of calculating engines invoked a new discourse of mathematical analysis.[42]

## Numerical analysis

Babbage foresaw that machine computation would give rise to new forms of mathematical analysis, and predicted that the progress of science would be impeded by their lack. In his letter to Brewster, Babbage argued that to exploit the facilities of the engine would require a mastery over the analysis that must *precede* computation by machinery.

In the context of tabulation by differences, he identified two elements of this analysis: how best to approximate functions by particular series, and the preparatory analysis to ensure that the approximation remains valid to the requisite accuracy within the restricted range of the function being tabulated. Neither element of analysis was new, and both were used by table-makers in the manual preparation of tables using the method of differences. However, the benefits of the engine for Babbage were two-fold: the stimulus to mathematics of systematizing the analysis of computational methods using differences, and the value of this analysis and of the engine in rendering practically useful abstract branches of mathematics. He further predicted new branches of analysis to optimize the efficiency of machine calculation, and illustrated the point by an example in which he manipulated a formula to show one version that required thirty-five multiplications and six additions to find its value, and a mathematically identical but alternative expression requiring only five multiplications and one addition.[43]

Finally, he asserted his conviction in the eventual dependence on machine calculation to relieve the burden of computation, whether or not the value of his calculating engines was recognized in his lifetime:[44]

If the absence of all encouragement to proceed with the mechanisms I have contrived, shall prove that I have anticipated too far the period at which it shall become necessary, I will yet venture to predict that a time will arrive, when the accumulating labour which arises from the arithmetical applications of mathematical formulae, acting as a constantly retarding force, shall ultimately impede the useful progress of the science, unless this or some equivalent method is devised for relieving it from the overwhelming incumbrance of numerical detail.

## Lardner and mathematics

In his enthusiastic championing of Babbage's engines, Lardner downplayed Babbage's mathematical aspirations.[45] In doing so he did Babbage's interests fatal damage. By identifying 'utility' as the practical utility of eliminating errors in the production of tables, he forced the engine advocates to defend the machine from a position of weakness. The utility of the engines as a solution to table-making was resoundingly rejected by experts in England and on the Continent—in 1842 by George Biddell Airy (Astronomer Royal at Greenwich), in 1844 by the Swedish astronomer Nils Selander, and in 1858 by Urban Jean Joseph Leverrier in France.[46]

Babbage, who was capable of incontinent savagery in his public criticism of those he disagreed with, was curiously silent in defence of the practical utility of his machines to table-making. It is unlikely that he simply disdained to reply to his critics, given his excesses in the past and the absence of any evidence of moderating restraint in his public outbursts. It is more likely that he was unwilling to defend his engines on the grounds of their limited practical benefits as formulated by Lardner.

Lardner's article has had a defining influence on historical perceptions of Babbage's motives, and the false emphasis he gave to the role of errors unbalanced subsequent interpretation.

## The utility of calculating engines

Victorian arguments for the utility of calculating engines are not easy to navigate. In much of Babbage's writing it is often difficult to unbundle compound statements of multiple benefits, or to establish relative importance in his arguments. He does not refer in any consistent or systematic way to the benefits of his machines and his writings do not present a unified or structured advocacy, nor do the arguments build on each other in a progressive way. The emphasis he gave to a particular feature tended to depend on his current preoccupation framed to meet the interests and proclivities of his audience.

The following map is offered as a navigation aid to arguments for the utility of calculating engines, articulated by Babbage and other contemporary advocates such as Lardner, Herschel, Francis Baily, and Georg and Edvard Scheutz. The map is part of a larger study of the utility, and helps us to locate Lardner's arguments in a broader landscape.[47]

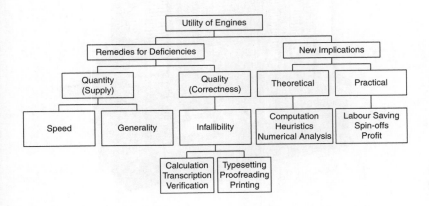

The terms in which the engine advocates framed utility can be seen to fall into two categories: remedies for known deficiencies in tables on the left, and new practical and theoretical implications of automatic computation on the right. The left-hand side of the table is concerned with how better to do, using machines, what was already being done in the manual production of printed tables. The supposed infallibility of machines ensured quality—that is, the unerring certainty of machinery would ensure the integrity of the processes involved in the manual production of tables: calculation, transcription, verification, typesetting, proof-reading, and printing.[48] The speed of the machine and the generality of the method of differences, as well as the generality of approximating within given domains all regular well-behaved mathematical functions by polynomials using Taylor expansions, would solve the problem of supply and meet the increasingly importunate needs for new tables by astronomers.

The metaphor of industrial production as a solution to the problem of supply was echoed by Lardner, who referred to the 'mechanical fabrication of tables', and also by Benjamin Herschel Babbage (Charles's son), who in 1872 described the difference engine as 'emphatically a machine for manufacturing tables'.[49] Even on the remedial side of the argument, the elimination of errors is only part of the story, and then only part of a yet larger picture of utility that extends beyond existing practice.

It was the right-hand side of the table that intrigued Babbage, and in this it was the new mathematical and theoretical implications of the engines that were at the centre of his early interests and dominated his work for decades after.

The analytical engine,
experimental part, 1871.

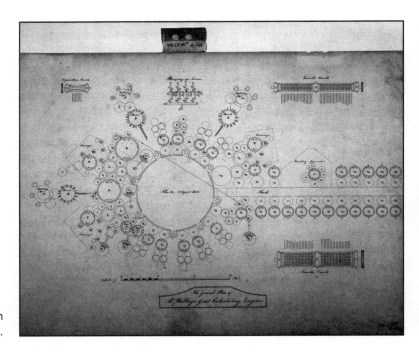

The analytical engine, design
drawing, 1840.

# The analytical engine

In his attempts to automate tabulation by the method of differences, Babbage was led from mechanized calculation to fully-fledged general purpose computation, and it was the designs for his 'analytical engine', conceived in 1834 but also unbuilt, that mark this essential transition. The analytical engine, while capable of tabulation, went well beyond this. It was a universal programmable computing engine capable of unimagined mathematical versatility. The jumping-off point of tabular errors was left behind, and the analytical engine, to which Babbage devoted most of his subsequent efforts, reaffirmed mathematics as a central and enduring motive for five decades of mechanical and logical design.

The designs for the analytical engine explicitly embody a raft of features that are recognizably present in latter-day electronic digital computers. It was programmable (using punch cards), had an internal repertoire of automatically executable functions (direct multiplication, division, addition, subtraction), and was capable, amongst other functions, of iterative looping, conditional branching, microprogramming, anticipating carriage of tens, and 'pipelining'. Its outputs included printed hardcopy, punched cards, and graph-plots. Its internal architecture anticipated the von Neumann model of 1945 that has dominated computer design since, and it featured, as does the von Neumann model, serial operation, a 'fetch execute cycle', memory, input/output, and the logical and physical separation of memory and central processor.[50]

Automatic multiplication was accomplished by accumulation of partial products. Division was achieved by tentative subtraction until a sign change was detected. The lost number was then restored by addition, followed by multiplication by 10 and repeated subtractions, until a further sign change. The number of subtractions between sign changes generated each digit of the quotient in turn. Division required conditional control, in this case detecting a sign change and taking one or another course of action depending on the outcome. These features were more than speculative suggestions: Babbage's technical archive contains designs for detailed mechanisms for these logical and systems functions.

Babbage's work embodied the earliest original exploration of the principles of general-purpose computation. He designed the first automatic computing machines, outlined computational method, explored the mathematical potential of computing engines, and formalized the logic of his designs in a generalized symbolic language.[51] His work up to 1871 can thus be seen as a sustained attempt to realize a machine that could fulfil the mathematical promise of a general-purpose computing engine.[52]

# Epilogue

Charles Babbage's name, more than any other, has become synonymous with the start, in the 19th century, of attempts to automate mathematics and computation

Difference engine prototype, 'the first printing calculator', built by Edvard Scheutz, 1843.

The Scheutzes' Difference Engine No. 3 (1859), used at the General Register Office for the 1864 life table.

using machines. His designs and partially built machines stand alone in originality, complexity, and logical and physical scale, as well as in practical and intellectual ambition. Others attempted calculating engines in the 19th century, although they worked in Babbage's shadow. The Swedish father-and-son team of Georg and Edvard Scheutz, stimulated by Lardner's account of Babbage's difference engine, undertook to build difference engines of their own for sale to observatories and institutions that produced tables. A prototype completed in 1843 was followed by two fully engineered machines in 1853 and 1859. Practical success was ambiguous and the venture was a commercial failure. The hoped-for economies from automated typesetting were not realized and, lacking the rigorous security devices of Babbage's designs, the Scheutz machines were temperamental.[53] Alfred Deacon in London, Martin Wiberg in Sweden (c. 1860), and Barnard Grant in the United

States (1876), all attempted difference engines in the second half of the century. Not all succeeded practically, and those who did had indifferent commercial success.

There were a few febrile twitches in the early 20th century. Between 1903 and 1909, an Irish auditor, Percy Ludgate, produced an 'analytical machine', an electrically driven programmable calculator of original design, apparently with no prior knowledge of Babbage's work. But this was an isolated episode and a developmental cul de sac. In Germany, Christel Hamman successfully used a difference engine of his own design for tabular interpolation, with results published in 1910 and 1911.[54] Despite the commitment of their originators and the ingenuity of implementation, the 19th- and early 20th-century initiatives failed to ignite a movement. Tabulation by differencing was eventually revived in the 1930s by Leslie Comrie who successfully implemented the technique using commercially manufactured calculating machines.[55] Electronic computers soon replaced mechanical and electro-mechanical calculators. The problems of tabulation became trivial and, as Babbage had predicted, machine computation for general mathematics and logic became indispensible to science and engineering.

The movement to build automatic calculating engines in the 19th century ultimately failed. In the late 1930s and 1940s the pioneers of the electronic era effectively reinvented the principles of automatic computation, largely in ignorance of the work of the 19th-century mechanists. The promised benefits of automatic calculation had eluded the reach of the great Victorian age of invention.

Florence Nightingale (1820–1910).

CHAPTER 11

# Vital statistics

*The measurement of public health*

M. EILEEN MAGNELLO

The twinned Victorian developments of industrialization and sprawling urbanization, along with a rapidly expanding population throughout the 19th century, produced overcrowded and unsanitary living conditions that led to endemic diseases, whilst the growth of factories, with toxic chemicals and carcinogenetic substances, compromised the health of the labouring class. The medically trained mid-Victorian vital statisticians engendered dramatic improvements in the overall standard of health and hygiene through their bureaucratic collection of a constellation of vital statistical data, which led to the implementation of the Public Health Acts and the reduction of mortality rates in Britain and in civilian and army hospitals.

Some of the earliest considerations for measuring public health in England began in the 17th century when John Graunt used his mercantile book-keeping methods, for which William Petty coined the term 'political arithmetic', in his *Natural and Political Observations upon the London Bills of Mortality* (1662).[1] This quantitative approach suggested, as Karl Metz has argued, that Graunt and Petty 'intended to use the mortality returns for the introduction of a health police in an attempt to prevent infectious diseases'.[2]

By the mid-18th century, Gottfried Achenwall, professor of law and politics at Göttingen University, had introduced the word *statistik* to 'examine inquiries respecting the population, the political circumstances, the production of the country and other matters of the state'.[3] This early modern German *staaten-kunde* did not need a numerical system since it 'corresponded to an essentially qualitative social structure in which everyone and everything possessed their proper place prescribed by tradition'.[4] The Scottish economist and first president of the Board of

Agriculture, Sir John Sinclair, who borrowed Achenwall's *statistik*, introduced the word 'statistics' into the English language in 1791, in his *Statistical Account of Scotland*. Sinclair wanted to measure the 'quantum of happiness' of the Scots, and thus emphasized the measurement of social phenomena rather than focusing on political matters.[5]

The collection of official statistics played a prominent role in all European countries from the late-18th to the early-19th centuries. The French use of *statistique* was allied to the 17th-century British political arithmetic involving numerical analyses, thus representing a shift from the Germanic qualitative system that described the state.[6] This Anglo-French use of statistics made it possible to address a variety of social, medical, and economic debates that shaped the development of vital and social statistics in Britain. In 1829, the Harley Street physician Francis Bisset Hawkins published his *Elements of Medical Statistics*, the first English textbook on vital statistics.[7] The systematic and bureaucratic quantification of the large scale of public health problems, which developed in a rapidly changing industrialized Britain in the mid-19th century, established the nexus that led to the creation and professionalization of vital statistics.

In this chapter, *vital statistics* refers to the description and enumeration used in census counts or in the tabulation of official statistics, such as marriage, divorce, and crime rates. This process is primarily concerned with average values and uses life tables, actuarial death rates, percentages, proportions, and ratios: probability is most commonly used for actuarial purposes. However, it was not until the 20th century that the singular form 'statistic', signifying an individual fact, came into use.

The vital statisticians' emphasis on *averages* as the primary unit of statistical measurement would lose its authority at the end of the 19th century, when a new mathematically based methodology began to take formation, largely due to Charles Darwin's ideas of biological variation and statistical populations of species. The decision to examine averages or measure variation was rooted in the philosophical ideologies that governed the thinking of statisticians, natural philosophers, and scientists throughout the 19th century. The emphasis on statistical averages was underpinned by the philosophical tenets of determinism and typological ideas of biological species, which helped to perpetuate the idea of an idealized mean. Determinism implied that there was supposed to be order and perfection in the universe; thus, variation was considered to be flawed and a source of error that should be eradicated, since it interfered with God's creation of His world. The typological concept of species, which was the dominant thinking of taxonomists and morphologists until the end of the 19th century, gave rise to the morphological concept of species, which was thought to have represented the ideal type. The presence of an ideal type was inferred from some sort of similarity; the morphological similarity then became the species criterion for typologists.[8]

These morphological ideas enabled some 19th-century biologists to classify species into 'types'; this could have had the effect of creating a proliferation of species, since any deviation from the type would have led to the classification of a new species. For many biologists interested in ideas of speciation (the multiplication of species), genuine change was possible only through the 'saltational origins' of new species, meaning that new species should have occurred by leaps or jumps in a single generation. Because Darwin's theory of evolution depended upon 'gradual' changes, it was incompatible with the typological definition of speciation. The adoption of Darwinian ideas of individual variation by Francis Galton, W. F. R. Weldon, and Karl Pearson led to a paradigmatic shift in statistical thinking and a different way of managing statistical data, and created an epistemic rupture between vital statistics and mathematical statistics.[9] The work of these late-Victorian mathematical statisticians is discussed in the next chapter.

These developments in statistics took place in the wider context of the Victorian culture of measurement, which occupied the activities of many scientists, mathematicians, and engineers in 19th-century Britain. The Victorians valued the precision and accuracy that instruments provided, since it gave them more reliable information. In the expanding industrial economy, it was essential to establish that the results were reproducible for an international market. Thus, engineers and physicists spent long hours in the laboratories recording and measuring electrical, mechanical, and physical constants for machines, apparatus, and other objects. Biologists and geologists collected as much information as possible on expeditions, voyages, and in the field, to create geographical maps, measure longitude and latitude, and classify new species of plants and animals. Statistics offered one way to quantify human measurements, especially for matters dealing with public health and hygiene, epidemics, heredity, and medicine. This chapter examines the innovative work of a number of Victorian vital statisticians, including Edwin Chadwick, William Farr, and Florence Nightingale.

## Populations, health, and vital statistics

Some of the developments surrounding the growth of vital statistics in the late-18th and early-19th centuries are linked to those commentators who were trying to determine the population of a country or of the world.[10] In 1798, the economist Thomas Robert Malthus was arguing in his celebrated work, *The Essay on the Principle of Population*, that unchecked human populations would always exceed the means of subsistence, and that human improvement depended on stern limits of reproduction. Malthus believed that populations would increase geometrically, whereas food supplies would increase arithmetically, and concluded that there would be a 'struggle for existence' in which the fittest individuals could survive and breed (a phrase that Darwin later adopted for his theory of natural selection). Although Malthus had assumed that food would remain a limited resource, one of the subsidiary effects of industrialization led to a greater availability of food. Once the rise of industry and subsequent employment of the working class replaced the idea that food was a limited national agricultural resource, Malthusian ideas were no longer a pressing concern.[11]

Thomas Robert Malthus (1766–1834).

Malthus found that no thorough investigation had been made into the cause of different rates of population growth: why some countries grew fast, and others not at all.[12] Whilst he had considered how population growth might constrain prosperity in the late-18th century, it was not until the mid-19th century, when the collection of population statistics in Europe and the United States had become extensive enough, that it was possible to contemplate the science of populations. The grandfather of the eminent Bertillion family of French demographers, Jean Paul Achille Guillard, first used the word *démographie* in 1855 to deal with the size, conditions, structure, and movement of populations, as well as the vital statistics of birth, marriage, and death, to describe those populations.

Advances in statistics were made in the 18th century, with the growth and application of mathematical probability, particularly from the work of Abraham De Moivre and Pierre-Simon Laplace, arising out of interests in games of chance and, more pragmatically, from insurance companies. Some of the earliest responses in Britain to Laplace and De Moivre's work came from a number of mathematicians who had been trained at Trinity College, Cambridge, in the early part of the 19th century, and who had been active in the Society for the Diffusion of Useful Knowledge. In 1830, the Society published an anonymous work entitled *On Probability*, attributed later to the Cambridge mathematician John William Lubbock and the legal counsel of the Home Office, John Elliot Drinkwater.

Following the developments in mathematical probability, the English began to use probability models for insurance offices and friendly societies for the construction of life tables; this had been first suggested by John Graunt and invented by the astronomer Edmond Halley. Some of the earliest teaching of probability took place in insurance offices. By 1857, the Institute of Actuaries was teaching a ten-week course on the 'Theory of life contingencies' for a fee of two guineas. An influential person in this transitional stage of probability in Britain was the mathematician and logician Augustus De Morgan, who published his *Essay on Probabilities* in 1838, which he used for life tables and insurance offices. His work played a key role in promoting probability for actuarial methods in Britain during the early-to-middle part of the 19th century.

By the end of the 18th century, attempts were being made to create a statistical office as an official body of the government. Two years after Sir John Sinclair produced the first volume of his *Statistical Account of Scotland* in 1791, the government established a Board of Agriculture in England to collect and publish 'some useful statements relating to the state of Agriculture in each county'.[13] At the beginning of the 19th century, a statistical office was set up in the Board of Trade to collect, arrange, and publish statements relating to the conditions and bearing upon the various interests of the British Empire.

Debates about the size of the population in the 1790s led English society to consider its military manpower and population resources. During the Napoleonic War years, the utilitarian philosopher Jeremy Bentham discovered that the government did not know how many paupers received relief, and that it could not even account for the amount of money in circulation. Registers of births, marriages, and deaths had been maintained by the Church of England since 1538, when Thomas Cromwell, Lord Chancellor to Henry VIII, instructed the clergy of every parish to keep registers of all baptisms, weddings, and funerals at which they officiated. The growth of ecclesiastical dissent throughout the 17th and 18th centuries meant, however, that the Established Church no longer covered the whole population and was thus inadequate as a source of national statistics. This lack of essential information indicated a basic instability in the affairs of the state and pointed to the need for a national system of keeping systematic records.

The idea of a national census first surfaced in 1753, but opposition from those who feared that the results would show that the population of England was smaller than commonly believed at the time, and could even encourage her warfaring enemies, was so vehement that the matter was dropped. Moreover, it was thought that a census would trespass on the liberty of the individual.

The first successful census was undertaken in England and Wales in 1801, after the Census Act had been approved by parliament in 1800.[14] This survey involved not only a house-to-house enquiry, but included clergy returns of baptisms and burials between 1700 and 1800, and marriages between 1754 and 1800. Despite this Act, there were no official requirements for individuals to register births, deaths, or marriages. Although registers were kept by Jews, Quakers, and many of the Free Churches and Chapels, they were regarded as an unacceptable source of national records, since they were outside the Established system.

Increasing concerns about those statistical problems that could be used to address economic and social issues were raised by the Belgian statistician and astronomer Adolphe Quetelet, the Cambridge-trained mathematician Charles Babbage, and Malthus, at the third meeting of the British Association for the Advancement of Science, held in Cambridge in 1833.[15] Quetelet brought his 'statistical budget' to this meeting, containing the 'proportion of crimes at different ages and in different parts of France and Belgium', but discovered that there was no suitable section of the Association for his discussion.[16] The situation prompted Malthus and his successor Richard Jones, professor of political economy at the East India Company College in Haileybury, Hertfordshire, to find a place for a private discussion of Quetelet's work. This subsequently led to the establishment of Section F of the British Association, which became known as the 'Statistical section' at its inception two years later.[17] At the 1833 meeting, Quetelet thought it was most unusual:[18]

a subject of wonder to every intelligent stranger, that in a country so intelligent as England, with so many illustrious persons occupied in statistical inquiries and where the state of population is the constant subject of public interest, the very basis on which all legislation must be grounded has never been prepared; foreigners can hardly believe that such a state of thing could exist in a country so wealthy, wise and great.

Indeed, many other European countries (Austria, Belgium, France, Italy, and the Netherlands) had already organized the systematic collection of vital statistics; once the British realized that they were lagging behind, it became a matter of national pride to organize a national system in line with their European neighbours.

Lambert Adolphe Jacques Quetelet (1796–1874).

The first statistical society in Britain had been founded in Manchester at the end of 1833, and its first task was to give empirical evidence for the Factory Commission. Just the year before, Jeremy Bentham had mentioned the need for a statistical society in London. The lack of an official requirement for registration was the driving force in the formation of the Statistical Society of London (now the Royal Statistical Society), which Babbage, Malthus, and Quetelet joined forces to establish in 1834. This society endeavoured to procure, arrange, and publish 'facts calculated to illustrate the condition and prospects of society'. Subsequent statistical societies were also formed in Bristol, Leeds, Liverpool, and Ulster. Following the recommendation of the British Association, the Statistical Society of London urged Parliament to set up a national registration system with a central office in London that would house duplicate copies of all records. After subsequent legislation in 1836, civil registration was introduced, the notification of births, marriages, and deaths was required, and the General Register Office (GRO) was established, giving England and Wales a system of demographic recording that was unique in Europe.

The Statists, as they were more commonly known, did not 'discuss causes nor reason upon probable effects'; instead, they collected and compared a set of facts which alone could form the basis of sound conclusions 'to benefit a social and political government'. Statistics was defined to be:

the observations necessary to the social and moral science, to the sciences of the Statists to whom the statesmen or legislator must resort for the principles on which to legislate or govern ... though the mathematician must lend aid to their pursuit.[19]

Although 'statistics' was still concerned with matters of the state, this Society helped to promote a change in emphasis from using statistics for the building of a social edifice, rather than for the creation of a political infrastructure.

This growing interest in the measurement of social phenomena was a consequence of a society that was undergoing rapid changes, precipitated by the transformation of the industrial age from an agrarian era; this, in turn, led to the urbanization of Britain and displaced many of the poor from the rural countryside to the growing industrial cities. Concomitant with these changes was an accelerated growth in the population throughout the 19th century.

In the expanding cities, the sheer magnitude of factory workers with their attendant work-related diseases and illnesses, the large numbers of unemployed receiving poor relief, and the introduction of the Education Acts that made it necessary to keep records of students, created a profusion of numerical information that could be codified and measured by these vital statisticians. Victorian Britain witnessed an explosion of industrial, technological, and social changes that engendered the coexistence of immense variation and apparent randomness in society. The confluence of these twinned statistical concepts became a source of diversification and quantification that was harnessed by the mid-Victorian vital statisticians to undertake statistical investigations of mass phenomena. Colossal amounts of data were collected by state agencies, private organizations, and various individuals interested in such social phenomena as poverty, disease, and suicide.

This early statistical movement drew on the early-modern Germanic use of *statistik*, where the emphasis had more to do with providing descriptive information about the state.[20] Many mid-Victorian statisticians regarded statistics as more than the mere collection of social data or a set of techniques; for this group, statistics embodied a separate academic discipline. It was, as Lawrence Goldman remarked, 'the centre-piece of the new study of man in society'.[21] These social measures had already been studied on the Continent in the 1830s by Quetelet, whose statistical thinking and methods greatly influenced the work of the Victorian vital statisticians.

Trained as an astronomer and meteorologist, Quetelet was one of the first statisticians to use statistical methods to quantify social phenomena by borrowing mathematical tools from astronomers. A student of Laplace, Quetelet's statistical ideas were also influenced by Malthus and the statistical work that the mathematician Joseph Fourier undertook in Paris in the early 1820s.[22] Following Laplace's discovery of the central limit theorem in 1810, Quetelet began to advocate a Laplacian empirical social science where empirical data conformed to the normal distribution.

Quetelet popularized the arithmetical mean in the 1830s when he discovered that astronomical error laws could be applied to the distribution of human features such as height and girth; this realization led in turn to his much celebrated construct of *l'homme moyen* (the average man). The regularities that Quetelet found in man and in meteors were comparable with the laws of physics, especially those of meteorology: he spoke of the social system in the same way that an astronomer spoke of the system of the universe. In this way, Quetelet aligned the average man with the centre of gravity.

Quetelet also noticed a similarity between the occurrences of regularities in phenomena in nature and in society: his realization of the inherent complexity underlying the measurement of social phenomena led him to devise a quantitative system of statistical laws, which he termed 'social physics', that ostensibly

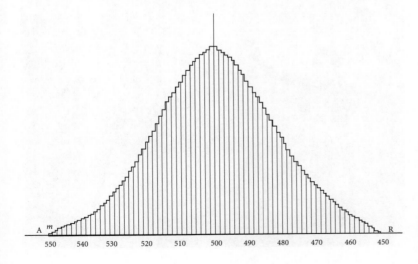

Quetelet's 'curve of possibility', later called the normal curve.

governed the varying rates of crime, marriage, and suicide.[23] Mean values were of any scientific value only when they represented a type, as deviations from this average were seen as flawed and a product of error. Quetelet was convinced that these mean values could be used to find the ideal type of society, politics, and morals. Since he thought that deviations from central values caused society's ills, he believed that a mean philosophical and political position should be able to resolve society's conflicts.

## The measurement of health

The introduction of the 1801 census in England and Wales led eventually to an increased knowledge about the number of deaths from various diseases, and brought about an awareness of the appalling sanitary conditions of towns. When the labouring poor flocked to the cities for factory jobs, they faced noxious toxins and dangerous working conditions. In Lancashire cotton mills, men who used mule-spinning machines (which gave the spinner greater control over the weaving process) were exposed to carcinogenic mineral oils and consequently developed cancer of the scrotum; tar and patent-fuel workers in South Wales faced similar health problems.[24] Miners and potters developed lung disease, those who inhaled dust from stones, flint, and sand got silicosis, whilst the match workers, who were exposed to phosphorus, were at risk of 'phossy jaw' (a disease that ate into the bone of the jaw, causing severe pain and eventually death). The most pernicious disease of the large urban cites was tuberculosis, which festered as the

A contemporary cartoon, illustrating the Victorians' obsession with public health.

population grew and poverty deepened.[25] Given the economic imperative in maintaining a healthy working class, intervention from the British government led to the creation of the Public Health Acts. This material provided a market for vital statisticians to measure health and illness, while the rise of epidemics led to the development of the discipline of epidemiology.

Industrialization not only dislocated the labouring classes from the farms in the countryside to the squalor of the cities, but it rapidly transformed the quality of food, creating intractable health problems. Overcrowding in the cities often led to inadequate housing conditions that lacked proper ventilation and sanitation, cesspools overflowed, and sewers ran directly into the Thames, posing major health risks to everyone. A key figure in the sanitary reform and in the use of statistics was the liberal-minded Edwin Chadwick who, as Secretary of the Poor Law Commission, was involved in central government's reorganization of the previous system of relief towards aid to the community's poor and destitute. It was largely his work that led to the first Sanitary Commission in 1838, followed by his *Inquiry into the Sanitary Conditions of the Labouring Population of Great Britain* in 1842.

Chadwick promoted the 1848 public health reform that led to the appointment of ministers of health in towns. Sanitary legislation led to authorities pulling down slum areas for redevelopment and establishing controls for standards of living. Public health reforms reduced significantly the occurrence of infectious disease and mortality, which fell during the last half of the 19th century. The success of the sanitary reforms gave new importance to the collection of statistics. When the chief question on the condition of Britain became the sanitary one, the function of statistics became the measurement of health.

Edwin Chadwick (1800–90).

Following the 1836 Births and Deaths Registration Act, Chadwick organized an amendment that required the cause of death to be stated when a death was registered, and urged for a better system in the collection of official vital statistics of births, deaths, and marriages. He then recommended the appointment of a Registrar General for the registration of births and deaths at the General Register Office. The post was created by parliament, and the novelist Thomas Henry Lister, the brother-in-law of the Secretary of State, was appointed by King William IV in September 1836.[26] Lister also implemented a new method of enumeration that enabled the population to be measured without duplication or omission. Although Lister 'laid the groundwork of the statistical project of the institution', his administrative skills were lacking.[27] As the historian Edward Higgs remarked, Lister has been regarded as 'a bit of a disaster as the administrative head of the GRO'.[28] His main criticism of Lister's stewardship of the Office was 'that he spent too much time on developing the GRO's scientific work, and not enough on ensuring that the Office was efficiently administered'.[29]

After the GRO was set up, there was a requirement to compile the statistical records. Despite the misgivings of the Treasury, Lister insisted on recruiting the medical statistician William Farr, who together with Edwin Chadwick helped to shape a medico-demographic programme at the GRO.[30] Lister and Farr planned the 1841 census so that it would produce the numbers at risk that could then be matched to the death statistics; this made it possible for Farr to calculate the mortality rates by age, occupation, area, and cause of death.[31]

William Farr (1807–83).

Farr, who studied medical statistics under the French clinician Pierre Charles Alexander Louis, developed a comprehensive *nosology* that listed secondary and tertiary causes of death, benefiting the medical profession and assurance societies alike. His work as Statistical Superintendent at the GRO in the 1840s was a landmark in the development of English preventive medicine and medical statistics: his methods and organization of vital statistics provided a template for all nations. Although his job was initially regarded as a minor clerical post, Farr turned it 'into a more important position than its creators could possibly have imagined'.[32]

It was during this time that Farr redefined statistics 'as a method of analysis rather than a social discipline in itself'.[33] The vital statisticians who worked at the GRO were taught 'shop arithmetic' until the 1870s when courses in vital statistics became available in London, and in 1878 Farr was an examiner in a course on the subject at the London School of Tropical Medicine.

As Statistical Superintendent, Farr began to work on the statistical studies of English life tables used by insurance actuaries. He regarded a life table as a *biometer*, because it gave the 'exact measure of the duration of life under given circumstances'.[34] The life table became the instrument of every British public health officer. The French *Annales Publique et de Médecine* served as a model for Farr's short-lived *British Annals of Medicine, Pharmacy, Vital Statistics and General Science*, which was the first British medical journal to use vital statistics: this information encouraged discussions of state medicine, public health, insurance, the census, and civil registration.[35] Due to Farr's influence, the *British Medical Journal* (*BMJ*) and the *Lancet* began to provide monthly reports on the vital statistics of Britain and foreign countries. As John Eyler has remarked, Farr's legacy with his colleague Thomas Rowe Edmunds was 'the creation of the modern discipline of vital statistics and using these statistics to assess public health and welfare'.[36]

Farr wanted to establish medicine on a firm basis through the classification and enumeration of disease, and to create a predictive social science based on numerical relationships. He went on to develop a series of 'statistical laws' used especially for his work on cholera and introduced the law of recovery, the law of epidemics, and the law of mortality. Farr's laws were not mathematically constant, as he hoped they might be: they were, instead, mathematical expressions of specific situations with results that varied as circumstances changed. They were thus not laws of epidemics as Farr thought, but rather they described various empirical distributions for different populations.

Beginning in the 1830s, a series of severe cholera outbreaks gripped London and spread rapidly to other parts of the country, killing thirty-two thousand people in just three months in 1831. Through various calculations, Farr discovered that the incidence of cholera was more prevalent on the coast where 50 in 10,000 died, than it was inland where the ratio was 17 per 10,000. From his findings, he deduced that the 'elevation in soil' (altitude) in London had a more consistent relationship with the mortality from cholera than with any other known factor.

The physician John Snow, one of the founding members of the London Epidemiological Society, showed instead that the link between mortality from cholera was due to polluted drinking water supplies, which had been contaminated in the water pumps with the bacteria *vibrio cholera*. Snow noticed that five hundred men, women, and children living within two hundred metres of a pump at the junction of Cambridge Street and Broad Street in Soho had died from cholera within ten days: it turned out that all those who died had drunk water from that pump. A few days later Snow persuaded the local officials to remove the handle from the pump, rendering it unusable, and the local epidemic began to subside. The science of epidemiology emerged in part through the recognition of associations between environmental factors and the development of diseases.[37]

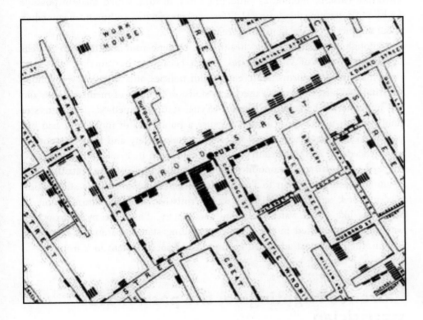

John Snow's cholera map.

Francis Bisset Hawkins, another founding member of the London Epidemiological Society, stressed the importance of detecting bias in medical observations. While Hawkins thought that a careful cultivation of vital statistics was the key to several sciences, because it would bring about a new philosophy of medicine, the medical statistician and professor of forensic medicine and hygiene at King's College, London, William Augustus Guy, advocated using statistical averages because he was convinced that this could help to illuminate the conditions of various human diseases. Guy learnt about using averages through the 'numerical method' developed by Pierre Louis, who thought that this method would confer status on the clinical physician, thus enabling him to create a science of medicine in the Parisian medical world. Louis's methodological innovation, which he introduced to evaluate the efficacy of different treatments used in traditional healing, was an arithmetical procedure for calculating the difference between two mean values for two groups and then subtracting the difference, but he erroneously assumed that a large difference signified a large effect in the treatment.

Louis's work was pursued by the German mathematician and physician Gustav Radicke, who realized that since the variation within each group was not taken into account, it was not possible to make meaningful comparisons by using Louis's method. Radicke thus proposed to measure this variation in each group with his 'index of dispersion'.[38] Although Radicke's procedure was a step in the right direction, it was not until 1908—when Karl Pearson's student, William Sealy Gosset (who used the pseudonym 'Student'), introduced his $z$-ratio as a statistical test for small samples—that this variation could be measured for small samples. This work enabled Pearson's successor, R. A. Fisher to transform and re-introduce Gosset's method as 'Student's $t$-test' in 1924, which made it possible for him to determine the statistical difference between two arithmetic means when small samples (of fewer than 20) were used.[39]

Although the British public health and hygiene movement led to a well-developed system of vital statistics, which became a bureaucratic device to implement social reforms and create health policies, this situation was unique to Britain. Whilst the French used vital statistics in areas of medicine, they did not implement a bureaucratic system of vital statistical methods for matters of public health in the 19th century, because a public health movement had not yet arisen in France, which was still an agrarian society and had thus not yet experienced the scale of health problems that arose in Britain's highly industrialized cities. Instead, economic, financial, and administrative statistics were of more pressing concern to French statisticians. In France little state action was pursued, sanitary legislation was permissive, and the introduction of amenities like piped water was slow and patchy. Thus, the move to action was most pronounced in the rapidly expanding state apparatus of Victorian Britain, then the most advanced economic and social system in Europe.

## Florence Nightingale: the passionate statistician

The statistical methods of Quetelet and Farr provided a conduit for health reforms for the mathematically inclined Florence Nightingale. Known to everyone as the 'Lady with the Lamp' who made nursing a respectable profession, it was, however, her role as the 'Passionate Statistician' that enabled her to have the greatest impact on nursing.[40]

Nightingale not only established nursing as a suitable career option for Victorian women, but revolutionized the profession through her use of statistics, since it was in her capacity as a statistician that she was able to introduce essential measures of sanitary reform in hospitals in the battlefield and in London. Using the new statistical methods and ideas of these vital statisticians, she persuaded various government officials of the importance of the lessons she learned in the Crimean war, showing that mortality rates could be reduced among the Army and in London once sanitary procedures became routine in hospitals.

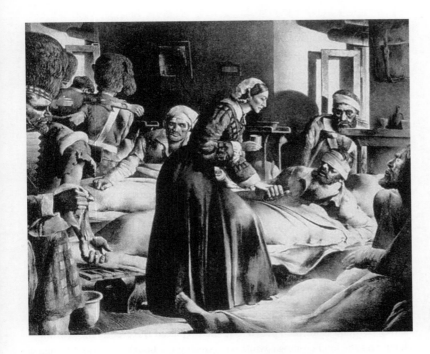

Florence Nightingale in a
Crimean hospital ward, c.1854.

Brought up in a liberal–humanitarian household, Nightingale's family were prominent members of the larger rising middle class. They were Christians who sought to improve social conditions for everyone and formed a part of the intellectual avant-garde who endorsed women's education.[41] Florence and her elder sister Parthenope benefited from their parents' ideas about education; they were fluent in French, German, and Italian and travelled with their family throughout Britain and the Continent. A graduate of Trinity College, Cambridge, their father William included Greek, Latin, and mathematics in their lessons.

By the time Florence was 9 years old, she was already organizing data in a tabular format.[42] As a young woman, she met a number of Victorian scientists at dinner parties, including Charles Babbage. Her fascination with numbers was so keenly developed at an early age that by the time she was 20 she began receiving two-hour instructions from a Cambridge-trained mathematician. In the mornings Florence would study material on the statistics of public health and hospitals, and eventually she accumulated a formidable array of statistical information on health and hospitals. Her enjoyment was so immense that she found the sight of a long column of figures to have been 'perfectly reviving'.

Nightingale regarded statistics as 'the most important science in the world', and maintained that 'to understand God's thought, we must study statistics for these are the measure of His purpose'.[43] She shared with Francis Galton the idea that the statistical study of natural phenomena was the 'religious duty of man'. Nightingale's ideology was rooted in the theology of the clergyman William

Derham, from whose ideas she developed her view that 'we learn the purpose of God by studying statistics'.[44]

Her mother did not approve of the hours that Florence spent studying mathematics, as she had been brought up with such advantages that she would fulfil her duty by getting married; but Florence had no interest in marriage. Much to her parents' consternation, she announced that she wanted to work as a nurse at Salisbury Hospital for several months. She then confessed that she had the idea of eventually setting up a house of her own to establish 'something like a Protestant Sisterhood, without vows, for women of educated feeling'.[45] Although she studied nursing practices in Paris, Rome, and Kaiserswerth (near Frankfurt), she eventually realized that 'there was no one to teach her to nurse' and that women would have to be trained by lay professionals.[46]

By 1852 her reputation in society, combined with her desire to professionalize the field of nursing, brought her to the attention of the ladies' committee of an Institution for Ill Gentlewomen, who offered her the position of Superintendent at this clinic for sick governesses, financed largely through charitable gifts. After successfully negotiating the terms of the work she was to undertake at 1 Harley Street, then becoming established as a pre-eminent area for medicine in London, she hired the Harley Street physician William Edward Stewart. In her new position, she installed a supply of hot water to all floors and a dumb-waiter to deliver hot food from the kitchen. She then replaced the grocer, the coal merchant, and the kitchen range, while reorganizing the bookkeeping and bringing the accounts into order.

A year after Nightingale improved the conditions in the hospital, Sidney Herbert, her life-long friend and Secretary at War, asked her to be 'Superintendent of the female nursing establishment in the English General Military Hospitals in Turkey' for the British troops fighting in the Crimean war, and to take a group of thirty-eight nurses with her. Given her years of training and advocacy for professional nursing, Herbert regarded Nightingale as the only person in Britain who had the unique qualifications for this insalubrious work.

Nightingale and her nurses worked alongside the pensioners and recovering soldiers who traditionally served the sick and wounded during a war. Herbert had responded to public outrage at the reports, expressed in *The Times*, of the suffering of the common soldier caused by the incompetence of the British Army commanders. He hoped that Nightingale's presence could pacify the public. Readers of *The Times* donated £7000 for her personal use, which was eventually used to improve hospital conditions, but it also inspired jealousy among Army doctors and officers. After the ship landed in Marseilles, Nightingale used some of the money to buy such goods as tinned foods, clothing, bed linen, and portable cooking stoves.

Once she arrived in the army base hospitals in Scutari in the Crimea, she found herself amidst utter chaos at the Barrack Hospital: there was no furniture, food, cooking utensils, blankets, or beds: rats and fleas were constant problems. Although she was able to get basins of milkless tea from the hospital, the same basins were used by the soldiers for washing, as well as eating and drinking. She drew the government's attention to problems that went far beyond her sphere of influence, and exposed the administrative incompetence and disorganization of the British military. Highlighting how the British failed to respond adequately to exigencies of war, *The Times* reported that the French were far better organized in medical matters.

Nightingale had already developed an exceptional capacity for large-scale organization and administrative reform, although some of the medical men, who did not want to relinquish any control in the Barrack Hospital, thought she lacked patience and the ability to compromise. With the appeal money from *The Times*, she was the only person with funds and the authority to rectify this bleak situation: she requested eating utensils, shirts, sheets, blankets, stuffed bags for mattresses, operating tables, screens, and clean linen, and wanted soldiers' wives to be employed as washers. Nightingale soon set up a laundry and a kitchen, with much of the food supplied by Fortnum and Mason.

During the first eighteen months of the Crimean campaign, she discovered in those men at the Barrack Hospital in Scutari an annual mortality rate of 60 per cent from such diseases as typhus, typhoid, and cholera—a rate even greater than the great plague in London. She found that, between the ages of 25 and 35, the mortality rate of 20 per 1000 in military hospitals was double that of civilian life. She wrote a report based on the army medical statistics and sent it as a confidential communication to the War Office and Army Medical Department.

From her findings, she requested the formation of a Royal Commission on the Health of the Army in November 1856. Due to Nightingale's influence William Farr was appointed a member: he offered her advice and they collaborated in the preparation of hospital statistics for her *Notes on Hospitals* (1859) and *Introductory Notes on Lying-in Institutions* (1871). One of the main outcomes of the statistical aspect of the Royal Commission was the creation of a department of Army Medical Statistics.

Nightingale popularized the use of pictorial diagrams for statistical information, with her development of *polar area graphs*, which were cut into twelve equal sectors, one for each month of the year, that revealed changes over time.[47] Her graphs dramatized the extent of the needless deaths amongst the soldiers during the Crimean war, and were used as a corrective tool to persuade the medical profession that deaths were preventable if sanitation reforms were implemented in hospitals.

In 1858 Nightingale returned to England and began examining London's hospital statistics. She found not just simple carelessness in the collation of statistical information, but a complete lack of scientific coordination. Hospital statistics gave little useful information on the average duration of hospital treatment or on the proportion of patients who recovered, compared with those who died. Such chaos in the management of hospital records was familiar territory for Nightingale, given her experience in the Crimea.

A few months after she returned, a meeting was to be held at the National Association for the Promotion of Social Sciences in Liverpool. She prepared a paper on the sanitary conditions of hospitals in London and Scutari, and these findings led to her book, *Hospital Statistics and Plans*. Her first recommendation was the adoption of the nomenclature of disease used by the Registrar General of

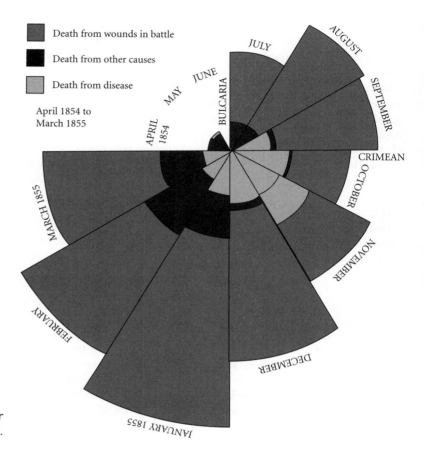

**Legend:**
- Death from wounds in battle
- Death from other causes
- Death from disease

April 1854 to
March 1855

BULGARIA

CRIMEAN

Months (clockwise from top): JULY, AUGUST, SEPTEMBER, OCTOBER, NOVEMBER, DECEMBER, JANUARY 1855, FEBRUARY, MARCH 1855, APRIL 1854, MAY, JUNE

Florence Nightingale's polar
area graph.

England and the use of standardized forms to collect hospital data: the total sick populations (the number of beds in use), the number of cases of diseases receiving medical or surgical treatment (listed by age, sex, and disease), the average stay in hospital (again classified by age and sex), and the annual proportion of recoveries compared with the beds occupied.

Nightingale wanted to establish a standardized form that would list causes of death, and create divisions for death by disease or following operations. Her hospital model forms were printed in 1859 and were adopted by St Bartholomew's, St Mary's, St Thomas's, and University College Hospitals, and by Guy's Hospital in 1861. Guy's advocated that each hospital should publish its own statistics annually. The results of these hospital reports were printed in the *Journal of the Statistical Society* in 1862. Although the forms were used for a period of time, the hospitals eventually found them too costly and time-consuming to administer. Due to her skills in deploying statistical methods in army sanitary reform and in London hospitals, she was the first woman to be elected a fellow of the Royal Statistical Society (in 1858) and an honorary foreign member of the American Statistical Association (in 1874).

In 1860 Nightingale met Quetelet, when he attended the International Statistical Conference held in London. At this conference she argued for the establishment of uniform medical statistics for hospitals. Although her forms were being used in Britain, she also wanted these forms to be adopted throughout Europe, thereby allowing comparisons to be made with various European hospitals. By then she had set up her Training School for Nurses at St Thomas's Hospital, where she not only emphasized the vocational side to nursing, but also invoked trust in God's mystical providence.

Later that year she became involved with the 1861 Census, when she persuaded the GRO to extend its scope by collecting statistics that would serve as a foundation for sanitary reform; she achieved this by including a census count of the sick and infirm who had previously been neglected. Rightly convinced of the importance of the relationship between health and housing, her second aim was to obtain complete data on the housing of the population, and, although some thought this would be too cumbersome, Nightingale retorted that it had already been done in Ireland.

## A plea for teaching statistics

Nightingale had long been aware that Members of Parliament had access to an enormous amount of statistical data, but made no use of this information. She was concerned that while these men had a university education, they had received no training about statistical methods, and she wanted to establish the teaching of statistics in universities. On 8 November 1872 she wrote to Quetelet to discuss the possibility of setting up a course of statistics at Oxford University, long established as the place where most future Members of Parliament pursued their studies.[48]

As we saw in Chapter 2, on learning of Quetelet's death in February 1874, she thought that the only fitting memorial to him was to establish the teaching of his statistics at Oxford. Her idea to establish a department of 'Applied Statistics' at Oxford was partly due to her association with her old friend Benjamin Jowett, Master of Balliol College and a classical scholar who had translated the major works of Plato and Aristotle into English. Nightingale had developed such a competent knowledge of Greek literature that Jowett even consulted her while writing the introduction to his translations of the dialogues of Plato.[49] From their first meeting in 1862, they developed a close friendship via their discussions about religion. Jowett felt considerable affection and devotion towards Nightingale and wanted to help her to set up a series of statistical lectures at Oxford.

However, it was not until 1891 that Nightingale returned to this project, and by New Year's Eve Jowett had a proposal for endowing a chair of statistics at Oxford. They consulted Francis Galton about the details, but Galton was not encouraging: he thought that it would be nearly impossible to set up a new professorship at Oxford or Cambridge, because the subject was not part of the undergraduate examinations, and the plan sank into oblivion. Nevertheless, by 1891 the role of economic statistics clearly fitted into the academic ethos of Oxford University, as witnessed by the appointment of the economist and statistician, Francis Edgeworth, to the Drummond professorship of political economy at All Souls College.[50]

Galton suggested instead that it would be more suitable to train applied statisticians by six lectures a year at the Royal Institution. Karl Pearson remarked later that he thought Nightingale had the better scheme, although the 'Royal Institution was valuable for announcing in a popular way the results of recent research, it was not an academic centre for training enthusiastic young minds to a new department in science'.[51] Nonetheless, Pearson thought that Nightingale's efforts to create such a chair should somehow be commemorated. In 1911, some twenty years later, he was looking for a new name for his newly established statistics department at University College, London, and he held that 'no fitter and worthy name occurred to me than that of Applied Statistics'. More than a century would pass before the Department of Biomathematics at Oxford was renamed the Department of Applied Statistics in 1988.

Nightingale was a practical statistician who showed how statistics could be a powerful tool when used judiciously. Her discerning examination of statistical data informed her that she could save the lives of the wounded and the sick by implementing essential and appropriate sanitary measures. By systematizing record-keeping practices at Scutari, she established the necessity of ensuring that hospital statistics were standardized across all hospitals in London and the rest of Britain. For Nightingale, statistics were a form of currency that would be of little value if they were simply accumulated and put aside; she thus demonstrated that their real importance lay in their subsequent analysis and interpretation. To her, statistics was not simply about collecting data or compiling numerical figures—the subject had important political and practical value that could lead to medical and social improvements, parliamentary reform, and, ultimately, the saving of human lives.

Francis Ysidro Edgeworth (1845–1926).

## Jevons and Edgeworth: the calculus of hedonics

The other side of the statistical pain-and-illness spectrum involved the measurement of pleasure. This was undertaken by two mathematically trained economists and statisticians, Stanley Jevons and Francis Edgeworth. Like so many other vital statisticians, Jevons was particularly interested in measuring social phenomena. In 1858, after five years in Australia where he had been working as an assayer for the British Royal Mint in Sydney, he produced an unpublished manuscript, *Social Statistics or the Science of Towns especially as regards London and Sydney*, which has been regarded as a 'pioneering exercise in urban sociology'.[52] Jevons's principal statistical innovations were linked to his work on index numbers, used in economics, and his statistical atlas, which contained a compilation of charts and diagrams that led to his contribution to time-series analysis.

Jevons thought that there was a direct relationship between a psychological state of well being and the act of engaging in economic activities, such as buying and selling, or borrowing and lending. For Jevons, economics was fundamentally quantitative, and thus amenable to mathematical analysis. His introduction of mathematics into economics in *The Theory of Political Economy* in 1871 helped to establish neoclassical economics, thereby distinguishing it from the classical economics of Adam Smith and David Ricardo.[53] Jeremy Bentham had used the *felicific calculus* as an algorithm to derive the degree or amount of happiness that a specific action is likely to cause; this algorithm was also known as the *hedonic calculus*, *hedonistic calculus*, and the *utility calculus*. Jevons used Bentham's utilitarianism to construct a mathematical theory of economics, adopting the hedonic calculus, which enabled him to transform sensations into quantities.[54]

One of Jevons's neighbours in Hampstead, north London, was Francis Edgeworth, whose book *Method of Ethics* (1877) was influenced by Jevons's *Theory of Political Economy*. In a similar vein, Edgeworth applied mathematics to ethics by using the calculus of variations for his hedonic measures to calculate the maximum amount of pleasure.[55] He developed his hedonical calculus more fully two years later.[56]

Like the mid-Victorian social statisticians, most of whom were influenced by the statistical ideas of Adolphe Quetelet, the economic statisticians developed their statistical methods by borrowing the mathematical tools from astronomers. While both groups introduced some new statistical ideas, neither developed a fully fledged methodology or a statistical school to promulgate such statistical methods. Perhaps part of the problem, as Stephen Sigler[57] observed, was that during this time sociology and economics lacked a single unifying theory, which made it difficult for practitioners to formulate their own statistical methodology. In contrast, as we shall see, the biological sciences offered Darwinism which, although still highly contentious in some circles, had the potential to create a unifying theory: Darwin supplied the crucial statistical material of biological variation for the next generation of mathematically trained statisticians to produce a new statistical methodology.

# Conclusion

The early-modern population arithmetic underpinned by the mercantile book-keeping methods used in John Graunt's *London Bills of Mortality* highlighted the conditions of the poor, and suggested the need to create health policies to prevent infectious disease. This work, and that taken up in all European countries in the 18th century, meant that economic, medical, and social issues could then be addressed numerically.

These developments precipitated an avalanche of social statistical data, which was amassed by the mid-Victorian vital statisticians who regarded their work as a distinct discipline within the social sciences. The market for vital statistics had been fuelled by the dramatic growth in the population of Victorian Britain, that registered eight million people in 1801 and swelled to forty million by 1901; moreover, the endemic scale of public health problems in the middle of the 19th century, combined with the appalling sanitary conditions in hospitals in the battlefield, and in the growing metropolis of London with its unacceptably high mortality rates, demanded a quantitative solution. The bureaucratic compilation of the vast amount of vital statistical data gathered in the mid-19th century enabled Chadwick, Farr, and Nightingale to identify the source of those problems, which led to the implementation of various public health policies to eradicate the unsustainably morbid living conditions of the Victorians. Farr's pivotal role in the professionalization of vital statistics led him to redefine statistics as a method of analysis, rather than as a discipline for the social sciences.

The role of biology remained influential for the next generation of late-Victorian statisticians, but the impetus for change was, instead, inextricably linked to evolutionary biology, one of the most pervasive ideas of Victorian life. Darwin's understanding of the crucial role of biological variation piqued Francis Galton's interest, which led him to devise a series of statistical methods to measure these individual differences. The concerns of such Darwinians as W. F. R. Weldon, about understanding species formation and finding empirical evidence of natural selection, shifted the direction of statistics—especially after Weldon accumulated a massive amount of zoological measurements that required the analysis of statistical data for problems that Galton's methods could not resolve. These circumstances led Karl Pearson to devise the infrastructure of his statistical methodology. With his colleagues and students he established the discipline of mathematical statistics in the closing years of the 19th century, as we shall see in the next chapter.

Karl Pearson (1857–1936).

# CHAPTER 12

# Darwinian variation and the creation of mathematical statistics

M. EILEEN MAGNELLO

Charles Darwin's prodigious work, which underscored the importance of biological variation for understanding the role of species formation and in establishing his theory of natural selection as the mechanism of evolution, helped to create a disciplinary and ideological bifurcation in statistics. While vital statisticians emphasized averages and regarded variation as a source of error, for the biometricians (Francis Galton, W. F. R. Weldon, and especially Karl Pearson) Darwinian biological variation provided the material from which they could erect a new kind of mathematically based statistical methodology. This development led to the creation of new techniques to measure and analyse statistical variation, correlation, regression, and goodness-of-fit.

As we saw in the previous chapter, the etymological and disciplinary trajectory of statistics began with the early-modern Germanic qualitative system for describing the state, which was then transformed to an Anglo-French numerical system of social statistics that was regarded as a separate academic discipline in the social sciences. This view changed in the mid-19th century when William Farr redefined statistics as a method of analysis. Statistics was to be transformed again at the end of the 19th century due, in part, to the growing professionalization of the sciences and medicine that were becoming more mathematical and, ultimately, to being anchored to the mathematization of Darwinism.

Charles Darwin (1809–83).

Charles Darwin's ideas about species formation (speciation), natural selection, and individual biological variation not only challenged the mid-Victorian vital statisticians' ideas of the role of statistical variation, but made it possible for them to analyse individual variation rather than ignore it. Moreover, the analysis of this Darwinian variation, which was largely due to the efforts of Karl Pearson along with his colleagues and students, crystallized the establishment and professionalization of the discipline of mathematical statistics in the early years of the 20th century.

This chapter is centred on a late-Victorian mathematical enterprise that led to the establishment of the academic discipline of mathematical statistics. Building upon some of the developments that emerged from the mathematical theory of probability in the late 18th century, via the work of such Continental mathematicians as Jacob Bernoulli, Abraham De Moivre, Pierre-Simon Laplace, and Carl Friedrich Gauss, this Victorian venture arose collectively through the work of Francis Galton, W. F. R. Weldon, and Karl Pearson, with contributions from Francis Ysidro Edgeworth and George Udny Yule.

Unlike vital statistics, mathematical statistics encompasses a scientific discipline that provides a methodology to analyse variation and is often underpinned by matrix algebra. This measurement and analysis of statistical variation is the lynch-pin of modern mathematical statistics, which deals with the collection, classification, description, and interpretation of data obtained by social surveys, scientific experiments, and clinical trials. Probability is used here for statistical tests of significance. This type of statistics is analytical, and can be used to make statistical predictions or to make inferences about the population; furthermore, it capitalizes on all the individual differences in a group by examining the spread and clustering of this statistical variation through such methods as the range and standard deviation.

The transition from the Victorian vital statistics that emphasized calculating average values to mathematical statistics that centred on the measurement of statistical variation, represented an ideological shift that occurred during the middle of the 19th century, when Charles Darwin began to study minute biological variation in plants and animals. When he suggested in 1859 that evolution proceeded by the accumulation of minute differences between individuals, he introduced the idea of continuous variation into biological discourse. Darwin had not only shown that variation is measurable and meaningful, by emphasizing statistical populations rather than focusing on one type, but had also discussed various types of correlation that could be used to explain natural selection.[1] Karl Pearson recognized the fundamental statistical concepts that underpinned Darwin's work, as 'every idea of Darwin, from variation, natural selection, inheritance to reversion, seemed to demand statistical analyses'.[2]

Inspired by Darwin's ideas on biological variation, his half-cousin Francis Galton began to devise statistical methods to measure it. Like Darwin, Galton worked as an independent scientist and never held a university post; he was, however, very active in many learned societies, including the British Association for the Advancement of Science, the Anthropological Institute, the Geographical Society, the Royal Institution, and the Royal Society. Inquisitive about the variation in humans, he looked for ways to measure the individual differences that Darwin made central to his work. And as Chris Pritchard has shown, George Darwin (Charles's son) also played a crucial role in Galton's development of statistics.[3]

Galton's creation of statistical scales, his ranking methods, the introduction of the median as a measure of central tendency, his semi-interquartile range (which was the first statistical method to measure Darwinian variation), and his work on correlation and regression heralded incipient changes in the development and growth of the statistics of variation. Subsequently, Galton's work captured the attention of the Darwinian zoologist W. F. R. Weldon, who graduated from Cambridge in 1881 with a first-class degree in the natural science tripos. Although Weldon first met Galton in Swansea in 1879 at the annual meeting of British Association for the Advancement of Science, they did not develop collegial relations for another decade.

## The biometrical species criterion

The first attempt to provide an empirical measure for the identification of species was considered by Galton in 1888; this was tested empirically some three years later by Weldon, when he employed Galton's method of correlation, in one of its earliest uses.[4] Although Weldon pursued Galton's idea, their work displayed two distinguishing features that were to affect their working relationship with Karl Pearson. First, Weldon was ready to break away from the essentialistic thinking derived from his morphological training at Cambridge, whereas Galton's belief that all frequency distributions should conform to the normal distributions suggested that he retained a strong typological commitment. Secondly, in the 1890s Weldon was considerably more interested in Darwin's ideas of species formation and theory of natural selection than Galton was. Weldon's aim to find a quantitative solution for these problems would eventually demand a statistical treatment that differed from the statistical methods that Galton had devised.

Weldon was looking for a way to find a working hypothesis for Darwin's variation, since the embryological and morphological methods he learnt at Cambridge did not enable him to examine the biological variation that Darwin emphasized. In the autumn of 1889 Weldon took up the Jodrell chair of zoology at University College, London (UCL); he went to Plymouth that summer to continue his work on the variation of marine organisms and began to use Galton's statistical methods in May.[5] By then, he had read Galton's *Natural Inheritance*, wherein Galton had shown that the deviations from the average size of organs in man, domesticated plants, and animals had followed the curve of error (the normal distribution). Weldon realized immediately that there was considerable material that remained unexplored, including 'species living in a wild state upon which natural selection acts'.[6]

With a wealth of new material to be examined statistically, Weldon embarked on this project in 1889, and one year later he showed that the variation in five pairs of organs in shrimp (*Crangon vulgaris*) was normally distributed.[7] With his statistical supposition of normality confirmed, Galton then suggested to Weldon that he ought to examine the relationship between two organs to see whether this would tell him if his collection had different groups of species or the same one with new species emerging.[8] Weldon was now prepared to test Galton's proposition that correlation could be used as a diagnostic criteria to identify membership in a species, using a 'biometrical species criterion'.[9] Weldon decided to investigate this numerical identification by using Galton's correlation, calling them 'Galton's functions', because he was examining relationships that could be described as *functional* (that is, when the size of one organ depends on the other). As a committed Darwinian, Weldon was undoubtedly supporting Darwin's idea of functional correlation.

To test the constancy of Galton's correlation, Weldon collected data from two thousand, nine hundred and eighty shrimp in the coastal cities of Plymouth, Southport, Sheerness, Roscoff in the north of France, and Helder in North Holland. He measured the total length of the carapace and the lengths of sinuous positions in samples of adult female shrimp, and found that the correlation values of the organs of the shrimp were nearly identical for each race (the values were all around 0.80). This encouraged him to think that he had found a criterion for the identification of species. Weldon was optimistic that a numerical measure that could be used for species of animals would be forthcoming, since 'the functional correlation between various organs ... could lead to the establishment of the great sub-divisions of the animal kingdom'.[10] In 1892, Weldon also sent a copy of this paper to Francis Edgeworth, who had just introduced the term 'coefficient of correlation'.[11]

Many of the statistical methods Galton devised, such as that used to calculate the median, could be used for subsets of data that were small enough to be counted. When Galton worked with a larger sample size, he often reduced the sample to one hundred because of the explanatory power of percentages. However, when Weldon needed to examine one thousand crabs in Naples and another thousand in Plymouth, Galton's methods became inadequate for Weldon's data. Since Galton's main interest was in anthropology and human heredity, he was not particularly interested in classifying and measuring vast quantities of zoological information. He was thus not drawn to problems of evolution or natural selection in the way that Weldon was; however, Weldon would soon find the perfect ally in Karl Pearson.

## Pearson's earliest statistical work

Third wrangler in the Cambridge mathematical tripos of 1879, Karl Pearson became professor of applied mathematics at University College, London, in 1884 where he taught mathematical physics to engineering students. As we saw in Chapter 3, in 1890 he also took up the Gresham chair of geometry at Gresham College in the City of London, where he first began to teach statistics. This

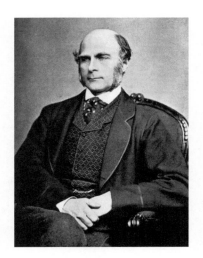

Francis Galton (1822–1911).

educational foundation offered series of free one-hour lectures, beginning at 6 p.m., to members of the public. As Gresham professor, Pearson was responsible for giving twelve lectures annually, and delivered thirty-eight lectures from 1891 to 1894.[12]

Pearson aimed to make geometry accessible for his audience of clerks, artisans, the industrial class, and others who worked in the City, and he wanted to introduce them to a way of thinking that would influence how they made sense of the physical world. Whilst his first eight lectures formed the basis of his *The Grammar of Science* (1892), the remaining thirty lectures dealt with Pearson's earliest ideas about mathematical statistics. As he could not lead the audience through the 'mazy paths of mathematical theory', he appealed to them by using graphs, geometrical figures, and illustrations, to teach statistics and deduction by easy arithmetic. Endorsing Sir Thomas Gresham's 16th-century ideal that lectures should address the application of science to practical matters, Pearson chose statistics as a topic for this group, since he thought that they would understand insurance, commerce, and trade statistics, and could relate to games of chance involving Monte Carlo roulette, lotteries, dice, and coins.

For Pearson, these problems were 'more readily deducible from the geometrical than from the numerical representation of statistics'.[13] The geometry of statistics was not only about the graphical representation of data, nor simply a fundamental process of statistical enquiry, but it was also a mode of ascertaining numerical truth and of statistical research.[14] Echoing Adolphe Quetelet's ambition for social statistics in the 1830s, Pearson proclaimed in his first Gresham lecture in statistics on 17 November 1891 that he was:[15]

now in a position to define statistics scientifically. Even [as we are] on the very threshold of this scientific realisation, it is clear that a great future awaits our present statistics and we may reasonably anticipate that the combination of statistics and analysis will create a science which will excel every other branch of mathematics, including astronomy, mechanics and physics.

Pearson could see that the physicist and the biologist were making greater use of statistical methods, and he was convinced that once this became more widely recognized it would be 'impossible to define statistics as having any special reference to human society, for it will pass from sociology to mathematical science'.[16]

Pearson's quantum leap from social and vital statistics to mathematical statistics facilitated the establishment of the academic discipline of mathematical statistics, and his dedication to this enterprise was to occupy much of his work for the rest of his professional life. His Gresham College lectures stimulated such an interest that his audience 'increased five to ten-fold in the first couple of years, and by 1893, nearly 300 people were attending his lectures'.[17] Pearson's success in attracting an audience of this size may have also played a role in encouraging him to develop his work in mathematical statistics.

Walter Frank Raphael Weldon (1860–1906).

# The analysis of Darwinian biological variation

Weldon first met Pearson in the autumn of 1891 at the Association for Promoting a Professorial University for London; this association endorsed a move to establish a university in which professors alone would teach and would largely control the university. The main idea was to unite all of the London lecturers so that the separate colleges would then become absorbed. The authorities supported a scheme that would have united King's College and University College to form what was widely regarded as a second-rate London University, to be named the Albert University: the academic rebellion was widespread and the project was defeated. The Association was, nevertheless, important in establishing the University of London Act of 1898, which reconstituted the University, admitting UCL—and several other London institutions—as a constituent college in 1900. Weldon was a tireless campaigner, and his commitment to Pearson marked the beginning of a lifelong friendship.

From 1891, until Weldon took up his post at Oxford in 1899, they saw each other almost daily and often several times a day. Since they both lectured from 1 to 2 p.m., they ate lunch together, which became the scene of many friendly debates where problems were discussed, solutions created, and even worked out on the back of the menu with pellets of bread. Members of staff who shared their table, such as Francis Oliver (Quain professor of botany), saw the tossing of pennies and arranging of crumbs, accompanied by pots of beer, and regarded such activities as a part of a regular lunch. From the beginning, Pearson and Weldon's relationship took on an emotional and intellectual intimacy.

Pearson regarded Weldon as 'one of the closest friends he ever had' and valued his opinions more highly than those of anyone else. He also saw in Weldon someone whose ideas meshed with those that he had been developing in his Gresham lectures. Because Weldon's impact on Pearson led him to change the direction of his career, Pearson acknowledged Weldon as the person 'who changed the whole drift of my work and left a far deeper impression on my life' than anyone else.[18] Although Pearson was to find himself on the threshold of creating a new kind of statistics at the end of 1891, it was not until the following year, when Weldon asked him for his advice on the data from his Naples crabs, that Pearson's early statistical ideas came to fruition.

Pearson delivered eighteen of his Gresham lectures on statistics and probability before seeing Weldon's data. Many of these lectures dealt with games of chance and mathematical probability, and some were influenced by the work of such statisticians as the Cambridge logician John Venn, Francis Edgeworth, and Stanley Jevons. By November 1892, Pearson was beginning to examine the zoological data that Weldon insisted was 'the keynote to the scientific measurement of the force of evolution'.[19]

During the Easter vacation of 1892, Weldon collected data from a thousand crabs in Plymouth Sound, and later that summer he and his wife, Florence Joy Tebb, collected twenty-three measurements from a thousand adult female shore crabs in the Bay of Naples. Like so many Victorian women who were married to scientists, Florence Tebb (who was educated at Girton College, Cambridge) helped him with the many laborious calculations of thousands of figures. One study involved taking various measurements in duplicate from eight thousand and sixty-nine crabs. There were no hand calculators, and the Brunsviga (a cheap, noisy, efficient, and sturdy calculating machine) was not readily available until 1895. After they had done some preliminary calculations at the end of that summer, Weldon sent a diagram of the frontal breadth (the forehead) of the carapace of these one thousand female crabs to both Galton and Pearson on 27 November.

Weldon found that all but one of the twenty-three characters he measured was normally distributed, thus confirming Galton's assumptions for these characters. When Weldon discovered that the distribution of the forehead of the carapace was not normally distributed, but was 'double-humped' (or bimodal), he thought that Galton would be interested in his results, because this bimodal curve might have meant that a new species had arisen.

In addition to doing some arithmetical calculations, Weldon tried to draw two normal curves inside it to help him to interpret the data; his attempt to break up the double-humped curve into two normal components was derived from Galton's belief that all distributions should be normally distributed. Weldon then concluded that either the crabs from Naples were two distinct races, or that they were in the process of creating a new species. He also seems to have been exploring Galton's claim that a new species could be due to the sudden appearance of 'sports' or sudden evolutionary jumps, saltations, or mutations, which could ostensibly have produced a new species (or what was known as 'instantaneous speciation'). By then, Weldon was convinced that the 'problem of animal evolution [was] essentially a statistical problem'.[20]

Weldon then asked Pearson whether he thought that the curve should be broken up into two groups. Pearson wanted to know if it would be possible to distinguish whether the species was really evolving into two different types, or whether the two groups were dimorphic (with two different varieties). He proceeded to look for another way to interpret the data without trying to normalize it, as Galton and Quetelet had done. Indeed, he was so committed to finding a statistical solution to this problem that he spent nearly his entire summer vacation in 1893 working on Weldon's crab measurements to find statistical resolutions of the data. Pearson and Weldon considered it important to make sense of the shape of the curve without distorting its original shape, as it might have revealed something about the creation of new species. For Pearson, 'the keynote to the most interesting problems in evolution [lay] in the non-symmetry of the frequency curves corresponding to the special organs in animals'.[21]

That Pearson would go on to create a new type of statistics was largely in response to the unshakable conviction, held by so many vital statisticians, mathematicians, and philosophers, that the normal distribution was the only feasible distribution for the analysis and interpretation of statistical data. Such was the tyranny of the normal curve that, by the end of the 19th century, most

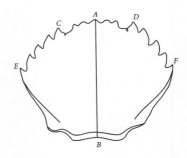

Carapace of a shore crab.

Weldon's hand-drawn double-humped curve.

statisticians assumed that no other could be used to describe empirical distributions, but this monolithic view was challenged by Pearson in the last decade of that century. Furthermore, the transition in measuring biological variation, rather than simply calculating averages, enabled Pearson and Weldon to create the tools that led to a statistical revolution when they translated Darwin's ideas about what kinds of natural processes occur in the world into statistical concepts.

Seven months later, Pearson showed Weldon's data to his Gresham College audience. He told them that if they applied the theory of chance to the existing biological problems of evolution, this would completely change their ideas of the problems of natural selection and heredity. At the very least, it would place the doctrine of evolution on a sound exact basis, eliminating the chaos and confusion that many 19th-century scientists held about evolution. Pearson's iconoclastic pedagogy encouraged his students to develop an awareness that they were living 'in an essentially critical period of science, when more exact methods and more sound logic were upsetting or modifying many of the whole statements, which had been taken for years as scientific gospel'.[22] He concluded that it was 'largely to Weldon that we owe this attempt to give an exact measure to the problem of evolution; and that Weldon's intense, laborious and careful measurements on the organs of shrimp and crab was the first step in the right direction'.[23]

## A new statistical methodology

Weldon wanted to measure the variation in an attempt to gain some understanding of evolution, to determine how a new species emerged, and then to detect empirical evidence of natural selection. He needed a statistical system that could measure the variation of his crabs, and one that could systematically handle large amounts of data; this was essential in the search for disturbances in the distributions from normality, to detect empirical evidence of natural selection. To help Weldon, Pearson had to create a formalized system of frequency distributions that could accommodate large sample sizes (more than one thousand), and to develop a system that did not rely on the normal distribution.

Until Pearson developed his new statistical system, the only procedures available for classifying a massive amount of data involved simple tabulation, the use of pie charts and various diagrams, or the reduction of data to smaller subsets, as Galton and various mid-Victorian vital statisticians had advocated. Some of these methods could be unwieldy, especially since the charts or tables were not standardized, and so comparisons or generalizations with other data sets could not be made. Once Pearson realized that a standardized system of frequency distributions was needed, so that everyone plotted data in the same way, it became possible to make comparisons and generalizations about data sets that had previously been impossible. Although the Victorian economist Alfred

Marshall envisaged that graphical methods ought to be standardized, it was Pearson who succeeded in providing the requisite methodology.[24]

On 18 November 1891, in his lecture on 'Maps and chartograms', Pearson showed his audience how to construct a frequency distribution by introducing the idea of a *histogram*, a term he coined to designate a 'time-diagram'; he explained that the histogram could be used for *historical* purposes to create blocks of time of 'charts about reigns or sovereigns or periods of different prime ministers'.[25] A *histogram* is a graphical version of a set of continuous data (such as time, inches, or temperature) that shows how many cases fall into contiguous rectangular columns. It is similar to a *bar chart*, except that the latter has gaps between the bars and uses discrete data (such as gender and political affiliation); both of these pictorial graphs are commonly used to depict statistical data. The next stage was to show his audience how to assemble frequency distributions for larger quantities of data, and how to construct these distributions.

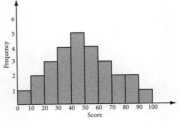

A histogram.

European Parliament election, 2004

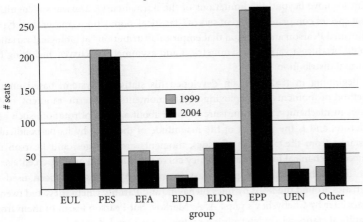

A bar chart.

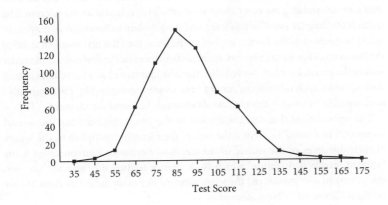

A frequency polygon.

Although Pearson used his histogram comprised of several blocks for continuous data, another way to depict the same set of data is to use a *frequency polygon*. This is a graph that joins with straight lines the mid-points of the bars of a histogram. Pearson was not necessarily looking for a specific type of curve, but he needed to know the shape of the curve that the data produced. This was the first step in Pearson's statistical innovations—not only had he provided the means to make sense of data that had hitherto not been analysed, but he was able to introduce rigour into scientific articles, and would go on to modernize the style of scientific writing in the 20th century.

## The method of moments

Pearson then adapted the mathematics of mechanics to construct a new statistical system for interpreting Weldon's data, since no such system existed at the time. This system enabled him to analyse data of all kinds of shapes, thereby permitting him to move beyond the limitations of the normal curve. Darwin's recognition that species comprised different sets of 'statistical' populations, rather than types, informed Pearson and Weldon that empirical distributions of biological variation could produce a variety of symmetrical and asymmetrical curves, including the normal distribution.

Beginning in 1892, Pearson developed his statistical system, based on the method of moments, by deploying higher moments. The term 'moment' originates in mechanics: it is a measure of force about a point of rotation, such as a fulcrum, and is the product of the magnitude of the force by its perpendicular distance from the point. In statistics, moments are averages, and Pearson replaced mechanical force by a frequency curve function, such as the percentage of the distribution within a given class interval.[26] Moments had been used in statistics around the 1850s by Jules Bienaymé and Pafnuty Chebyshev, and twenty years later by Chebyshev's pupil A. A. Markov, but Pearson learnt of them from graphical statics while determining moments on a loaded continuous beam.

Pearson used the method of moments to show his audience how to calculate the mean, by determining the point about which the lever balances on the fulcrum. The mean is the 'balance point' of this lever, and is equivalent to the centre of gravity (or mass) in mechanics. If a force is applied to the lever, then the first moment is called the 'moment of force': calculations are made to determine the first moment in order to find the mean. Pearson showed that 'the first moment of any set of lines at unit distance from each other is the sum of their lengths multiplied by their respective distances from a parallel straight line about which he found the moment'.[27]

The second time that force is applied to the lever, this becomes the second moment. The second moment is the sum of their lengths multiplied by the *square* of their distances or the square of the standard deviation: Pearson called it the 'squared standard deviation' which R. A. Fisher termed the 'variance' in 1918; thus, Pearson also introduced the term *standard deviation* in his Gresham lecture on skew curves on 23 November 1893.

The third moment is the sum of their lengths multiplied by the *cube* of their distances, and is used to find a measure of the skewness of a distribution; it indicates whether the lever balances on the fulcrum: an unbalanced lever is analogous to an unbalanced normal distribution or a frequency distribution that is skewed. The fourth moment is found by multiplying the lengths by the fourth power, and measures how flat or peaked is the curve of the distribution; for the fourth moment Pearson coined the word *kurtosis* (from the Greek word for bulginess). It has three further components:

- if data is clustered or peaked around the mean, he called the peaked-ness *leptokurtic*;

- if it spreads out across the distribution, the curve is *platykurtic*, for it resembles the shape of a platypus;

- if it produces a normal curve, this is termed *mesokurtic*.

When William Sealy Gosset ('Student') was teaching statistics in the 1920s, he used an illustration of a platypus to show a platykurtic curve, and two long-tailed kangaroos for a leptokurtic curve.[28]

* In case any of my readers may be unfamiliar with the term "kurtosis" we may define meso-kurtic as "having $\beta_2$ equal to 3," while platykurtic curves have $\beta_2 < 3$ and leptokurtic $> 3$. The important property which follows from this is that platykurtic curves have shorter "tails" than the normal curve of error and leptokurtic longer "tails." I myself bear in mind the meaning of the words by the above *memoria technica*, where the first figure represents platypus, and the second kangaroos, noted for "lepping," though, perhaps, with equal reason they should be hares!

Student's illustrations of Pearson's platykurtic and leptokurtic curves.

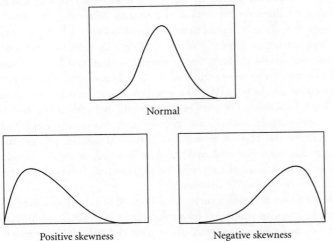

Normal

Positive skewness

Negative skewness

Skewness.

When a distribution is skewed, the mean tends to lie closer to the tail (see the positive and negative skewed curves above). Pearson's first coefficient of skewness enabled him to calculate the asymmetry, by taking the difference between the mean and the mode and dividing the result by the standard deviation:

$$\text{skewness} = \{\text{mean} - \text{mode}\} \, / \, (\text{standard deviation}).$$

Pearson continued with his analogy of mechanics to show his audience how to calculate the standard deviation and the covariance. He showed them that the standard deviation corresponds to the 'moment of inertia'—an important concept in mechanics, being a measure of the beam's ability to resist buckling or bending—and that the covariance corresponds to the 'product moment of dynamics', which is concerned with the effect of force on the motion of objects.

From the method of moments, Pearson thus established four parameters for curve fitting: the *mean* to show how the data clustered, the *standard deviation* to show how it spread, *skewness* to show whether there was a loss of symmetry, and *kurtosis* to show whether the shape of the distribution was peaked or flat. These four parameters describe the essential characteristics of any distribution: the system is parsimonious and elegant, making these statistical tools essential for the interpretation of any set of statistical data, whatever shape the distribution takes.

By the autumn of 1893, Pearson was convinced that, for the first time in the history of biology, there was a chance that the science of life could become an exact mathematical science. He subsequently announced his statistical methods and theory in a series of twenty-three papers on the 'Mathematical contributions to the theory of evolution', published mainly by the Royal Society from 1893 to 1916. In particular, between 1893 and 1895, he developed a general formula to use for subsets of various types of frequency curves and defined a series of types of curves.

Although Pearson was not the first person to devise statistical methods to analyse asymmetrical curves, he was the first to have provided a systematic treatment for them via his method of moments. Some six years earlier, on 1 September 1887, John Venn had raised the problem of curve-fitting asymmetrical distributions when he wrote to *Nature* to challenge Quetelet's assumption that all distributions should conform to the normal distribution.[29] Venn found that some distributions of barometric readings and thermometric observations were asymmetrical, and he hoped that this would encourage discussions on the matter. Edgeworth endeavoured to work out the probability by using Poisson's higher-order approximation for some asymmetrical data that he found in life tables of men between the ages of 50 and 100.[30] Pearson wanted to show that the asymmetrical point binomial could be used to fit asymmetrical curves with a more generalized form of treatment for asymmetrical curves than others had offered with the method of moments.[31]

When Pearson devised his curves (later referred to as the *Pearsonian family of curves*), he provided a variety of theoretical curves in varying graduations, which could then be superimposed onto an empirical curve to determine which curve gave the best 'fit'. The more important ones that remain an essential part of theoretical statistics today include: the gamma curve (Type III), which he went on to use for finding the exact chi-square distribution; the family of asymmetric curves (Type IV), created for Weldon's data; Type V, the normal curve; and Type VII, now known as 'Student's distribution for *t*-tests', which was introduced by Gosset and formalized by Fisher. One of Pearson's students, Churchill Eisenhart, recognized in 1974 that:[32]

Pearson's family of curves did much to dispel the almost religious acceptance of the normal distribution as the mathematical model of variation of biological, physical and social phenomena.

Incidentally, the generalized method of moments, which is used in a number of disciplines today, emerged as an important method of estimation and inference in econometrics in the early 1980s, and by the 1990s it was regarded as a flexible tool of application for a large number of econometric and economic models.[33]

After Pearson examined the asymmetric curves that had been derived from Weldon's crab data in Naples, he decided that an objective method of measuring the goodness of fit was a desideratum for distributions that do not conform to the normal curve. Pearson's earliest consideration of determining a measure of the goodness-of-fit test came out of his lecture on 22 November 1893, when he asked, 'Can you always fit a normal curve to a set of data?'[34] At a time when many believed that the normal distribution was the only distribution to fit an empirical curve, Pearson's task was to show this group that this was 'not always' the case, since not all data would fit into the normal curve. In his last set of Gresham lectures, in May 1894, Pearson discussed various procedures for goodness-of-fit testing, when he introduced the sixth moment as an empirical measure of a goodness-of-fit test for asymmetrical distributions. By this time he had provided the infrastructure of his statistical methodology, and, as mentioned in Chapter 3, he soon began to teach statistics at University College. He introduced his second measure of a goodness-of-fit test at UCL in the spring of 1894, which he thought provided a 'fairly reasonable measure of a goodness of fit'.[35]

Whilst Pearson continued to work on curve fitting and finding a goodness-of-fit test throughout the 1890s, his work was interrupted by Francis Galton who needed some help with correlation. The interruption proved beneficial, as Pearson was able to expand the corpus of his statistical methods by devising twenty-two methods of correlation, of which eleven continue to be used today.[36] Before becoming acquainted with Galton's idea of correlation, Pearson was convinced that formal mathematics could be applied only to natural phenomena that were determined by causation, but Galton's ideas of correlation enabled Pearson to make an epistemological shift from seeing the world in terms of causation and replacing that with correlation, especially in the biological sciences. Pearson went on to become an ardent anti-causalist who thought that the universe was not controlled by laws of causation, in its narrowest sense, but that variation instead played a bigger role in helping to explain phenomena in the universe.[37]

In 1895, Pearson worked out the mathematical properties of the product–moment correlation coefficient (which measures the relationship between two continuous variables) and simple regression (used for the linear prediction between two continuous variables).[38] By the end of the 19th century Pearson's student, George Udny Yule, had introduced a novel approach to interpreting correlation and regression with a conceptually new use of the *method of least squares*, which Yule used to locate the regression line and to fit a line to a set of points. This methodology, which was based on the theory of errors, had been devised in the early 19th century by Gauss, Laplace, Legendre and others to determine, for example, the shape of the earth.[39]

In 1896, Pearson introduced a higher level of mathematics into statistical theory, the coefficient of variation, the standard error of estimate, multiple regression, and multiple correlation. At the beginning of the 20th century, he began to develop a classificatory system (or scales of measurement) for different types of variables to measure correlation. Although most of the variables he studied in the 1890s were continuous (such as height and weight), he realized by 1900 that some biological variables were discrete (such as eye colour). Pearson created two subdivisions for discrete variables consisting of 'nominal' and 'ordinal'. Moreover, he realized that data could be ranked and could take on such characteristics as binary (0, 1) and dichotomous (two mutually exclusive values) categories, requiring the use of very specific methods of correlation.[40]

Pearson's ongoing work throughout the 1890s on curve fitting signified that he needed a criterion to determine how good the fit was. Although many other 19th-century scientists attempted to find a goodness-of-fit test, they did not give any underlying theoretical basis for their formulas. This Pearson managed to do.[41] In 1900 he devised his chi-square goodness-of-fit test. The overriding significance of this test meant that statisticians could use statistical methods that did not depend on the normal distribution to interpret their findings. Indeed, the exact chi-square goodness-of-fit test represented Pearson's single most important contribution to modern statistical theory, and it elevated the practice of mathematical statistics substantially.

Thus we see that, inspired and supported by Weldon, Pearson's major contributions to statistics were:

- introducing standardized statistical data management procedures to handle very large sets of data;
- challenging the tyrannical acceptance of the normal curve as the only distribution on which to base the interpretation of statistical data;
- providing a set of mathematical statistical tools for the analysis and interpretation of statistical variation;
- professionalizing the discipline of mathematical statistics.

## The Pearson and Galton fallacy

Despite the indispensable role of W. F. R. Weldon that led to Pearson's decision to change careers from a mathematical physicist to a biometrician (or mathematical statistician), Pearson has long been erroneously viewed as a disciple of Galton, who followed in his footsteps and who merely expanded what Galton started. Consequently, it has been falsely assumed by various scholars that Pearson's motive to create a new statistical methodology arose from problems of eugenics.[42] As I have argued elsewhere, Pearson not only managed the Drapers' Biometric Laboratory and the Galton Eugenics Laboratory separately, since they occupied separate physical spaces, but he maintained separate financial accounts, established very different journals, and created two completely different methodologies.[43] Moreover, he took on his work in the Eugenics Laboratory reluctantly, and primarily as a personal favour to Galton. Pearson thus emphasized to Galton that the sort of sociological problems that he was interested in pursuing for his eugenics programme were markedly different from the research that was conducted in the Biometric Laboratory.[44]

The principal methodology in the Drapers' Biometric Laboratory was underpinned by Pearson's seminal statistical work on curve-fitting and goodness-of-fit testing for distributions of various shapes, in addition to a series of correlational methods and statistical regression that were accompanied with a higher level of mathematics. The methodology in Pearson's Eugenics Laboratory involved instead the use of family pedigrees and actuarial death rates. Pearson's approach in the development and deployment of the principal methods in the two laboratories differed in a number of ways. His biometric methods were used for the *analysis* of data and were generalizable; moreover, they were innovative, sophisticated, and rigorous. In contrast, the pedigrees in the Eugenics Laboratory were used initially by Galton, and were a tool for *collecting* data and were not generalizable. Although the family pedigrees were not statistically rigorous, they were visually impressive. Adopting and thus endorsing Galton's methodological procedures to address a range of medical and social problems, *vis-à-vis* family pedigrees, may have also reflected Pearson's ambivalence towards this enterprise.[45]

As a result, historians have tended to gloss over crucial differences in the techniques and the methodologies in the two laboratories. This historiographical tendency to link *in toto* Pearson's work in the Galton Eugenics Laboratory with his work in the Drapers' Biometric Laboratory remains the most problematic aspect in the historiography of Pearsonian statistics. Moreover, the general historiographical trend has been to over-emphasize Pearson's work on correlation and regression to the neglect of all other statistical methods, techniques, tools, and instruments that played a central role in the many different projects that he undertook in both laboratories.[46]

Juxtaposing Pearson alongside Galton and eugenics has distorted the complexity and totality of Pearson's intellectual enterprises, since there was virtually no relationship between his research in pure statistics and his agenda for the eugenics movement. This long-established but misguided impression can be attributed to

- an excessive reliance on secondary sources containing false assumptions;

- the neglect of Pearson's voluminous archival material;

- the use of a minute portion of his more than six hundred published papers;

- a conflation of some of Pearson's biometric and crainometric work with that of eugenics;

- a blatant misinterpretation and misrepresentation of Pearsonian statistics.

Conversely, some statisticians who have written on Pearson have invariably assumed that the impetus to Pearson's statistics came from his reading of Galton's *Natural Inheritance*.[47] This view, however, fails to take into account that Pearson's initial reaction to Galton's book in March 1889 was actually quite cautious. It was not until 1934, when Pearson was 77 years old, that he reinterpreted the impact that Galton's book had on his statistical work in a more favourable light—long after Pearson had established the foundations of modern statistics.

There has also been a tendency to overemphasize the role that Pearson's iconoclastic and positivistic book, *Grammar of Science*, played in the development of Pearson's statistical work, whilst neglecting other influential factors in his life.[48] This book, which contains Pearson's first eight Gresham lectures, was written when he was 34 years old and represents his philosophy of science as a young adult. It does not reveal everything about Pearson's thinking and ideas, especially those in connection with his development of the modern theory of mathematical statistics. Thus, it is not helpful to view this particular book as a guide to what he was to do throughout the remaining forty-two years of his working life.

Finally, nearly all historians of science have failed to take into account that Pearson's and Galton's ideas, methods, and outlook on statistics were as different from each other as two people's could possibly have been. Whilst Pearson's main focus was goodness-of-fit testing, Galton's emphasis was correlation (though Galton never even used Pearson's product–moment correlation); Pearson introduced higher-level mathematics for doing statistics, and his work was more mathematically complex than Galton's; Pearson was interested in very large data sets (more than one thousand), whereas Galton was more concerned with data sets of around one hundred (owing to the explanatory power of percentages); Pearson undertook long-term projects over several years, whilst Galton wanted faster results. Moreover, Galton thought that all data had to conform to the normal distribution, whereas Pearson emphasized that empirical distributions could take on any number of shapes. Pearson's method of curve fitting and reporting data represents a radical departure from Galton's interpretations of observational data: in a Galtonian system much information would be lost by normalizing the asymmetrical distributions, whereas in a Pearsonian system all the information would be used.

Galton's contribution to the corpus of Pearsonian statistics has to do principally with his recognition of the explanatory significance of using correlation instead of causation, when two or more variables may be related to each other, and providing Pearson with the framework that led him to develop a battery of correlational methods for variables that used different scales of measurement. Moreover, Pearson's transformation of Galton's hereditarian ideas of biological regression to a purely statistical concept gave statisticians a tool for determining whether statistical predictions could be made for two linearly related variables.

These uni-dimensional and mono-causal accounts of Pearson's work undermine the complexity of Pearson, who was a far more multi-faceted person than has been conveyed by a number of historians and philosophers of science. The historiographical tendency to link Galton's work in his Eugenics Record Office, which Galton created in 1904, is unwarranted, because it overlooks the totality and complexity of Pearsonian statistics and completely ignores the crucial role Weldon played in Pearson's switch from being a competent but unpromising elastician to a resoundingly successful mathematical statistician.

## Conclusion

The Victorian creation of a mathematically based statistical methodology hinged on the measurement and analysis of a different statistical unit of measurement from the one used by the mid-Victorian vital, social, and economic statisticians. Influenced by Adolphe Quetelet, who borrowed some of the statistical tools from astronomy to devise his *homme moyen*, these mid-Victorian statisticians espoused Quetelet's arithmetic mean, whilst mortality tables and actuarial death rates were also used by vital statisticians. Since social and economic statistics lacked a single unifying theory of their discipline, this made it difficult for them to develop their own statistical methodology. Yet, the biological variation that Darwin made central to his work, combined with his theory of natural selection as the mechanism of evolution, offered the biological sciences the potential for a unified science. The highly statistical nature of Darwin's ideas of individual biological variation and populations of species provided the statistical framework that made it possible for Galton, Weldon, and especially Karl Pearson, with his colleagues and students, to create a new and rigorous mathematically based discipline of statistics.

Biological Darwinism precipitated a paradigmatic transition in the measurement and analysis of statistical data when individual *variation* became the principal unit of statistical investigations. The shift in the locus of statistical units of measurement from averages to variation brought about a new way of analysing and managing statistical data on a larger scale than had been previously possible. Francis Galton initiated this analysis: his ideas of correlation provided a new way of understanding the way in which two or more variables may be related to each other that did not imply causation. Furthermore, his ideas about correlation and regression led Pearson to develop a battery of correlational methods for variables that used different scales of measurement.

Although a number of social and economic statisticians made great use of methods derived from astronomy for their statistical studies, these were of limited use for Pearson, Weldon, and the biometricians. While astronomers were concerned with observations made under *stable* conditions, the conditions of the objects studied by the biometricians (plants, animals, and human) were *unstable*, because of the highly variable nature of biological organisms: environmental and ecological changes constantly affect the growth and development of plants and animals, whilst genetic variation and mutations give human beings enormous diversity. Moreover, any attempt to quantify and study human society creates a constellation of possible explanations. Thus, each and every individual variation was to be measured, rather than ignored or regarded as unwarranted: statistical investigations could be used to illuminate an understanding of nature's most complex group of biological organisms.

The challenge of finding an empirical criterion for the process of species divergence (or speciation) in marine organisms and later in plants, and empirical evidence of natural selection from Weldon's data, enabled Pearson to introduce the method of moments as the building blocks for his emergent discipline of mathematical statistics. This made it possible for him to move beyond the restrictions of the normal distribution for the analysis of biological data. Such work enabled his creation of a battery of correlational methods, and ultimately to devise the exact chi-square goodness-of-fit test.

With this new outlook, Pearson was able to establish the Biometric School at University College, London, in 1893, which became the Drapers' Biometric Laboratory in 1903 and, eight years later, UCL's Department of Applied Statistics. These institutional developments, along with the establishment (with Weldon and Galton) of the journal *Biometrika*, enabled Pearson and his students to promulgate his statistical methods and theory to an international community of empirical scientists. These methods continue to be used in the 21st century in the biological, medical, social, and behavioural sciences, as well as in industry, economics, psychometrics, and education.

LIBRARY OF USEFUL KNOWLEDGE.

THE

# DIFFERENTIAL AND INTEGRAL

# CALCULUS

CONTAINING

DIFFERENTIATION, INTEGRATION, DEVELOPMENT, SERIES, DIFFERENTIAL EQUA-
TIONS, DIFFERENCES, SUMMATION, EQUATIONS OF DIFFERENCES, CALCULUS
OF VARIATIONS, DEFINITE INTEGRALS,—WITH APPLICATIONS TO
ALGEBRA, PLANE GEOMETRY, SOLID GEOMETRY,
AND MECHANICS.

ALSO,

ELEMENTARY ILLUSTRATIONS OF THE DIFFERENTIAL AND
INTEGRAL CALCULUS.

BY

## AUGUSTUS DE MORGAN, F.R.A.S. AND C.P.S.,

OF TRINITY COLLEGE, CAMBRIDGE,
PROFESSOR OF MATHEMATICS IN UNIVERSITY COLLEGE, LONDON.

Ταῦτα δὲ τοῖς μὲν πολλοῖς καὶ μὴ κεκοινωνηκότεσσι τῶν μαθημάτων οὐκ
εὔπιστα φανήσειν ὑπολαμβάνω· τοῖς δὲ μεταλελαβηκότεσσι, καὶ περὶ τῶν
ἀποστημάτων καὶ τῶν μεγεθέων, τᾶς τε γᾶς, καὶ τοῦ ἀλίου, καὶ τᾶς
σελήνας, καὶ τοῦ ὅλου κόσμου, πεφροντικότεσσι, πιστὰ διὰ τὰν ἀπόδειξιν
ἐσσεῖσθαι. Διόπερ ὠήθην καὶ τινας οὐκ ἀνάρμοστον εἴη ἔτι ἐπιθεωρῆσαι
ταῦτα.—ARCHIMEDES.

PUBLISHED UNDER THE SUPERINTENDENCE OF THE SOCIETY FOR THE
DIFFUSION OF USEFUL KNOWLEDGE.

LONDON: BALDWIN AND CRADOCK,
47, PATERNOSTER-ROW.

MDCCCXLII.

The title page of De Morgan's *The Differential and Integral Calculus.*

## CHAPTER 13

# Instruction in the calculus and differential equations in Victorian and Edwardian Britain

I. GRATTAN-GUINNESS

The history of the calculus in Britain in the 19th century is unique, in that it is a case of a subject re-learnt; for the British abandoned Newton's version and took up the very different ones due to Leibniz and Euler, and to Lagrange. After this re-programming, they dutifully handled not only the calculus itself but also related topics, especially differential equations and the calculus of variations. They also showed an enthusiasm for operator methods that had been inspired on the Continent by Lagrange's theory, but had lost their appeal there by 1830. Most of this activity was at the level of teaching and writing textbooks; the British did not produce significant research work in these areas until the 1890s, but then the rise was rapid.

By the second decade of the 19th century, three traditions of the calculus were in place in Britain. The dominant version had been Isaac Newton's theory of fluxions and fluents, but this was largely confined to British mathematicians and had developed little for many decades. Then the two Continental versions of Leibniz and Euler, and of Lagrange, became known and adopted: the publication in 1799–1805 of the first four volumes of Laplace's *Traité de Mécanique Céleste* was the main stimulus of change, because of the high status of celestial mechanics.[1]

Especially for applications, preference was given to the differential and integral calculus as set out by Leibniz, modified by Euler with his addition of the differential coefficient (the forerunner of our derivative), and extended in the mid-18th century to multivariate calculus. The calculus of variations had developed alongside in its algebraic form, using the '$\delta$' due primarily to Lagrange.[2]

But among those mathematicians more concerned with the pure calculus, especially Robert Woodhouse at Cambridge[3] and his student successors Charles Babbage and John Herschel with their 'Analytical Society' (from 1812 onwards), the favour went to Lagrange's version of the calculus. This reduced the subject to a branch of algebra, where any mathematical function $f(x + h)$ was assumed to be expandable as a power series in $h$ for all values of $x$, with the 'derived functions' defined from the coefficients of the powers of $h$.

Lagrange's theory helped to develop two new algebras, first among the French and then in England with Babbage and Herschel in the vanguard: functional equations (which became known as 'the calculus of functions'), where the functions themselves were the objects of study rather than any values that they took; and differential operators ('the calculus of operations'),[4] an algebra based upon $D = d/dx$, where higher orders of differentiation were construed as powers of $D$.

In each version of the calculus the integral was specified ('defined' is too strong a term) as some sort of inverse of the fluxion, differential, derivative, or differential operator. In particular, in the calculus of operations, integration was understood to be $1/D$, with both it and the differential operator $D$ being open to manipulations emulating those of common algebra; an example is given in Box 13.1.

---

## Box 13.1: Solving a differential equation

To solve the ordinary differential equation

$$\frac{d^2 y}{dx^2} - 2\frac{dy}{dx} - 3y = 0,$$

or, in operator terms, $D^2 y - 2Dy - 3y = 0$,

we 'separate the symbols' of operation from those of quantity to give

$$(D^2 - 2D - 3)y = 0 \quad \text{or} \quad (D + 1)(D - 3)y = 0.$$

Thus, since a solution of $(D + 1)\, y_1 = 0$ is $y_1 = C_1 e^{-x}$, and since $y_2 = C_2\, e^{3x}$ similarly satisfies $(D - 3)\, y_2 = 0$, the general solution of the original equation is

$$y = C_1 e^{-x} + C_2 e^{3x}.$$

From the late 1810s Cauchy created a fourth tradition of founding the calculus at the École Polytechnique in Paris. It was based upon a *theory* of limits, that itself was grounded in the careful study of convergent infinite sequences of values, and not just the modestly developed *notion* that his predecessors (such as Newton) had grasped. By these means he set up mathematical analysis as we recognize it, covering also the theory of functions and infinite series.

Cauchy defined the derivative as the limit of the difference quotient, and the integral as the limit of a sequence of partition sums, and in both cases he allowed for the possibility that the limit did not exist. As one offshoot, in 1822 he produced a number of counter-examples—such as the function $f(x) = e^{-1/x^2}$ which does not have a Taylor series expansion at $x = 0$—that refuted the fundamental assumption underlying Lagrange's foundational approach to the calculus.

Cauchy's new version (which was not motivated by those counter-examples) was bad news for all predecessor theories, and to some extent it was so received; in particular, his colleagues at the École Polytechnique, an engineering school, were annoyed. Its adoption in Europe was steady but very gradual, even among those mathematicians who stressed rigour. An important stimulus was the series of lecture courses given by Karl Weierstrass at Berlin University for nearly thirty years, starting in the late 1850s. The principal achievements in the foundations of real-variable analysis that we owe to him and his students are refinements of the theory of limits, including definitions of the irrational numbers and clear distinctions between such concepts as the least upper bound and the upper limit of a sequence of values; also included were an examination of the refinements involved in handling multiple limits, especially the various modes of convergence of series of functions and forms of continuity of a function, and some new important general theorems, such as his own approximation theorem.

Students came to hear these lectures from across Europe, and even from the USA (which was not then an important mathematical country)—but it seems that no British students attended.[5] What was happening there? This question is addressed below, mainly at the educational level, covering not only the foundations and principal parts of the calculus and real-variable mathematical analysis, but also differential equations. British contributions to foundational aspects at the research level were very modest compared with Continental developments, and many of them arose from the distinguished contributions to mechanics and mathematical physics.[6] Further, complex-variable analysis received only very patchy attention in Britain until the 1890s, and is noted only in a later section of this chapter.

# Developments in early Victorian Britain and earlier

While Newton's version of the calculus petered out around the mid-1820s, Euler's version long continued to retain its high status among those figures concerned with applications. Here Britain was strong, all century long, from figures such as William Whewell and George Airy in several areas of mechanics from the 1820s onwards, through men such as William Rowan Hamilton, James MacCullagh, and George Green at the beginning of our period, via William Thomson, George Gabriel Stokes, P. G. Tait, and James Clerk Maxwell in the middle, to G. H. Darwin, G. F. FitzGerald, Joseph Larmor, J. J. Thomson, and Lord Rayleigh towards the end of it. These scholars worked in many parts of mechanics and mathematical physics, and thereby used, and on occasion contributed to, mathematical analysis. Many (perhaps too many) of these contributors worked, or had studied, at the University of Cambridge, with its notorious mathematical tripos and the Smith's prize essay competition;[7] but at least its graduates came to master a wide range of techniques, in which Euler's calculus was joined by differential equations and the calculus of variations.[8] In addition, the textbooks in those branches of applied and applicable mathematics usually took the calculus as it was already known, almost always in the form just described. An important textbook author was E. J. Routh, who supported his teaching as a Cambridge coach with a popular book on dynamics in editions from the 1860s onwards, followed in the 1890s by books on statics and particle dynamics.

Up to this time, one could tell broadly similar stories about Continental countries. But to a much greater extent than any other country, Lagrange's version of the calculus retained popularity in Britain, even after Hamilton independently rediscovered one of Cauchy's counter-examples to it in 1833. In particular, the new Lagrange-inspired algebras were pursued, especially that of differential operators.

George Boole became a major figure in this area from the 1840s onwards, but he was just one among a substantial cohort of around twenty practitioners who kept these methods alive up to the mid-1870s.[9] Most of these men were not major mathematicians, and indeed not mathematicians by profession; nevertheless, as well as Boole, other notable figures included Robert Murphy, Duncan Gregory, and Arthur Cayley.

Some attention was also paid to difference equations, then called 'the calculus of finite differences', where forward differencing '$\Delta$' and summation '$\sum$' were handled as inverse operators and were linked to the differential calculus, as shown in Box 13.2.

As the French had already perceived, the method could be extended to solve linear differential equations, which were written in the form $f(D)\, y = X$, where $f(D)$ is the appropriate differential function (to us, functor) for the equation, and $X$ contains all the non-operational terms; solutions were sought by manipulating $f^{-1}(D)$. Four differential equations drew special attention: Laplace's equation, and the ordinary equation in an angular variable obtained from it by imposing spherical polar coordinates and separating the variables, another ordinary equation also due to him concerning the shape of the earth, and Riccati's equation. Boole seems to have hoped to find one all-embracing method to cover solutions of at least linear equations; but his own contributions themselves rendered such a hope unlikely, and a variety of methods emerged.

## Box 13.2: Difference equations

Exploiting the analogy between exponentiation and repeated differentiation, we can write Taylor's theorem as

$$f(x + h) = \sum_{k=0}^{\infty} \frac{h^k}{k!} \left( \frac{d}{dx} \right)^k f(x) = \sum_{k=0}^{\infty} \frac{h^k}{k!} (D)^k f(x)$$

$$= \sum_{k=0}^{\infty} \frac{(hD)^k}{k!} f(x), \text{ for } h > 0.$$

Now, since $e^x = \sum_{k=0}^{\infty} x^k / k!$, we obtain

$$f(x + h) = e^{hD} f(x),$$

resulting in the curious abbreviation:

$$\Delta f(x) = f(x + h) - f(x) = (e^{hD} - 1) f(x).$$

Some authors even wrote $\Delta = e^{hD} - 1$, inaugurating operator algebra.

---

In 1840, Boole[10] came up with the best method to date for solving linear differential equations with constant coefficients; we present this in Box 13.3. He also found solutions for ordinary equations with certain forms of variable coefficients.

One book clearly expressed the aspirations for these methods in its title: R. Carmichael's *Treatise on the Calculus of Operations, Designed to Facilitate the Processes of the Differential and Integral Calculus and the Calculus of Finite Differences* (1855).[11] Their popularity is partially explained by the enthusiasm for algebras of all kinds displayed throughout the century by the British, especially English and Irish mathematicians (see Chapter 15). The low level of professionalization of mathematics in Britain may also have encouraged the use of the methods; they made good sport for gentlemen of leisure, unencumbered by lots of difficult technical theorems of the kind that attended the theories of special functions (Bessel, Legendre, elliptic, and so on). As Niccolò Guicciardini has put it: 'The main reason for the success of these techniques is that they were easy to learn and offered immense possibilities of dull proliferation'.[12] Furthermore, as we saw in Chapter 7, the journals that became available proved to be sympathetic venues for publication, with *The Cambridge and Dublin Mathematical Journal* featuring numerous papers on these methods.[13] By contrast, operator methods on the Continent seemed largely to have died out, or at least died down, by about 1830.

# Box 13.3: Boole's method for solving differential equations

In order to solve the linear differential equation,

$$\frac{d^n y}{dx^n} + A\frac{d^{n-1} y}{dx^{n-1}} + B\frac{d^{n-2} y}{dx^{n-2}} + \dots + R\frac{dy}{dx} + Sy = X,$$

we replace it by $n$ first-order linear equations

$$\left(\frac{d}{dx} - a_i\right) y_i = (D - a_i)\, y_i = f(x), \text{ for } i = 1, 2, \dots, n,$$

and then sum linear combinations of the solutions.

Taking the generalized Product Rule

$$D^n(uy) = \sum_{k=0}^{n} \binom{n}{k} D^k(u) D^{n-k}(y)$$

and the fact that $D^k(e^{ax}) = a^k e^{ax}$, and letting $u = e^{ax}$, we obtain

$$D^n(e^{ax}y) = \sum_{k=0}^{n} \binom{n}{k} a^k e^{ax} D^{n-k}(y) = e^{ax}[(D+a)^n(y)]$$

$$(\,\text{by the Binomial Theorem}).$$

So

$$(D+a)^n(y) = e^{-ax}[D^n(e^{ax}y)].$$

Of course, if $n = -1$, then $D^n = \int dx$ and

$$(D+a)^{-1}(y) = e^{-ax}\int e^{ax} y\, dx.$$

Thus, first-order linear equations of the form $(D - a_i)\, y_i = f(x)$ can be solved by

$$y_i = (D - a_i)^{-1} f(x) = e^{a_i x}\int e^{-a_i x} y\, f(x)\, dx.$$

This was research activity, but quite a few of the results filtered down to some British textbooks. During the century there appeared in Britain thirty-three textbooks on, or covering, the calculus; ten on differential equations (usually containing some general theory and certain solution methods), which replaced the Newtonian fluxional equations, and ten on the calculus of variations; the two main publishers were Cambridge University Press and Macmillan.[14] As with the operator gentlemen, most authors were not distinguished figures. A notable Cambridge author was Isaac Todhunter, known for his textbooks (one of which is described below), and also for his substantial historical writing in these areas.[15] Boole produced influential textbooks on differential equations (1859) and difference equations (1860), with Todhunter preparing a posthumous edition of the former in 1865.

Some authors handled the calculus of differential operators, and a number of calculus books also covered parts of the theory of differential equations (ordinary differential equations more than partial ones). Several books also treated related topics, such as the calculus of variations, special functions and integrals, infinite series, and difference equations.

A good example is provided by John Hymers, a fellow of St John's College, Cambridge. His *Treatise on Differential Equations* (1839) contained an efficient presentation, mainly of specific solutions to various basic types of ordinary and partial equations; he made very little use of operator methods; the same can be said of the substantial survey of difference equations and related topics that, unusually but agreeably, he adjoined to his book. In the second 'enlarged' edition of 1858, the book's structure and range was more or less maintained; however, nearly half of the eighty extra pages were now devoted to sections on the method of 'separation of symbols' for linear differential equations in all three calculi.

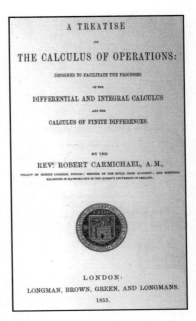

Carmichael's *Treatise on the Calculus of Operations*.

## A limited use of limits

For foundations the authors often drew upon either the Euler or the Lagrange version of the calculus. Several authors affirmed the importance of limits; however they did not seem to advocate a Cauchy-like theory but used only the notion, following Newton's approach and maybe also its revival by William Whewell, particularly in his *Doctrine of Limits* of 1838. Unfortunately, many gave space to the falsehood, espoused by Whewell and others, that 'what is true up to the limit is true at the limit'. So the British record of teaching the theory of limits in no way resembles the traditions of *Cours d'Analyse* or *Lehrbücher der Analyse* that one found on the Continent, in which at least parts of mathematical analysis, after the manner of Cauchy and Weierstrass, were gradually publicized.

The pioneer in raising the status of limits in Britain was Augustus De Morgan.[16] In his early 30s he developed a theory of limits, giving the first formal definition of a limit in the English language: 'When, under circumstances, or by certain suppositions, we can make $A$ as near as we please to $P$,... then $P$ is called the limit of $A$'.[17] Curiously, this definition appeared not in a work on calculus, but

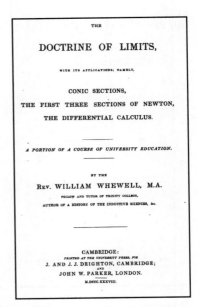

Whewell's *Doctrine of Limits*.

in his *Elements of Algebra* (1835), the branch of mathematics where De Morgan felt limits to lie.

De Morgan then largely followed Cauchy in his sizeable textbook on *The Differential and Integral Calculus* (1842) and in many related topics, though with few mentions of his mentor; he did not even note the counter-examples to Taylor's series. He founded the differential calculus by defining the 'differential coefficient' as the limiting value of the difference quotient $\Delta y/\Delta x$ as the denominator goes to zero. But he followed the popular but unhappy international practice of symbolizing it with '$dy/dx$', which was to be taken as a *whole symbol*, not a ratio.[18] The integral was defined both as the anti-differential and as the limiting value of a sequence of sums, after which some of the principal 'processes' were rehearsed.[19]

Among other topics in his book, De Morgan included basic difference calculus, some elementary ordinary differential equations, various series expansions (including some using differential operators), basic calculus of variations and its use in mechanics, numerical analysis, and the evaluation of many definite integrals (including a pioneering but seemingly uninfluential passage[20] on Cauchy's evaluations using complex-variable methods). Of special note is a chapter on methods of summing divergent series;[21] and although Cauchy had decreed such series to be illegitimate mathematics, De Morgan maintained this as one of his research interests, perhaps under the influence of his former Cambridge tutor and Analytical Society alumnus George Peacock (who incidentally *was* aware of Cauchy's counter-examples). Although not disorganized, the book is not easy to follow, with several topics repeated in later chapters. Nevertheless it presented a great deal of material that was not familiar in early Victorian Britain.

A typical example of a later use of limits in a textbook, by one of the better-known authors of the period, is Isaac Todhunter's *A Treatise on the Differential Calculus and the Elements of the Integral Calculus* (1852). Todhunter declared his allegiance to limits right at the start, and cited several works, including French ones (Cauchy appearing via the work of his disciple, the Abbot Moigno, published in the 1840s) in which such theory was deployed. But he neither presented a theory of limits itself, nor did he clearly draw upon it in his uses of limits! His foundation of the calculus was basically the same as De Morgan's, including the reading of '$dy/dx$'; he also used '$\phi'(x)$', but sparingly.

At nearly four hundred pages Todhunter's account was quite comprehensive at its level, with many examples. Topics included the handling of implicit functions, functions of several variables and functions of functions, a Cauchy-like development of Taylor series (though no mention of the counter-examples to Lagrange, whose own series was presented), optimization, and the basic use of the calculus on planar curves (including curve tracing, determining asymptotes, and evolutes and involutes). The final chapters included an account of differentials in the traditional Leibnizian sense, integration as the converse of differentiation and also as a sum, integration by parts, and some applications to areas and volumes.

That Todhunter's book was popular among its target audience is evidenced by its eight editions, produced between 1852 and 1878; indeed, it remained the foremost introductory calculus textbook for much of the Victorian period.[22]

A TREATISE ON THE

DIFFERENTIAL CALCULUS

WITH NUMEROUS EXAMPLES.

BY

I. TODHUNTER, M.A.
FELLOW OF ST JOHN'S COLLEGE, CAMBRIDGE.

THIRD EDITION, REVISED.

MACMILLAN AND CO.
Cambridge:
AND 23, HENRIETTA STREET, COVENT GARDEN,
London.
MDCCCLX.

Todhunter's *A Treatise on the Differential Calculus*, 3rd edition.

But despite its regular revisions, the book's view of the calculus stayed stuck in the 1840s, even as the century approached its final quarter. The fact that the fifth edition of 1871 opined that there had not been 'any real improvements' in the 'methods of explaining and developing the principles'[23] of the calculus in the twenty years since its first publication, gives some indication of how insular and out of date Victorian British mathematics had become by this time. The story of the general theory of differential equations seems to be similarly unexceptional; but by contrast there were many Victorian British contributions to solutions of particular equations and to related topics, especially potential theory. Many of these solutions came from British enthusiasm for the $D$-calculus,[24] but the majority arose within their impressive contributions to mechanics and mathematical physics, a survey of which appears in Chapter 8. It is strange that British strength in the applications of the calculus contrasted so greatly with their comparative apathy to the pure side.

## Some signs of change

For four decades no British author wrote anything that significantly updated either De Morgan or Todhunter's textbooks. Even at Cambridge, the majority of the fellows, tutors, and students remained largely ignorant of recent mathematical developments outside Britain. As one figure recalled:[25]

The college lecturers could not read German, and did not read French. One of the most eminent of them in the eighteen-nineties used to speak of the discoverer of the Gamma-function as "Yewler", and the founder of the theory of functions as "Corky".

In contrast, by late in the century Continental analysts had not only absorbed Weierstrassian sophistications, such as multiple limits theory, theories of irrational numbers, and modes of convergence of series of functions, but were also taking account of Georg Cantor's point-set topology, generalizing the Riemann integral to measure theory, developing functional analysis and linear integral equations, and continuing to elaborate the theories of differential equations, functions, and series.

Andrew Russell Forsyth (1858–1942).

At last, by the end of the century, some Britons (almost all Cambridge graduates) were beginning to participate in such research—and to write textbooks that showed some awareness of these and related developments—with respect to both technical developments and foundational issues. One of the first such authors was the Trinity College Fellow A. R. Forsyth, later to succeed Arthur Cayley as Sadleirian professor at Cambridge. His *Treatise on Differential Equations* of 1885 (with several later editions), followed by a compendious survey *Theory* in six volumes from 1890 to 1906, was a major contribution. Both works raised the standard of books in English on this subject, the former book and the first volume of the latter receiving the compliment of German translations, in 1912 and 1893, respectively. In the preface of his *Treatise*, Forsyth showed his awareness of recent Continental developments, while confessing that his own use

The sixth edition (1956) of Forsyth's *A Treatise on Differential Equations*, first published in 1885.

of them was restricted. He also indicated that most of the problems originated in Cambridge, from examination papers or articles in journals.

Forsyth's book of 1885 contrasts interestingly with arguably its most distinguished predecessor, Boole's *Treatise on Differential Equations* of a quarter-century earlier. It carried the same title, was about the same length, and was also published by Macmillan. Forsyth praised Boole's book in his preface, along with some Continental volumes. Each book covered roughly the same range of first-order and second-order ordinary and partial differential equations, including simultaneous equations. Each author handled the general theory of solutions, with Forsyth somewhat more elaborate overall, but Boole more emphatic about singular solutions.[26] Neither man presented existence theorems, and nor did they dwell much upon applications, Boole to geometry, Forsyth to mechanics. But both handled in some detail important equations, such as Riccati's equation and those defining the special functions via power series; Forsyth also devoted a chapter to the hypergeometric series.

Some of the differences arose from the new results in Forsyth, for which he gave references. One difference that marked a sign of the times lay in their handling of D-methods: Boole gave them eighty pages, while Forsyth allowed them around twenty, mainly covering their least controversial context of factorizing differential polynomials. Again, Forsyth gave more space than Boole to the theoretical aspects of partial first-order and second-order equations that had been studied by French mathematicians such as Lagrange, Laplace, Paul Charpit, Gaspard Monge, and A. M. Ampère, between 1780 and 1810, and by Carl Jacobi in the 1830s. These aspects included methods of transforming an equation into another kind, or a solution into another solution, and finding conditions under which there could be established subsidiary equations, or intermediate solutions of second-order equations.

Although Forsyth's *Treatise* was later to be judged 'the most successful book on differential equations that has ever appeared in any language',[27] it was not without its fair share of flaws. Writing in 1971, the geometer Leonard Roth wrote that:[28]

this work has done more than anything else to retard the true development of the subject; for over two generations it has continued to put wrong ideas into people's heads concerning the nature and scope of the theory and, thanks to the author's forceful and authoritative style, in this it has been overwhelmingly successful.

Recognizing the incompleteness of his account, Forsyth covered far more material in his six-volume *Theory*, which was divided into four parts, each with its own index. The first part was dedicated to exact equations and Pfaff's problem on manipulating total differential equations. The second part (comprising two volumes) treated non-linear ordinary equations of the first and second order, including some existence theorems, types of singularity, forms of singular solution, and systems of equations. The third part contained a selection of properties of linear ordinary equations, especially those that related to the theory of functions. The final part (in two volumes) was devoted to partial equations. The coverage included some existence theorems, cases where the general solution could be determined, aspects of Lie theory, Hamiltonian methods as applied to dynamics (a rare intrusion of applied mathematics), general integrals of equations of second and higher order, an integral solution that he called 'Laplace-transformation' but is not the theory now carrying that name, second-order equations admitting finite forms of solution, certain methods of solution that admitted intermediate solutions, some transformation of equations themselves into related forms, and a few forays into equations of the third and higher orders. At around two thousand seven hundred pages (= 350 + 750 + 500 + 1100), his *Theory* eclipsed all previous English works on differential equations.

While Forsyth's volumes were appearing, British authors produced two rather similar calculus textbooks that showed some awareness of Continental developments: Horace Lamb's hefty *Elementary Course of Infinitesimal Calculus* (1897), and George Gibson's rather less extensive *Elementary Treatise on the Calculus with Illustrations from Geometry, Mechanics and Physics* (1901). Both books were based on lecture courses given at provincial universities: Manchester (Lamb) and Glasgow (Gibson). Taking some note of the modern theory of functions, they presented a theory of limits, and defined notions such as the continuity of functions and the absolute and conditional convergence of infinite series; however, neither author gave a definition of irrational numbers or mentioned Cauchy's counter-examples. Offering '$dy/dx$' as a whole symbol as usual, they also made some use of the $D$-calculus, mainly for factorizing differential polynomials. Still uncommon for the time, both books contained indexes and many examples, the latter containing many drawn from mechanics and some physics.

Gibson followed up with *An Introduction to the Calculus based on Graphical Methods* (1904), stressing the difference quotient rather more than the differential coefficient, and including a chapter on 'graphical integration' and instruments such as intergraphs. Slightly later, an exceptionally popular textbook was Sylvanus P. Thompson's *Calculus Made Easy*, which presented the Euler version in an easy-reading style. First published in 1910, this book is still in print today.

As British awareness of Continental developments slowly increased, their work was heralded, appropriately, by the English translation of a German textbook. Axel Harnack had written one on the calculus and mathematical analysis in 1881, related to his teaching to engineers at Darmstadt and Dresden Polytechnika, but intentionally going far beyond that level to include substantial parts of real-variable and complex-variable analysis. G. L. Cathcart at Trinity College, Dublin, became aware of the book in 1888 and began negotiating with Harnack about an English translation, including possible revisions desired by the author. But then Harnack suddenly died; so Cathcart worked on the book as printed, publishing his translation (which included a fine index) under the modest title *An Introduction to the Study of the Differential and Integral Calculus* (1891): Harnack's original title had used *Elemente*.

An excellent mathematician, Harnack followed a level of rigour and thoroughness that had long been known, especially in Germany and France, but was largely unfamiliar in Britain. He began with the real number system, without offering one of the new definitions of irrationals. He then handled, sometimes in detail, all the main features of Weierstrassian multiple-limit analysis when presenting its processes: uniform continuity, differentiating under the integral sign, uniform, absolute, and conditional convergence of infinite series, and continuous and discontinuous functions (though only an occasional use of Fourier series), infinite products, and multiple and repeated integrals. He even included a note on Georg Cantor's point-set topology, which was not regular fodder either in Germany in 1881 or in Britain a decade later. He used limits to define the 'differential quotient', and also the integral as the limiting value of a sequence of sums.

## Waking up to complex analysis

For complex-variable analysis, Harnack started out from complex numbers and variables before passing to functions. He presented especially analytic and elliptic functions. Weierstrass's complex function theory seems to have been a greater influence on him than that of Bernhard Riemann, and Cauchy's version was largely confined to the evaluation of definite integrals.

These features of his book are especially noteworthy, for if the British were largely uninterested in matters concerning real-variable analysis at this time, they remained almost totally ignorant of the subject of complex analysis until the very last decade of the 19th century. Indeed, while Forsyth was preparing the second volume of his *Theory*, on linear differential equations, he realized that:[29]

[since its theory] is centred around the study of the singularities of their coefficients and the behaviour of the solutions when the independent variable is taken along paths in the complex plane ... all this would be unintelligible to a senior wrangler who had never heard of poles or branch-points or contour-integrals.

In order to rectify this situation, Forsyth 'suspended his work on differential equations to teach his colleagues the language which was current throughout the rest of the mathematical world'.[30] The result was his *Theory of Functions of a Complex Variable* (1893).

The scale of this book was enormous. In over six hundred pages, Forsyth gave the first thorough-going treatment in English of the theory of uniform functions, power series, contour integration, conformal mapping, and topological surfaces. Perhaps for the first time, the book presented in a single volume the Cauchy-style methods of the French analysts together with the theories of Riemann and Weierstrass, even including discussions of recent work on automorphic functions by Klein and Poincaré.

Unfortunately, like his *Treatise on Differential Equations*, this book was deeply flawed. In a contemporary review, the American W. F. Osgood pointed

to the loose form in which theorems are often stated and proofs given; it only too often happens that the ideas on which the proofs rest are lacking in rigour, or that important matters are overlooked.[31]

The principal reason for this was that Forsyth was not a specialist in complex analysis, and his mathematical skills, formidable though they were, did not extend to this particular area. It would appear that:

despite his intentions and his absorption of the material, he never comes within reach of comprehending what modern analysis is really about: indeed whole tracts of the book read as though they had been written by Euler.[32]

But then, that was not the point. The book's prime purpose had been to acquaint the British with half a century of Continental analytical developments as quickly as possible, and in this regard, if nothing else, it was abundantly successful. In a Cambridge where 'one at least of the four professors used to refer to the complex variable $x + iy$ as a "semi-imaginary quantity"',[33] Forsyth's book 'burst in 1893 with the splendour of a revelation'.[34] Despite all its many flaws,

from the day of its publication in 1893, the face of Cambridge was changed: the majority of the pure mathematicians who took their degrees in the next twenty years became function-theorists.[35]

It is for this reason that Forsyth's *Theory of Functions* was described as having 'had a greater influence on British mathematics than any work since Newton's *Principia*'.[36] For better or worse, it was this book that ignited the British love affair with modern analysis that would last well into the 20th century.

# Epilogue: an Edwardian renaissance

By the first decade of the new century, spurred on by their new-found interest (and, as it turned out, ability) in the theory of functions, British authors were writing advanced textbooks and monographs on the calculus and analysis, in a rapid (and 'Edwardian') transformation of a Victorian slumber.[37] Not surprisingly, almost all of these mathematicians had been undergraduates at Cambridge in the 1890s, around the time of the publication of Forsyth's *Theory of Functions*. We conclude this chapter with a review of the main works, in chronological order of publication, focusing upon real variables.

Forsyth's student, E. T. Whittaker, produced his first book in 1902 when he was in his early 30s: *A Course of Modern Analysis: An Introduction to the General Theory of Infinite Series and of Analytic Functions; With an Account of the Principal Transcendental Functions*. In the first part he rehearsed a collection of topics in 'the processes of analysis', real and complex, in a rather odd order: chapters on complex numbers; absolute convergence of infinite series, starting with the definition of limit; 'fundamentals of analytic functions', including the Taylor series; 'uniform convergence and infinite series'; Cauchy's theory of residues for evaluating definite integrals; series expansions; Fourier series; and asymptotic expansions. The second part treated 'the transcendental functions' in a rather more systematic fashion, giving a fine survey of the main special functions: beta functions, gamma functions, Legendre functions, hypergeometric functions, Bessel functions, Mathieu functions, and elliptic functions. While most of these topics were known about in Britain, this book provided both a valuable update and gathered much information in one place, especially the second part.

Edmund Taylor Whittaker (1873–1956).

Also in his early 30s, Bertrand Russell published *The Principles of Mathematics* in 1903, after a rather complicated gestation.[38] It was largely concerned with Cantor's set theory and the mathematical logic of Giuseppe Peano and his followers in Turin, the latter topic then almost unknown in Britain. The main thesis of the book was that this logic, which Russell himself had enriched with a logic of relations, sufficed to express all mathematics, not just its proof methods but also its 'objects'. Starting out with cardinal integers, defined as classes of equipollent sets, he proceeded to the real line, the calculus and geometry, and some parts of mechanics—not 'all' mathematics, then, but quite a lot from rather few assumptions. However, he also found his now famous paradox of set theory, that the set of all sets not belonging to themselves belongs to itself *if and only if* it does not. This *double* contradiction messed up his mathematical pyramid at its base, and he had no satisfactory solution to offer.

In 1906, three years after moving to the University of Sydney from his native Scotland, H. S. Carslaw produced his *Introduction to the Theory of Fourier Series and Integrals and the Mathematical Theory of the Conduction of Heat*. In the first part he presented the series via careful accounts of rational and irrational numbers, the convergence of infinite series and the basic theory of functions, the definition of the Riemann integral, and functions defined in terms of integrals, before arriving at Fourier series and integrals, where he treated a wide range of properties. In the second part he closely followed Fourier's book on heat diffusion, which had been translated into English in 1878; after forming the diffusion equation, and providing series solutions for the various finite bodies such as the ring, rod, and sphere he used the integrals to solve for heat diffusion in infinite bodies. The only novelty relative to Fourier was a chapter on the role played by Green's functions in solutions. This was the first extended account in English of Fourier analysis, which had played such a prominent role in the development of mathematical analysis on the Continent.

William and Grace Chisholm Young are notable, among other reasons, as the first husband-and-wife partnership in the history of mathematics. They left Britain in 1899 to study in Göttingen and find out what mathematics really was. There, Felix Klein, Mrs Young's former dissertation supervisor, alerted them to set theory, which in recent years had attracted a lot of interest; their book *Theory of Sets of Points* (1906) was the first textbook on point-set topology in any language. They led up to the definition and application of the 'Lebesgue integral', to use the name that Young himself had proposed when he found that Henri Lebesgue had anticipated him in defining an integral of such generality.

Ernest William Hobson (1856–1933).

In his *Theory of Functions of a Real Variable and the Theory of Fourier Series* (1907), E. W. Hobson ran through much of the material in Whittaker's and Carslaw's books at great length, in a volume of nearly eight hundred pages. In addition to the usual extensive account of functions, set theory was prominent, and he also covered both the Riemann and Lebesgue integrals as part of his preparation for Fourier series, where in effect he updated Carslaw's account.

In 1908, Thomas Bromwich devoted a large volume to an *Introduction to the Theory of Infinite Series*, perhaps the first such book in English by a British author, and based upon some years' teaching at University College, Galway. In this book, he gave a detailed account of criteria and tests for the convergence of series of constants and variable terms, and then considered series of complex-variable terms. He explained absolute and conditional convergence, and the uniform convergence of series of functions, and noted the recent Continental insights about summability and formal power series. In a suite of appendices he handled definitions of irrational numbers, and several special functions that were defined from series or made much use of them.

Finally, at a deliberately more elementary level than the books just summarized, G. H. Hardy, also in his early 30s, put out *A Course of Pure Mathematics* in 1908. Despite its title, it was confined to real-variable mathematical analysis. He presented real variables and their functions, which dominated much of the later survey, but he also included some trigonometry and analytical geometry. In his preface he mentioned that he had left out uniform convergence, double series, and infinite products, and suggested that mastery of his book would allow the reader to move on to Bromwich's, of which he had recently read the proofs.

Apart from the Youngs' book, all the texts above appeared in later editions from their authors, Carslaw's in separate volumes on Fourier analysis and on heat diffusion. Whittaker's and Hardy's books became particularly well known; subsequent editions of the former were undertaken in collaboration with G. N. Watson,[39] while the latter became noticeably less elementary with

Thomas John l'Anson Bromwich (1875–1929).

From the Preface to Hardy's *A Course of Pure Mathematics*.

successive editions. Indeed, it was the influence of later incarnations of both of these books, particularly Hardy's *Course* with its focus on introducing undergraduates to rigorous analysis, which received much of the credit for transforming the mathematics curriculum at British universities in the first half of the 20th century.

Most of these volumes were published by the Cambridge University Press, which in 1903 also started a series of short self-study books entitled the 'Cambridge Tracts in Mathematics and Mathematical Physics', under the co-direction of Whittaker. Parts of modern mathematical analysis featured in several numbers, where Hardy and W. H. Young were among the authors.

The Youngs and Hardy were also playing noteworthy roles in research. In 1911, the latter would begin his collaboration with J. E. Littlewood, one of the longest and most fruitful partnerships in the history of mathematics, producing around a hundred joint papers of enormous influence covering a wide range of analytical topics. Their research, and that of the 'school' of young analysts which grew up around them, would establish mathematical analysis as an acknowledged forte of British mathematics. In the space of a few short years, in this one field, British mathematicians had progressed from near ignorance to original research.

By the first decade of the 20th century, Britain was no longer isolated from contemporary developments in real and complex mathematical analysis, either in research or in education. This was a remarkable transformation from the Victorian period when their contributions had been surprisingly modest, especially when compared with those from France, Germany, Italy, and Scandinavia. The change from around 1900 was substantial and rapid, but it was also well overdue.

THE FIRST SIX BOOKS OF

# THE ELEMENTS OF EUCLID

IN WHICH COLOURED DIAGRAMS AND SYMBOLS

ARE USED INSTEAD OF LETTERS FOR THE

GREATER EASE OF LEARNERS

BY OLIVER BYRNE

SURVEYOR OF HER MAJESTY'S SETTLEMENTS IN THE FALKLAND ISLANDS

AND AUTHOR OF NUMEROUS MATHEMATICAL WORKS

LONDON

WILLIAM PICKERING

1847

The title page of Oliver Byrne's 1847 edition of Euclid's *Elements* in colour.

## CHAPTER 14

# Geometry

*The Euclid debate*

AMIROUCHE MOKTEFI

The Victorian period witnessed a wide debate on the teaching of geometry in schools and colleges. Until the mid-1860s, Euclid's *Elements* was used with almost no rival as a text-book for the purpose. Then rival manuals were proposed, Euclid's text was challenged, and an anti-Euclid association was created. The fear of chaos and other reasons prevented radical changes, however, and it was only in the early 1900s that Euclid's dominance finally ceased in British schools.

As in several other mathematical disciplines, revolutionary developments occurred in geometrical research in the 19th century. The discovery of non-Euclidean geometries alone wholly changed the scene, but only a handful of British mathematicians made noteworthy contributions to such developments. Geometry was an essential subject in Victorian education, but it was not a discipline in which British mathematicians excelled. Contrary to the situation in most countries on the Continent, the teaching of pure geometry in Britain was almost entirely based on Euclid's *Elements*. From the late 1860s onward, this dominance was widely criticized and an anti-Euclid association was even founded. The debate that followed placed many of Britain's ablest mathematicians and teachers in opposition to each other. Euclid's *Elements* resisted its rivals throughout the Victorian period and the book continued to keep its privileged place in the school curriculum until the early 1900s.

In this chapter we describe how the debate began, evolved, and ended. We see how Euclid benefitted from the Oxbridge dominance in Victorian education in general, and from Cambridge's leading position in mathematics in particular. No significant change was possible without the approbation of Oxford and

Cambridge Universities, and their preference for liberal education over vocational training gave Euclid a comfortable status that made the task of his challengers difficult. One had eventually to wait for the introduction of practical mathematics to make the reform possible and allow important improvements in geometrical teaching.

## Euclid: 'a very English subject'

The first English Euclid: Henry Billingsley's edition of 1570.

At the beginning of the 19th century, Euclid's *Elements* was the standard text for the teaching of geometry in British schools and colleges. This domination was recorded by Dionysius Lardner in the preface of his Euclid textbook, first published in 1828 with a twelfth edition in 1861. He attributed this supremacy to Euclid's merits over his numerous rivals:[1]

Two thousand years have now rolled away since Euclid's Elements were first used in the school of Alexandria, and to this day they continue to be esteemed the best introduction to mathematical science. They have been adopted as the basis of geometrical instruction in every part of the globe to which the light of science has penetrated; and, while in every other department of human knowledge there have been almost as many manuals as schools, in this, and in this only, one work has, by common consent, been adopted as an universal standard...

This unprecedented unanimity in the adoption of one work as the basis of instruction has not arisen from the absence of other treatises on the same subject. Some of the most eminent mathematicians have written, either original Treatises, or modifications and supposed improvements of the Elements; but still the "Elements" themselves have been invariably preferred. To what can a preference so universal be attributed, if not to that singular perspicuity of arrangement, and that rigorous exactitude of demonstration, in which this celebrated Treatise has never been surpassed?

Euclid's dominance should be looked at in context, however, as by the beginning of the 19th century the teaching of mathematics itself was often limited, and sometimes non-existent, in British schools. But in mathematics Euclid had a privileged and unrivalled position in the curriculum, and this authority increased with the growing attention given to mathematics in schools by the mid-century, and with the establishment of the Cambridge mathematical tripos (see Chapter 1). Geometry did indeed fit very well to the ideal of a liberal education supported by Oxford and Cambridge Universities, whose training had no vocational or professional aim, clergymen excepted. The university was defined as a place to form the minds and characters of the gentlemen attending it. The influential William Whewell, professor of moral philosophy at the University of Cambridge, distinguished between 'permanent studies' and 'progressive studies', the former being more essential and to be mastered first. He considered geometry, in the Euclidean style, as a permanent subject, unlike algebra and calculus:[2]

But still, these progressive portions of Mathematics cannot take the place of the permanent portions, in our Higher Education without destroying the value of our system. Wherever Mathematics has formed a part of a Liberal Education, as a discipline of the Reason, Geometry has been the branch of mathematics principally employed for this purpose. And for this purpose Geometry is especially fitted. For Geometry really consists entirely of manifest examples of perfect reasoning: the reasoning being expressed in words which convince the mind, in virtue of the special forms and relations to which they directly refer.

Indeed, Euclidean geometry was considered worth learning for the sake of its logic, at least as much as it was for the sake of its geometry. Thanks to its formal structure, where theorems were proved only by deduction from what has already been proved or assumed, Euclid's system was studied as an instrument for training the mind in correct reasoning. Euclid was also a classical text, having crossed the centuries almost unchanged. Such a 'long-established demonstrated science' naturally deserved to be an important ingredient of liberal education, provided that it was taught using Euclid's own treatise: *The Elements*.

*The Elements* was written by Euclid around 300 BC. It contains thirteen books, to which two more books were later added by subsequent authors. Later versions of the Greek text survived and were translated into Latin and Arabic, before the printing revolution made it available in modern European languages in the 15th and 16th centuries. The first English translation, by Henry Billingsley, appeared in 1570 with a preface by John Dee, and many hundreds of further editions followed.[3,4] By the mid-19th century, the use of textbooks was essential for the purpose of geometrical teaching and examining, and Euclid's text, with its various and increasingly numerous editions, dominated the scene. An Oxford guidebook of 1861 explains how geometry should be learned with the use of textbooks:[5]

Hints for the Mathematical Class-course must rather be given in the form of directions as to the choice of books than in that of remarks on the books or subjects themselves. It is hoped that, by using from the first the books most fitted for the Schools, a great saving of labour which is now very often thrown away may be effected, and that the utmost may thus be made of the limited time at [one's] disposal ...

Geometry must, of course, be commenced by acquiring a thorough knowledge of Euclid, which will be best read in Mr. Pott's octavo edition. The VIth book should not be read until the principles of ratio and proportion enunciated in the Vth are entirely mastered. It is not usual to read more than the first twenty propositions of the XIth Book; and the XIIth may be said to be completely superseded by other methods. In Mr. Pott's Appendix is a short but useful chapter on Transversals.

Here the author is referring to Robert Potts' edition, first published in 1845.[6] Like most of the 19th-century popular editions, this edition was based on the 1756 translation of the Scottish mathematician Robert Simson.

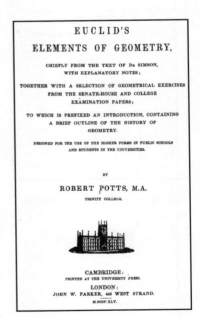

Potts's edition of Euclid's *Elements*.

Isaac Todhunter (1820–84).

Todhunter's edition of Euclid's *Elements*.

Another standard edition was produced by Isaac Todhunter, Britain's most popular mathematical textbook writer in the second half of the 19th century.[7] Todhunter was awarded a B.A. (1842) and an M.A. (1844) from University College, London, before entering St Johns College, Cambridge, where he graduated as senior wrangler in 1848. He remained at St Johns, where he served as a mathematical tutor, lecturer, and examiner. In 1862, he was elected a Fellow of the Royal Society, and was also a founding member of the London Mathematical Society in 1865. Although he made some interesting mathematical contributions and wrote authoritative histories of mathematical subjects, his reputation was (and still is) mainly based on his numerous textbooks in many disciplines (calculus, geometry, trigonometry, mechanics, algebra, etc.). His works were widely used in schools and colleges, and were regularly re-edited and translated into foreign languages. His best-seller, *Algebra for Beginners*, sold nearly six hundred thousand copies.[8]

While studying at University College, Todhunter was taught by Augustus De Morgan, one of Britain's leading mathematicians and teachers of the time. De Morgan was a fervent admirer of Euclid, a passion that he inculcated to his pupils. Todhunter's *Elements of Euclid for the Use of Schools and Colleges* appeared first in 1862 and, like Potts' edition, was based on Simson's 1756 text. It contained only eight books (I –VI, XI and XII), being those that were usually read in British universities. One special feature of Todhunter's edition was the splitting of geometrical demonstrations into their constituent parts, a technique that had been used by De Morgan thirty years earlier. Todhunter's standards of editorship guaranteed wide success for the book in Britain and abroad. It continued to be used long after his death, and attained more than twenty editions and half a million copies.[9]

However, not all textbooks were as original and popular as those of Potts or Todhunter. We can clearly observe the over-production of Euclid textbooks by looking at the mathematical reviews in popular journals and magazines. Hugh MacColl,[10] the mathematical reviewer for *The Athenaeum* in the last decades of the 19th century, wondered more than once about the *raison d'être* of the numerous redundant Euclid editions that he noticed in the journal; as he wrote in his 1891 review of a Euclid edition by A. E. Layng, headmaster of Stafford School:[11]

Why so many Euclids and all running in the old grooves? Mr. Layng's is carefully compiled and well arranged, but since the same praise is due to so many others, why add to the interminable list?

All through the Victorian period, hundreds of Euclid editions were published and used for geometrical teaching. From the end of the 1860s, however, alternative manuals that aimed to replace Euclid's text began to appear. By that time, several local and foreign reports had been issued on the state of British education, and most agreed on the backward state of geometry teaching. It was argued that Euclid-based teaching appealed to the pupils' memory, rather than to their intelligence. Thus, one could study Euclid for years and learn its content by rote, but still know almost nothing about geometry, especially if the textbook being used did not contain exercises. Consequently, several textbooks appeared to challenge the dominance of Euclid's *Elements*. Although none succeeded in replacing, or even competing with, Euclid's text, the contest was growing, and a wide-ranging debate on the suitability of Euclid's text as a manual for the teaching of geometry in British schools and colleges was slowly, but unavoidably, taking place within the community of mathematicians and mathematics teachers.

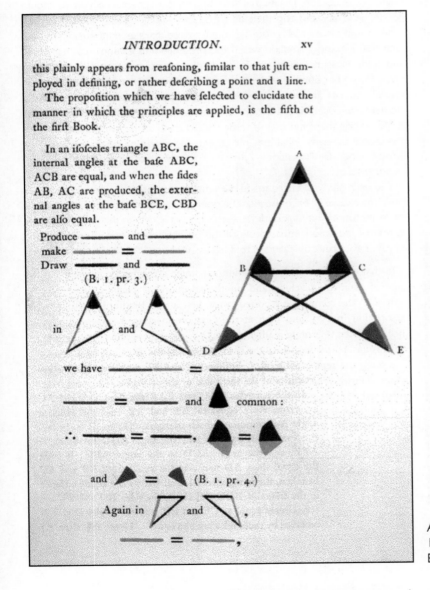

A page from Oliver Byrne's 1847 colour edition of Euclid's *Elements*.

# Euclid and his modern rivals

One of Euclid's most notorious opponents was James Joseph Sylvester. In his opening address to the mathematical and physical section of the 1869 meeting of the British Association for the Advancement of Science in Exeter, he pleaded strongly for the abandonment of Euclid in schools and colleges:[12]

[N]o one can desire more earnestly than myself to see natural and experimental science introduced into our schools as a primary and indispensable branch of education: I think that that study and mathematical culture should go on hand in hand together, and that they would greatly influence each other for their mutual good. I should rejoice to see mathematics taught with that life and animation which the presence and example of her young and buoyant sister could not fail to impart; short roads preferred to long ones; Euclid honourably shelved or buried "deeper than did ever plummet sound" out of the schoolboy's reach; morphology introduced into the elements of Algebra; projection, correlation, and motion accepted as aids to geometry; the mind of the student quickened and elevated and his faith awakened by early initiation into the ruling ideas of polarity, continuity, infinity, and familiarisation with the doctrine of the imaginary and inconceivable...

The early study of Euclid made me a hater of geometry, which I hope may plead my excuse if I have shocked the opinions of any in this room (and I know there are some who rank Euclid as second in sacredness to the Bible alone, and as one of the advanced outposts of the British Constitution) by the tone in which I have previously alluded to it as a school-book...

In that year, the British Association appointed a committee to report on the state of geometry teaching and on the adequacy of Euclid's *Elements* as a textbook. The committee reported twice: first at the Bradford meeting of the Association in 1873, and then at the Glasgow meeting in 1876. Although it included Britain's leading mathematicians, the committee came to no significant conclusions, partly because of the conflicting opinions of its members.

Among the representatives of schoolmasters, this committee also included James Maurice Wilson, mathematics master at Rugby School, who had recently published a successful anti-Euclid textbook. Wilson was senior wrangler in 1859 and was elected to a fellowship at St John's College, Cambridge, in the next year.[13] At Rugby School, with the support of his headmaster Frederick Temple, he undertook the writing of a manual that would supersede Euclid's treatise. For this purpose, he consulted geometrical textbooks from the Continent and the United States, where Euclid had long been abandoned. A first edition of his book, *Elementary Geometry*, that corresponded to Euclid's first two books, appeared in April 1868.

In the preface of his book, Wilson listed four major objections to the use of Euclid for teaching geometry:[14]

James Maurice Wilson
(1836–1931).

- Euclid's demonstrations are artificial, because he sacrificed simplicity and naturalness, and required some unnecessarily strict rules such as the exclusion of hypothetical constructions;

- the invariably syllogistic form of his reasoning is unsuitable for beginners, as it makes geometry 'unnecessarily stiff, obscure, tedious and barren';

- the length of Euclid's demonstrations 'exercises the memory more than the intelligence' of the learner;

- Euclid is unsuggestive as he 'places all his theorems and problems on a level, without giving prominence to the master-theorems, or clearly indicating the master-methods'.

With regard to these faults, Wilson pleaded for the abandonment of Euclid's text, and for the use of his own, for the purpose of geometrical teaching:[15]

In common then with some of our ablest mathematicians, and with many who are engaged in teaching mathematics, I am of opinion that the time is come for making an effort to supplant Euclid in our schools and universities. Already the fifth book has practically gone; and in consequence the study of the sixth book has become somewhat irrational.

For the improvement of our Geometrical teaching in England two things seem to be wanted. First, that Cambridge, and the Government Examiners, should follow the example of Oxford and London, and examine not in Euclid only, but in the Geometry of specified subjects, according to a programme for each examination. Secondly, that textbooks should be written to illustrate what is required. This book is the first part of such a text-book of Elementary Geometry.

Thanks to favourable circumstances Wilson's textbook gained some success, and Wilson himself was invited to present his views before the London Mathematical Society and the Royal High School in Edinburgh. His growing notoriety allowed him to publicize the problem of geometrical teaching, and to act as a spokesman for the mathematical schoolmasters opposed to the use of Euclid's manual.[16]

Meanwhile, the need for an anti-Euclid association was promptly being felt. The idea seems to have been first expressed in May 1870 by Rawdon Levett, the mathematics master of King Edward's School (Birmingham),[17] in a letter that appeared in the journal *Nature*:[18]

There are many engaged in the work of education in this country, besides those who have come prominently forward in the matter, who feel strongly that Geometry as now taught falls far short of being that powerful means of education in the highest sense which it might easily be made. They find themselves, in the majority of cases, compelled to use in their classes a textbook which should long ago have become obsolete.

ELEMENTARY GEOMETRY

*PART I.*

ANGLES, PARALLELS, TRIANGLES, EQUIVALENT FIGURES, WITH THE APPLICATION TO PROBLEMS.

COMPILED BY

J. M. WILSON, M.A.

FELLOW OF ST JOHN'S COLLEGE, CAMBRIDGE, AND MATHEMATICAL MASTER OF RUGBY SCHOOL.

London and Cambridge:
MACMILLAN AND CO.
1868

[*All Rights reserved.*]

Title page of J. M. Wilson's *Elementary Geometry*.

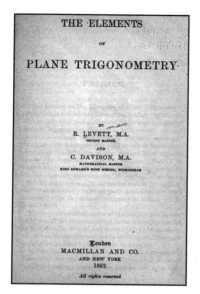

Title page of Rawden Levett's *Elements of Plane Trigonometry.*

Thomas Archer Hirst (1830–92).

We have lately had instances in abundance of the power of combined action. If the leaders of the agitation for the reform of our geometrical teaching would organise an Anti-Euclid Association, I feel sure they would meet with considerable and daily-increasing support.

We of the rank and file do not feel strong enough to act alone, and yet think we might do something to help forward the good cause by co-operating with others.

The immediate objects of such an association should be in my opinion (1) To collect and distribute information connected with the subject; (2) To induce examining bodies to frame their questions in geometry without reference to any particular textbook.

After a short meeting at Wilson's house in Rugby, a founding meeting was organized at University College, London, on 17 January 1871, and was attended by twenty-six 'members' and a 'few gentlemen' who were not yet members. It led to the creation of the Association for the Improvement of Geometrical Teaching (AIGT), with Thomas Archer Hirst as its first president. Wilson and Joshua Jones (King William's College) were elected vice-presidents, while Levett and E. F. M. MacCarthy (King Edward's, Birmingham) were secretaries.

Hirst fitted the presidency very well because of his long involvement in scientific institutions and organizations, which made him a leading player in the scientific milieu of his time. Elected a fellow of the Royal Society in 1861, he had been a founder-member of the influential X-Club since 1864, promoting science in England independently from religion. In addition to his organizational skills, Hirst was also a renowned mathematics teacher. He first taught mathematics at Queenwood College near Salisbury, becoming professor of mathematical physics at University College in 1865, replacing Augustus De Morgan two years later as professor of mathematics, before resigning this position in 1870. Hirst also gave a highly praised series of geometry lectures to the Ladies Educational Association of London around this time.[19]

In addition to his position as a co-founder (in 1865), first vice-president, treasurer, and later president of the London Mathematical Society, Hirst acted as secretary of the British Association for the Advancement of Science from 1866 to 1870. He was assistant registrar of the University of London from 1870 to 1873, at the time of his election as the first president of the AIGT, a position he would keep until 1878. In the meantime, in 1873 he became director of the Royal Naval College in Greenwich (as mentioned in Chapter 3), staying on until his retirement in 1883.

Thanks to his continental education (in Germany, France, and Italy), Hirst was acquainted with those educational systems where Euclid had long been abandoned. It was at Hirst's suggestion that the word 'improvement' was substituted for the word 'reform' in the original name of the association, as proposed by Levett. Hirst explained that 'reform' might be a merely ephemeral object, while, after the present immediate necessity has been satisfied, there would always be room for 'improvement'.[20]

Syllabus of Hirst's lectures on geometry to the Ladies' Educational Association, 1870.

At this first meeting, the participants resolved to appoint a committee who would prepare a syllabus of geometry for the use of teachers, textbook authors, and examiners. The Association's *Syllabus of Plane Geometry* appeared in 1875, and was regularly revised and reissued in subsequent years. The Association did not initially resolve how to produce its own textbook, but its first report (in 1871) contained a list of modern geometry manuals to be preferred to Euclid's *Elements.* The list included eight English texts by J. R. Morell, B. Peirce, E. M. Reynolds, H. Wedwoods (twice), J. M. Wilson, R. Wormell, and R. P. Wright. Also included were ten French texts (by Amiot, Bos, Ch. Briot, Briot and Vacquant, M. Pronchet, Legendre, Vincent and Saigey, Vernier, Roucher, and De Comberousse), six German titles (by Baltzer, Becker, Fischer, Grabow,

Meyer, and Schumann), one Italian text (by Sannia e D'Ovidio) and one Spanish title (by Don Juan Cortazar).[21]

It was not until the arrival of a new president, Robert Baldwin Hayward, in 1878 that the Association began work on its own alternative textbook, *The Elements of Plane Geometry*. This eventually appeared in two parts (1884 and 1886), covering the material of Euclid Books I–II and III–VI, respectively. The AIGT regularly circulated its outputs, notably its syllabus and the textbook, to examining boards at British universities and to learned societies, but it attained only a limited success. By 1888, however, Oxford and Cambridge did accept proofs differing from Euclid's in their geometry examinations, provided that they did not violate Euclid's numbering or logical order.

Although the Association received considerable support from schoolmasters dissatisfied with the teaching requirements, and much sympathy from many younger mathematicians of the time, it also had to face the hostility of virulent opponents, one of whom was Britain's leading mathematician, Arthur Cayley, Sadleirian professor of mathematics at Cambridge. In 1936, Arthur W. Siddons reported the following anecdote regarding Cayley's opposition to the reformers:[22]

I remember Levett telling me that on one occasion he and others went up to Cambridge for a meeting of the Mathematical Board at which they hoped the Board would agree to the abolition of Euclid. I think this meeting must have been in 1887. Cayley was in the Chair and dominated the meeting. One remark of Cayley's Levett quoted to me, "the proper way to learn geometry is to start with the geometry of *n* dimensions and then come down to the particular cases of 2 and 3 dimensions". Cayley opposed the abolition of Euclid and Levett told me after the meeting a member of the board said to him, "We cannot go against Cayley".

Siddons also reported another anecdote, told by Edward Mann Langley in the 15th general report (January 1889) of the AIGT, about Cayley's opposition to the reform at the Cambridge Senate:[23]

One or two uncompromising opponents had been met with. Professor Cayley was the most formidable; and he was such an ardent admirer of Euclid that he overshot the mark, and his opposition told in favour of the Association. In the course of the discussion he (Mr. Langley) having drawn attention to the fact that the authorised treatise was an inconsistent one—a mixture of Euclid and Simson—in which use was not made of the Corollaries introduced by Simson in proving subsequent propositions, Professor Cayley suggested striking out Simson's additions and keeping strictly to the original treatise; upon which a member of the Senate whispered that perhaps to study it in the original Greek would be better still.

These anecdotes show clearly both Cayley's influence in Cambridge and how he used it to plead strongly for the retention of Euclid's treatise, going as far as to object to any change in the Euclidean text. Not all of Euclid's advocates were opposed to any change, however; indeed, the majority were in favour of a gradual improvement of geometrical teaching, using an amended version of Euclid's text.

This was, for example, the position of Augustus De Morgan, whose early writings called for deep modifications in Euclid's text, notably in Books V and X, to make them more intelligible to students. De Morgan did not appreciate Wilson's rival manual: in an anonymous review for *The Athenaeum*, he pointed out Wilson's limited understanding of Euclid's logical structure, and objected to the theory of direction used in the treatment of angles and parallels. De Morgan claimed his clear preference for an amended version of Euclid:[24]

We feel confidence that no such system as Mr Wilson has put forward will replace Euclid in this country. The old geometry is a very English subject, and the heretics of this orthodoxy are the extreme of heretics: even Bishop Colenso has written a Euclid. And the reason is of the same kind as that by which the classics have held their ground in education. There is a mixture of good sense and of what, for want of a better name, people call prejudice: but to this mixture we owe our stability. The proper word is *postjudice*, a clinging to past experience, often longer than is held judicious by after times. We only desire to avail ourselves of this feeling until the book is produced which is to supplant Euclid; we regret the manner in which it has allowed the retention of the faults of Euclid; and we trust the fight against it will rage until it ends in an amended form of Euclid.

De Morgan's views were echoed by his student, Todhunter, in his 1873 collection of essays, *The Conflict of Studies*, where he expounded his educational views.[25] The fifth essay was devoted to 'Elementary geometry' and contained a defence of Euclid's *Elements* and the rejection of his rivals' manuals, notably Wilson's.

Todhunter shared this position with another famous mathematics teacher, Charles L. Dodgson, lecturer at Christ Church, Oxford, but better known by his pen-name of Lewis Carroll. In *Euclid and his Modern Rivals*, first published in 1879 and enlarged in 1885, Dodgson collected together, discussed, and then rejected, thirteen textbooks written to supersede Euclid (by W. Chauvenet, W. D. Cooley, F. Cuthbertson, O. Henrici, A. M. Legendre, E. Loomis, W. D. J. R. Morell, B. Peirce, E. M. Reynolds, W. A. Willock, J. M. Wilson, R. P. Wright, and the Syllabus of the AIGT). The book was written as a drama in four acts, involving Minos (the judge), Radamanthus, Euclid's ghost, and Herr Niemand (a German Professor who spoke on behalf of Euclid's rivals). None of the rival textbooks was retained by Dodgson, and most were severely criticized. Only Legendre's and Peirce's works were praised, but were considered as inadequate for beginners.

Dodgson explained first that it is necessary to employ one (and only one) textbook and its sequence and numbering of propositions. Then he argued that there are strong *a priori* reasons for retaining Euclid's manual. Finally, he showed that none of the manuals offered as substitutes can replace, nor even compete with, that of Euclid. Dodgson concluded that Euclid's (and only Euclid's) text should be maintained, and that his logical sequence and numbering of propositions should be used for the purpose of teaching geometry:[26]

Charles Dodgson
(Lewis Carroll) (1832–98).

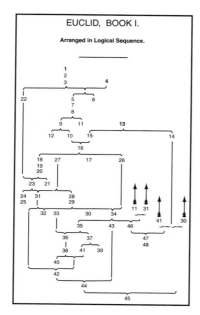

EUCLID, BOOK I.

**Arranged in Logical Sequence.**

Schematic image of the propositions in Book I of Euclid's *Elements*, from *Euclid and his Modern Rivals*.

*Euc[lid]*. 'The cock doth craw, the day doth daw,' and all respectable ghosts ought to be going home. Let me carry with me the hope that I have convinced you of the importance, if not the necessity, of retaining my order and numbering, and my method of treating straight Lines, angles, right angles, and (most especially) Parallels. Leave me those untouched, and I shall look on with great contentment while other changes are made—while my proofs are abridged and improved—while alternative proofs are appended to mine—and while new Problems and Theorems are interpolated.

In all these matters my Manual is capable of almost unlimited improvement.

Thus, Dodgson did not exclude the possibility of improving the original text. He did himself introduce a few changes in his own Euclid edition, first published in 1882,[27] and even introduced a new axiom of parallels in a late treatise, *Curiosa Mathematica—Part I: A New Theory of Parallels* (1888).[28]

# The passing of Euclid

The failure of the AIGT, in its early years, to remove Euclid's *Elements* from the British curriculum is not surprising. As we have seen, mathematicians and mathematics teachers were far from unanimous in their support for the Association's aims. The members of the Association themselves, although they agreed on the necessity of improving geometrical teaching, did not agree on the extent of the change. Most were, in fact, in favour of progressive improvement too.

This caution may be explained by a fear of potential chaos in mathematical teaching and examination if Euclid were abandoned, a fear that most teachers and examiners, both Euclid's opponents and proponents, shared. The 'harmony' of examination systems required the use of a single sequence and a single numbering of propositions. Until then, Euclid's sequence and numbering had been used for the purpose. The radical and sudden introduction of numerous manuals, with different numberings and sequences, would break that uniformity. In addition, the substitute manuals produced were as yet far from perfect, and none could make a convincing claim for its sequence to be used in preference to that of Euclid. Consequently, the retention of Euclid's order and numbering left no room for significant changes.

Thus, despite the attacks of the AIGT, Euclid's *Elements* remained the standard text for geometrical teaching in Britain, though not in other western countries. In 1900, David E. Smith recorded that:[29]

So complete as a specimen of logic was Euclid's treatment of elementary geometry, that it has been used as a text-book, with slight modifications, for over two thousand years. This use has not, however, been general. Indeed, it has needed the exertions of men like Houël in France and Loria in Italy, and other Continental writers, to recall from time to time the merits of Euclid to the educational world. But in England Euclid still holds a sway that is practically absolute.

By this time, the AIGT had undergone several internal changes. As we saw in Chapter 7, following the suggestion of E. M. Langley, the Association launched its own journal, *The Mathematical Gazette*, on the lines of the French *Journal des Mathématiques Elémentaires*. The first issue appeared in April 1894, with Langley as the founding editor. *The Mathematical Gazette* replaced the Association's annual reports, which already included papers on various mathematical subjects and educational issues.[30] In March 1897, the Association for the Improvement of Geometrical Teaching changed its name and became the Mathematical Association, to express more appropriately its expanded educational desire to include all the disciplines of elementary mathematics.

Contrary to geometrical education which remained almost unchanged, geometrical research witnessed important changes in the 19th century, notably through the discovery of non-Euclidean geometries, although British mathematicians played no role in their early development. Despite some works of Girolamo Saccheri, Johann Heinrich Lambert, Adrien-Marie Legendre, Ferdinand Schweikart, and Carl Friedrich Gauss in the 18th and early 19th centuries, the discovery of non-Euclidean geometries is generally credited to the mathematicians János Bolyai and Nikolai Lobachevsky in the 1820s and 1830s. However, these works gained little recognition from the mathematical community. It needed much time and further development by Bernhard Riemann, Eugenio Beltrami, and Felix Klein for these new geometries to reach a wide audience. In Britain, only a few mathematicians were ready to accept these ideas and what they might tell us about the nature of mathematical and physical space. Although Cayley noticed Lobachevsky's work in 1865,[31] it was William Kingdon Clifford and Hermann Helmholtz who made non-Euclidean geometries known to the British mathematical community in the 1870s.[32]

However, the introduction of these geometries and the development of projective and algebraic geometries, to which Cayley mainly made important contributions, did not significantly affect Euclid's dominance in British schools. The Euclid debate focused on purely pedagogical and administrative issues. Most mathematicians, even those sympathetic to the AIGT's views, were either not interested in, or not familiar with, such non-mathematical concerns, and focused rather on formal matters in relation to Euclid's work. For instance, Bertrand Russell, who in 1897 published an *Essay on the Foundations of Geometry*,[33] contributed to the Association's journal in 1902 with a short note in which he explains that Euclid's logical merits had been highly overestimated:[34]

It has been customary when Euclid, considered as a text-book, is attacked for his verbosity or his obscurity or his pedantry, to defend him on the ground that his logical excellence is transcendent, and affords an invaluable training to the youthful powers of reasoning. This claim, however, vanishes on a close inspection. His definitions do not always define, his axioms are not always indemonstrable, his demonstrations require many axioms of which he is quite unconscious. A valid proof retains its demonstrative force when no figure is drawn, but very many of Euclid's earlier proofs fail before this test.

In its early years, the AIGT, and later the Mathematical Association, had been concerned almost completely with deductive geometry alone, as it was taught on Euclidean lines in British schools and universities, particularly in Oxford and Cambridge. In one way the Association's aims were modest, because it remained within the liberal and non-utilitarian paradigm and did little for vocational education: indeed, the Association paid almost no attention to practically orientated geometry, and more generally to practical mathematics, the teaching of which was progressively growing for the needs of engineering and science education. Ironically, it was this 'practical movement', under the leadership of John Perry,[35] that ultimately led to deep reforms in mathematical education, and stopped Euclid's hegemony in the early years of the 20th century.

From 1875 to 1878, after some early teaching experience at Clifton College in Bristol and a short post as an assistant to Lord Kelvin at Glasgow, Perry worked as a professor in civil engineering at the Imperial College in Tokyo. On his return to Britain, he conducted pioneering technical education at Finsbury Technical College in London from 1882, and later at the Royal College of Science, where he was appointed professor of mathematics and mechanics in 1896. In addition to his textbooks—notably, a popular *Calculus for Engineers* (1896)—Perry wrote regularly in journals, preaching at length for the reform of mathematical education. He advocated the introduction into schools of the practical methods that he taught successfully in his own classes, prior to the deductive methods, notably the Euclid-based geometrical teaching:[36]

John Perry (1850–1920).

I think it very important to try to get a view of our system of teaching mathematics which is not too much tinted with the pleasant memories of one's youth. Like all the men who arrogate themselves the right to preach on this subject, I was in my youth a keen geometrician, loving Euclid and abstract reasoning. But I have taught mathematics to the average boy at a public school, and this has enabled me to get a new view...

The framers of educational methods took in their youth to abstract reasoning as a duck takes to water, and of course they assume that a boy who cannot in one year understand a little Euclid must be stupid. In truth, it is a very exceptional mind, and not, perhaps, a very healthy mind, which can learn things or train itself through abstract reasoning; nor, indeed, is much ever learnt in this way. Do we philosophise about swimming before we know how to swim; or about walking or jumping, or cycling or riding a horse, or planing wood, or chipping or filing metals, or about playing billiards or cricket? Is it through philosophy that we learn a game of cards, or to read or to write? No; we first learn by actual trial; we practice as our mind lets us; we philosophise afterwards—perhaps long afterwards. Then if we are too clever or stupid, we insist on teaching a pupil from the point of view which we have at the end of our studies, and we refuse to look at things from the pupil's point of view.

On 14 September 1901, Perry gave a talk at the Glasgow meeting of the British Association, where he condemned the current situation of mathematical teaching in British schools, and called for immediate and deep changes. This talk and the discussion that followed were reported on by R. F. Muirhead in the October 1901 issue of the *Mathematical Gazette*[37] and stimulated reactions by E. M. Langley, Charles Godfrey, and A. W. Siddons in the following issue.[38] Langley retained three main ideas from the Glasgow debate: the necessity of stopping the domination of examinations over the school system, the urgency of abandoning Euclid's book for the purpose of teaching geometry, and the need to introduce experimental or intuitional methods, in addition to the already existing deductive methods, in mathematical (and especially geometrical) teaching.

The British Association appointed a committee to consider the issue of the teaching of elementary mathematics, with Andrew R. Forsyth in the chair and Perry as secretary. The committee reported at the Belfast meeting in September 1902. It drew up general suggestions, but insisted on the necessity of introducing practical mathematics into schools. With regard to the retention of Euclid, the committee recommended more freedom in the choice of textbooks (Euclid included), and asked the examining bodies to relax their requirements in order to allow teachers to proceed in such a way as to fit with the needs of their pupils:[39]

In the opinion of the Committee, it is not necessary that one (and only one) text-book should be placed in the position of authority in demonstrative geometry; nor is it necessary that there should be only a single syllabus in control of all examinations. Each large examining body might propound its own syllabus, in the construction of which regard would be paid to the average requirements of the examiners…

In every case, the details of any syllabus should not be made too precise. It is preferable to leave as much freedom as possible, consistently with the range to be covered; for in that way the individuality of the teacher can have its most useful scope. It is the competent teacher, not the examining body, who can best find out what sequence is more suited educationally to the particular class that has to be taught.

In January 1902 the Mathematical Association appointed a 'Teaching Committee' in order to keep in touch with the reform movement. The Committee reported first on geometry, in the May 1902 issue of the *Mathematical Gazette*. The report asked for the introduction of experimental courses, but surprisingly did not attack the Euclidean order.[40] In fact, the members of the Committee took the diplomatic route, considering it more profitable to suggest moderate changes without violating the Euclidean order.

This decision shows how secure Euclid's position still was, and how far the members of the committee were from anticipating the imminent end of Euclid's dominance. Indeed, although several universities had relaxed their requirements on geometry examinations by allowing other proofs, Euclid could never be completely overthrown in British mathematical education as long as the University of Cambridge maintained its formal allegiance to the *Elements*. After all, as we saw in Chapter 1, Cambridge was still the centre of British mathematics at this time,[41] and Cayley was able to exploit his distinguished professorial position to prevent the abolition of Euclid there. But after Cayley's death in 1895, his successor in the Sadleirian professorship, Andrew Forsyth, proved to be a key figure in the agitation that would lead ultimately to the abandonment of Euclid in the early years of the 20th century.

Forsyth was Chair of the British Association Committee mentioned above, and in 1903 became President of the Mathematical Association. He was also an influential member of the Syndicate appointed by the University of Cambridge in December 1902 to consider examinations in mathematics. On 9 May 1903, the Syndicate issued a general report that recommended, at Forsyth's instigation, the acceptance of *any* systematic proof, Euclidean or otherwise. To be effective, however, the recommendation needed the agreement of the Senate. A. W. Siddons, a member of the Syndicate, later recollected:[42]

The great question now was, would the recommendations be passed by the Senate; we feared that flysheets would appear and that every country parson who ever took pupils would come up and vote against the abolition of Euclid. However, a storm was brewing at Cambridge about the report of another Syndicate; and under the shadow of that storm, owing to Professor Forsyth's careful steering, our report sailed through without a division and Euclid as a textbook ceased to dominate English education.

The recommendations were adopted by the Senate on 11 June 1903, and the new regulations came into force the following year.[43]

With the capitulation of Cambridge, Euclid's rivals had won the decisive battle. The victory was as sudden as it was unexpected, however, and the rivals were not yet prepared for radical change. For a time, many schools continued to teach geometry in the old way, since Euclid was not forbidden by the new regulations—in Forsyth's words, it was 'only disestablished, not abolished'.[44] Thus, the *Elements* was now merely one option competing with a variety of new manuals designed to meet the needs of the reform.[45] But since the book no longer fitted in with the practical geometry courses now in fashion, the disappearance of Euclid from the British classroom became inevitable.

Still, in deductive geometry Euclid's ghost continued to haunt schoolmasters for a long time. The period of confusion that followed the reforms encouraged the need for a single sequence in geometry, and in the early years of the 20th century British mathematics teachers repeatedly wondered whether Euclid's sequence should be restored.[46] In the event, that never happened: Euclid's hallowed position as the prime geometry textbook had gone for good.

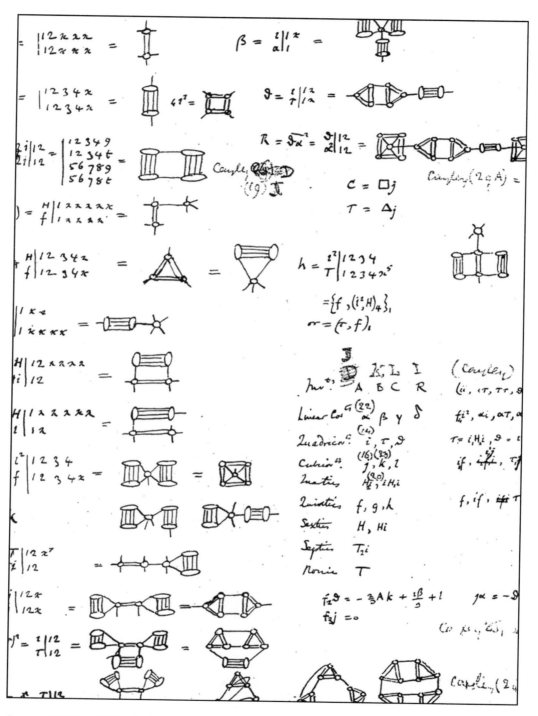

Some of William Clifford's binary quantics.

CHAPTER 15

# Victorian algebra

*The freedom to create new mathematical entities*

KAREN HUNGER PARSHALL

Geometry and mathematical physics may have interested a majority of mathematicians in Victorian Britain, but algebra also served to focus their mathematical attention. In the century's first half, algebraic work centred on the development of the so-called 'symbolical algebra' and the creation of new algebras, while in its second half, the theory of invariants dominated and the abstract theory of groups witnessed key developments. Underlying much of this research was the philosophical question of how free mathematicians were to create new mathematical entities. The Victorian British response was, ultimately, 'quite'.

Over the course of the 19th century, and especially during the Victorian period proper, a number of British mathematicians began successfully to explore new and original lines of algebraic research. In the opening decades of the century, that research centred on the debate—sparked by mathematicians such as Francis Maseres and William Frend, and then carried on by Robert Woodhouse, Augustus De Morgan, George Peacock, and others—about the nature of the negative and imaginary numbers and, indeed, about the very nature of algebra. This avenue of inquiry, partly philosophical and partly mathematical, resulted at the beginning of the Victorian era and into the 1850s in the development of what was called 'symbolical' algebra, as well as in the calculus of operations, the largely algebraic study of the manipulation of symbols. The latter area, in particular, matured in the changed educational atmosphere of a Cambridge that had increasingly embraced the analytical—that is, algebraic—agenda of the Cambridge Analytical Society, beginning in the 1810s. By the time that Victoria came to the throne in 1837, the former members of the Society and their supporters had succeeded in reorienting the Cambridge tripos in this more algebraic direction and, in so doing, had provided a generation of mathematicians—by no means all of whom were associated with Cambridge—with a taste for things algebraic.[1]

This new algebraic inclination almost immediately manifested itself, in the 1840s and 1850s, in significant British developments in what would become by the end of the century the theory of algebras over the real and complex fields. British mathematicians approached the new mathematical entities that they encountered in their work from a point of view that can be characterized, albeit in a 19th-century as opposed to a 20th-century sense, as both constructive and structural; at least initially, they emphasized the creation of new algebras which obey certain recognized rules of addition and multiplication and analysed the properties that these new constructs satisfy. To a lesser degree, a few British mathematicians, but most notably Arthur Cayley, followed this same kind of structural approach in exploring the abstract theory of groups, beginning in the 1850s.

Perhaps the area that most prominently put British algebraic researchers on the international mathematical map, however, was invariant theory. Building on an 1841 paper by George Boole, the British—as opposed to the Continental—style of invariant theory developed at the hands primarily of Cayley and James Joseph Sylvester, starting in the early 1850s. Their work attracted the attention of George Salmon in Ireland, of some of Sylvester's American students in the late 1870s and early 1880s, and of others back in Britain up to the end of the Victorian era. This chapter traces these widely ranging algebraic developments from the 1830s to 1900.

## The development of 'symbolical algebra'

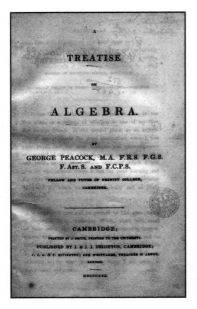

Peacock's *Treatise on Algebra*.

Although the initial debate over the foundations of algebra and its proper role in (especially) the Cambridge mathematical curriculum flared famously in the closing decade of the 18th and the opening two decades of the 19th century, the status of algebra came once again to the fore in 1830 following the publication in that year of George Peacock's *Treatise on Algebra*.[2] Along with Charles Babbage and John Herschel in the 1810s, Peacock had aimed, through the founding of the Cambridge Analytical Society, to replace once and for all the Newtonian approach to the calculus with the analytic approach exemplified in their translation of Silvestre Lacroix's text on the differential and integral calculus. By 1830, Peacock had formulated an interpretation of algebra that proved equally thought-provoking to his contemporaries.

In an effort 'to articulat[e] what algebra was, and analyz[e] what it could therefore achieve', Peacock gave, according to Menachem Fisch, 'a new exposition of what algebra *is* and how it should be accounted for' that hinged on the distinction between what Peacock termed 'arithmetical' and 'symbolical' algebra.[3] Peacock viewed 'arithmetical' algebra as 'the science, whose operations and the general consequences of them should serve as the guides to the assumptions which become the foundation to symbolical Algebra'.[4] In other words, the arithmetic of the positive whole numbers determined the properties of arithmetical algebra and,

[t]hough the science of…arithmetical algebra does not furnish an adequate foundation for the science of symbolical algebra, it necessarily suggests its principles, or rather its laws of combination; for in as much as symbolical algebra, though arbitrary in the authority of its principles, is not arbitrary in their application, being required to include arithmetical algebra as well as other sciences, it is evident that their rules must be identical with each other, as far as those sciences proceed together in common.[5]

Peacock termed this identification the 'principle of the permanence of equivalent forms', and although he distinguished between the two algebras—arithmetic and symbolic—their results were linked. As Fisch has put it,[6]

[i]n the last analysis [Peacock's algebra] is generalized arithmetic pure and simple with the additional license to generalize both symbols and operations. Traditional generalized arithmetic (e.g. Frend's) only generalized operations, and was therefore constrained by the limiting inventory of arithmetically admissible quantities.

Contrary to the views of some of his contemporaries and of some later historians, 'Peacock did no more…than to generalize the entire system'.

Contemporaneous reactions to the interpretation of algebra that Peacock expressed in his *Treatise* varied. William Rowan Hamilton responded in 1833 with his own exposition of 'algebra as the science of pure time'. In his view,[7]

[t]he Study of Algebra may be pursued in three very different schools, the Practical, the Philological or the Theoretical, according as Algebra itself is accounted an Instrument [or an Art, in Hamilton's terminology], or a Language, or a Contemplation.

For Hamilton, '[t]he felt imperfections are of three answering kinds':[8]

The Practical Algebraist complains of imperfection when he finds his Instrument limited in power; when a rule, which he could happily apply to many cases, can be hardly or not at all applied by him to some new case…The Philological Algebraist complains of imperfection, when his Language presents him with an Anomaly; when he finds an Exception disturb the simplicity of his Notation, or the symmetrical structure of his Syntax; when a Formula must be written with precaution, and a Symbolism not universal. The Theoretical Algebraist complains of imperfection, when the clearness of his Contemplation is obscured; when the Reasonings of his Science seem anywhere to oppose each other, or become in any part too complex or too little valid for his belief to rest firmly upon them; or when…he cannot rise to intuition from induction, or cannot look beyond the signs to the things signified.

To Hamilton's way of thinking, Peacock had approached algebra as an art from a primarily practical point of view, whereas he sought to deal with it as a science from a theoretical viewpoint (see the next section).

Another contemporary, Augustus De Morgan, entered into the debate and ultimately offered his own resolution of how best to interpret algebra. In the second part of his paper 'On the foundations of algebra', read before the Cambridge Philosophical Society in November 1841, De Morgan proposed to separate the 'laws of operation from the explanation of the symbols operated upon or with'—that is, he established algebra in terms of a set of axioms that it should obey.[9] The operations were addition, subtraction, multiplication, and division, and although the axioms laid out were essentially the axioms for what we would now term a 'field' (without an explicit statement of associativity), De Morgan was interested only incidentally in the establishment of axioms. He likened the relationship of axioms to the algebra they underlie to working a jigsaw puzzle of the map of a country. As he explained:[10]

a person who puts one of these together by the backs of the pieces, and therefore is guided only by their forms, and not by their meanings, may be compared to one who makes the transformations of algebra by the defined laws of operation only: while one who looks at the fronts, and converts his general knowledge of the countries painted on them into one of a more particular kind by help of the forms of the pieces, more resembles the investigator and the mathematician.

Determining the meaning of the axioms in a particular context was the mathematician's real goal, to De Morgan's way of thinking.

De Morgan's axiomatization moved algebra more markedly away from the arithmetic of the whole numbers by its full incorporation of subtraction in the context of 'literal symbols, *a*, *b*, *c*, &c. [that] have no necessary relation except this, that whatever any one of them may mean in any one part of a process, it means the same in every other part of the same process'.[11] A further move away from arithmetic and toward a more formal approach to algebra was made initially by a number of Cambridge graduates who had been reared mathematically in the algebraically charged 1830s. These students, among them Duncan Gregory and Robert Ellis, were products of the reform of the Cambridge mathematics curriculum at the hands of Peacock, Babbage, and Herschel, and were thus not only thoroughly imbued with the analytic (that is, algebraic) approach to the calculus, but also profoundly influenced by the contemporaneous debate about the meaning of algebra. They had cut their mathematical eye teeth on texts like Peacock's *Treatise*, and they were equally aware of more targeted texts like De Morgan's *Elements of Algebra Preliminary to the Differential Calculus*, which came out initially in 1835 and in a second edition in 1837. In De Morgan's book, in particular, they were confronted with the view that:[12]

[t]he very great difficulty of the differential calculus has always been a complaint; and it has frequently been observed that no one knows exactly what he is doing in that science until he has made considerable progress in the mechanism of its operations...I have long believed the reason of this to be that the fundamental notions of the differential calculus are conventionally, and

Duncan Gregory (1813–44).

with difficulty, excluded from algebra, in which I think they ought to occupy an early and prominent place.

Young students of mathematics exposed to this new way of treating the calculus began to explore—from the kind of algebraic point of view that Peacock and De Morgan espoused—the behaviour of the actual notation in the context of their own explorations of the meaning of symbolic algebra. In his paper 'On the real nature of symbolical algebra', read before the Royal Society of Edinburgh in May 1838, Gregory considered these questions of notational behaviour in a foundational light, inspired by Peacock's work. In particular, he analysed operations *per se* in terms of their properties and in so doing defined five distinct classes. For example, if $F$ and $f$ operate on $a$ and if $F$ and $f$ satisfy the four relations

$$FF(a) = F(a), ff(a) = F(a), Ff(a) = f(a), \text{ and } fF(a) = f(a)$$

relative to $a$, then $F$ and $f$ are in the first class, or were what Gregory termed 'circulating operations'. As he noted, the arithmetic operations of addition and subtraction belong to this class, while operations that satisfy the relations

$$f_m(a)f_n(a) = f_{m+n}(a) \text{ and } f_m(a)f_n(a) = f_{mn}(a)$$

are in the third class—that is, the same class as the exponentiation operator.[13] As Gregory put it in the introduction to his paper, his efforts to categorize operations and thereby

to investigate the real nature of Symbolical Algebra, as distinguished from the various branches of analysis which come under its dominion, took its rise from certain general considerations, to which I was led in following out the principle of the separation of symbols of operation from those of quantity.[14]

In pursuing this line of research, Gregory and his Cambridge mathematical *confrères* had observed, for example, the similarities between the law of exponents and the behaviour of the composition of functions. Thus, they noted that

$$x^m \cdot x^n = x^{m+n}, \frac{d^m}{dx^m}\left(\frac{d^n u}{dx^n}\right) = \frac{d^{m+n}}{dx^{m+n}} u \text{ and } f^m(f^n(x)) = f^{m+n}(x).$$

How then, they asked, can symbols of operation be manipulated, and to what extent can they be manipulated independently of the symbols on which they operate?[15] The answers to these and related questions form the algebraic calculus of operations, and engaged such aspiring mathematicians as Ellis and Gregory, who in 1837 founded the *Cambridge Mathematical Journal* for the encouragement of young researchers like themselves. As discussed in Chapters 7 and 13, this journal quickly came to serve as the principal publication outlet for research in this new area, which became a sort of British mathematical cottage industry in the 1840s and early 1850s.

# The creation of new kinds of algebras

It was in this same spirit of algebraic experimentation and exploration—as well as in the spirit of his own philosophical musings on algebra as the science of pure time—that Ireland's William Rowan Hamilton began to consider the complex numbers in the summer of 1829 and into the 1830s.[16] From a philosophical point of view, he found the usual conception of the complex numbers as sums of a real and an imaginary part problematic. They were hybrids, part real and part imaginary. How could two such different things legitimately be added? Drawing on John Warren's *A Treatise on the Geometrical Representation of the Square Roots of Negative Quantities* (1828), Hamilton succeeded in resolving this problem in 1833, with his representation of the complex numbers as ordered pairs of real numbers.

As we saw in Chapter 5, Hamilton wrote the complex number $a + b\sqrt{-1}$, for real numbers $a$ and $b$, as the ordered 'couple' $(a, b)$—that is, he identified complex numbers with points in the real plane. He then specified, among other things, the binary operations of addition and multiplication on pairs of couples:

$$(a, b) + (c, d) = (a + c, b + d) \text{ and } (a, b) \times (c, d) = (ac - bd, bc + ad).$$

Hamilton thus *defined* the usual binary operations on the new construct, the ordered couple, in the real plane. Was it then possible to represent points in 3-dimensional space analogously as ordered triples with suitably defined binary operations—that is, (in terminology developed later) is it possible to create a hypercomplex number system (or algebra) that is 3-dimensional over the reals?[17]

Working on and off for a decade to answer this question, Hamilton thought he would be able to answer it in the affirmative by actually creating such an algebra. In 1843, however, he unexpectedly discovered not a 3-dimensional but a 4-dimensional algebra—the *quaternions*—and, even more surprisingly, established the non-commutativity of its multiplication. Quaternions are 'numbers' of the form

$$(a, b, c, d) = a + bi + cj + dk,$$

where $a$, $b$, $c$, and $d$ are real and where $i$, $j$, $k$ satisfy the relations

$$ij = -ji = k, \ jk = -kj = i, \ ki = -ik = j \text{ and } i^2 = j^2 = k^2 = -1.$$

As in the two-dimensional case, Hamilton defined addition component-wise, but the relations, as Hamilton specified them, resulted in a multiplication satisfying, like that of the ordinary integers, the condition that every non-zero element has a multiplicative inverse, but unlike it, not satisfying the commutative law.

Hamilton had *created* a new number system with its own set of operational rules, a number system that does not share the usual properties of arithmetic. In so doing, he fundamentally challenged the notion of algebra as Peacock had conceived of it with his principle of the permanence of equivalent forms.

The plaque on Brougham Bridge, Dublin, celebrating Hamilton's quaternions.

Here as he walked by on the 16th of October 1843 Sir William Rowan Hamilton in a flash of genius discovered the fundamental formula for quaternion multiplication $i^2 = j^2 = k^2 = ijk = -1$ & cut it on a stone of this bridge

Hamilton's friend, John Graves, expressed the misgivings of many of Hamilton's contemporaries when he confessed that 'I have not yet a clear view as to the extent to which we are at liberty arbitrarily to create imaginaries, and to endow them with supernatural properties'.[18] It was, nevertheless, not long before other new algebras were discovered which had even more surprising 'supernatural properties'.

Arthur Cayley and John Graves almost immediately extended Hamilton's ideas to the 8-dimensional octonions, the multiplication of which is both non-commutative and (as Hamilton later discovered) non-associative. By 1844, Hamilton had discovered what he called the *biquaternions*, the 4-dimensional algebra with relations on $i, j$, and $k$ as defined for the quaternions, but with coefficients taken from the complex numbers rather than from the reals.[19] These biquaternions, or 'complex quaternions' (in more modern terminology), behave even worse than the quaternions and the octonions relative to the 'expected' properties of multiplication. It is easy to see, for example, that the product of the two non-zero complex quaternions

$$(-\sqrt{-1}, 1, 0, 0) = -\sqrt{-1} + i \text{ and } (\sqrt{-1}, 1, 0, 0) = \sqrt{-1} + i$$

is 0; in other words, the complex quaternions contain zero divisors (non-zero elements whose product is 0). These discoveries piqued a fair amount of interest through the 1840s, with De Morgan and Charles Graves creating triplets with specified operations (spurred on by the work of Hamilton on quaternions and Warren on the representation of the complex numbers) and Thomas Kirkman and James Cockle isolating higher-dimensional examples—they called them *pluquaternions* and *tessarines*, respectively—of hypercomplex number systems over the reals.[20]

Later in the century, William Kingdon Clifford and James Joseph Sylvester considered hypercomplex number systems from a more general point of view. By the 1870s, Clifford had come to focus his mathematical attention not only on Hamilton's system of quaternions, but also on the type of algebras formulated by Hermann Grassmann in his *Ausdehnungslehre* of 1844 and 1862. In a paper published in the inaugural volume of Sylvester's *American Journal of Mathematics* in 1878, Clifford showed how to interpret the quaternions and other examples in Grassmann's setting, as well as how to use quaternions to define more general algebras. For example, influenced by the work of Grassmann and by the 1870 treatise on *Linear Associative Algebra* by the American mathematician Benjamin Peirce, Clifford formed a 16-dimensional algebra by taking what would today be called the tensor product of two quaternion algebras, and he generalized this construction to consider new algebras which were tensor products of $m$ copies of the quaternions.[21]

Also inspired to some extent by Peirce's work, but by the 1881 version of it published posthumously in the *American Journal*, Sylvester focused between 1882 and 1884 on the quaternions and their generalizations. His initial research impetus came from his aroused interest in matrices. He asked whether it is possible to define a system of matrices that satisfy a set of relations analogous to those satisfied by the quaternions. He quickly realized that, for $\theta = \sqrt{-1}$, the matrices

$$I = \begin{pmatrix} 1 & 0 \\ 0 & 1 \end{pmatrix}, v = \begin{pmatrix} 0 & 1 \\ -1 & 0 \end{pmatrix}, u = \begin{pmatrix} 0 & \theta \\ \theta & 0 \end{pmatrix} \text{ and } w = uv = \begin{pmatrix} -\theta & 0 \\ 0 & \theta \end{pmatrix}$$

satisfy the necessary relations and form a basis of $2 \times 2$ matrices for the quaternions, when viewed as an algebra over the reals. This led him, quite naturally, to ask whether the construction can be generalized to larger square matrices. Letting $\rho = \sqrt[3]{-1}$, he answered that question in the affirmative by defining

$$U = \begin{pmatrix} 0 & 0 & 1 \\ \rho & 0 & 0 \\ 0 & \rho^2 & 0 \end{pmatrix} \text{ and } V = \begin{pmatrix} 0 & 0 & 1 \\ \rho^2 & 0 & 0 \\ 0 & \rho & 0 \end{pmatrix},$$

and showing that the matrices

$$I, U, V, U^2, UV, V^2, U^2V, UV^2, U^2V^2$$

form a basis of a 9-dimensional algebra over the reals with the desired properties. He dubbed this new algebra the *nonions*.[22]

Sylvester tried to codify this and other results he had obtained on matrices into a general theory during the course of his lectures at the Johns Hopkins University in 1882, but more specifically in what he intended to be the first in a series of 'Lectures on the principles of universal algebra', published in the *American Journal* in 1884. There, Sylvester treated the collection of square matrices (generally over the reals) as an algebra with the operations of multiplication, addition, and scalar multiplication, and he explored some of the now-standard concepts associated with the theory of algebras, such as spanning sets and the characteristic equation and its roots. Although he focused specifically on the $2 \times 2$ and $3 \times 3$ cases, he asserted, in an inductive style typical of much of then-contemporaneous British algebraic research, general results for $n \times n$ matrices and envisioned a purely algebraic development of the subject.[23]

As Sylvester acknowledged in his 'Lectures', the first step toward the realization of an algebra of matrices had already been taken by his friend Cayley in the 1850s. Prior to Cayley's work, matrices had arisen in two distinct contexts: solving simultaneously systems of linear equations in more than two unknowns, and linearly transforming the variables of homogeneous polynomials in any number of variables. It was in the first setting that Cauchy had isolated the notion of a *determinant* and that Sylvester had subsequently coined the term *matrix*. As Sylvester defined it in 1850, a matrix is 'a rectangular array of terms, out of which different systems of determinants may be engendered, as from the womb of a common parent'.[24] By the time the word 'matrix' came to be associated with the concept of a determinant, a theory of determinants as distinct algebraic entities had developed independently of the context of solving systems of linear equations. Many British mathematicians, among them, Sylvester, Cayley, William Spottiswoode, and Charles Dodgson (Lewis Carroll) contributed to this new line of research.[25]

Cayley and Sylvester were also well aware of the second context in which matrices arose, given their evolving work on a theory of invariants (see below) in the 1850s. In his *Disquisitiones Arithmeticae* (1801), Gauss had considered how ternary quadratic forms with integer coefficients, expressions of the form

$$a_1 x^2 + a_2 y^2 + a_3 z^2 + 2a_4 xy + 2a_5 xz + 2a_6 yz,$$

are affected by a linear transformation of their variables. In order to derive the new binary ternary form, he applied the linear transformation

$$x = \alpha x' + \beta y' + \gamma z', \quad y = \alpha' x' + \beta' y' + \gamma' z', \quad z = \alpha'' x' + \beta'' y' + \gamma'' z',$$

which he denoted by the square array

$$\begin{matrix} \alpha, & \beta, & \gamma \\ \alpha', & \beta', & \gamma' \\ \alpha'', & \beta'', & \gamma''. \end{matrix}$$

Then, in calculating the composition of two such transformations, he effectively gave an explicit example of matrix multiplication.

In his 1858 paper, 'A memoir on the theory of matrices', Cayley extended this idea of what could then, in the light of Sylvester's new terminology, be termed 'matrix multiplication' to the notion of 'matrix addition' and to the analysis of the explicit properties of both operations. As Cayley put it:[26]

matrices (attending only to those of the same order) comport themselves as single quantities; they may be added, multiplied or compounded together &c: the law of addition of matrices is precisely similar to that for the addition of ordinary algebraical quantities; as regards their multiplication (or composition), there is the peculiarity that matrices are not in general convertible.

In exploring how these operations behave, Cayley recognized matrices as independent algebraic entities that satisfy all of the properties that would later constitute an algebra.

He actually recognized even more. '[I]t is nevertheless possible', he went on to say in his 1858 paper, 'to form the powers (positive or negative, integral or fractional) of a matrix, and thence to arrive at the notion of a rational and integral function, or generally of any algebraical function, of a matrix'.[27] The stunning example that Cayley gave was the so-called Cayley–Hamilton theorem: *An n × n matrix M satisfies the equation*

$$\det(x\mathbf{I} - \mathbf{M}) = 0,$$

*where I is the n × n identity matrix.* Typically, Cayley stated his theorem in general but verified it only in the 2 × 2 and 3 × 3 cases. This British work on specific kinds of algebras, as well as on the algebra of matrices, ultimately coalesced with parallel developments on the Continent into a free-standing theory of algebras in the work, among others, of the Scots mathematician

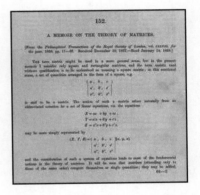

Cayley's matrix paper of 1858.

Joseph Henry Maclagan Wedderburn, beginning in the last years of Victoria's reign.[28]

## Another 'new' mathematical construct: the abstract group

Although it did not attract the attention in Victorian Britain that it did on the Continent, the abstract theory of groups had caught at least Cayley's eye by the 1850s. In a short two-part paper published in 1854, he began to explore groups from what can be characterized as a structural approach—by means of group multiplication tables—but his inspiration seems to have come from the calculus of operations.

Cayley set the stage by defining in a general setting 'a symbol of operation', $\theta$,

which may, if we please, have for its operand, not a single quantity $x$, but a system $(x, y, \ldots)$, so that $\theta(x, y, \ldots) = (x', y', \ldots)$, where $x', y', \ldots$ are any functions whatever of $x, y, \ldots$. In particular, $x', y'$, &c may represent a permutation of $x, y$, &c., $\theta$ is in this case what is termed a substitution.[29]

He proceeded to interpret the multiplication (or composition) of symbols, to note that—depending on what the operands were—this multiplication need not be commutative (as in the case of his own octonions), and to provide an abstract definition of a group. In his words:

[a] set of symbols, $1, \alpha, \beta, \ldots$ all of them different, and such that the product of any two of them (no matter in what order), or the product of any one of them into itself, belongs to the set, is said to be a *group*.

As the title of his paper announced, however, Cayley was interested particularly in 'a theory of groups, as depending on the symbolic equation $\theta^n = 1$', and he duly noted:

that if the entire group is multiplied by any one of the symbols, either as further or nearer factor, the effect is simply to reproduce the group; or what is the same thing, that if the symbols are multiplied together so as to form a table . . . that as well each line as each column of the square will contain all the symbols . . . [and] that the product of any number of the symbols . . . is a symbol of the group.

Without using the modern terminology, he then proceeded to analyse groups of composite order in terms of their possible multiplication tables via generators and relations and, given his characteristic case-by-case analysis, to demonstrate the existence of two fundamentally different groups of order 4—the cyclic group and the Klein four group—as well as of the two different groups of order 6 defined by the equation $\theta^6 = 1$—the cyclic group and the group of permutations on three letters. In a third part of the same paper published in 1859, Cayley analysed the 'next' case of $\theta^8 = 1$ and considered the ramifications of some of his findings for higher orders.[30]

By the late 1850s, groups had also come to interest one of Cayley's mathematical acquaintances, the Anglican clergyman Thomas Kirkman. Kirkman had noted with interest the announcement in 1858 of a prize question on the theory of groups of substitutions in the *Comptes Rendus* of the Paris Academy of Sciences. The Grand Prix would be awarded in 1860, and Kirkman spent the years from 1858 to 1860 working on his submission.[31]

Although the Academy ultimately found none of the three submissions—by Camille Jordan, Émile Mathieu, and Kirkman—suitably original to justify the awarding of the prize, Kirkman made a brief announcement of his ideas at the 1860 meeting of the British Association for the Advancement of Science held in Oxford, and then submitted them for publication to the *Memoirs of the Manchester Literary and Philosophical Society* in 1861.[32] There, drawing from the work of Cauchy, he gave a widely ranging account of permutation groups. In addition to laying out the basic definitions, Kirkman treated (to use modern terminology) such fundamental concepts as left and right cosets, the normalizer of a group, extensions of a group, and the direct product of groups. He also gave a new construction of the projective groups of degree $q + 1$, where $q$ is a power of a prime. In particular, he constructed the projective group of degree 9 and order 1512 (the first of what are now called the *Ree groups*) some thirty years before Frank Nelson Cole claimed to have discovered this new finite group.

Kirkman continued his researches on group theory into 1863, giving group-theoretic interpretations of some of his combinatorial work and presenting in his paper 'Hints on the theory of groups' what has been termed 'probably the first systematic treatment in English of the elements of group theory'.[33] Around the same time, Sylvester worked briefly on the theory of substitutions, publishing a series of short notes in the *Comptes Rendus* and the *Philosophical Magazine* in 1861, based on work that he had started in the summer of 1860.[34]

Following these British results of the 1850s and early 1860s, little of import for the history of the theory of groups issued from the British Isles. In 1878, however, and in response to a call from Sylvester to help him launch his new *American Journal of Mathematics*, Cayley submitted four short notes. He opened the first of them by recapitulating some of his ideas on group theory from the 1850s, particularly the notion of the group multiplication (or 'Cayley') table. He next averred that 'the general problem' in group theory 'is to find all the groups of a given order $n$'.[35] This articulated what became a research agenda—taken up particularly by a number of American mathematicians at the end of the 19th century—as well as the 20th-century quest for the classification of all finite simple groups. He closed with a statement of what is now termed 'Cayley's theorem':

the general problem of finding all groups of a given order $n$, is really identical with the apparently less general problem of finding all the groups of the same order $n$, which can be formed with the substitutions upon $n$ letters.

As if these were not enough group-theoretic gems, another of his short notes contained one more, his diagrammatic graphical representation (or 'Cayley

William Burnside (1852–1927).

graph') for a group that has now become an essential tool in combinatorial and geometrical group theory.[36]

While these and other nuggets of group-theoretic work issued from the mathematicians of Victorian Britain, it certainly cannot be said that it was a strong indigenous research tradition that resulted in 1897 in William Burnside's influential, book-length codification of then-known group-theoretical results, *Theory of Groups of Finite Order*. Although he acknowledged Cayley's work, Burnside deemed that the subject had been developed mostly in France in the first half of the 19th century and then in France and Germany thereafter. As he saw it, '[t]he subject is one which has hitherto attracted but little attention in this country', but he expressed his hope that 'by means of this book, I shall succeed in arousing interest among English mathematicians in a branch of pure mathematics which becomes the more fascinating the more it is studied'.[37] If Victorian Britain was not a major contributor to group theory, it did witness the creation of one of the 19th century's major areas of mathematical research, invariant theory.

## The evolution of the theory of invariants

In an 1841 paper, entitled 'Exposition of a general theory of linear transformations', George Boole explored the relationships between a homogeneous polynomial of degree $n$ in two variables (what the British invariant theorists would later call a 'binary $n$-ic') and a linear transformation of that form.[38] In particular, he examined the specific example of a binary quadratic form

$$Q = ax^2 + 2bxy + cy^2$$

and a non-singular linear transformation of the variables $x$ and $y$ given by

$$x = mx' + ny' \text{ and } y = m'x' + n'y',$$

where $m$, $n$, $m'$, and $n'$ are real numbers with $mn' - m'n \neq 0$.[39]

Boole next applied a formal process involving partial differentiation to generate what is now termed the *discriminant* of Q. Specifically, he calculated $\partial Q/\partial x = 2ax + 2by$ and $\partial Q/\partial y = 2bx + 2cy$, equated each of these expressions to 0, and eliminated the variables from the resulting system of linear equations in $x$ and $y$. The result of this elimination, denoted by $\theta(Q)$, is $b^2 - ac$. Boole repeated the process on $R = Ax'^2 + 2Bx'y' + Cy'^2$, the binary quadratic form that results from applying the given linear transformation to Q, to get $\theta(R) = B^2 - AC$. From there, it was easy to show that $\theta(R) = (mn' - m'n)^2 \theta(Q)$, so $\theta(R)$ and $\theta(Q)$ are equal up to a power of the determinant of the linear transformation.[40] In the terminology that Sylvester later developed, Boole had discovered an *invariant* of the binary quadratic form Q—that is, an expression in the coefficients of Q that remains unaltered, up to a power of the determinant, under the action of a non-singular linear transformation on Q.

Arthur Cayley, who had read Boole's paper in 1844, fairly quickly devised new calculational procedures involving generalized determinants for generating invariants of binary forms. In particular, relative to the binary quartic form

$$ax^4 + 4bx^3y + 6cx^2y^2 + 4dxy^3 + ey^4, \tag{1}$$

Boole's partial differentiation method of 1841 yielded a rather complicated invariant $K$. Using a new method in 1844, Boole discovered another invariant of (1),

$$J = ace + 2bcd - ad^2 - eb^2 - c^3,$$

while Cayley's so-called *hyperdeterminant method* of 1845 produced yet another new invariant,

$$I = ae - 4bd + 3c^2.$$

As Boole quickly discovered, using a brute force calculation, these three invariants satisfy the relation $K = I^3 - 27 J^2$.

Cayley saw even more. To use modern terminology, he recognized that the invariants $I$ and $J$ form a minimum generating set of invariants for the binary quartic, and that the elements in a minimum generating set need not be independent—that is, they might satisfy an algebraic dependence relation.[41] As he foresaw, the task of devising a successful theory for detecting and analysing these dependence relations—or *syzygies*, as Sylvester would later dub them—would 'present many great difficulties'.[42]

After these initial results, Cayley moved on to other mathematical topics, particularly in algebraic geometry, but by early February 1850 Sylvester was hard at work on determinant-theoretic ideas that were growing from his discussions with Cayley.[43] In his 1850 paper 'On the intersections, contacts, and other correlations of two conics expressed by indeterminate coordinates', as well as in his correspondence with Cayley well into the autumn of 1850, Sylvester considered two conics $U$, $V$ in three variables $x$, $y$, $z$:

$$U = ax^2 + by^2 + cz^2 + 2d'yz + 2b'xz + 2c'xy = 0$$

and

$$V = \alpha x^2 + \beta y^2 + \gamma z^2 + 2\alpha' yz + 2\beta' xz + 2\gamma' xy = 0.$$

He then asked under what algebraic conditions these conics intersect.[44] His solution involved an analysis of the roots of an equation that he denoted by $\square(U + \lambda V) = 0$, where $\square(U + \lambda V)$ is the determinant

$$\begin{vmatrix} a + \alpha\lambda & c' + \gamma'\lambda & b' + \beta'\lambda \\ c' + \gamma'\lambda & b + \beta\lambda & d' + \alpha'\lambda \\ b' + \beta'\lambda & d' + \alpha'\lambda & c + \gamma\lambda \end{vmatrix};$$

this is precisely the determinant that Sylvester would later call the *discriminant* and identify as an invariant.[45]

By the end of 1851, Sylvester and Cayley had continued to push this whole line of thought toward a full-blown theory of invariants. They had developed a new language in which they discussed their evolving ideas, and they had isolated a number of questions and constructs that came to shape their research. On 5 December 1851, Cayley wrote a letter to Sylvester that amounted to the moment of conception of their theory of invariants. Cayley's message was short and sweet:[46]

Every Invariant satisfies the partial diff[erentia]l equations

$$\left( a\frac{d}{db} + 2b\frac{d}{dc} + 3c\frac{d}{dd} + \cdots + nj\frac{d}{dk} \right) U = 0$$

$$\left( b\frac{d}{db} + 2c\frac{d}{dc} + 3d\frac{d}{dd} + \cdots + nk\frac{d}{dk} \right) U = \tfrac{1}{2} nsU$$

(s the degree of the Invariant) & of course the two equations formed by taking the coeff[icien]ts in a reverse order. This will constitute the foundation of a new theory of Invariants.

By 1852, Sylvester had also made his first major contribution to that theory in his formidable paper, 'On the principles of the calculus of forms'.[47] In this work, among many other things, Sylvester studied the properties of a curve in $x$ and $y$. For reasons of mathematical efficiency, he wanted to express the curve in the simplest possible form without loss of generality—that is, he wanted to determine a rotation of the axis system, or linear transformation, to render the curve in its *canonical form*. The problem of determining the canonical form of a quantic was thus key to his approach to questions about forms and their properties, and he· found it explicitly by examining an associated form derived from it—that is, an invariant. His invariant-theoretic approach allowed him to turn a *geometrical* question into an *algebraic* one, but to what extent could this kind of algebrization of geometry be carried out—that is, to what extent was a *theory* of invariants lurking in the shadows? Sylvester's paper on the calculus of forms shed some light on this vision thus far only obscurely seen.

In its second instalment, Sylvester duly recorded his debt to Cayley for a calculational technique Cayley had first articulated in his letter of December 1851, but swiftly staked his claim that '[t]he method by which I obtain these equations and prove their sufficiency is my own ...'.[48] Given a binary $n$-ic form

$$\phi = a_0 x^n + na_1 x^{n-1}y + \tfrac{1}{2}n(n-1)a_2 x^{n-2}y^2 + \cdots + na_{n-1}xy^{n-1} + a_ny^n,$$

define the following two formal differential operators:

$$\mathfrak{X} = (a_0\partial_{a_1} + 2a_1\partial_{a_2} + 3a_2\partial_{a_3} + \cdots + na_{n-1}\partial_{a_n}) \qquad (2)$$

as before, but now

$$\mathfrak{Y} = (a_n\partial_{a_{n-1}} + 2a_{n-1}\partial_{a_{n-2}} + 3a_{n-2}\partial_{a_{n-3}} + \cdots + na_1\partial_{a_0}). \qquad (3)$$

If $I$ is a homogeneous expression in the coefficients of $\phi$, and if $\mathfrak{X}(I) = 0$ and $\mathfrak{Y}(I) = 0$, then $I$ is an invariant of $\phi$. Moreover, the partial differential operators $\mathfrak{X} - y\partial_x$ and $\mathfrak{Y} - x\partial_y$ suffice analogously to detect covariants—that is, expressions in the coefficients and the variables of a given binary form that remain invariant (up to a power of the determinant) under a linear transformation of the variables.

While in this and other papers in the early years of the 1850s, Sylvester may have taken the lead in laying out many of the initial ideas of invariant theory, Cayley quickly extended and refined those results. Sylvester's 1852 papers 'On the principles of the calculus of forms' and his massive 1853 assault, 'On a theory of syzygetic relations of two rational integral functions...', examined forms and their invariants as well as the thorny problem of detecting syzygies—that is, algebraic dependence relations between elements in a minimum generating set of covariants.[49] Cayley's more focused and systematic 1854 'Introductory memoir upon quantics' and his 1856 'Second memoir upon quantics' adapted the differential operators $\mathfrak{X}$ and $\mathfrak{Y}$ to the context of covariants and, in so doing, went far toward articulating that theory more fully.[50] These four papers—two by Sylvester and two by Cayley—represent the birth of the British approach to invariant theory, an area and a theoretical point of view on which both friends would work to develop and perfect for the rest of their active mathematical careers.

That theory received its first systematic treatment at the hands of their friend and co-worker, the Irish mathematician George Salmon, in his 1859 *Lessons Introductory to the Modern Higher Algebra*.[51] There, Salmon laid out the concepts and defined the research agenda that would remain central to the theory as it developed into the 1880s. For a binary quantic (or binary form)

$$a_0x^m + a_1\binom{m}{1}x^{m-1}y + \cdots + a_m\binom{m}{m}y^m$$
$$T\downarrow \qquad\qquad\qquad\qquad (4)$$
$$A_0X^m + A_1\binom{m}{1}X^{m-1}Y + \cdots + A_m\binom{m}{m}Y^m,$$

where $T: x \to aX + bY$, $y \to a'X + b'Y$ is a non-singular linear transformation, Cayley, Sylvester, and ultimately their fellow invariant-theorists Percy MacMahon, Edwin Bailey Elliott, and others, wanted to find all homogeneous polynomials in the coefficients and variables of (4) that remain 'invariant' under $T$. In other words, if $K(a_0, a_1, \ldots, a_m; x, y)$ is a homogeneous polynomial in the coefficients and the variables of (4), then $K$ is a *covariant* if

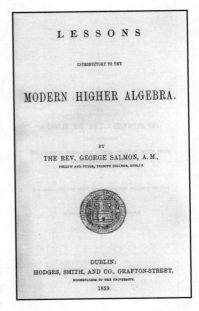

LESSONS

INTRODUCTORY TO THE

MODERN HIGHER ALGEBRA.

BY

THE REV. GEORGE SALMON, A. M.,

FELLOW AND TUTOR, TRINITY COLLEGE, DUBLIN.

DUBLIN:

HODGES, SMITH, AND CO., GRAFTON-STREET,
BOOKSELLERS TO THE UNIVERSITY.

1859.

Salmon's *Lessons Introductory to the Modern Higher Algebra*.

$$K(A_0, A_1, \ldots, A_m; X, Y) = \Delta^l K(a_0, a_1, \ldots, a_m; x, y),$$

where $\Delta = \det(T)$ and $l \in \mathbf{Z}^+$. It is not difficult to show that if $K$ is a covariant, then $K$ has a constant degree of homogeneity in $x$, $y$, denoted by $\mu$ and called the *order* of $K$, and a constant degree of homogeneity in the coefficients $a_0, a_1, \ldots, a_m$, called the *degree* and denoted by $\theta$. Moreover, given a monomial of $K$, its *weight* is defined to be the sum of the products of the subscripts and their corresponding superscripts; for example, the monomial $a_3^2 a_4^3 x^6 y^2$ has weight $(3 \times 2) + (4 \times 3) + (1 \times 6) + (0 \times 2) = 24$, where $x$ has arbitrarily been assigned the subscript 1 and $y$ the subscript 0. With these definitions in place, it can be shown that every covariant $K$ of degree $\theta$ and order $\mu$ has constant weight $\frac{1}{2}(m\theta + \mu)$, where $m$ is the 'degree' of (4).

Now consider again $\mathfrak{X}$ and $\mathfrak{Y}$ as in (2) and (3) above, and take a perfectly general (homogeneous) polynomial $A_0$ in the coefficients of (4), of degree $\theta$ and weight $\frac{1}{2}(m\theta - \mu)$, such that $\mathfrak{X}A_0 = 0$. Setting $A_j = (\mathfrak{Y}A_{j-1})/j$ for $1 \le j \le \mu$, define

$$K = A_0 x^\mu + A_1 x^{\mu-1} y + \cdots + A_{\mu-1} xy^{\mu-1} + A_\mu y^\mu.$$

Such a construct $K$ is a homogeneous polynomial of weight $\frac{1}{2}(m\theta + \mu)$ and order $\mu$. In fact, as Cayley had argued in his 1854 'First memoir upon quantics', $K$ so defined is always a covariant of (4) and *every* covariant of (4) of order $\mu$ and degree $\theta$ can be expressed in this way. From this, Cayley also deduced that, for given $m$, $\theta$, and $\mu$, determining the number of linearly independent covariants reduces to determining the number of partitions of $\frac{1}{2}(m\theta - \mu)$ and of $\frac{1}{2}(m\theta - \mu - 1)$, with parts taken from the integers $0, 1, 2, \ldots, n$ and repetitions allowed. Thus, a key question in invariant theory reduced to a problem in partition theory, and so Cayley and Sylvester embarked on combinatorial researches to tackle these and related issues (see Chapter 17).[52]

In his 1856 'Second memoir', Cayley used much of this evolving theory in explicitly calculating minimum generating sets of covariants for quantics of successive degrees. In particular, he erroneously argued—and Salmon initially perpetuated the error—that a minimum generating set for the binary quintic form has infinitely many elements. Still, he and Sylvester continued in their search for distinct (that is, both linearly and algebraically independent) covariants for given binary forms. They literally took on what was, as far as they knew, a problem of infinite proportions, but in so doing, they created what their contemporaries like Salmon recognized as a new area of mathematical research.[53]

Much to their consternation, that problem took on more finite proportions in 1868, when Paul Gordan proved that Cayley had been in error in his work on the binary quintic. Working in the parallel invariant-theoretic tradition that had grown up in Europe, particularly from the 1840s on, in the work of Gotthold Eisenstein, Otto Hesse, Siegfried Aronhold, Alfred Clebsch, and Gordan himself, Gordan proved that, given a binary form, a minimum generating set of invariants for that form is always finite.[54] This result at once established Gordan's reputation and called into question the techniques of the British school. It thus became incumbent upon the British to vindicate their work by showing that their techniques, too, were strong enough to provide a proof of Gordan's theorem. Moreover, since Gordan's methods did not lend themselves to efficient direct computation of minimum generating sets—the prime objective of the British school—the British were still very much in the game.

Cayley began the rehabilitation process in 1871 with his paper, 'A ninth memoir on quantics'. There, he used Gordan's theorem in the context of the British approach to calculate explicitly the covariants in a minimum generating set of a binary quintic. In that paper, too, he expressed his 'hope that a more simple proof of Professor Gordan's theorem will be obtained—a theorem the importance of which, in reference to the whole theory of forms, it is impossible to estimate too highly'.[55] Sylvester, largely out of the mathematical fray in the late 1860s and early 1870s, took up this challenge in earnest in November 1876 from the professorial post he had just assumed at the Johns Hopkins University.

Introducing his graduate students to the area of invariant theory in his lecture courses and in the mathematical seminar that he ran, he succeeded in bringing his students up to the research level. In particular, Fabian Franklin, who earned his Ph.D. under Sylvester's direction in 1880, worked closely with Sylvester in 1878 and 1879 as a sort of 'human calculator', using the British combinatorial techniques to calculate minimum generating sets for binary quantics of degrees up to and including the tenth.[56]

Paul Gordan (1837–1912).

Work like that of Sylvester and his students in Baltimore, as well as new research in England by James Hammond and Percy MacMahon, may have vindicated the British calculational techniques, but the British were still unsuccessful in pushing their methods through to a proof of Gordan's finiteness theorem. Throughout the 1870s and 1880s, Sylvester continued to work toward a British-style proof of that theorem, being repeatedly convinced that he had found one, only to realize some flaw.

The rules of the game changed fundamentally in 1889 when the young David Hilbert entered the scene. In a paper entitled 'Über die Endlichkeit des Invariantensystems für binäre Grundformen' and published in the *Mathematische Annalen*, Hilbert gave a simpler, more modern proof of Gordan's result.[57] Cayley immediately responded with his own paper in the *Annalen*, in which he thought he had not only greatly simplified Hilbert's argument, but had done it along British lines.[58] It could have been that long-sought vindication of British techniques, except for one thing: it was wrong. Neither Cayley, nor Sylvester, nor any other invariant-theorist working within the British framework, ever succeeded in producing a British-style proof of Gordan's finiteness theorem.

David Hilbert (1862–1943).

By 1893, Hilbert had provided an existence theorem for finite bases of covariants of quantics in any finite number of variables, completely outstripping with his very general non-calculational techniques the work of the British and of Gordan and his contemporaries on the Continent.[59] In the advanced textbook codification of invariant theory that he published in 1895, the Oxford mathematician E. B. Elliott highlighted British techniques, but also outlined enough of the Continental pre-Hilbertian techniques to prove what had been—until Hilbert—the 19th century's key theorem in invariant theory: Gordan's finiteness theorem.

Edwin Bailey Elliott (1851–1937).

Unfortunately, unbeknownst to its author, Elliott's book was obsolete before it was even published.[60]

## Epilogue

If, as the members of the Cambridge Analytical Society claimed in the early years of the 19th century, British mathematicians had been severely hampered by their adherence to the Newtonian calculus, the Victorian era witnessed a revitalization and reorientation of mathematical research that stemmed, at least in part, from the introduction of analytic (that is, algebraic) techniques. By the 1830s, mathematicians in Cambridge, Ireland, London, and countryside hamlets had begun to explore the implications of approaching mathematical questions algebraically, as well as the extent to which they could define new algebraic entities and create new areas of algebraic research. Throughout the Victorian era, they focused particularly on questions about what we now term *algebras* and *invariants*, but they did not totally neglect group theory which so engaged their contemporaries in France, Germany, and elsewhere. They formed a band of 'brothers in algebra', united in the face of criticism by those like William Thomson,[61] who wished in 1864:

that the CAYLEYS would devote what skill they have to such things [as applied mathematics] instead of to pieces of algebra which possibly interest four people in the world, certainly not more, and possibly also only the one person who works.

These algebraists announced their research findings at the Royal Societies in Dublin, Edinburgh, and London, at meetings of the British Association for the Advancement of Science, and (after 1865) before their fellow members of the London Mathematical Society. They also published them in the pages of journals at home, but perhaps more importantly abroad. In so doing, they contributed in lasting and historically significant ways to the stock of algebraic knowledge.

# A Syllogism worked out.

---

That story of yours, about your once meeting the sea=serpent, always sets me off yawning;

I never yawn, unless when I'm listening to something totally devoid of interest.

---

### The Premisses, separately.

### The Premisses, combined.

### The Conclusion.

That story of yours, about your once meeting the sea=serpent, is totally devoid of interest.

Lewis Carroll's method of diagrams.

## CHAPTER 16

# Victorian logic

*From Whately to Russell*

## I. GRATTAN-GUINNESS

The subject of logic provided two modest but distinctive streams of work to the landscape of Victorian mathematics and philosophy. The first began to flow in the late Georgian period: much of it was discursive in character, but from the 1840s a notable movement made use of various algebraic principles; our interest will lie largely there, using the name 'algebraic logic' for the movement. The second stream started very late in the Victorian epoch, so that we note only its first stages. Again, mathematics provided much of the inspiration, but this time mathematical analysis and set theory formed the main stimuli for 'mathematical logic'. This chapter reviews both traditions, stressing the major differences between them and considering the striking fact that these developments happened at all.

Most intellectual topics develop more or less continuously over time, although progress may on occasion be very slow. It is rare that a topic stops entirely or regresses, at least in reputation. Yet this was the fate of logic in Europe, especially from the 17th century. By then the subject was held to cover specifying concepts, judging the truth-value of premises involving them, and deducing consequences from the premises. The last category was regarded as especially characteristic of logic, and so became a main target of criticism.

The best known logical theory was that of syllogistic forms, due to Aristotle, and a main cause of criticism was a general decline in the reputation of Aristotelianism. Part of that attack was the advocacy of induction as a philosophy of science (by Francis Bacon, to mention one influential writer). Alternative sources included the advocacy of signs in general, of which words provided an important example. In his *Essay Concerning Human Understanding* of 1690, John Locke introduced the word 'semiotike' for such sources of knowledge, but the word 'logic' was applied to the case of words—rather misleadingly, as that word also continued in its traditional interpretation concerning deduction.[1]

## Whately: a sudden and surprising revival of logic

Richard Whately (1787–1863).

The situation changed suddenly and unexpectedly in the mid-1820s. The cause was the book *Elements of Logic* (1826) by the theologian Richard Whately. His concerns included logic as an art and as a science, valid methods of drawing inferences, fallacies, and the relationships between logic and language(s). No particular response had followed its first appearance in 1823, as an article in the *Encyclopaedia Metropolitana*; maybe the high price of the encyclopaedia gave it a small readership anyway. However, even though no significant new techniques or interpretations of logic were advocated, either there or in the book version, the reactions to the book were rapid and substantial.

Within less than a decade Whately had issued his fifth revised edition, and several other authors had published articles and books. The most important of these was by George Bentham (nephew of Jeremy) in his *Outline of a New System of Logic* (1827). He extended Aristotle's system of syllogisms by taking traditional forms such as 'All *A*s are *B*s' and adjoining further ones like 'All *A*s are some *B*s' and 'Some *A*s are no *B*s'. He described his book as a reply to Whately, who then ignored it in later editions of his work.

In the course of this revival and later, both deductive and inductive logics were developed, the latter partly in connection with the prevailing inductivist philosophy of science, which in turn was a stimulus for some of the studies of probability theory. The account below is restricted to deductive logic, and to cases where mathematical methods, or imitations of them, were prominent. At first, reasoning and proofs in mathematics were not major themes, and no mathematician played a significant role, but from the late 1830s this situation was to change.

## De Morgan: algebraization and the logic of relations

As we saw in Chapter 3, Augustus De Morgan was based for most of his career at University College, London, as professor of mathematics. His research centred upon algebras, both the traditional sort and new ones, or at least new applications. His first substantial foray into logic was an effort to make logical sense of Euclid's *Elements*, a pioneering exercise in the West, in an article of 1833 and especially in his 1839 pamphlet *First Notions of Logic (Preparatory to the Study of*

*Geometry*). This text reappeared as the initial chapter of his book *Formal Logic* (1847), which also contained his first uses of algebra in logic. He followed these up with a suite of five papers published with the Cambridge Philosophical Society between 1846 and 1862, and a few other notes, papers, and pamphlets.[2]

One of De Morgan's first standpoints, stated in the opening lines of his pamphlet, was to *detach* from logic discussion of the truth-values of candidate premises; instead, he assumed them to be true and regarded logic as the means by which true conclusions could be inferred from them. Maybe he came to this stance from his understanding of the role of axioms of Euclid; at all events, most of his successors also followed it, making a break with the past, including Whately.

Another valuable feature, stated in the opening paragraphs of De Morgan's first paper, was an explicit mention of a 'universe of possible conceptions' within which logical discourse may take place. Within that domain he added to the syllogistic repertoire by admitting 'numerically definite syllogisms', which involved propositions such as '45 *X*s are each of them one of 70 *Y*s'. This was essentially the same kind of extension as that made by Bentham, and it led him into a priority dispute, not with Bentham but with the Scottish logician William Hamilton, who had come up with the 'quantification of the predicate' by working with propositions like 'All *X*s are some *Y*s'. The dispute was ironic, in that neither author seemed to know of Bentham's book (although the situation for Hamilton is ambiguous); it had indeed sold very poorly, and in the meantime its author had embarked on a distinguished career as a botanist.

Augustus De Morgan (1806–71).

Novel with De Morgan was the symbolization of the syllogistic forms by systematically using capital letters for terms (predicates) or their associated classes ('aggregates') and lower-case ones for their contraries or complements, and deploying round brackets and full stops to symbolize the various kinds of deduction. Typically for an algebraist, he drew upon properties such as symmetry and antisymmetry; the semiotic *tour de force* was a 'zodiac' of twelve combinations of brackets and stops from which trios could be formed in various ways to represent valid syllogisms.[3]

Again, the algebraic laws named after him arose initially in connection with classes: 'The contrary of an aggregate is the compound of the contraries of the aggregates; the contrary of a compound is the aggregate of the contraries of the components' (1858).[4] He associated inference in logic with elimination in common algebra, and also considered geometrical schemes such as Euler diagrams (to be described later) as replacements of it. On philosophical issues, he tried to address the relationship between logic and language by distinguishing the (logical) form of a proposition from its (extra-logical) matter, but with only partial success.

De Morgan also pointed out some limitations of Aristotle's logic: for example, that the valid inference 'man is animal, therefore the head of a man is the head of an animal' lay outside it. More importantly, in 1860 he devoted the fourth of his suite of articles to the logic of two-place relations, a quite fundamental extension of logic that had escaped the attention of logicians for the previous two millennia.[5] One of the features that he emphasized was the inverse (any, one, or many) of a relation, and properties of compounded relations such as (son of uncle of) $\neq$ (uncle of son of). The mathematical ally of this theory is the study of functional equations, such as $f(x+y) = f(x)f(y)$, where the function $f$ is the unknown, and inverses and compounding of functions are major features: De Morgan had published the first systematic account of this newish algebra in 1836, as an article in the *Encyclopaedia Metropolitana*.[6]

In the following year De Morgan published another long article in the same encyclopaedia, this time on probability theory.[7] Especially in his early years he saw certain connections to logic, such as the degree of belief in the truth of premises used in syllogisms, and the validity of deductions; in certain ways he thereby revived the topic of probability logic.[8] The full title of his book of 1847 was *Formal Logic: or, the Calculus of Inference, Necessary and Probable*, and he devoted three of its chapters to probability theory and induction.

## Boole, 1847: a new algebra for logic

George Boole (1815–64).

Until 1847 De Morgan had the algebraization of logic more or less to himself;[9] but then another voice was heard. The motivation to publish came from the priority dispute between De Morgan and Hamilton, but the logic offered was of a very different character, as were the links to probability theory.

George Boole was a schoolmaster in and around Lincoln, making a very respectable career as a mathematician despite being self-taught in the subject.[10] Researching in differential equations, he specialized in solutions using the newish algebra of differential operators, where differentiation was symbolized by $D$ and integration as its inverse $D^{-1}$; it enjoyed a following in Britain, and Boole was the master.[11]

This algebra also influenced Boole's own 'mathematical analysis of logic', the title of his 1847 book, which apparently came out on the same day as De Morgan's *Formal Logic*. Boole based his logic upon the mental act of forming classes $x, y, \ldots$ of objects from some given universal class 1, or alternatively upon the classes themselves. In both interpretations these objects satisfied an algebra with three laws: distributivity and commutativity[12] (as with differential operators), and also an 'index law', a name that for operators denoted the law of powers ($D^m D^n = D^{m+n}$), but in logic referred to the novel law $xx = x$, with its attendant law $x(1-x) = 0$, where the 'multiplication' denotes the repetition of the mental act or the intersection of classes, and '$-$' signifies mental exception (the universe 1 except for $x$) or class complementation. Curiously, he did not state that $x + (1-x) = 1$—that is, the union of a class and its complement is the universal class.

Upon this basis, Boole developed an algebra of logic in which he cast one or more premises into algebraic equations, chose one of the classes as subject ($z$, say), and determined as a logical consequence the relationship between $z$ and the other classes involved in the formulation of the premise(s). The deduction was obtained by means of an expansion theorem for any logical function $\phi(x)$ of $x$ (or $\phi(xy)$ of $x$ and $y$, and similarly for more variables), which he proved by developing $\phi$ in a MacLaurin series, as modified by the index law. For two variables it read:

$$\phi(xy) = \phi(00)(1-x)(1-y) + \phi(01)y(1-x) + \phi(10)x(1-y) + \phi(11)xy.$$

This type of theorem took over the role traditionally assigned to rules of inference: in a novel philosophical move, Boole did not require the lines of the deduction to be interpretable as logic. Together with his double interpretation of $x$ as mental acts and as classes, there are traces here of model theory, which was barely evident elsewhere in mathematics at that time.

Several features of Boole's algebra were new for the time; for example, the cancellation law was lost (if $xy = xz$, then it does not necessarily follow that $y = z$). Also, in forming the consequence in terms of the subject variable, the coefficients were capable of division and could have some exotic values. If 0/0 was the coefficient for $xy$ (say) then it would be replaced by 'an arbitrary elective symbol' $v$, and if the coefficient of $x(1-y)$ (say) was 1/0 (or any value other than 1 or 0) then $x(1-y)$ was to be set equal to 0 as a necessary condition for the solution to be obtainable at all. Thus, for example (one of Boole's), given the single premise relating $x$, $y$, and $z$,

$$\phi(xyz) = x(1-z) - y + z = 0,$$

the expansion theorem for $z$ as the subject yielded

$$z = (0/0)xy + (1/0)x(1-y) + 1(1-x)y + 0(1-x)(1-y),$$

so that the consequence read

$$z = vxy + (1-x)y \text{ as long as } x(1-y) = 0;$$

this means that 'the class $Z$ consists of all the $Y$s that are not $X$s, and an *indefinite* remainder of $Y$s that are $X$s', provided that all $X$s are $Y$s.


# Boole, 1854: some extensions and limitations of his logic

In 1849 Boole took up a post as professor of mathematics at the newly founded Queen's College, Cork, a post that he held until his sudden death fifteen years later. In Cork he wrote a second and larger book on logic, *An Investigation of the Laws of Thought*, which appeared in 1854. The title made explicit his vision of logic as expressing, *normatively*, the manner in which correct reasoning was to be

AN INVESTIGATION

OF

THE LAWS OF THOUGHT,

ON WHICH ARE FOUNDED

THE MATHEMATICAL THEORIES OF LOGIC
AND PROBABILITIES.

BY

GEORGE BOOLE, LL.D.

PROFESSOR OF MATHEMATICS IN QUEEN'S COLLEGE, CORK

LONDON:
WALTON AND MABERLEY,
UPPER GOWER-STREET, AND IVY-LANE, PATERNOSTER-ROW.
CAMBRIDGE: MACMILLAN AND CO.
1854.

Boole's *Laws of Thought*.

effected. The account was broadly the same as before, although with some clarifications (such as the status of the universe of discourse, and the possibility of deploying several of them in a logical analysis), and important new lemmas (such as the elimination theorem $\phi(0)\ \phi(1) = 0$, and its analogues for more variables). He also showed further cases of emulating the methods of manipulating ordinary linear equations; for example, a collection of logical premises could be reduced to the single equation $\sum_r e_r V_r = 0$, on multiplying each of the corresponding logical functions $V_r$ by an arbitrary constant $e_r$ and continuing as usual.

Among interesting differences between the books, Boole laid greater emphasis on the interpretation as classes than as mental acts, and also showed a marked increase in ambition: in the first book several of the examples were syllogisms, but now the Aristotelian doctrine was confined to the last of the fifteen chapters devoted to logic. They were followed by six chapters on probability theory, where he built upon the perception that compound events could be construed as Boolean combinations of simple ones.

While the scope and methods of Boole's logic were impressive and new, there were also omissions and oversights. The lack of a symbol for 'not' sometimes made his system clumsy, and also rendered him silent on the important proof-method of *reductio ad absurdum*. His three laws should have been joined by that for associativity, which he used on occasion. He also failed to note that some of his examples admitted singular solutions (such as $x = 0$, $y = 0$); this was surprising in a mathematician who extolled such solutions of differential equations. The symbol '=' covered only identity, and so did not capture the full linguistic repertoire of 'is'. He did not take up De Morgan's logic of relations, although this appeared in 1860 when Boole was not working on logic.

Perhaps the most striking silence is that Boole did not adapt his algebra to set up the calculus of propositions. Instead he offered a rather strange theory of 'secondary propositions', in which a proposition $X$ is true for a part $x$ of the pertaining universe 1 of time; if it is always true, then '$x = 1$', while if it is never so, then '$x = 0$'.

The prominent place given to Boolean algebra in computing today might lead us to expect that Boole took an interest in the computers of his day, the 'engines' of Charles Babbage, also a British pioneer in differential operators (and especially functional equations). But this was not at all the case. His logic was oriented towards analysing thought, with a belief in the creative power of the mind: he was convinced of its ability to perceive the general in the particular case, a view that guided his practice as a schoolmaster. Thus, the mechanical repetition of actions was not among his interests—and conversely, Babbage did not take up logic.[13]

Partly linked to Boole's conception of the mind, there was an important religious element in his logic. He was a Unitarian, a stance with which he associated his '1', the Universe. For example, in his *Laws of Thought* he devoted a chapter to the logical analysis of two arguments; any context would have sufficed, but he chose two purported proofs of the existence of the one God. Further, in the last lines of the concluding chapter of the book, on 'The constitution of the intellect', he even used the Dissenter phrase 'Father of Lights' to denote the Godhead.

## Jevons as a critic of Boole

After his *Laws of Thought* Boole went back largely to mathematics, producing two influential textbooks on differential and difference equations. He also published some papers on probability theory, and some of his logic manuscripts belong to that period. However, he did not publish on logic; in particular, as mentioned above, he did not take up De Morgan's logic of relations. Both he and De Morgan innovated in logic, but in quite different ways and drawing upon different recently developed algebras; De Morgan worked largely within syllogistic logic with an important extension into relations, while Boole put forward new principles that produced a logic of greater range than any predecessor (setting aside relations). They corresponded quite amicably,[14] but neither one engaged much with the other's theory.

William Stanley Jevons (1835–82).

Boole's first serious British reader was Stanley Jevons, who is better remembered for his contributions to neo-classical economics and the philosophy of science. He also took up logic as another off-beat topic of interest: he had been a student of De Morgan at University College, London, but he studied Boole's logic as presented in *Laws of Thought*. He produced a short book on *Pure Logic* in 1864, in which he took exception to some of Boole's principles, while welcoming the thrust of the enterprise. They corresponded shortly before his book appeared and Boole died.[15]

One feature of Boole's system concerned the union $x + y$ of two classes, which he defined only if they had no elements in common. Perhaps he wished to avoid multi-sets, where elements can belong to a class more than once;[16] but for Jevons the restriction was unacceptable, especially in a theory purporting to capture thought. In their exchange they focused upon Jevons's equation $x + x = x$; for Boole it was quite mistaken since $x + x$ could not be defined, although the *equation* $x + x = 0$ had the solution $x = 0$.

Another point of discord was the extent of the analogy between logic and algebra. Boole had taken it quite far, not only in his laws but also in the use of concepts such as $+$, $-$, $=$, 1, 0, and 0/0. Jevons wanted to reduce this link, distinguishing pure logic from its applications to (for example) mathematics. Working with 'terms' $A$, $B$, ... and their complements $a$, $b$, ... relative to the universe (De Morgan's notational pattern), he laid out principles such as

$$A + A = A, \quad AA = A, \quad A = A(B + b) = AB + Ab, \quad \text{and} \quad Aa = 0.$$

He symbolized propositions such as 'all $A$s are $B$s' as '$AB = A$', instead of Boole's '$A(1 - B) = 0$'. Avoiding all expansion theorems, he drew inferences from a premise $P$ (or several of them) by various means, including taking the disjunction of all the conjunctive propositions containing the components of $P$ (such as $ABc$,

$Abc, \dots$), choosing one of them as subject ($A$, say), and laying down a suite of rules for deleting those propositions that contradicted $P$ in some way. This left the rest to make up the consequence, in the initial form $A = bCd + Bcd + \dots$ (say), that may be subject to simplification.

It was more towards automation than mathematics that Jevons oriented his logic: in 1870 he produced a little machine that imitated his main rule of inference and thereby generated the consequence from a given collection of premises involving propositions to four terms.[17] He also wrote popular books on logic, omitting all algebraic versions from them, but the reception of his algebraic logic was not much greater than either of its predecessors. However, his reading of $x + x$ was generally to prevail, as we now see.

## Later Victorian logicians

After Jevons, the principal further developments of algebraic logic passed to C. S. Peirce in the USA and Ernst Schröder in Germany: they developed De Morgan's logic of relations and Boole's algebra of logic together. Independently of Jevons, Peirce also redefined $x + x$, and with his student O. H. Mitchell he introduced universal and existential quantification of individuals with respect to predicates (respectively, 'for all $x$, …' and 'there is an $x$ such that…'); they saw them as extensions of conjunction and disjunction, and symbolized them appropriately by '$\prod$' and '$\sum$'. The use of 'quantification' presumably echoed its employment by Hamilton, although the associated theory is much more powerful; the cognate noun 'quantifier' is due to Peirce.[18]

Meanwhile, in Britain a few Victorian mathematicians and philosophers took up aspects of these new algebras, with Boole's work making rather more impact than De Morgan's. An unusual example occurred around 1866, soon after Boole's death, when the chemist Benjamin Brodie attempted to develop a similar algebra for chemical operations on substances. '$x + y$' denoted the creation of a unit of hydrogen and one of oxygen separately, while '$xy$' was their compound creation; the axiom was $xy = x + y$. While the effort was unsuccessful (a consequence was that $0 = 1$), Brodie's correspondence with mathematicians shows the reception of Boole's algebra at the time.[19]

An edition of *The Mathematical and Other Writings* by the mathematician R. L. Ellis appeared in 1863, containing some notes on Boole; their researches had overlapped in the recently developed algebras. Ellis had also informed Boole and other Britons of Leibniz's contributions to logic, including the index law; at that time, much of Leibniz's work in logic was still unpublished.[20] Another friend of Boole was the mathematician and educator Robert Harley, who produced a notable obituary of Boole in 1866, and also commented upon the logic. In the early 1870s A. J. Ellis submitted two papers to the Royal Society on aspects of Boole's logic and on De Morgan's logic of relations, and their relationship to algebras of the time such as quaternions.[21]

Two Scots contributed notably in the late 1870s. The schoolmaster Hugh McColl, who taught in Boulogne, published a suite of papers with the London Mathematical Society on 'the calculus of equivalent statements', generally following Boole but with a Jevonsian inclination to limit the links between mathematics and logic. His most substantial innovation came in his first paper (1877), where he proposed that the calculus of propositions, considered as units, satisfies the laws of Boole's algebra; this reading became one of its main applications, replacing Boole's theory of secondary propositions. MacColl (as he spelt his name later) is better remembered for his second phase of work in logic, from the late 1890s onwards, when he pioneered forms of modal logic.[22] The second Scot, Alexander MacFarlane, published *Principles of the Algebra of Logic* (1879) and some related papers. He tried to work modified versions of Boole's algebra and De Morgan's logic of relations, but in a less comprehensive way than Peirce was already developing.[23]

# Venn and Johnson: algebraic logic at Cambridge

All figures considered so far were pursuing these algebraic logics as research topics, with little opportunity for teaching them. The revival of logic had indeed led to an increase in logic teaching in higher education, but most of it was traditional (though not necessarily Aristotelian) and prosodic, with a modest use of symbols. However, Boole's logic began to gain some attention in Cambridge University, where courses in elementary and advanced logics were offered within the moral sciences tripos. The first course was long taught by the Reverend John Venn, who distilled it into the first edition of his book *Symbolic Logic* (1881), the origin of this name; some related papers also appeared around then, especially with the Cambridge Philosophical Society.

Venn followed Boole, even to the extent of his reading of $x + x$, although interestingly he ignored the religious connotation of the logic. He said surprisingly little on the logic of relations, or on the propositional calculus (where he disliked Boole's temporal interpretation while also down-playing McColl's proposal). But he drew upon his extensive knowledge of the history of logic, including non-symbolic and non-British.

One interest was the systems of diagrams that some logicians had introduced. Venn's own main innovation here was a diagrammatic representation of classes, based upon drawing circles (or convex shapes) so that *all* intersections were shown, and marking the ones appropriate to the argument at hand; he had to replace convexity by more complicated shapes when more than four classes were involved.[24]

Unfortunately his name is usually associated with a different method of illustrating a logical argument, by drawing circles in various states of overlap, inclusion, or disjointness relative to a universe, which was due to Leonhard Euler in 1768; in his book Venn described both Euler's diagrams and his own

John Venn (1834–1923).

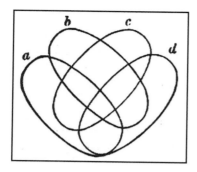

Venn's diagram for four classes.

approach.[25] He published a second edition in 1894, following more or less the same chapter structure but with many additions made, both logical and historical; the main difference was that he followed Jevons's definition of $x + x$ because 'the voting has gone this way'.

In the preface to this edition Venn thanked the help given by W. E. Johnson, who taught the course in 'advanced logic' at Cambridge when students wished to take it. An infrequent publisher, some of his ideas appeared in 1892 in a long paper on 'the logical calculus' in the philosophical journal *Mind*.[26] He focused on the propositional calculus including quantification, referring to all main predecessors except De Morgan. He did not pursue connections with mathematics in detail; for example, when introducing the contradiction and tautology as logical constants, he explicitly rejected '0' and '1' and used instead 'falsism' and 'truism'. He elaborated a system of notations that has remained deservedly unpopular. The main fruits of his thoughts on logic, deductive and inductive, did not appear in print until the three-volume book *Logic* was published between 1921 and 1924, by which time many of them were at best nostalgic to readers, at least those with a mathematical interest.

## Lewis Carroll as a logician

Some other figures treated aspects of algebraic logic from time to time, and interestingly a few of them were women. They included Boole's widow Mary, mainly on philosophical and educational applications of his logic, Constance Jones who also included aspects of mathematical logic, Sophie Bryant who was interested mainly in the operator reading of Boole's logic, and Lady Victoria Welby, a disciple of Peirce who was mainly concerned with semiotics. But the most important female contributor to logic in the Victorian period was Peirce's American student Christine Ladd-Franklin, who pioneered the 'antilogism' method of handling syllogisms in the 1880s.[27] And one man stood out, as the fool (in the Shakespearean sense): the Oxford mathematics tutor C. L. Dodgson.

Writing under his pen-name of 'Lewis Carroll', Dodgson published *The Game of Logic* in 1886 and the first part of his *Symbolic Logic* in 1896: materials for its second part were not published until 1977.[28] As usual, the aim was to deduce logical consequences from given premises: the framework was basically Aristotelian, although generalized from syllogisms to 'soriteses' in which more than two premises were considered; among predecessors, Boole and Venn had handled them, but Carroll's examples went up to as many as fifty premises.

Carroll used various methods to draw inferences. One was algebraic, and somewhat similar to De Morgan's, with lower-case letters to denote classes, primes to indicate the complements relative to some universe, subscripts 0 and 1 to indicate emptiness and non-emptiness of classes, and concatenations for intersection. He then symbolized propositions so that, for example, $xy'$ denoted 'some $x$s are not $y$s' and $x_1y_0$ meant 'all $x$s are not $y$s'. He also laid down rules for eliminating classes, or their associated terms, from some or all of the given premises to obtain the logical consequences.

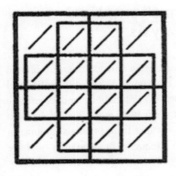 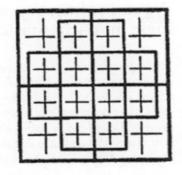

Examples of Carroll's nested boxes.

Carroll also devised a geometrical procedure, using a variant of Venn diagrams, with boxes and nested sub-boxes instead of circles to represent classes. He also placed grey or red counters in (or across) them, to show that the corresponding classes were respectively empty or not empty, and he read off consequences from some or all of the premises by studying the corresponding (sub-)array of counters.[29] In another procedure, the most novel part of his theory (but published only posthumously), he laid out premises and the expected consequence C in a tree format, effecting the deduction by assuming the negation of C and obtaining a contradiction, as in the *reductio* proof method used in mathematics.[30]

Carroll's logic is nice and amusing, especially when conclusions such as 'No wise young pigs go up in balloons' can safely be drawn, but it is not outstanding. However, he also explored the mysterious world of 'hypothetical propositions'; in particular, he wrote a witty but wise paper in *Mind* in 1895, where he clearly showed the importance of distinguishing implication from inference (although he did not offer that interpretation).[31]

Of still higher order are his *Alice* books (1865, 1871) and some of his other entertainments, such as *The Hunting of the Snark*. These works deserve their fame, but the quality of his logical insights in them is not widely recognized. Yet all sorts of issues in logic are there: among others were meaning by convention, connections between logic and philosophy, signs and their referents (if any), and identity. Especially brilliant is the disappeared Cheshire cat, who has left behind only his grin—a masterpiece of phenomenological logic, involving its distinction between the moments of a whole (such as the smile of the cat) and its parts (such as the lips of the cat).[32]

The best appreciation to date of this aspect of Carroll's logic appeared in 1918, in a book of amusing short essays on logic and its philosophy by the historian of mathematics, Philip Jourdain, to which he attached an extensive appendix of passages from Carroll's writings as illustrative texts. The book was called *The Philosophy of Mr. B\*tr\*nd R\*ss\*ll*, in honour of his former logic teacher at Cambridge University, who indeed contributed a few chapters himself.[33] But the logic that they had in mind was not the algebraic logic of De Morgan, Boole, and Carroll, but that to which Russell had contributed—a Victorian logic that was completely different from anything seen so far.

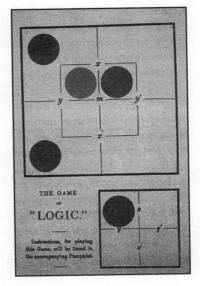

Lewis Carroll's *The Game of Logic*, showing the board and counters.

## Rigour, set theory, and mathematical logic in mathematical analysis

We must return briefly to the early 19th century, when a major emergent branch of mathematics was mathematical analysis, incorporating the differential and integral calculus with the theories of mathematical functions and of infinite series. The chief architect of this fusion was Augustin-Louis Cauchy, with his unifying glue of a proper *theory* of limits, as distinct from previous notions; in particular, he stressed that in all these contexts a sequence of values may not have a limit at all. This approach greatly increased the level of rigour in mathematics of the time, to which he adjoined two further principles: that mathematical functions must always be single-valued, and that definitions of concepts, and proofs of theorems, must be laid out in as much detail as possible.

Cauchy's approach did not meet with rapid or universal approval: for example, in late-1830s Britain De Morgan's adhesion to it was only partial.[34] But it gradually gained ground from the late 1850s onwards, especially at Berlin University in the teaching of Karl Weierstrass, who brought further refinements both to the theory of limits itself and its applications across mathematical analysis. Further, from the early 1870s Georg Cantor was inspired by it to create set theory: the point-set topology concerning the distribution of points on the line, the theory of infinitely large cardinal and ordinal numbers, and the doctrine of 'order-types' (the different ways in which the members of finite and infinite sets could be ordered).[35]

Logic was not explicitly deployed in these foundational studies until the late 1880s, when Giuseppe Peano at the University of Turin sought to enhance Weierstrass's aspiration for rigour by formalizing and symbolizing both the mathematical and the logical notions involved. With Cantor's set theory at centre stage, he stated the axioms or laws of the required 'mathematical logic'—his name for the logic of propositions and 'propositional functions' (his word for predicates)—including universal and existential quantification over members of a set. Then he developed the mathematical theory at hand in a highly formalized and symbolic manner. Helped by an impressive roster of followers at Turin, the basic accounts were published in the book *Formulaire Mathématique* (five editions under his editorship, with variant titles, from 1895–1908), while research papers in this tradition appeared in his journal *Rivista di Matematica* (eight volumes under his editorship, with variant titles, from 1891–1906).[36]

## Russell and Whitehead: mathematical logic at Cambridge

In the first half of the 1890s, ignorant of such developments (like most of his compatriots), Bertrand Russell took the Part I of the mathematics tripos at Cambridge University and then Part II of the moral sciences tripos (though, it seems, not the logic courses). After winning a research fellowship at his college (Trinity), he combined his specialities into an ambitious programme in the philosophy of mathematics, starting with *An Essay on the Foundations of Geometry* (1897) and moving on to mathematical analysis and related topics. One of these was Cantor's set theory, which was little known in Britain; Russell came across it principally when he reviewed for *Mind* the book *De l'Infini Mathématique* (1896) by the French philosopher Louis Couturat. Logic also crossed his desk sometimes, especially in the book *Universal Algebra* (1898) by his former Trinity College tutor A. N. Whitehead, where a version of Boole's logic was presented.

At this time Russell's adopted philosophy was the standard Cambridge creed, a highly idealist tradition called 'neo-Hegelianism',[37] which may have helped to prevent him from seeing any clear-cut standpoint from which he could frame a suitable position. In 1899 he abandoned this philosophy, in a trend initiated by his friend, the philosopher G. E. Moore; they went for an opposite extreme, positivism, where abstract objects were not at all welcome.

Soon afterwards the main conversion occurred for Russell, during the last year of the Victorian era, and specifically between 9.30 and 13.30 on Friday 3 August 1900. Russell and Whitehead were attending the First International Congress of Philosophy in Paris, and for that session Couturat had organised a suite of four lectures by the 'Peanists', as the group was affectionately known. Peano and one of his main followers spoke, and Couturat read papers by two others in their absence. We can even identify the magic moment of this morning: Peano spoke first, on legitimate forms of definition in mathematics, and in the discussion period he had a dispute with Schröder, during which Russell realized Peano's superiority.

From then on Peano was Russell's father figure, especially in the rapid sequence of developments that followed.[38] During the rest of 1900 Russell wrote much of a new version of a book on the philosophy of mathematics that he had been trying to produce for years. He also noticed, with understandable surprise, that the Peanists had not included a logic of relations in their mathematical logic; so he furnished one, in a paper published in 1901 in the *Rivista*. He followed this with a treatment of well-ordering, which was Cantor's fundamental order-type because it was the one satisfied by the sets of finite and infinite positive ordinal integers.

Bertrand Russell as a B.A. in mathematics at Trinity College, Cambridge, 1893.

## Logicism with Russell and Whitehead

Early in 1901 Russell also found the philosophical standpoint underlying his new approach, and maybe from the following situation. When the Peanists clothed a mathematical theory in their logical garb, every now and then they included two columns listing the symbols that they were going to use in the exegesis to come, one for mathematical concepts and the other for logical ones; symbols for

Alfred North Whitehead
(1861–1947).

concepts in set theory would sometimes be in one column, sometimes in the other, and sometimes even in both.

So where is the division between mathematics and logic? Whether from this motive or another, Russell decided that there was *no* division, but that mathematics was *part* of this mathematical logic (including the logic of relations); it supplied not only the methods of reasoning, but also the 'objects' required in a mathematical theory. This philosophical line guided the rest of Russell's logical career, a position that came later to be called 'logicism'. Executing it in detail in symbolic Peanese would be a formidable task, but during 1901 and 1902 Whitehead's interest in the project blossomed into a formal collaboration.

But, also in 1901, bad news stymied this excellent progress. While considering one of Cantor's proof-methods (namely, the power-set argument), Russell formed the set of all sets that do not belong to themselves, and found that it belongs to itself *if and only if* it does not do so. This was a genuine paradox, a double contradiction, rather than the single contradiction that, for example, licenses the *reductio* proof-method in mathematics. Moreover, it lay at the heart of his logicism, for it needed only sets and membership in its formulation.[39] So the logicist enterprise would need substantial modification.

Russell did not complete his 1900 manuscript on the philosophy of mathematics until well into 1902,[40] and the solution to his paradox that he offered in an appendix to *The Principles of Mathematics* (1903) was stop-gap—in fact, as he found later, leaky. It took until 1907 before he and Whitehead constructed a system that could ground logicism, and even then important weak points were evident. The main outcome was their vast three-volume *Principia Mathematica* (1910–13), published by Cambridge University Press, for which W. E. Johnson acted as referee.[41] But that story belongs to their period of Edwardian pain, after the few months of late Victorian pleasure.

## Algebraic logic ≠ mathematical logic

We close this chapter with two reflections on the practice of logic in Victorian Britain. The first concerns the several, and fundamental, differences between what we have termed algebraic logic and mathematical logic.[42] The algebraists applied mathematics (usually an algebra of some kind) to logic, while the 'mathematicals' (as we shall call them) applied their logic to mathematics. The algebraists often noted, and even exploited, structural properties such as symmetry and duality, and Schröder was to make it an explicit method; the mathematicals usually ignored them. The algebraists, especially Boole, tended to hide the details of a derivation in the workings of their algebra; the mathematicals wanted to exhibit them as precisely as possible, in the spirit of Cauchy and Weierstrass. The algebraists handled logic as a qualitative theory: so did the mathematicals, but they also covered the quantitative side in proffering definitions of numbers integral, rational, and irrational, and proceeding on to other branches of mathematics, especially mathematical analysis and geometry. Several of the algebraists also studied probability theory (then another small mathematical topic), to some extent through concerns with inductive as well as deductive logic; none of the mathematicals followed suit.

On collections the mathematicals drew heavily on Cantorian set theory, whereas the algebraists followed the tradition of part-whole theory where, for example, the class of European men is a part of the class $M$ of men, and no distinction is made between membership of a European man to $M$ and the inclusion of its unit class within $M$.[43] This distinction has major consequences on both technical and philosophical grounds. For example, while algebraic logicians studied aspects of Cantor's set theory (such as Peirce on continuity and Schröder on finitude), they could not express the mathematical contexts *fully* in their systems, in ways available to the Peanists and to Russell and Whitehead. The distinction also affected the reading of universal and existential quantification.

Concerning language, the algebraists focused on nouns and adjectives, whereas the Peanists (and especially the logicists) looked closely at six little words: 'all, every, any, a, some, the'. (Nobody bothered much about adverbs.) An important part of the motivation of the mathematicals lay in the deep concern about whether an expression had no, one, or more referents; by contrast, the algebraists were concerned with the semiotics of their logic. (Peirce revived this word, in this and other contexts.)

Some of the differences between the two styles of logic had been sensed or anticipated as early as 1858 by De Morgan. In the kind of *bon mot* of which he was a fine expert, he coined the term 'mathematical logic', not in Peano's later and precise sense described above, but in this context:[44]

As joint attention to logic and mathematics increases, logic will grow up among the mathematicians, distinguished from the logic of the logicians by having the mathematical element properly subordinated to the rest. This *mathematical logic* ... will commend itself to the educated world by showing an actual representation of their form of thought—a representation the truth of which they recognise—instead of a mutilated and onesided fragment, founded upon canons of which they neither feel the force nor see the utility.

The algebraic logic of De Morgan, Boole, and others has dominated this chapter, with the mathematical logic of Russell and Whitehead supplying the end. The gap between these two kinds of symbolic logic is exemplified by the fact that when in 1901 Russell added to mathematical logic a logic of relations, he could have drawn upon a substantial body of algebraic theory by De Morgan, Peirce, and especially Schröder; but instead, he made hardly any use of it.

## Symbolic logics as marginal enterprises

Our second reflection is the fact that the community of algebraic logicians (De Morgan, Boole, Jevons, etc.) was small and its work was of no great general concern. Their logic was too mathematical for the philosophers and too philosophical for the mathematicians. Boole, for example, was highly regarded by his mathematical contemporaries as a specialist in differential equations, but to them (apart from De Morgan and Jevons) his logic was a curiosity.

However, Victorian Britain provided quite a large proportion of the personnel involved worldwide: far more than France, for example, which was a much more important country for mathematics. One reason may lie in the popularity of algebras in Britain, especially in England and Ireland.[45] In addition to the continuing worries there about the legitimacy of negative numbers, unique in level for that time, there was especially the strong interest taken in developing and even inventing new algebras: differential operators, functional equations, probability theory (to which all the remarks of this section also apply), quaternions, determinants, matrices, and invariants, as well as algebras for logic. The only kinds of algebra where Britons did not take major roles were the largely German-speaking and French enterprises of Lie and abstract algebras, where the dominating case in the Victorian period was group theory. Maybe one attraction of algebras was that the student did not require a massive amount of technical preparation, unlike the major Continental preoccupations of real- and complex-variable analysis. This feature did not make these algebras 'easier' mathematics to pursue than other branches, but rather meant that the difficulties lay more in conceptual issues. In algebraic logic some of these are very subtle; for example, most of our authors slipped between terms, classes, predicates, and attributes without always noticing (and we have not attempted any clean-up above).

Mathematical logicians formed another small set, and for similar reasons; for example, the Peanists were hardly known in Britain before Russell and Whitehead came across them in Paris. *Principia Mathematica* has always been very little read—unfortunately, as it contains a lot of interesting mathematics, especially in the second and third volumes. From the 1900s onwards mathematical logic largely eclipsed algebraic logic, but its place in mathematics, and in philosophy in general, remained modest, and not only in Britain.

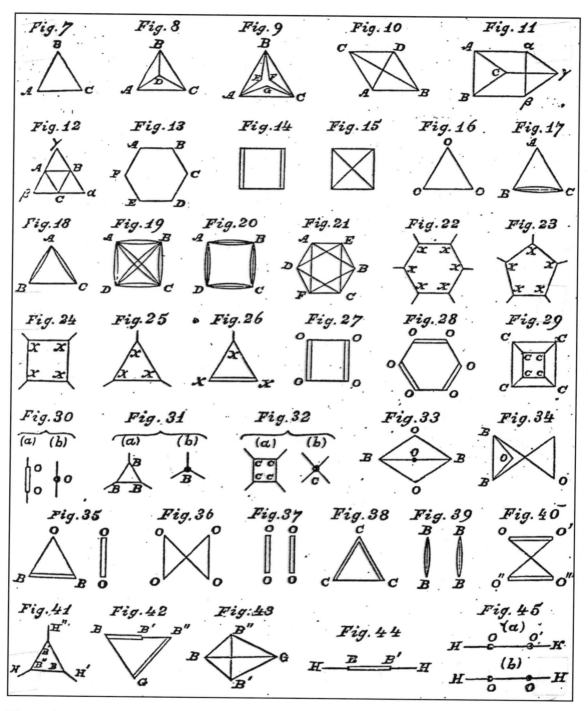

Sylvester's chemical trees, from his lengthy paper in the first issue of the *American Journal of Mathematics*.

CHAPTER 17

# Combinatorics

*A very Victorian recreation*

ROBIN WILSON

I n this chapter we explore the area of combinatorics (or combinatorial analysis), the mathematics of arranging and counting things. In Victorian times, Britain was a leading venue for such pursuits, indulged in by professional mathematicians and enthusiastic amateurs alike. We outline five specific topics that illustrate the range of Victorian combinatorics, meeting along the journey such illustrious mathematicians as Augustus De Morgan, Arthur Cayley, James Joseph Sylvester, and Sir William Rowan Hamilton, and such amateurs as Thomas Kirkman, Wesley Woolhouse, and Alfred Kempe.

Many areas of mathematics are so technical that amateur mathematicians steer well clear of them. However, there are a few topics, such as number theory and combinatorics, where the concepts are easy to grasp and the questions are easy to pose even if many of them are exceedingly difficult to answer. In these areas, well-meaning amateur mathematicians have long shown remarkable persistence in producing solutions, sometimes with considerable success, in order to show their superiority over the professionals. Even when their attempted 'solutions' are incorrect, their efforts may exhibit remarkable skill and ingenuity.

Inevitably, many of these amateurs are cranks, dabblers whose attempts are simplistic or nonsensical. But in Victorian times several of them were lovers of mathematics who had previously taken degrees in the subject before moving on to distinguished careers in the church, the law, or the army, while a literary example, to be found in Thomas Pynchon's *Gravity's Rainbow*, was the elderly Brigadier Ernest Pudding who, when pensioned off, went to sit in the study of an empty house in Devon…there to go at a spot of combinatorial analysis, that favourite pastime of retired Army officers, with a rattling intense devotion.[1]

In this chapter we investigate five areas in which Victorian mathematicians, professional and amateur alike, exhibited their skills: the study of so-called 'Steiner triple systems' and the 'schoolgirls problem' of Thomas Kirkman, cycles on polyhedra and Sir William Rowan Hamilton's 'Icosian game', Cayley and Sylvester's investigations into trees and chemical molecules, the contributions of Major Percy MacMahon and others on partitions of integers, and finally, the celebrated 'four-colour problem' on the colouring of maps.

## Triple systems

Wesley Stoker Barker Woolhouse (1809–93).

Our first area of Victorian combinatorics is now known as the study of *Steiner triple systems*. But, as we shall see, the Swiss mathematician Jakob Steiner contributed virtually nothing to their development, and the credit should rightly go to a Lancashire clergyman, the Revd. Thomas Penyngton Kirkman, who made substantial contributions to the subject. Kirkman had studied mathematics, science, philosophy, and classics at Trinity College, Dublin, before entering the church, eventually becoming rector of the tiny parish of Croft-with-Southworth, near Warrington, where he stayed for over fifty years. His parochial duties were undemanding and left him time to have seven children and to do much mathematics, becoming a Fellow of the Royal Society in the process. In 1847 Kirkman wrote an important paper about these triple systems. But what are they, and how did he learn about them?

The story can be traced back to the annual *Lady's and Gentleman's Diary* (see Chapter 7), which, according to its title page, was:

DESIGNED PRINCIPALLY FOR THE AMUSEMENT AND INSTRUCTION
OF
STUDENTS IN MATHEMATICS:
COMPRISING
MANY USEFUL AND ENTERTAINING PARTICULARS,
INTERESTING TO ALL PERSONS ENGAGED IN THAT
DELIGHTFUL PURSUIT.

In 1844 the *Diary* appointed a new editor, Wesley Woolhouse, a keen amateur mathematician who had been Deputy Secretary of the Nautical Almanac. Woolhouse challenged his readers with a combinatorial problem, which he may have learned from James Joseph Sylvester who had recently written on a similar topic:[2]

*Prize Quest.* (1733); *by the Editor.*[3]
Determine the number of combinations that can be made out of *n* symbols, *p* symbols in each; with this limitation, that no combination of *q* symbols, which may appear in any one of them shall be repeated in any other.

This is a problem about arranging things: we are asked to arrange a number ($n$) of symbols into groups of $p$ elements in such a way that a particular condition is satisfied. Unfortunately, the problem can be interpreted in various ways, as illustrated by the attempted solutions sent in. In fact, no-one managed to produce a satisfactory solution to the question, and in 1846 Woolhouse duly presented his readers with a simplified challenge,[4] corresponding to the special case $p = 3$ and $q = 2$:

How many triads can be made out of $n$ symbols, so that no pair of symbols shall be comprised more than once amongst them?

Here is an example for $n = 7$, where the seven symbols are the numbers 1–7 and the triads, or triples, are arranged vertically:

| 1 | 2 | 3 | 4 | 5 | 6 | 7 |
|---|---|---|---|---|---|---|
| 2 | 3 | 4 | 5 | 6 | 7 | 1 |
| 4 | 5 | 6 | 7 | 1 | 2 | 3 |

The condition to be satisfied is that no pair of numbers may occur together more than once. In this example, each pair appears exactly once—for example, the numbers 3 and 5 appear together in the second triple, while the numbers 2 and 6 appear together in the sixth triple.

Such arrangements of numbers are now usually called *Steiner triple systems*; they were studied in the 1830s by the German geometer Julius Plücker, and it is conceivable that James Joseph Sylvester may have been familiar with Plücker's work before communicating it to Woolhouse. Such systems are now used in agriculture, in the design of experiments. For, suppose that you wish to compare seven varieties of wheat. If your fields are too small to include all seven types of wheat, you can plant varieties 1, 2, and 4 in the first field, varieties 2, 3, and 5 in the second, and so on. With the above arrangement, you can then directly compare each pair of varieties in one of the seven fields.

Notice that the above system is easy to construct, since from the first triple (1, 2, 4) we can obtain each successive triple by adding 1 to each number, following 7 by 1; so, once we have found a suitable starting triple, the rest follows. Such systems are called *cyclic systems*, and we shall meet them again.

Another triple system is shown below, for $n = 9$. Again, each pair of numbers appears in exactly one triple—for example, the numbers 3 and 8 appear together in the eighth triple:

| 1 | 1 | 1 | 1 | 2 | 2 | 2 | 3 | 3 | 3 | 4 | 7 |
|---|---|---|---|---|---|---|---|---|---|---|---|
| 2 | 4 | 5 | 6 | 4 | 5 | 6 | 4 | 5 | 6 | 5 | 8 |
| 3 | 7 | 9 | 8 | 9 | 8 | 7 | 8 | 7 | 9 | 6 | 9 |

Thomas Penyngton Kirkman (1806–95).

For which values of $n$ do triple systems exist? In a ground-breaking paper entitled 'On a problem in combinations', presented to the Literary and Philosophical Society of Manchester on 15 December 1846 and subsequently published in *The Cambridge and Dublin Mathematical Journal*,[5] Kirkman used a simple counting argument to show that triple systems can occur only when $n$, the total number of symbols, is one of the numbers in the sequence

$$7, 9, 13, 15, 19, 21, 25, \ldots$$

—these numbers are all of the form $6k + 1$ or $6k + 3$, where $k$ is an integer. He also showed (and this was much more difficult) that for each number in this sequence one can construct at least one such triple system, and he gave an explicit construction for doing so. In later papers he developed these ideas further.

Before proceeding, let us revisit the triple system with $n = 9$:

| 1 | 4 | 7 | | 1 | 2 | 3 | | 1 | 2 | 3 | | 1 | 2 | 3 |
|---|---|---|---|---|---|---|---|---|---|---|---|---|---|---|
| 2 | 5 | 8 | | 4 | 5 | 6 | | 6 | 4 | 5 | | 5 | 6 | 4 |
| 3 | 6 | 9 | | 7 | 8 | 9 | | 8 | 9 | 7 | | 9 | 7 | 8 |

Here we have rearranged the twelve triples so that the system splits into four parts with all the nine numbers appearing in each part. In our agricultural context, we can think of each part as representing a *season*: in the first season, we plant three fields with varieties 1, 2, 3 in the first; 4, 5, 6 in the second; and 7, 8, 9 in the third; in the next season we plant them with varieties 1, 4, 7 / 2, 5, 8 / 3, 6, 9; and so on for the remaining two seasons. After four seasons, we will have compared each pair of varieties exactly once.

Such a division into seasons can happen only when $n$, the number of symbols, is divisible by 3, which means that it must have the form $6k + 3$. This includes the above case of $n = 9$, and also includes the case $n = 15$, as we now see.

An entertaining feature of *The Lady's and Gentleman's Diary* was that it always included a number of queries sent by the readers. In 1850 there were queries by Mr Lugg on the origin of April Fools' Day, by the Revd. Hope on the three sons of Noah, by Mr Herdson on the saltiness of the sea, and by Kirkman on triple systems; he had thought of this problem while working on his paper of 1847:

Fifteen young ladies in a school walk out three abreast for seven days in succession: it is required to arrange them daily, so that no two shall walk twice abreast.[6]

If there had been only nine young ladies, we could have used the triple system above, with the four seasons corresponding to four days: on the first day, 1 walks with 2 and 3, 4 walks with 5 and 6, and 7 walks with 8 and 9; on the second day, 1 walks with 4 and 7, and so on—and no two young ladies walk together more than once.

The problem of the fifteen young ladies, now known as *Kirkman's schoolgirls problem*, also appeared in the *Educational Times* 'thus versified by a lady':[7]

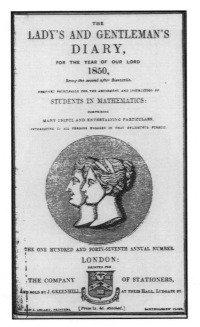

*The Lady's and Gentleman's Diary of 1850.*

A governess of great renown
Young ladies had fifteen,
Who promenaded near the town,
Along the meadows green.

But as they walked
They tattled and talked,
In chosen ranks of three,
So fast and so loud,
That the governess vowed
It should no longer be.

So she changed them about,
For a week throughout,
In threes, in such a way
That never a pair
Should take the air
Abreast on a second day;
And how did the governess manage
it, pray?

A solution of Kirkman's schoolgirls problem, listing the five triples for each day, is as follows:

| Monday: | 1–2–3 | 4–5–6 | 7–8–9 | 10–11–12 | 13–14–15 |
|---------|-------|-------|-------|----------|----------|
| Tuesday: | 1–4–7 | 2–5–8 | 3–12–15 | 6–10–14 | 9–11–13 |
| Wednesday: | 1–10–13 | 2–11–14 | 3–6–9 | 4–8–12 | 5–7–15 |
| Thursday: | 1–5–11 | 2–6–12 | 3–7–13 | 4–9–14 | 8–10–15 |
| Friday: | 1–8–14 | 2–9–15 | 3–4–10 | 6–7–11 | 5–12–13 |
| Saturday: | 1–6–15 | 2–4–13 | 3–8–11 | 5–9–10 | 7–12–14 |
| Sunday: | 1–9–12 | 2–7–10 | 3–5–14 | 6–8–13 | 4–11–15 |

Notice that, as before, any two schoolgirls walk together exactly once; for example, schoolgirls 3 and 10 walk together on *Friday*.

The schoolgirls challenge proved to be more successful than the 1844 Prize Question, and two solutions appeared in the *Diary* for 1851: one by Kirkman himself, and one apparently obtained independently by Mr Bills of Newark, Mr Jones of Chester, Mr Wainman of Leeds, and Mr Levy of Hungerford; how they all produced exactly the same solution is unclear. Kirkman claimed that his own solution was 'the symmetrical and only possible solution', but he was wrong; another symmetrical solution, different from his, had been obtained a few months earlier by Arthur Cayley.

Kirkman later presented variations on his schoolgirls problem, such as the following one:[8]

Sixteen young ladies can all walk out four abreast, till every three have *once* walked abreast; so can thirty-two, and so can sixty-four young ladies; so can $4^n$ young ladies.

The first person to treat the fifteen schoolgirls problem in a systematic way, rather than by organized guesswork, was another Victorian clergyman, the Revd. Robert Anstice, who had studied mathematics in Oxford. Anstice's achievement was to find a *cyclic* solution:[9] it is given below, with the fifteen schoolgirls denoted here by 0 – 6 in Roman type, **0** – **6** in bold face, and a separate symbol for infinity. Notice that once we have the arrangement for Monday, we can then obtain the arrangement for each successive day by adding 1, always following 6 by 0, and leaving infinity unchanged.

| Monday: | ∞–0–**0** | 1–2–**3** | 1–**4**–**5** | 3–**5**–**6** | **4**–**2**–**6** |
|---|---|---|---|---|---|
| Tuesday: | ∞–1–**1** | 2–**3**–4 | 2–**5**–**6** | **4**–**6**–**0** | **5**–**3**–**0** |
| Wednesday: | ∞–2–**2** | 3–4–**5** | 3–**6**–**0** | **5**–**0**–1 | **6**–**4**–**1** |
| Thursday: | ∞–3–**3** | 4–5–**6** | 4–**0**–1 | **6**–**1**–2 | 0–**5**–**2** |
| Friday: | ∞–4–**4** | 5–**6**–**0** | 5–**1**–**2** | **0**–2–3 | 1–**6**–**3** |
| Saturday: | ∞–5–**5** | **6**–**0**–1 | **6**–**2**–**3** | 1–3–**4** | 2–**0**–**4** |
| Sunday: | ∞–6–**6** | **0**–1–**2** | **0**–3–**4** | 2–4–**5** | 3–**1**–**5** |

At this point we revisit the brilliant but eccentric James Joseph Sylvester, whose chequered career, on both sides of the Atlantic, culminated in his election as the Savilian professor of geometry in Oxford at the age of 69. Sylvester believed that *he* had originated the schoolgirls problem, and said so in his own inimitable, but somewhat incomprehensible, way:[10]

... in connexion with my researches in combinatorial aggregation ... I had fallen upon the question of forming a heptatic aggregate of triadic synthemes comprising all duads to the base 15, which has since become so well known, and fluttered so many a gentle bosom, under the title of the fifteen school-girls' problem; and it is not improbable that the question, under its existing form, may have originated through channels which can no longer be traced in the oral communications made by myself to my fellow-undergraduates at the University of Cambridge ...

Kirkman naturally thought little of Sylvester's priority claim, and replied:[11]

No man can doubt, after reading his words, that he was in possession of the property in question of the number 15 when he was an Undergraduate at Cambridge. But the difficulty of tracing the origin of the puzzle ... is considerably enhanced by the fact that, when I proposed the question in 1849, I had never had the pleasure of seeing either Cambridge or Professor Sylvester.

Then, after citing his own paper, Kirkman concluded:

No other account of it has, so far as I know, been published in print except this guess of Prof. Sylvester's in 1861.

However, Sylvester did devise an interesting extension of the schoolgirls problem. The total number of possible triples of fifteen schoolgirls is 455, which is 13 × 35, and, as reported by Cayley,[12] Sylvester asked:

Can we arrange all these 455 triples into 13 *separate* solutions to the problem? that is, can we arrange 13 weekly schedules so that each possible triple of schoolgirls appears just once in the quarter-year?

Kirkman claimed a solution in 1850, but he was wrong—and indeed, a correct solution was not to be found for over a hundred years, when R. Denniston[13] of Cambridge used a computer to construct one in 1971. Around the same time, the general problem of solving the schoolgirls problem for larger numbers of school-girls (21, 27, 33, . . . ) was solved by Dijen Ray-Chaudhuri and Rick Wilson,[14] and independently a few years earlier by Lu Xia Xi, a schoolteacher from Inner Mongolia. However, Sylvester's extension of the problem for these higher num-bers remains unsolved to this day.

Where does Jakob Steiner fit into this story? In 1853, he wrote a short note on triple systems,[15] a topic that he had probably encountered through his study of Plücker's work. In this note, Steiner correctly observed that triple systems with $n$ symbols can exist only when $n$ has the form $6k + 1$ or $6k + 3$. He then asked for which numbers $n$ such triple systems can be constructed, completely unaware that Kirkman had completely solved this problem six years earlier; this lack of awareness probably arises from the fact that *The Cambridge and Dublin Mathematical Journal*, although well known in Britain, was little known on the Continent. The situation was further complicated when M. Reiss[16] solved Steiner's problem using methods very similar to those of Kirkman, causing the latter to complain sarcastically:[17]

. . . how did the Cambridge and Dublin Mathematical Journal . . . contrive to steal so much from a later paper in Crelle's Journal, Vol. LVI., p. 326, on exactly the same problem in combinations?

Poor Kirkman was unlucky. Not only is he rarely credited with his fundamen-tal contributions to triple systems, but he also failed to receive credit for his invention of *Hamiltonian cycles*, as we now see.

# Polyhedra

Sir William Rowan Hamilton has frequently been described as a child prodigy who knew Latin, Greek, and Hebrew at the age of 5, and then apparently learnt Persian, Syriac, Sanskrit, and other languages by the age of 13 (although there is some doubt about this—see Chapter 5). He became Astronomer Royal of Ireland while still an undergraduate, and was knighted at the age of 30.

Before we come to Hamilton's work on polyhedra, we consider some examples. A *polyhedron* is a solid shape whose faces are polygons—triangles, squares, pentagons, etc.; for example, a cube is bounded by six square faces and a dodecahedron is bounded by twelve pentagons. In some polyhedra the faces are of different shapes; for example, a *truncated octahedron* has both square faces and hexagonal faces.

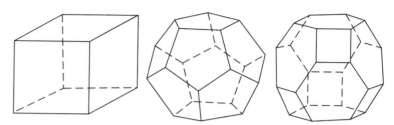

Cube, dodecahedron, and truncated octahedron.

The Victorians were interested in problems that can be stated as follows:

A fly decides to visit all the corners of a cube, and then return to its starting point. What route should it take?

One such cyclical route is shown below on a flattened cube—just follow the solid line. Similarly, there are cycles that visit all the vertices of a dodecahedron and return to the starting point.

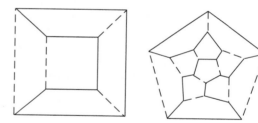

Routes on a cube and a dodecahedron.

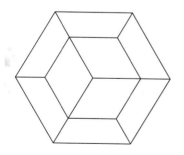

Kirkman's 'cell of a bee'.

Are there such cyclical routes for *all* polyhedra? In 1856, Kirkman gave an example to show that, for some polyhedra, the answer is 'no'. Using his own word *q-acron* for a polyhedron with *q* vertices, he observed that:[18]

if we cut in two the cell of a bee…we get a 13-acron…The closed 13-gon cannot be drawn.

To explain why this is, we can colour the vertices so that each of its edges has a black end and a white end. Since any cycle has to alternate black and white vertices, the numbers of black and white vertices must be the same. But there are six black vertices and seven white vertices, so a cyclical route cannot be found.

In the following year, Hamilton became interested in drawing cycles on a dodecahedron—these arose out of some investigations he was carrying out in algebra, called the *icosian calculus*. Following his work on quaternions, where he was interested in symbols *i*, *j*, and *k* satisfying the equations $i^2 = j^2 = k^2 = ijk = -1$ (see Chapters 5 and 15), he now considered three symbols *i*, *k*, and *l* satisfying the equations $i^2 = k^3 = l^5 = 1$, where *i*, *k*, and *l* are linked by the equation $l = i \times k$. Writing $m = i \times k^2$ (which is the same as $l \times k$), he obtained the following equation containing twenty symbols:

$$l^3 m^3 l m l m l^3 m^3 l m l m = 1.$$

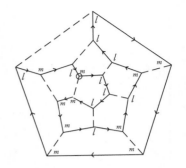

Tracing a cycle on a dodecahedron.

Hamilton then showed that we can interpret this equation in terms of a cycle through the twenty vertices of a dodecahedron: to see how, start anywhere and think of *l* as *turn right* and *m* as *turn left*, giving

right, right, right, left, left, left, right, left, . . . .

Hamilton was so pleased with his idea that he converted it into a game called *A Voyage Around the World*, in which the twenty vertices of the dodecahedron are labelled with the consonants *B*, *C*, *D*, . . . , *Z*, representing places (Brussels, Canton, Delhi, . . . , Zanzibar) and the object is to 'travel around the world' and return to the starting point; one solution is to follow alphabetical order:

*B C D F G H J K L M N P Q R S T V W X Z B.*

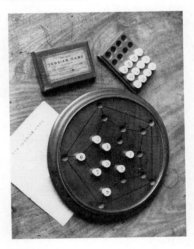

Hamilton proudly sold his game to a games manufacturer for £25—a wise move as it did not sell. The game had twenty numbered pegs to be placed in cyclic order into the twenty holes of a board. The instructions for his 'Icosian game' contain various puzzles of the form:

given five initial points, such as *B, C, D, F, G*, in how many ways can you complete the cycle?

Hamilton's Icosian game.

Such were Sir William's importance and influence that these cycles are now named *Hamiltonian cycles*, rather than being more justly credited to Kirkman, who had preceded Hamilton by several months and who had discussed cycles on general polyhedra, and not just on a dodecahedron.

## Trees

In the 1840s, Arthur Cayley found himself unable to obtain a fellowship in Cambridge without taking holy orders, while many academic jobs were closed to James Joseph Sylvester, who was Jewish. Sylvester took employment in a London actuarial firm, while Cayley worked as a lawyer at Gray's Inn; in their spare time they wrote several hundred mathematical papers and quickly established themselves as the leading pure mathematicians in England (see Chapter 15).

In the late 1850s, Cayley and Sylvester became interested in tree structures. These are branching diagrams that contain no cycles: familiar examples include

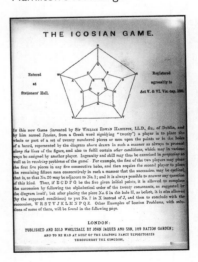

family trees and certain chemical molecules such as alkanes (or paraffins). Earlier, in the 1840s, trees had been used by Gustav Kirchhoff in his investigations into currents in electrical networks.

Cayley was primarily concerned with counting rooted trees, trees where everything branches down from one special vertex, called the root, placed

---

## Box 17.1: Cayley's approach to counting rooted trees

In order to count rooted trees, Cayley used an 18th-century device called a *generating function*,

$$1 + A_1 x + A_2 x^2 + A_3 x^3 + \cdots,$$

where $A_n$ is the number of rooted trees with $n$ branches; this has the advantage that you need to consider only one expression at a time, instead of a whole string of numbers all at once.

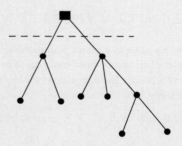

Slicing off the root, he obtained several smaller rooted trees; this leads to the following recurrence relation from which each successive number $A_i$ can be found iteratively: knowing $A_1$ you can deduce $A_2$; knowing $A_1$ and $A_2$ you can deduce $A_3$; and so on:

$$1 + A_1 x + A_2 x^2 + A_3 x^3 + \cdots = (1-x)^{-1} \times (1-x^2)^{-A_1} \times (1-x^3)^{-A_2} \times \cdots.$$

Using this, Cayley obtained the following numbers of rooted trees; for example, there are 48 rooted trees with 6 branches, and 719 rooted trees with 9 branches.

| $n$ | 1 | 2 | 3 | 4 | 5 | 6 | 7 | 8 | 9 | 10 | ... |
|---|---|---|---|---|---|---|---|---|---|---|---|
| number | 1 | 2 | 4 | 9 | 20 | 48 | 115 | 286 | 719 | 1842 | ... |

inappropriately at the top. His interest in such trees originated in a problem of Sylvester on the differential calculus and, in a paper of 1859,[19] he first drew the small rooted trees with a given number of terminal vertices, and demonstrated that just 3 rooted trees have 3 terminal vertices, and 13 rooted trees have 4. Cayley's approach to counting rooted trees is outlined in Box 17.1.

Counting unrooted trees, where there is no special root-vertex, is a much harder task, but Cayley found a way of achieving it, by starting at the middle of the tree and gradually working outwards. He obtained the numbers below; for example, there are 106 unrooted trees with 9 branches.

The trees with up to four branches.

| $n$ | 1 | 2 | 3 | 4 | 5 | 6 | 7 | 8 | 9 | 10 | ... |
|---|---|---|---|---|---|---|---|---|---|---|---|
| number | 1 | 1 | 2 | 3 | 6 | 11 | 23 | 47 | 106 | 235 | ... |

What was the purpose of all this? In the 1850s and 1860s the chemical theory of valency came to be explained, and Cayley's motivation for counting trees was to enable him to enumerate those chemical molecules (paraffins, alcohols, etc.) that have a tree structure. A list of Cayley's writings on trees includes both mathematical and chemical papers.

Sylvester was also interested in chemistry. He had been fascinated by the recent use of tree-like pictures to depict molecules, by the celebrated chemist Edward Frankland, and convinced himself that there was a connection between such 'graphic formulae' and certain algebraic ideas from invariant theory. Indeed, in a paper in 1878, he waxed eloquent:[20]

Chemistry has the same quickening and suggestive influence upon the algebraist as a visit to the Royal Academy, or the old masters may be supposed to have on a Browning or a Tennyson. Indeed it seems to me that an exact homology exists between painting and poetry on the one hand and modern chemistry and modern algebra on the other...

Some of Sylvester's drawings from this paper are certainly chemical, while others are more related to algebra.

The motivation for Sylvester's interest arose partly from his appointment in 1876 as the first professor in mathematics at the newly-founded Johns Hopkins University in Baltimore, USA, and he was faced with the problem of giving an inaugural lecture to a mixed audience. Indeed, in the above paper, which appeared in the first issue of the *American Journal of Mathematics* (which he had just founded), he describes in a long sentence how he came across this supposed connection between chemistry and algebra:[21]

Casting about, as I lay in bed one night, to discover some means of conveying an intelligible conception of the objects of modern algebra to a mixed society, mainly composed of physicists, chemists and biologists, interspersed only with a few mathematicians...I was agreeably surprised to find, of a sudden, distinctly pictured on my mental retina a chemico-graphical image...

Two chemical trees.

When Sylvester summarized the results of his lengthy paper in *Nature* in 1878, he described the link between chemical atoms and algebraic expressions called binary quantics, saying:[22]

Every invariant and covariant thus becomes expressible by a graph.

This was the first appearance of the word *graph*, in the sense of graph theory.

Some of Sylvester's chemical work was carried out with William Kingdom Clifford, whose researches at University College, London, were cruelly cut short by his untimely death in 1879. Shortly after, the whole cottage industry of English invariant theory was superseded by the more powerful Continental methods of Paul Gordan and David Hilbert. As we saw in Chapter 15, Cayley and Sylvester tried hard to regain the initiative, but without success.

Before leaving the study of trees, we mention one further tree-counting result. In 1889, Cayley wrote a short note[23] in which he calculated the number of ways of connecting a number of given points to form a tree; for example, if there are 4 points, they can be joined in 16 ways. Referring back to an earlier result of C. W. Borchardt on determinants, Cayley asserted that if there are $n$ points, then the number of such trees is $n^{n-2}$. Unfortunately, his proof was less than adequate: his explanation was restricted to the case $n = 5$, and it is unclear how his method can be extended to other values of $n$. A complete proof had to wait until several years after his death.

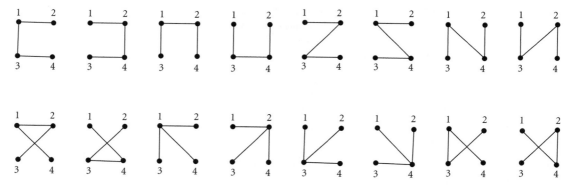

The 16 ways of joining 4 points to form a tree.

## Partitions

Our next topic is *partitions*, or 'divulsions of integers' as Leibniz called them when he introduced them in a letter to Jacob Bernoulli.[24] Such partitions of integers were studied in depth by Leonhard Euler in a chapter of his 1748 *Introductio in Analysin Infinitorum*.

Let $p(n)$ be the number of partitions of $n$—that is, the number of ways of splitting $n$ into smaller natural numbers; for example, $p(4) = 5$, corresponding to the partitions

$$4 = 3 + 1 = 2 + 2 = 2 + 1 + 1 = 1 + 1 + 1 + 1;$$

The order of the summands does not matter—for example, $2 + 1 + 1$ is considered the same as $1 + 2 + 1$. We can draw up a table of values of these partition numbers $p(n)$, for different values of $n$; examples include $p(5) = 7$, $p(10) = 42$, and $p(200) = 3{,}972{,}999{,}029{,}388$. But how can one show that $p(200)$ has this value?

To find the partition numbers Leonhard Euler used an iterative formula. In his 1748 investigations he proved that if $p(x)$ is the generating function for the numbers $p(n)$—that is,

$$p(x) = p(0) + p(1)x + p(2)x^2 + p(3)x^3 + \cdots$$
$$= 1 + x + 2x^2 + 3x^3 + 5x^4 + 7x^5 + \cdots$$

—then

$$p(x) = [(1 - x) \times (1 - x^2) \times (1 - x^3) \times \cdots]^{-1}.$$

A couple of years later Euler obtained his celebrated *pentagonal number formula*:

$$(1 - x) \times (1 - x^2) \times (1 - x^3) \times \cdots$$
$$= 1 - x - x^2 + x^5 + x^7 - x^{12} - x^{15} + \cdots;$$

here the exponents $1, 2, 5, 7, 12, 15, \ldots$ all have the form $\frac{1}{2} k(3k \pm 1)$.

Multiplying these results together gives

$$p(x) \times (1 - x - x^2 + x^5 + x^7 - x^{12} - x^{15} + \cdots) = 1,$$

and on taking the coefficient of $x^n$ and rearranging the equation, Euler deduced that

$$p(n) = p(n - 1) + p(n - 2) - p(n - 5) - p(n - 7)$$
$$+ p(n - 12) + p(n - 15) - \cdots;$$

this yields $p(n)$ by iteration and is still the most efficient way of finding it. Euler himself used his iterative formula to find all the partition numbers up to $p(66) = 2{,}323{,}520$.

In the 1840s and 1850s a number of English mathematicians worked on partitions, using methods of finite differences and largely unaware of Euler's work. In particular, spectacular work was done by Sylvester, who used Cauchy's residue theorem from complex analysis to obtain an expression for the coefficient of $x^n$ in the expansion of an arbitrary rational function.

Another amateur mathematician interested in partitions was Major Percy MacMahon, who had been invalided out of the Army while fighting with the Punjab Frontier Force. An important mathematical figure in the late 19th and early 20th century, he wrote over a hundred papers. From 1894 to 1896 MacMahon was President of the London Mathematical Society, and in his presidential address he described Sylvester's result as 'incomparably the finest contribution

Percy Alexander MacMahon (1854–1929).

that has ever been made to combinatory analysis'.[25] Sylvester himself was equally enthusiastic, asserting that:[26]

partitions constitute the sphere in which analysis lives, moves, and has its being; and no power of language can exaggerate or paint too forcibly the importance of this till-recently almost neglected (but vast, subtle and universally permeating) element of algebraical thought and expression.

MacMahon is now mainly remembered for his contributions to combinatorics, at a time when few others were working in this area. Using Euler's iterative

### TABLE IV*: $p(n)$.

| | | | | |
|---|---|---|---|---|
| 1... 1 | 51... 239943 | 101... 214481126 | 151... 45060624582 |
| 2... 2 | 52... 281589 | 102... 241265379 | 152... 49686288421 |
| 3... 3 | 53... 329931 | 103... 271248950 | 153... 54770336324 |
| 4... 5 | 54... 386155 | 104... 304801365 | 154... 60356673280 |
| 5... 7 | 55... 451276 | 105... 342325709 | 155... 66493182097 |
| 6... 11 | 56... 526823 | 106... 384276336 | 156... 73232243759 |
| 7... 15 | 57... 614154 | 107... 431149389 | 157... 80630964769 |
| 8... 22 | 58... 715220 | 108... 483502844 | 158... 88751778802 |
| 9... 30 | 59... 831820 | 109... 541946240 | 159... 97662728555 |
| 10... 42 | 60... 966467 | 110... 607163746 | 160... 107438159466 |
| 11... 56 | 61... 1121505 | 111... 679903203 | 161... 118159068427 |
| 12... 77 | 62... 1300156 | 112... 761002156 | 162... 129913904637 |
| 13... 101 | 63... 1505499 | 113... 851376628 | 163... 142798995930 |
| 14... 135 | 64... 1741630 | 114... 952050665 | 164... 156919475295 |
| 15... 176 | 65... 2012558 | 115... 1064144451 | 165... 172389800255 |
| 16... 231 | 66... 2323520 | 116... 1188908248 | 166... 189334822579 |
| 17... 297 | 67... 2679689 | 117... 1327710076 | 167... 207890420102 |
| 18... 385 | 68... 3087735 | 118... 1482074143 | 168... 228204732751 |
| 19... 490 | 69... 3554345 | 119... 1653668665 | 169... 250438925115 |
| 20... 627 | 70... 4087968 | 120... 1844349560 | 170... 274768617130 |
| 21... 792 | 71... 4697205 | 121... 2056148051 | 171... 301384802048 |
| 22... 1002 | 72... 5392783 | 122... 2291320912 | 172... 330495499613 |
| 23... 1255 | 73... 6185689 | 123... 2552338241 | 173... 362326859895 |
| 24... 1575 | 74... 7089500 | 124... 2841940500 | 174... 397125074750 |
| 25... 1958 | 75... 8118264 | 125... 3163127352 | 175... 435157697830 |
| 26... 2436 | 76... 9289091 | 126... 3519222692 | 176... 476715857290 |
| 27... 3010 | 77... 10619863 | 127... 3913864295 | 177... 522115831195 |
| 28... 3718 | 78... 12132164 | 128... 4351078600 | 178... 571701605655 |
| 29... 4565 | 79... 13848650 | 129... 4835271870 | 179... 625846753120 |
| 30... 5604 | 80... 15796476 | 130... 5371315400 | 180... 684957390936 |
| 31... 6842 | 81... 18004327 | 131... 5964539504 | 181... 749474411781 |
| 32... 8349 | 82... 20506255 | 132... 6620830889 | 182... 819876908323 |
| 33... 10143 | 83... 23338469 | 133... 7346629512 | 183... 896684817527 |
| 34... 12310 | 84... 26543660 | 134... 8149040695 | 184... 980462880430 |
| 35... 14883 | 85... 30167357 | 135... 9035836076 | 185... 1071823774337 |
| 36... 17977 | 86... 34262962 | 136... 10015581680 | 186... 1171432692373 |
| 37... 21637 | 87... 38887673 | 137... 11097645016 | 187... 1280011042268 |
| 38... 26015 | 88... 44108109 | 138... 12292341831 | 188... 1398341745571 |
| 39... 31185 | 89... 49995925 | 139... 13610949895 | 189... 1527273599625 |
| 40... 37338 | 90... 56634173 | 140... 15065878135 | 190... 1667727404093 |
| 41... 44583 | 91... 64112359 | 141... 16670689208 | 191... 1820701100652 |
| 42... 53174 | 92... 72533807 | 142... 18440293320 | 192... 1987276856363 |
| 43... 63261 | 93... 82010177 | 143... 20390982757 | 193... 2168627105489 |
| 44... 75175 | 94... 92669720 | 144... 22540654445 | 194... 2366022741845 |
| 45... 89134 | 95... 104651419 | 145... 24908858009 | 195... 2580840212973 |
| 46... 105558 | 96... 118114304 | 146... 27517052599 | 196... 2814570987591 |
| 47... 124754 | 97... 133230930 | 147... 30388671978 | 197... 3068829878530 |
| 48... 147273 | 98... 150198136 | 148... 33549419497 | 198... 3345365983698 |
| 49... 173525 | 99... 169229875 | 149... 37027355200 | 199... 3646072432125 |
| 50... 204226 | 100... 190569292 | 150... 40853235313 | 200... 3972999029388 |

MacMahon's table of partition numbers.

formula, he obtained correct values of $p(n)$ for all values of $n$ up to 200; the value of $p(200)$, given earlier, took him a whole month to calculate. His table of partition numbers, shown opposite, appeared in the fundamental Hardy–Ramanujan paper on partitions mentioned below. In 1915, while working as Deputy Warden of the Standards with the Board of Trade, MacMahon wrote a two-volume classic text *Combinatory Analysis* covering many aspects of the subject and including his best-known result, his powerful 'master theorem', which he used to solve a range of counting problems involving permutations and latin squares.

In the 1880s a useful geometrical contribution to the study of partitions was made by N. M. Ferrers of Caius College, Cambridge, and further developed by Sylvester; this approach related 'conjugate partitions' such as $5 + 5 + 4 + 3 + 3$ and $5 + 5 + 5 + 3 + 2$ and enabled one to deduce certain partition results from others.

All of these advances led, ultimately, to a spectacular exact formula for $p(n)$ by Hardy and Ramanujan,[27] one of the most startling results in the whole of mathematics. In 1918, they proved that $p(n)$ is the integer nearest to the product of two complicated expressions $A_q$ and $\phi(q)$, where $\phi(q)$ involves $\pi$, square roots, derivatives, and exponentials, and $A_q$ involves exponential sums, Legendre symbols, and 24th roots of unity!

```
•  •  •  •  •      5
                   +
•  •  •  •  •      5
                   +
•  •  •  •         4
                   +
•  •  •            3
                   +
•  •  •            3
                 ┌────
5 + 5 + 5 + 3 + 2 │ 20
```

Conjugate partitions.

## Map colouring

No discussion of Victorian combinatorics would be complete without a mention of map colouring, and in particular the celebrated *four-colour problem*.

In October 1852 Francis Guthrie, a former student at University College, London, whom we met briefly in Chapter 6, was colouring a map of England, and noticed that he needed only four colours to colour it so that neighbouring counties were coloured differently. *Is this true for all maps?*, he wondered. His brother Frederick approached his teacher, Augustus De Morgan, Professor of Mathematics at University College. De Morgan immediately became fascinated with the problem, and wrote to Sir William Rowan Hamilton, saying:[28]

A student of mine asked me today to give him a reason for a fact which I did not know was a fact—and do not yet. He says that if a figure be anyhow divided and the compartments differently coloured so that figures with any portion of common boundary *line* are differently coloured—four colours may be wanted but not more…Query: cannot a necessity for five or more be invented.

Although Hamilton was not interested in De Morgan's 'quaternion of colours', De Morgan was sufficiently intrigued by the problem to write about it to various friends, including the philosopher William Whewell, Master of Trinity College, Cambridge. In fact, it was in the middle of an unsigned book review (actually by

Francis Guthrie (1831–99).

De Morgan) of Whewell's *Philosophy of Discovery* that the problem first appeared in print, in characteristically bizarre language:[29]

…Now, it must have been always known to map-colourers that four different colours are enough. Let the counties come cranking in, as Hotspur says, with as many and as odd convolutions as the designer chooses to give them; let them go in and out and roundabout in such a manner that it would be quite absurd in the Queen's writ to tell the sheriff that A. B. could run up and down in his bailiwick: still, four colours will be enough to make all requisite distinction…

On 13 June 1878, a few years after De Morgan's death in 1871, Arthur Cayley revived the question at a meeting of the London Mathematical Society,[30] and in the following year, a 'proof' that four colours are sufficient was given by Alfred Kempe,[31] a London barrister who had formerly studied mathematics with Cayley at Cambridge, and who had attended the meeting (see Box 17.2); he later became treasurer of the Royal Society. Kempe's paper was commissioned for the *American Journal of Mathematics* by its editor-in-chief, James Joseph Sylvester.

Alfred Bray Kempe (1849–1922).

---

## Box 17.2: Kempe's 'proof' of the four-colour theorem

Kempe's proof was essentially as follows.

We assume that the result is false, so that there are maps that need more than four colours. Of these, we take such a map *M* with the *smallest* possible number of colours—so this map *M* cannot be four-coloured, but any map with fewer countries can be. Using a result called *Euler's polyhedron formula*, one can easily show that every map has a country with at most five neighbours, such as a triangle, square, or pentagon. We now look at these in turn.

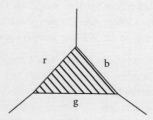

If the map has a triangle, we shrink it to a point. There are then fewer countries, so (by our assumption) we can colour the resulting map with four colours. Now put the triangle back again—it is surrounded by three countries, which take up only three colours, so there is a spare colour for the triangle. This gives the required contradiction.

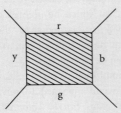

If, next, there is a square, we follow Kempe and look at just two of the colours — say, the red–green part of the map. Are the reds and greens above the square connected to those reds and greens below it? There are two cases: either they are not connected, or they are connected.

- In the first case, we can interchange the reds and greens above the square without affecting those below the square. The two countries above and below the square are now both green, and so can colour the square red.

- In the second case, interchanging the colours leaves us no better off than before—but now we look at the blues and yellows on the left and right of the square. Those on the right are separated from those on the left by the red–green chain, so we can interchange the blues and yellows on the right without affecting those on the left. The square can then be coloured blue.

In each case we can four-colour the resulting map, giving the required contradiction.

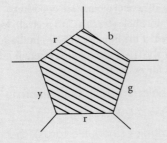

Finally, if there's a pentagon, Kempe claimed that you can do two simultaneous interchanges of colour to get rid of the two reds. This gives just three colours around the pentagon, so the colouring can again be completed, giving the required contradiction. This deals with all cases and completes the proof.

Percy John Heawood
(1861–1955).

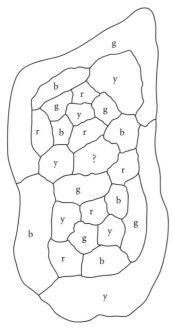

The map that Heawood constructed to illustrate the flaw in Kempe's argument.

Kempe's argument was widely accepted—by Cayley, Sylvester, Tait, and others—and so it was a great surprise, eleven years later, when Percy Heawood of Durham wrote a paper[32] in which he pointed out a fundamental error in Kempe's proof. Heawood had first learned about the problem in 1880, while an undergraduate in Oxford, and he became hooked, writing papers on it over many years; the last appeared when he was aged 90.

Heawood's paper was a bombshell, as no-one (least of all, Kempe) was able to patch up the mistake. Heawood managed to salvage enough from Kempe's argument to show that *all maps can be coloured with five colours*—itself a worthwhile result—and also tried to extend the result to other mathematical surfaces, as we see below.

Where did Kempe go wrong? Recall that when dealing with the pentagon, Kempe tried to carry out two colour interchanges at once. Heawood constructed a map to show why this is not allowable in general: either interchange may be carried out separately, but doing both at the same time can result in two neighbouring countries being re-coloured the same.

The error proved extremely difficult to patch up—in fact, it took a further eighty-six years to do so. The task was finally completed by Kenneth Appel and Wolfgang Haken of the University of Illinois in 1976. Their proof was exceedingly complicated, making extensive use of a computer to test many thousands of cases, but the basic underlying ideas can be traced back to those of Kempe.

To conclude this chapter, we return briefly to Heawood's 1890 paper, and to his attempted extension of the four-colour problem to other surfaces. Using a map projection, we can see that colouring maps on the plane is the same as colouring maps on a sphere (such as a globe)—but what about colouring maps on other mathematical surfaces, such as a torus? Here the magic number turns out to be seven: using an appropriate version of Euler's polyhedron formula, we can easily prove that seven colours are always enough, and Heawood constructed a torus map that actually needs all seven colours (see opposite).

If we now introduce more holes (such as in a pretzel), we need more colours. In fact, as Heawood showed, if the number of holes in the surface is $g$, then we can colour any map with $[\frac{7}{2} + \frac{1}{2}\sqrt{(1 + 48g)}]$ colours, where $[x]$ denotes the integer part of $x$ (so $[3] = [\pi] = 3$); for example,

- if there is one hole (a torus), then any map can be coloured with $[\frac{7}{2} + \frac{1}{2}\sqrt{49}] = 7$ colours;
- if there are two holes (a double-torus), then any map can be coloured with $[\frac{7}{2} + \frac{1}{2}\sqrt{97}] = [8.424\ldots] = 8$ colours.

What Heawood failed to show is that there are always maps that need this number of colours—and this took a further seventy-eight years to prove. Proving the *Heawood conjecture*, as it came to be known, turned out to involve twelve completely separate cases, and was finally settled in 1968, mainly by the German mathematician Gerhard Ringel and the American Ted Youngs.

# Conclusion

It might be supposed that the above combinatorial advances in Britain over the second half of the 19th century would set the stage for further developments in the ensuing years, but this was not to be. Cayley, Sylvester, and Kirkman all died in the 1890s, while Kempe wrote no more on the subject, and publications such as *The Lady's and Gentleman's Diary* and *The Educational Times* ceased publication. The strong connections between combinatorics and recreational mathematics began to disappear as research mathematicians were appointed in greater numbers at British universities, especially after the First World War, and professional mathematicians gradually took over from the amateurs.

Meanwhile, the major research in combinatorics moved to Europe, with important contributions to graph theory by Julius Petersen in Denmark and Dénes König in Hungary, and major developments on enumeration by the Hungarian George Pólya. Across the Atlantic, an American school was similarly beginning to develop, leading to major combinatorial advances spearheaded by such influential figures as G. D. Birkhoff, Oswald Veblen, and Hassler Whitney.

But back in Britain, with the notable exceptions of MacMahon's *Combinatory Analysis* (1915–16) and Hardy and Ramanujan's (1918) paper on partitions, British mathematicians showed little interest in combinatorial topics for several decades as their interests moved towards analysis, number theory, and other areas.

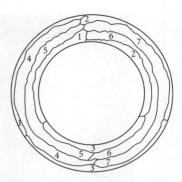

Heawood's drawing of a seven-coloured map on a torus.

Karen Parshall's *James Joseph Sylvester: Jewish Mathematician in a Victorian World*.

## CHAPTER 18

# Overstating their case? Reflections on British pure mathematics in the 19th century

JEREMY GRAY

In this chapter I argue that the British pure mathematicians of the 19th century have been overrated, to the detriment of historical writing on the subject. To do so, I first give an overview of their work that locates them in a more European context, and then turn to more overtly historiographic questions.

The literature in the history of mathematics, as in any branch of history, is partisan. Even when no claims are made for the importance of some hitherto neglected person, group, or activity, there is a natural tendency to boost the qualities of those one is writing about or, if it is more appropriate, to denigrate them. The same is true at the level of historiography: certain approaches or historians meet with approval, others with criticism. This chapter is no different, except that its target is perhaps unfamiliar. It may even be unwelcome in the generally warm-hearted, if not simply uncritical, spirit that pervades much of the history of mathematics. I shall argue that the British pure mathematicians of the 19th century have been overrated, to the detriment of historical writing on the subject.

Now that we have two thorough accounts of the lives and works of both Cayley and Sylvester, by Tony Crilly and Karen Parshall, respectively,[1] it may be that a more accurate period in the study of pure mathematics in Victorian Britain is about to begin. But until that happens, there is a perception among mathematicians today that is at odds with the real achievements of these men and their lesser contemporaries, a legacy of overstatement and undue simplification that must be confronted, and indeed too high an opinion of these Victorians in the opinions of the men themselves (although Cayley may perhaps be exempted from that charge), bound up with an insularity to which the British are ever prone.

The first part of this chapter looks at the work of these Victorian pure mathematicians with a view to putting them in a more European context. The second part turns to more overtly historiographic questions.

## Part I: Victorian pure mathematicians

## The first generation

It is the very earliest generation of Victorians who have the most to answer for. The men who did most to advance mathematics and science in England in the 1830s, 1840s, and 1850s were Charles Babbage, Augustus De Morgan, and William Whewell, and they make a fascinating trio.

There is a good case to be made for Babbage as a man of vision and restless energy that led him to try to reform the whole organization of scientific life in Britain, to attack the Royal Society for becoming self-serving and indifferent to science, to improve the collection and use of statistics, and most notably to design and attempt to build a successful programmable computer. This case has often been made.

What is less often said, and at best grudgingly allowed, is that there is at least as strong a case to be made that Babbage was a self-important and spectacular under-achiever with an unduly high opinion of himself. Although we live in an era when computer projects generate vast cost over-runs and generate much valid criticism, we do not seem to take this hard-won wisdom back to the past. The British Treasury has a well-earned reputation for meanness, and no one would claim that its economic advisors were (or are) soft-hearted, or even that they had a shrewd sense of the nature and value of public investment. But their willingness to pay out quite a considerable sum of money (£17,000) and then to pay no more is not indicative of their folly. Not only was the analytical engine not finished, it probably never could have been, not just because it stretched engineering to the limit, but because its designer made the fatal mistake of always seeking to improve it before ever getting it to work. All this is well documented in studies by Anthony Hyman and Doron Swade (see also Chapter 10).[2]

More significantly, because his influence in Cambridge was enormous, there stands before us William Whewell, Master of Trinity College from 1841 until his death in 1866. Whatever his merits, and whether his over-bearing, even tyranni-cal, manner was what Trinity needed, he was quite blind to mathematics as a research subject: in his opinion it could only ever be an aid to science. In his *Of a*

*Liberal Education in General, and with Particular Reference to the Leading Studies of the University of Cambridge*, Whewell distinguished between 'permanent' and 'progressive' studies:[3]

The Studies by which the Intellectual Education of young men is carried on, include two kinds; which, with reference to their subjects, we may describe as Permanent, and Progressive Studies. To the former class belong those portions of knowledge which have long taken their permanent shape;—ancient languages and their literature, and long-established demonstrative sciences. To the latter class belong the results of the mental activity of our own times: the literature of our own age, and the sciences in which men are making progress from day to day.

Permanent mathematics included Euclidean geometry and Newtonian mathematics as presented in Newton's *Principia*. The heavily analytical areas of mathematics were progressive, producing results quickly but with diminished understanding of how the results were obtained. But since, in Whewell's view, the goal of Cambridge mathematical teaching was 'not to produce a school of eminent mathematicians, but to contribute to a Liberal Education of the highest kind', the fundamental effect of his views was to subordinate mathematics to general education and to general science. Both were sure ways to sterilize the subject.

In Whewell's day, mathematics was at a strikingly low ebb in Cambridge. It is hard for a subject to grow from a weak position, but it is worth noting the time that it took to do so in Britain. In comparison, Norway produced Abel and Lie, both better mathematicians than anyone in Britain in the whole of the 19th century. There can be many reasons for that disparity, but I shall argue that British insularity and a disdain for mathematical analysis are significant factors.

Among those responsible for the climb back, if that is not overstating the case, was Augustus De Morgan, the founding father of the London Mathematical Society. De Morgan is not a major name in European mathematics and science. He was a respectable minor figure, but another with an unduly high opinion of himself and his work. At the very least, anyone who could applaud his own rather superficial approach to divergent series, while failing adequately to respond to Cauchy's profound, extensive, and clear theory of convergent series, has the matter backwards—a point we shall return to.

Indeed, what is striking about so many British mathematicians is their belief that they were the equals of their Continental peers, when no such comparison can be entertained. This was not the opinion of Thomas Archer Hirst, who made a serious attempt to learn geometry from the Europeans and was respected by them, nor of Olaus Henrici, who was educated in Germany and made the journey in the opposite direction, teaching mathematics in London without ever falling for the national myth.

That said, self-belief is a complicated matter. It is likely that few (if any) do really first-rate work without a high degree of self-belief. Something must carry them through the lonely months and years when they have little to show for their ideas, and a faith in their own capabilities and originality is surely essential. It becomes unfortunate, and even destructive, when it is not matched by a lively sense of what other people are doing, whether they are seen as rivals or as colleagues in some great scientific quest.

Throughout the Victorian period the British seem to have been ambiguous about the merits of their universities and the ethos they sustained, at times prone to self-doubt, at others surely buoyed up by Britain's growing industrial wealth and the extent and riches of its Empire. As we shall see, they retained some of Whewell's blindness to mathematics and adhered to a crippling utilitarianism.

## The second generation

The top wranglers of the late 1830s and early 1840s have some claim to comprising the most illustrious of them all. Sylvester, the 2nd wrangler in 1837, was the first of them, but was unable to obtain a degree from Cambridge or a fellowship there because he was a Jew and could not subscribe to the Thirty-Nine Articles of the Church of England, as Cambridge then required. He was followed by Stokes (senior wrangler in 1841), Cayley (senior wrangler in 1842), and William Thomson (2nd wrangler in 1845).

Cayley's reputation is certainly high, but is oddly hard to capture. Many would probably call him the greatest English mathematician apart from Newton, and if asked for evidence would point to his great productivity—thirteen volumes of his *Collected Mathematical Papers*—and the large number of mathematical objects and discoveries that carry his name: these are handily listed in Crilly[4] and include the Cayley–Bacharach theorem, the Cayley–Hamilton theorem, the Cayley–Klein metric, Cayley lines, Cayley numbers, Cayley–Plücker coordinates, Cayley tables, Cayley trees, Cayley graphs, Cayley's absolute, Cayley's theorems on the transformation of elliptic functions and on permutation groups, and Cayley's $\Omega$-process. This list does scant justice to Cayley's work on invariant theory, which occupied him on and off for most of his working life, and nor does it reflect his selfless efforts reporting on Continental work on various domains of mathematics in order to make it accessible to an English audience. (Cayley was fluent in several languages.)

Crilly's own assessment of Cayley's achievements is much subtler, and we return to it later. But the list is ominous. The Cayley–Bacharach theorem indicates the problem. The subject is the theory of algebraic curves, and the theorem says that:

if $n \geq l$, $n \geq m$, and $n < l + m$, then a curve of degree $n$ through

$$N = \{lm - (l + m - n - 1)(l + m - n - 2)\}/2$$

points, common to a curve of degree $l$ and a curve of degree $m$, must pass through the remainder.

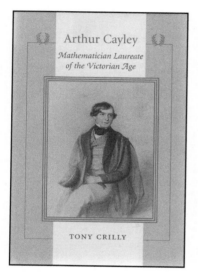

Tony Crilly's *Arthur Cayley: Mathematician Laureate of the Victorian Age.*

It is a good theorem, but the trouble is that Cayley's contribution to it is wrong.[5] In a paper of 1886, Bacharach drew attention to the falsity of this claim and showed that the correct statement is:

any curve of degree $r$ through $mn - N$ points of intersection of a curve of degree $m$ and a curve of degree $n$ goes through the remaining $N$ points, provided that, if these $N$ points lie on a curve of degree $m + n - 3$, then an arbitrary curve of degree $r$ through the $mn - N$ points must not simultaneously pass through the remaining $N$ points.[6]

The Cayley–Hamilton theorem says that *every matrix satisfies its own characteristic equation* (see Chapter 15). Cayley discovered it for $2 \times 2$ matrices, satisfied himself that it is true for $3 \times 3$ matrices, and then published it. As Leopold Kronecker knew very well, to prove the theorem in general required new conceptual tools, and higher standards of rigour.[7]

The Cayley–Klein metric and Cayley's absolute conic come up in projective geometry. Cayley's idea was that if one introduced a purely imaginary conic into projective space, one could then capture Euclidean geometry as a sub-geometry (in some sense) of projective geometry. This is true, but he did not see, as Klein did, that if one introduces a non-degenerate real conic into projective space, then one gets non-Euclidean geometry in the same way, thus clarifying the status of non-Euclidean geometry and also helping to explain how geometries and their transformations can be organized in an insightful way. Cayley never understood how coordinates arise in projective geometry, and he remained baffled by non-Euclidean geometry all his life.

The Cayley lines are an ingenious and non-trivial contribution to the further study of Pascal's hexagram, a fascinating configuration in projective geometry. The Cayley numbers, also known as *octonions*, are another laudable discovery, as are the Cayley–Plücker coordinates for lines in projective three-dimensional space. A Cayley table is a multiplication table for groups, but one can doubt whether any good group theory is done by studying them. Cayley trees are graphs, and although they are a valuable topic in modern graph theory it can be doubted whether Cayley's use of them to analyse chemical structures was much more than a mathematician's exercise (see Chapter 17). Cayley's theorem on the transformation of elliptic functions is but one theorem among many in that rich and technically complicated subject, and Cayley was never deterred by complications. His theorem that every abstract group is isomorphic to a permutation group is helpful, if elementary—but then the whole subject of group theory was in its infancy in Cayley's day—and Cayley's $\Omega$-process is a workhorse in invariant theory.

This snapshot or report card is no fairer an assessment of a life's work than any other brief account, and while it highlights several real achievements, it also throws up several problems. Let me stress that these achievements are real: many a mathematician would be glad to have them to his or her name, and the equivalent work at any time since would give its author a good reputation. There are more: Cayley's work on scrolls and the geometry of algebraic surfaces is worth remembering. His discovery of the twenty-seven lines on a cubic surface

(Cayley found that there were lines, but Salmon enumerated them[8]) proved to be the start of an inexhaustible mine of investigation. Certainly, Cayley's name is attached to one or two ideas that figure more prominently in the undergraduate syllabus than in advanced mathematics, and there is nothing unusual or wrong about that. But the lack of rigour, the failure to push examples to see what they can yield, and the failure to separate the general case from its exceptions, are all weaknesses in Cayley's work. Most destructively, they are weaknesses in his most significant domain and the subject of his life-long collaboration with Sylvester—the theory of invariants.

Before turning to that, we shall find it helpful to consider Sylvester and his career in somewhat the same way, while admitting that no one makes quite the same claims for Sylvester's brilliance that they do for Cayley. Sylvester is the less luminous part of the double star of Cayley–Sylvester.

Sylvester was a Jew who had to make his own way in the Anglican world of Victorian England and, as Parshall's biography of him shows very clearly, he flourished much better at the newly founded Johns Hopkins University in Baltimore than he ever did in his native land. Like Cayley, who for a time had a successful career as a barrister, Sylvester too led a non-mathematical life and became a well-regarded actuary, prominent in the Institute of Actuaries, which he helped to establish in 1848. But it is his mathematics and the position that he occupies in the history of mathematics that are our concerns. His achievements are more tightly focused on the theory of invariants than Cayley's were. Outside that field he is remembered for his law of inertia in the study of matrices, for his work on the theory of partitions, and for a number of smaller discoveries.

We must now turn to the *theory of invariants*, the main accomplishment of Cayley and Sylvester and their principal influence on pure mathematics in Britain in their lifetimes; here it is disturbingly hard to find out what they did, and Parshall's biography of Sylvester is a significant help (see also Chapter 15). Everyone who writes about the subject historically starts with the familiar observation that, if a non-degenerate conic section is defined by

$$ax^2 + bxy + cy^2 + fx + gy + h = 0,$$

then the sign of $b^2 - 4ac$ tells you whether the conic is an ellipse, a parabola, or a hyperbola. The crucial observation is that under linear changes of the variables the quantity $b^2 - 4ac$ is multiplied by a square (the square of the determinant of the matrix defining the change of variables). It seems to have been George Boole who drew attention to this as a clue to a subject for further investigation:

- Are there analogous quantities for quadratic expressions in several variables?

- Are there analogous quantities for polynomial expressions of degree higher than 2 in two or more variables?

The quantity $b^2 - 4ac$ is called an *invariant* because of the way it transforms, and because it is not altered at all by transformations of determinant 1. For more complicated expressions than those of the second degree, it turns out that there are important expressions that transform in interesting ways, but which also involve the variables; these are called *covariants* of the original expression. For example, a curve $C$ of degree 3 or more has inflection points, and these are not altered by a linear transformation. A curve called the *Hessian* of $C$ meets the curve $C$ in its inflection points, and the equation of this Hessian curve is a covariant of the equation defining $C$. This observation illustrates several things: that there is a story that starts with $b^2 - 4ac$ (Boole was right!), that the new invariants and covariants have geometric significance, that the subject becomes very complicated very quickly, and that other mathematicians were already involved.

These observations generate their own sets of implications. We may start with the complexity. Cayley and Sylvester were happy to embark on vast series of calculations in which Cayley's $\Omega$-process was a vital tool: the amount of work involved is daunting even to contemplate. The metaphor to which both Crilly and Parshall recur, that of the Victorian explorer (perhaps a botanist) who journeys through new territories and brings back exotic discoveries, is apt in several ways, not least in conveying the difficult and time-consuming nature of the activity. Both historians use the metaphor mainly to convey their sense of the achievements and limitations of the work of Cayley and Sylvester, but it is also helpful in enabling them to finesse the problem of describing in detail what those men did—and that is the problem: the botany metaphor breaks down in several ways.

It is not merely that few find the expression for a covariant of an equation of degree 6 exciting, unlike the strange plants, animals, and insects brought back by the explorer. It is also that few *mathematicians* find these covariants interesting. The historian of mathematics is not the only one who finds the complexity of the work of Cayley and Sylvester difficult to do justice to: the mathematician is another. Better tools are needed than a capacity for accurate hard work, and reassurance is also needed that the quest is not hopeless. One might, for example, wish to know whether there are only finitely many invariants and covariants of a polynomial of a given degree. Since this collection is closed under addition and multiplication it forms a 'ring', in modern terminology, so what is at stake is whether the covariants and invariants of a polynomial expression of a given degree are generated (as a ring) by a finite set of them, and if so, how this set can be found. What, for example, is the number of its generators?

Successful work on these problems is largely associated with other names. A better tool for the study of the subject is the symbolic method of Aronhold and Clebsch in Germany in the 1850s and 1860s. The finiteness result for forms of any degree, but in only two variables, is due to Paul Gordan in 1868; its extension to any number of variables was the work of Hilbert in the 1880s. The discovery that the generators of this ring themselves satisfy certain polynomial relations is, happily, the discovery of Cayley and Sylvester and has gone ever since by the name that Sylvester gave it: the theory of *syzygies*.

Aronhold's general method dates from 1849, two years before Sylvester proclaimed a similar thing, and neither Cayley nor Sylvester ever fully caught

Karen Parshall's *James Joseph Sylvester: Life and Work in Letters*.

up with it. Parshall notes that 'the British—and especially Sylvester—continued to develop their techniques in isolation from the German approach, and vice versa'.[9] Gordan's finiteness theorem, which Cayley had not expected because he believed that the generating set of covariants for a binary quintic would be infinite, further vindicated the methods that the Germans were using. But it was poorly adapted to the computation of minimum generating sets, and so Cayley and Sylvester continued to look for a proof of Gordan's theorem using their approach.

There is nothing wrong with that, except that they failed. And while their laborious discovery of numerous syzygies was well regarded by their European contemporaries, their erroneous conjectures were also noted. As Crilly described Hermite's discovery of the skew invariant of degree 18 of a binary quintic: 'Its appearance surprised Cayley and ran counter to his previously held belief that the degrees of invariants in the case of the quintic were multiples of four'.[10] Crilly also noted that Salmon then computed this expression explicitly: the formula for it ran to five quarto pages!

But it is not Cayley's mistaken hunch that should be criticized. It is the general insufficiency of method that brought Cayley and Sylvester down. It was not sharp enough, it was not deep enough. It rested on too naïve a desire to generalize from low-dimensional cases. This worked for the Cayley–Hamilton theorem, but not for the more substantial subject of invariant theory.

At the University of Oxford the situation was famously dire: there was virtually no first-class mathematician there, other than Henry Smith (see Chapter 2). Joan Richards has suggested that this is because Oxford invested heavily in logic rather than mathematics, and that in turn may have been because of the profound battles fought in Oxford over religion.[11] Whatever the force of this observation, Smith was a significant mathematician, the author of a six-part *Report on the Theory of Numbers* and of original work on the subject—one essay famously (though posthumously) won Smith a prize of the Paris Academy of Sciences. He also discovered the Cantor set before Cantor, and modified its construction to produce a set that was still nowhere dense but, unlike the Cantor set, not of measure zero (to lapse into anachronism). He also sorted out one of Cayley's murkier contributions to the analysis of complicated singularities of plane algebraic curves. Smith's reputation in his day and since is a topic worth returning to.

The remaining pure mathematician of note in mid-Victorian Britain was William Kingdon Clifford. With Clifford, things might have been different, but he died in 1879 at the age of 33. The observation that mathematicians do their best work when young is just a cliché, but 33 is a respectable age for a mathematician and by that stage Clifford had done very little. His best work was the invention of biquaternions, the discovery of what became the Clifford–Klein space forms (the flat torus, in his case), and some remarks concerning Riemannian geometry that he never followed up.[12] These are good pieces of work, but, as I shall argue below, overestimating their importance blinds us to the real historical situation that Clifford's case actually exemplifies.

# The third generation

Matters began to change in the 1880s, with the arrival of Charlotte Scott, Andrew Forsyth, Ernest Hobson, James Harkness, Frank Morley, Henry Baker, and William Burnside. Of these, Scott, Harkness, and Morley left for jobs in America, while Burnside worked in isolation at the Royal Naval College in Greenwich. This left Hobson, Forsyth, and Baker to tackle Cambridge. Hobson produced the first substantial book in English on modern real analysis (his 1907 *Theory of Functions of a Real Variable*), but Hardy records that Hobson went to his grave feeling that he had failed to get the British to catch up with their Continental peers, and indeed that was left as a task for Hardy and Littlewood.[13] Forsyth wrote an ambitious book on complex analysis, and in its day and in this country it was a great success. In his obituary of Forsyth, E. T. Whittaker said that it 'had a greater influence on British mathematics than any work since Newton's *Principia*' (see Chapter 13).[14]

But this is what the German-trained American mathematician W. F. Osgood had to say about it:[15]

But we cannot stop with pointing to the loose form in which theorems are often stated and proofs given; it only too often happens that the ideas on which the proofs rest are lacking in rigour, or that important matters are overlooked.

A detailed summary of several pages followed, finding much to criticize and some to praise, before he concluded:

It will be seen . . . that the book is not one that can safely be put into the hands of the immature student for a first introduction to the study of functions. But the student who is already familiar with its elements and who has acquired some degree of critical power will find its pages incentive to valuable work in this wide field.

In the end it didn't matter, because Forsyth never made the transition from Cayley-style symbol crunching (at which he excelled) to the conceptual style of thinking required in 20th-century mathematics. Moreover, he was driven out of Cambridge by a moralizing campaign of truly Edwardian hypocrisy, and never found his way again in mathematics, even though he eventually became a professor at Imperial College, London.

Henry Frederick Baker was one of four men who were bracketed senior wrangler in 1887, and he became a fellow of St John's College in the next year. His research career began in the theory of invariants, where he mastered the Clebsch–Aronhold symbolism; he went on to study with Klein in Göttingen, and his first book, *Abelian Functions*, was published in 1897. He presented this topic as the best guide 'to the analytical developments of pure mathematics during the last seventy years' and, very much under the influence of Klein, based his approach on the methods of Riemann, while also presenting the arithmetic theories of Kronecker and of Dedekind and Weber and the more geometrical ideas of Brill, Noether, Clebsch, and Gordan. The geometry of algebraic surfaces was the subject of his next book, *Multiply Periodic Functions* (1907). Baker went on to be Cayley's true successor at Cambridge, a geometer of international distinction, if not exactly of the highest rank.

The greatest success of the whole bunch turned out, perhaps to his own surprise, to be Burnside, who single-handedly brought group theory to Britain and did lasting work of international calibre. The reader is referred to the two-volume edition, *The Collected Papers of William Burnside*,[16] and to its introductory essays, for this interesting (but by no means singular) case of first-rate original mathematics being done in isolation from the principal structures for the support of advanced work in the subject.

Something should also be said about the situation in Scotland and Ireland, both then part of Britain (see also Chapters 4 and 5). Scotland had long had a number of universities and an independent mathematical tradition that could boast of Maclaurin, Simson, and Stirling in the 18th century. In the 19th century, Glasgow rose to prominence as a thriving centre of industry and technology, and its university's physics department was dominated for fifty years by William Thomson. Peter Guthrie Tait, his co-author for the *Treatise of Natural Philosophy* (1867), was the senior wrangler in 1852 and became professor of natural philosophy at Edinburgh in 1860, a post he held for almost forty years.

Despite his addiction to quaternions, Tait did substantial work in physics, independently of his long collaboration with Thomson, and he also worked on numerous mathematical topics. As we saw in Chapter 8, he produced a prodigious classification of knots as part of an idea that he and Thomson entertained that atoms might be knotted pieces of the aether. Despite their eminence, and perhaps because of the contrast between the well-developed mathematical skills of the high Cambridge wranglers and the generally much lower level of teaching in Scotland, neither Tait nor Thomson succeeded in breaking the position that Cambridge held as the centre of British mathematical life. Indeed, after noting that Thomson was the only Glasgow graduate of the entire 19th century to become a major physicist, David Wilson[17] goes as far as to say that Thomson's name contributed to the reputation of the mathematical tripos at Cambridge and helped to draw students there and away from Glasgow.

In the opening years of the 19th century, Trinity College in Dublin did better than the insular English at keeping abreast of Continental developments in mathematics, but as the century progressed the situation seems to have got worse. Where once Dublin had Hamilton and MacCullagh, it declined to Salmon, a splendid textbook writer until he turned permanently to theology. In 1849 Cork acquired Boole, and he did almost all of the work for which he became famous in that charming backwater, but the situation in Ireland was not propitious for the advance of mathematics.

That leaves us with the London Mathematical Society. Here I side with the views of John Heard[18] rather more than those of Rice and Wilson,[19] since I do not think that the Society was particularly energetic, international, or successful in the period. Nowhere, it seems, was there an institution in Britain with the vision, or the capacity, to promote the growth of mathematics.

# Analysis

To start to get a proper perspective on pure mathematics in Victorian Britain, we need to look at what was happening in Continental Europe, and in particular at work in mathematical analysis.

Mathematical analysis was undoubtedly the largest and most important branch of mathematics in the 19th century. Real analysis was reformulated entirely by Cauchy, advanced by Dirichlet, systematically extended by numerous subsequent writers, and invigorated by the spirit of Fourier and the subtle theory of Fourier series. Complex analysis was one of the triumphant creations of the 19th century. It is a stream with two sources: Abel, Jacobi, and the whole theory of elliptic functions was one, and the other was fed from the springs labelled Cauchy, Riemann, and Weierstrass. It was the core subject at Berlin, the central university of the time, but was equally important at the École Polytechnique in Paris. There is an intimate connection, emphasized by Riemann, between complex function theory and harmonic function theory, whence there are applications in physics. There are also strong connections to algebraic geometry and to number theory. Put simply, but without exaggeration, in the Victorian period analysis was by far the major domain of mathematics.

The British contribution to all of this exactly demonstrates their marginal status throughout the century. The remarkable work of George Green, which languished until it was picked up by William Thomson (later Lord Kelvin), was put to use by Thomson himself, Stokes, Maxwell, and Smith. But British mathematicians, by not becoming analysts in any significant way, marginalized themselves.

The same could be said of other branches of mathematics, especially number theory, where Smith was the only strong British mathematician, and group theory before Burnside; but analysis is the crucial example because of its central and fundamental role in 19th-century mathematics. The failure to contribute to its profound developments in the period not only indicates the failings of British pure mathematics, but indicates a failure to appreciate the true nature of the subject. The achievements of Cayley and Sylvester, real and original though they undoubtedly were, did not enable mathematicians of high calibre to follow them into the heartland of mathematics.

## Part II: Historiographic issues

If much scholarship has overstated the merits of Cayley and Sylvester, it is worth asking whether it has misled us in other ways. Are there better ways to think about the British and mathematics in the Victorian era?

Inevitably and rightly, pursuing the history of mathematics is to some extent a personal journey. Novice historians probably start with some of the received opinions of the tribe—famous names, and some associations of these with ideas that have survived: Cayley and matrices, Cayley and group theory, Cayley, Sylvester, and invariant theory, and Clifford and Riemannian geometry—that sort of thing. On the way, they learn more about these people: about them as mathematicians, about their mathematics, about the mathematics of their time, about other mathematicians (some of whom they may not have heard of), about other branches of mathematics, not all of which they may have studied (or really understand, or even like). But it is hard to shake off those first associations: Cayley and Sylvester are big names, and Clifford would have been if he had lived long enough. De Morgan one greets as an old friend that one has rather lost touch with. Henry Smith is more of a distant relative, Burnside is a singularity, and Baker and Forsyth are unknowns inhabiting the interim between Cayley and Sylvester in the mid-19th century and Hardy and Little-wood in the early 20th.

If the novice historian has things to unlearn, as well as many things to learn, the seasoned practitioners have currents of their own to navigate. Three instincts inhabit the minds of reputable historians: to discover what happened and tell us what it was, to discover that we have had it all wrong and to put us right, and to confirm that we had it more or less right all along. The last motive is not often stressed by historians, eager to show that they are independent spirits and discoverers just as scientists are, but it is there just the same and it comes out when particular exaggerations gain sweeping currency.

In addition, there are methodological questions that animate historians. They become concerned that present views about a topic might have become imposed illegitimately on the past, be they intellectual, sociological, value judgements, or pieces of psychology. The professional mathematician writing history tends to find it easier to see the old mathematics through present-day lenses and courts anachronism, while being not greatly interested in contextual matters such as the structure of universities and the academic profession during the historical periods under discussion.[20] Opinions about what is important in mathematics now can be taken back uncritically to the past,[21] teaching and textbook writing may be ignored or discussed at length, with motivations ascribed or misrepresented.

To descend from the lofty heights of reflections on the nature of history (or, if you prefer, to rise from its dusty plains) to the particular question of how we might think about the history of mathematics in Victorian Britain, we may raise a number of specific issues. They overlap, so it may be helpful to list them separately first:

- the issue of pure mathematics versus applied mathematics, or of pure mathematics as the 'real' mathematics;

- the utilitarianism of the Victorians, amounting to philistinism, together with their priorities, successes, and failures;

- the work done so far on the history of mathematics in Victorian Britain, and the work that remains to be done.

# (Pure) mathematics

These issues are at work whenever someone produces a history of mathematics, and they work on us when we read them. What the parade of names that goes: De Morgan, Cayley, Sylvester, Clifford, Forsyth, Baker, Burnside, Hardy, and Littlewood implies is the arrival of pure mathematics in Britain. Compare it with the names: Airy, Green, Stokes, Thomson, Tait, Maxwell, Rayleigh, Lamb, and Love; this is the line of British applied mathematics, or mathematical physics, and perhaps at some point it becomes classical 19th-century physics. So we have a curious overlapping collection of feelings, some of which get articulated as thoughts and which may strike some people as prejudices. For example, there is a tendency, bravely opposed by Ivor Grattan-Guinness, to talk about the history of pure mathematics, but to speak as if it were simply the history of mathematics, unqualified by any adjective. People who do this may be articulating a prior preference for pure over applied mathematics—or they may have a feeling (itself 20th-century in origin) that there is a real distinction here, even if it is only 20th-century in scope. In either case, a certain value judgement follows almost immediately. Historians have to consider what mathematics they are writing about, where they draw the line between mathematics and (say) physics or statistics, and if indeed there is a line to draw.

# British philistinism

It is generally admitted that there was a striking degree to which British culture in the 19th century disliked mathematics, except as an aid to science. There is an astonishing degree of philistinism and rampant utilitarianism about British life in the period, perhaps more in the first half than in the second half. When William Thomson regretted that Cayley was not to be drawn into 'useful' work it was a forceful point, because Thomson was a very distinguished and successful scientist.

It was another matter when John Couch Adams intervened. For not only was Thomson up in bustling Glasgow, but Adams, who had let the discovery of Neptune slip through British fingers, accomplished work of value, but little of distinction, during his thirty-three years as Lowndean professor of astronomy and geometry at Cambridge itself. But British utilitarianism went much deeper than that: installing a research culture of any kind, first in Cambridge and then elsewhere, was a remarkably difficult thing to do.

Consider the battle between Huxley and Sylvester. Huxley had spoken and written in 1869 on the need for renewed education in science, calling for a hands-on pedagogy that stressed the inductive character of scientific knowledge and its use of immediate observation. In the published address, he contrasted this with the state of mathematics, which 'knows nothing of observation, nothing of experiment, nothing of induction, nothing of causation!'. Sylvester, who had been elected President of the British Association for the Advancement of Science in that year and was wondering what to say, took the occasion of their meeting to reply, emphasizing the observational, comparative, inductive, and indeed imaginative and inventive, characteristics of mathematics. This is the speech in which he expressed his wish to see Euclid buried 'deeper than did e'er plummet sound', and where he extolled the charms of algebra; but he could not speak from personal experience of the charms of mathematical analysis, and nor did he.

The clash between Huxley and Sylvester reveals their different experiences of mathematics. For Huxley it was a mere aid to thought and, in the areas in which he worked, not a sophisticated tool. It belonged, like too much of school education, to the lifeless domains of rote learning. For Sylvester, mathematics was a living and inspiring subject of research. It was not an easy climate in which to be a mathematician, if Huxley could so confidently and gratuitously dismiss it as he did. The historian of mathematics is inexorably led from the mathematicians to their contexts, the expectations placed upon them, their opportunities and their obstacles, large and small.

Britain in the 19th century was the world's major industrial and military nation. What were, or should have been, the intellectual, cultural, and scientific priorities of the world's richest nation? France arranged matters very differently, and so did the states of what became the unified Germany, perhaps because they had the trauma of defeat at the hands of Napoleon. We can certainly argue today that Germany overtook Britain because it got right the relationship of mathematics and science to industry and technology, but in the 1860s we could not have argued, with much prospect of convincing one's audience, that Britain was doing it wrong. The British may have proceeded in a particularly utilitarian way, but it brought results that had notable successes throughout the century. But, like the fat man at the sweet counter, Britain learned as little as possible from every setback until it was too late.

I suggest that there is a particular spin on the question of pure mathematics in an affluent country—indeed, a country that then led the world—that has resonances for today. After all, although one can imagine a relatively poor country trying to force its students (who are expensive to produce) into useful work, it is easier to imagine an affluent country tolerating a few eccentrics. England was a curious mixture of the affluent and the penny-pinching in this period, and it often relied on the good work of the privately wealthy (see Chapter 9).

But I also think that it would be interesting to look at pure mathematics in Britain in the 19th century in the way that one looks at the arrival of mathematics in countries not known for their mathematical prowess. The game of catch-up with the leading nations is always interesting to watch. In Japan, Portugal, and Poland a few decades later, the pure mathematicians got going first—but not in Britain. In the British case it was complicated by the fact that Britain was the world's most powerful nation. Probably, the British were lulled by this into a falsely high estimate of their own worth, and the European interest in what those from Britain were doing derived partly from the fact that they were British. Still, it is interesting to observe that when American mathematicians started playing catch-up, they did not bother much with British mathematics but headed straight for the research-orientated institutions of Germany. The research university was a creation of the 19th century, and it prospered much better in the soil of France and Germany than it did in Britain, which came late to the idea.

## Historians (plus and minus)

Looking at what has been done by historians on the mathematics and mathematicians of the Victorian period, we get a curious set of feelings. We now have biographies of Cayley and Sylvester, and we can quickly get a good idea of what they did. We also have books on Green and Thomson, but of Clifford, Forsyth, and Baker, and of Airy, Stokes, Rayleigh, and Tait we know much less. To our knowledge, there is no multi-tome Victorian survey of the Cambridge physicists by an acolyte—nothing to compare with the three-volume biography of Helmholtz by Koenigsberger.[22] Sylvanus P. Thompson's life of Kelvin from 1910 is certainly informative in the eulogistic manner that is to be expected, but is more superficial.[23] The 1882 biography of Maxwell[24] does not help in this respect, being full of letters and such things as Maxwell's poems, while almost free of mathematics when it comes to the sections on Maxwell's scientific work.

The problem is augmented by two factors. One is that Thomson, Stokes, Rayleigh, and Maxwell were working physicists: they pursued questions in physics, they conducted experiments, and they made the simplifications and approximations that are necessary to bring the natural world into a comprehensible condition. The place of their mathematics in all of this is hard to elucidate, and is made harder by the second factor: historians of science do not like to dwell on the mathematics. The specifically mathematical accomplishments of these men are excluded so that the reader may be conveyed at once to the matters of scientific insight and dispute. This makes it hard, if not impossible, to discover whether these eminent physicists actually did any new mathematics.

The case of Stokes is particularly acute. Richly informative though it is, David Wilson's comparative study of Kelvin and Stokes does not describe any mathematics in any detail.[25] Stokes's 1849 account of the convergence of Fourier series, the first in English, is described at best only in passing in the historical literature. His version of uniform convergence is described in various places.[26] The best account I know of the history of the Navier–Stokes equation is in Olivier Darrigol's *Worlds of Flow*,[27] where the prior discovery of this equation by Navier, Cauchy, and Poisson is naturally described. The phenomenon of Stokes sectors in the complex plane was discovered by Stokes in the course of a delicate asymptotic analysis of problems to do with diffraction; happily, this part of the story has caught the attention of Michael Berry.[28]

For Thomson the situation is better, inasmuch as we have Sharlin and Sharlin's biography,[29] even though it tends to keep the mathematics at bay; the study by

Andrew Warwick's
*Masters of Theory.*

Smith and Wise[30] is much stronger in every respect. For Maxwell, things are better because there is a wealth of literature on him and the Maxwellians.[31] But it seems that we may have entered the world of physics here, by which we mean that Thomson's and Maxwell's mathematics, although ingenious, is not in and of itself sufficiently original to be noted in a history of mathematics, as opposed to a history of the applications of mathematics (assuming that this distinction is valid). Maxwell's original mathematics is not on a par with his physics; his contributions to graphical statics were described by Scholz,[32] and his strikingly Riemannian presentation of complex analysis is worth mentioning in a history of that subject, if only to underline how little the subject was appreciated in the Cambridge of the day.[33]

This gap in the historical literature raises another issue: the awkward relationship between the history of science and the history of mathematics. Whatever the reasons for it, a disciplinary boundary seems to be deeply entrenched institutionally in British and American universities, one that the remarkable breadth of the history of science and the rise of the biological sciences makes it ever harder to cross. The example of Stokes's work on asymptotic analysis should haunt us. Who knows what other sharp analyses of mathematical topics were carried out by these Cambridge physicists in pursuit of genuine scientific questions? The secondary literature is quiet, and it is necessary to go back to the originals.

To look more closely at the divide between the history of science and the history of mathematics, we can compare Crilly's biography of Cayley and Parshall's of Sylvester with Andrew Warwick's *Masters of Theory.*[34] Warwick starts off in Cambridge at the start of the 19th century, and as matters specialize he follows the physicists in the end. They too had their struggles to establish their subject, to install research, and to reform teaching and the syllabus, although they were not as oppressed as the pure mathematicians, whom they at best tolerated. Now, was Warwick following the main road and taking the natural path along which his story developed, or did he leave out the emergence of pure mathematics because that story does not matter to present-day historians of science? Or was he tacitly endorsing the preferences and prejudices of the Cambridge scientists and tutors that he wrote about? Turn it around: should we recognize that Cayley and Sylvester actually are minor figures in British scientific intellectual life, that writing about them is frankly a bit 'nerdy', a bit like train spotting? Does your answer to that question depend on your opinion of the quality of the mathematics that Cayley and Sylvester did?

The special role of Euclid's *Elements* in any account of British mathematics in the 19th century has often been noted and discussed (see Chapter 14). In this connection, it is natural to wonder about the impact of the discovery of non-Euclidean geometry, published by Bolyai and Lobachevskii in the 1830s but left languishing until first Riemann, then Beltrami, then the translations of Jules Hoüel, and then the work of Poincaré rescued it. The record is striking and has, I think, been misunderstood. Philip Kelland was intriguingly open-minded about non-Euclidean geometry in 1864, but this contrasts sharply with Cayley's failure to grasp it in 1865, and his oddly defensive attitude to Euclidean geometry. Then there is Clifford's embrace of it in the 1870s, followed by very little until it became a generally accepted subject at Cambridge, where A. N. Whitehead was the first to lecture on it in 1893.

There is not much to explain here. The story resembles most communities' responses: neglect, doubt, and finally acceptance, but it was more muted because of the very strong hold of Euclid's *Elements* on the British community. What Cayley and one or two others wanted was mathematics, and Euclidean geometry in particular, as an exemplar of certainty, one that by its example held open the door to other certainties in philosophy and, most importantly, religion. In contrast, what Clifford wanted to do was to shut that door, as has been noted recently by Barrow-Green and Gray.[35]

All this was well described to us many years ago by Joan Richards in her book *Mathematical Visions*.[36] But the singular nature of the British, and the blinkers that they wore when looking into geometry, makes them unsuitable as illustrations of the way that the 19th-century mathematical community as a whole responded to non-Euclidean geometry. The British response resembled that of other countries, but it was much weaker, and Joan Richards went too far (in my view) in holding up the British response as perhaps typical. It was atypical, because it was worse.

Recently there have been many substantial biographies of 19th-century mathematicians. In addition to the valuable biographies of Cayley and Sylvester already mentioned, we are fortunate to have lives of Monge by Taton,[37] Cauchy by Belhoste,[38] Liouville by Lützen,[39] Hadamard by Maz'ya and Shaposhnikova,[40] Abel and Lie by Stubhaug[41] (with one on Mittag-Leffler promised from the same author), and perhaps some others. The Germans are largely lacking: only two on Cantor spring to mind, by Dauben, and Purkert and Ilgauds,[42] three of several slight ones on Gauss by Dunnington, Bühler, and Wussing,[43] one on Riemann by Laugwitz[44] along with an old one, and a small study of Klein by Tobies.[45] There are other ways to write history, of course, and it is not my intention to enter into a discussion about the pros and cons of biography versus any other mode of writing history. But we may wonder whether a certain imbalance has not opened up.

Crilly and Parshall made it quite clear that whatever the indefatigable Cayley and Sylvester may have done with their invariant theory, they were corrected and eventually surpassed by Hermite and by the Germans, Aronhold, Clebsch, and Gordan, not to mention Hilbert who left them all behind. But we simply do not have good accounts of most of what these other people did, except for Hilbert, another hero. And 'hero' is the word.

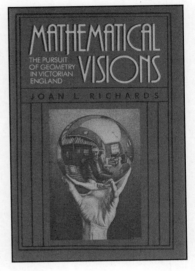

Joan Richards'
*Mathematical Visions*.

## The arrival of pure mathematics

Let us move to our conclusions. The recent work on the history of mathematics in Britain does not support our harsh account. We are not saying that De Morgan, Cayley, Sylvester, Clifford, Smith, and the others, when they are written about, are presented uncritically. We are not even saying that my opinion of their merits—somewhat overdrawn here for polemical purposes—should be accepted anyway. We are not saying that they should be written about, still less thought about, according to this or that fashion in the history of science. Even if the books were poor—and in fact they are good—it is better to have full-length studies of Cayley and Sylvester than none, as is the case for some major British scientists.

Cayley and Sylvester are also worth talking about because we now have two good books about them. Sylvester, it is fair to say, was the more vibrant of the two: a Jew, with a career in America as well as here—maybe even a man with more impact in his day, while always being something of an outsider. Cayley was the insider. Both were fine algebraists. But I cannot help wondering whether there is an irony in Crilly's conclusion about Cayley: he calls him 'the mathematician laureate of the Victorian age', and we think that is exactly right. He was the very model of what the British thought a pure mathematician ought to be. In his time the Poet Laureate was Tennyson, a major poet; but how poor most of the Poets Laureate have been!

I suggest that we should not look at the story of pure mathematics in Britain in the 19th century as a success story, but as a particular kind of failure—or, if you prefer, of partial success against great odds. It was a case of under-development in the richest country in the world, and it requires us to challenge the self-perceptions of the British at the time, and of our own as readers and writers of history. If the names of Cayley and Sylvester have cropped up so often in this chapter, it is in fair part because there are not many other people to talk about. Henry Smith would be the best counter-example to this assertion, but his relative lack of recognition in his day, compounded by being at the 'wrong' university, kept him at the margins. Britain was not Germany, France, or Italy, but it might look better, I agree, if historians of mathematics and historians of science filled in some of the larger outstanding gaps.

# NOTES, REFERENCES, AND FURTHER READING

*In the Notes for each chapter, works cited by author and year only are fully identified in the introductory material for that chapter.*

## INTRODUCTION (ADRIAN RICE)

Until now, although no single book has considered the broad topic of mathematics in Victorian Britain, several sources have covered various aspects of the subject.

For information on the mathematical climate in pre-Victorian Britain, Niccolò Guicciardini's *The Development of Newtonian Calculus in Britain 1700–1800* (Cambridge, 1989) and Helena M. Pycior's *Symbols, Impossible Numbers, and Geometric Entanglements: British Algebra through the Commentaries on Newton's Universal Arithmetic* (Cambridge, 1997) are excellent studies.

The history of the development of certain subjects by Victorian mathematicians is featured in several publications, including Michael J. Crowe's *A History of Vector Analysis: The Evolution of the Idea of a Vectorial System* (Dover, 1985), M. Eileen Magnello's *Victorian Values: The Origins of Modern Statistics* (Icon Press, 2006), and Joan L. Richards' *Mathematical Visions: The Pursuit of Geometry in Victorian England* (Academic Press, 1988). Victorian geometry also figures in the first half of June Barrow-Green and Jeremy Gray's paper 'Geometry at Cambridge, 1863–1940', *Historia Mathematica* 33 (2006), 315–56, while Victorian studies on logic are detailed in the earlier sections of Ivor Grattan-Guinness's *The Search for Mathematical Roots, 1870–1940* (Princeton Univ. Press, 2000).

Aspects of Victorian British work on mathematical physics, particularly with respect to that emanating from the University of Cambridge, are contained in Peter M. Harman (ed.), *Wranglers and Physicists: Studies on Cambridge Physics in the Nineteenth Century* (Manchester Univ. Press, 1985), and more recently in Andrew Warwick's *Masters of Theory: Cambridge and the Rise of Mathematical Physics* (Univ. of Chicago Press, 2003). Another worthy history of the study of mathematics in the Cambridge of this period is Alex D. D. Craik's

*Mr. Hopkins' Men: Cambridge Reform and British Mathematics in the 19th Century* (Springer, 2007).

Finally, there are a number of biographies of several Victorian British mathematicians that contain valuable material. These include: Thomas L. Hankins' *Sir William Rowan Hamilton* (Johns Hopkins Univ. Press, 1980); Tony Crilly's *Arthur Cayley: Mathematician Laureate of the Victorian Age* (Johns Hopkins Univ. Press, 2006); Karen Parshall's *James Joseph Sylvester: Jewish Mathematician in a Victorian World* (Johns Hopkins Univ. Press, 2006); Robin Wilson's *Lewis Carroll In Numberland: His Fantastical Mathematical Logical Life* (Allen Lane, 2009); Des MacHale's *George Boole: His Life and Work* (Boole Press, 1985); Anthony Hyman's *Charles Babbage, Pioneer of the Computer* (Princeton Univ. Press, 1982); Dorothy Stein's *Ada: A Life and a Legacy* (MIT Press, 1985); Crosbie Smith and M. Norton Wise's *Energy and Empire: A Biographical Study of Lord Kelvin* (Cambridge, 1989); D. M. Cannell's *George Green: Mathematician and Physicist, 1793–1841* (Athlone Press, 1993); Basil Mahon's *The Man Who Changed Everything: The Life of James Clerk Maxwell* (Wiley, 2003); Mark Bostridge's *Florence Nightingale: The Woman and Her Legend* (Viking, 2008); and Theodore M. Porter's *Karl Pearson: The Scientific Life in a Statistical Age* (Princeton Univ. Press, 2004).

## NOTES

1. *The Athenæum*, 2 Nov. 1839, p. 825, col. 1.

2. Guicciardini (1989), pp. 139–42.

3. Charles Babbage, *Reflections on the Decline of Science in England, and on Some of its Causes* (B. Fellowes, 1830), p. 1.

4. Babbage (note 3), p. ix.

5. I. Grattan-Guinness, *The Fontana History of the Mathematical Sciences: The Rainbow of Mathematics* (Fontana, 1997), p. 280.

6. Guicciardini (1989), pp. 82–91.

7. Pycior (1997).

8. A. Rice, 'Inexplicable? The status of complex numbers in Britain, 1750–1850', *Around Caspar Wessel and the Geometric Representation of Complex Numbers* (ed. J. Lützen) (Royal Danish Academy of Sciences and Letters, 2001), pp. 147–80.

9. Rice (2001), pp. 157–59.

10. D. F. Gregory, 'On the real nature of symbolical algebra', *Trans. Roy. Soc. Edinburgh* **14** (1838), 208–16, on p. 211.

11. G. Peacock, *A Treatise on Algebra*, Vol. 2, 2nd ed., J. & J. J. Deighton (1845), p. 449.

12. Crowe (1985).

13. Rice (2001), pp. 174–75.

14. A. Cayley, 'A memoir on the theory of matrices', *Phil. Trans. Roy. Soc.* **148** (1858), 17–38. Sylvester's original definition of a matrix is contained in: J. J. Sylvester, 'On a new class of theorems', *Phil. Mag.* (3rd series) **37** (1850), 363–70.

15. K. H. Parshall, 'Towards a history of nineteenth-century invariant theory', *The History of Modern Mathematics*, Vol. 1 (ed. D. Rowe and J. McCleary) (Academic Press, 1989), 157–206.

16. Cayley (note 14), p. 18.

17. Wilson (2009), pp. 115–23.

18. The Scottish philosopher Sir William Hamilton (1788–1856) is not to be confused with the Irish mathematician Sir William Rowan Hamilton (1805–65).

19. P. Heath, 'Introduction', A. De Morgan, *On the Syllogism, and Other Logical Writings* (ed. P. Heath) (Routledge and Kegan Paul, 1966), vii–xxxi, on p. xv.

20. M. E. Baron, 'A note on the historical development of logic diagrams: Leibniz, Euler and Venn', *Mathematical Gazette* **53**, No. 384 (1969), 113–25.

21. J. Venn, 'On the diagrammatic and mechanical representation of propositions and reasonings', *Phil. Mag.* (5th series) **10**, No. 59 (July 1880), 1–18.

22. See Grattan-Guinness (2000).

23. A. N. Whitehead and B. Russell, *Principia Mathematica*, Vol. 2 (Cambridge, 1912), p. 86.

24. See D. Swade, *The Cogwheel Brain: Charles Babbage and the Quest to Build the First Computer* (Little, Brown, 2000).

25. A. Bromley, 'Charles Babbage's Analytical Engine, 1838', *Annals of the History of Computing* **4** (1982), 196–217.

26. A. A. Lovelace, 'Sketch of the Analytical Engine invented by Charles Babbage Esq.', in *The Works of Charles Babbage*, Vol. 3 (ed. M. Campbell-Kelly), New York Univ. Press (1989), 158–70. See also J. Fuegi and J. Francis, 'Lovelace & Babbage and the creation of the 1843 "Notes"', *Annals of the History of Computing* **25** (4) (2003), 16–26.

27. M. Norton Wise, 'Treatise on natural philosophy', *Landmark Writings in Western Mathematics 1640–1940* (ed. I. Grattan-Guinness), Elsevier (2005), 521–33.

28. I. Grattan-Guinness, 'Why did George Green write his essay of 1828 on electricity and magnetism?', *American Math. Monthly* **102** (1995), 387–96.

29. See Cannell (1993).

30. J. Lützen, *Joseph Liouville 1809–1882: Master of Pure and Applied Mathematics* (Springer, 1990), p. 139.

31. James J. Cross, 'Integral theorems in Cambridge mathematical physics, 1830–1855', in Harman (1985), pp. 113–48, on p. 144.

32. Crowe (1985), p. 176.

33. Warwick (2003), pp. 363–95.

34. F. Nightingale, *Notes on Matters Affecting the Health, Efficiency, and Hospital Administration of the British Army, Founded Chiefly on the Experience of the Late War* (London, 1858).

35. M. Bulmer, *Francis Galton: Pioneer of Heredity and Biometry* (Johns Hopkins Univ. Press, 2003).

36. M. E. Magnello, 'Karl Pearson's mathematization of inheritance: from ancestral heredity to Mendelian genetics (1895–1909)', *Annals of Science*, **55** (1998), 35–94.

37. M. E. Magnello, 'Karl Pearson's paper on the chi-square goodness of fit test', *Landmark Writings in Western Mathematics 1640–1940*, (ed. I. Grattan-Guinness) (Elsevier, 2005), pp. 724–31.

38. See Magnello (2006).

39. A. C. Rice, R. J. Wilson and J. H. Gardner, 'From student club to national society: The founding of the London Mathematical Society in 1865', *Historia Math.* **22** (1995), 402–21.

40. It is worth noting that the Moscow Mathematical Society has some claim to be older than the London Mathematical Society, since it was formed in 1867 as a reorganization of a Mathematical Circle that originated at Moscow University in 1864. However, it remained a relatively small body for many years and, unlike its British counterpart, had no immediate influence on the subsequent formation of similar learned societies.

41. H. Gispert, 'The effects of war on France's international role in mathematics, 1870–1914', in *Mathematics Unbound: The Evolution of an International Mathematical Research Community, 1800–1945* (ed. K. H. Parshall and A. C. Rice) (American and London Mathematical Societies, 2002), pp. 105–21, on pp. 106–7.

42. D. E. Smith and J. Ginsburg, *A History of Mathematics in America Before 1900* (Arno Press, 1980), pp. 105–11.

43. H. Gispert and R. Tobies, 'A comparative study of the French and German Mathematical Societies

before 1914', *L'Europe mathématique—Mathematical Europe* (ed. C. Goldstein, J. Gray, and J. Ritter) (Éditions de la Maison des Sciences de l'Homme, 1996), pp. 408–30.

44. W. Burnside, 'On the theory of groups of finite order', *Proc. Lond. Math. Soc.* (2) **7** (1909), 1–7, on p. 1.

45. K. Hannabuss, 'Henry Smith', *Oxford Figures: 800 Years of the Mathematical Sciences* (ed. J. Fauvel, R. Flood, and R. Wilson) (Oxford, 2000), pp. 203–17, on p. 206.

46. E. T. Whittaker, 'Andrew Russell Forsyth', *Obituary Notices of Fellows of the Royal Society* **4** (1942), 209–27, on p. 218.

47. See Maria Panteki, *Relationships between algebra, differential equations and logic in England: 1800-1860*, Ph.D. thesis, Middlesex Univ., 1992.

48. See Crowe (1985).

49. E. Seneta, 'Early influences on probability and statistics in the Russian empire', *Archive for History of Exact Sciences* **53** (1998), 201–13.

50. Hannabuss (2000), p. 216. Smith's presidential address was originally published as: H. J. S. Smith, 'On the present state and prospects of some branches of pure mathematics', *Proc. Lond. Math. Soc.* (1st series) **8** (1876), 6–29.

51. *Proceedings of the Fifth International Congress of Mathematicians (Cambridge, 22–28 August 1912)* (ed. E. W. Hobson and A. E. H. Love) (Cambridge, 1913), p. 46.

## CHAPTER 1 (TONY CRILLY)

For an overview of mathematical education at Cambridge in the Victorian era, modern studies include Peter M. Harman (ed.), *Wranglers and Physicists: Studies on Cambridge Physics in the Nineteenth Century* (Manchester Univ. Press, 1985), Andrew Warwick's *Masters of Theory: Cambridge and the Rise of Mathematical Physics* (Univ. of Chicago Press, 2003), and J. Smith and C. Stray (eds.) *Teaching and Learning in Nineteenth Century Cambridge* (Boydell Press, 2001).

An authoritative book written by a Cambridge don of the period is W. W. Rouse Ball's *A History of the Study of Mathematics at Cambridge* (Cambridge, 1889).

Useful general histories and guides to primary sources are D. A. Winstanley's book-ends *Early Victorian Cambridge* (1940/1955) and *Later Victorian Cambridge* (1947). Modern scholarship on the history of Cambridge University is afforded by P. Searby's *History of the University of Cambridge 1750–1870*, Vol. 3 (1997), and C. N. L. Brooke's *History of the University of Cambridge 1870–1990*, Vol. 4 (1993). A shorter history of the university is E. S. Leedham-Green's *Concise History of the University of Cambridge* (1996). (All of these works are published by Cambridge University Press.)

From the social history perspective, M. M. Garland's *Cambridge before Darwin: The Idea of a Liberal Education*

*(1800–1860)* (Cambridge, 1980) and R. McWilliams-Tullberg's *Women at Cambridge* (Gollancz, 1975/1998) are invaluable. A classic is S. Rothblatt's *The Revolution of the Dons: Cambridge Society and Victorian England* (Cambridge, 1969/1981). Biographies of mathematicians educated at Cambridge throw light on the mathematical tripos. Modern biographies include Tony Crilly's *Arthur Cayley: Mathematician Laureate of the Victorian Age* and Karen Parshall's *James Joseph Sylvester: A Jewish Mathematician in the Victorian World* (both published in 2006 by Johns Hopkins Univ. Press).

Reminiscences and diaries by former students are also available. Special mention might be made of Romilly's diary, which is a remarkable record of academic life at Cambridge for the first half of the Victorian period: J. T. Bury (ed.), *Romilly's Cambridge Diary 1832–1842* (Cambridge, 1967) and M. E. Bury and J. D. Pickles (ed.), *Romilly's Cambridge Diary 1842–1847* and *Romilly's Cambridge Diary 1848–1864*, Cambridge Records Society **10** (1994), and **14** (2000).

### NOTES

1. The Senate House examinations were the final examinations, held every January, on the basis of which the University of Cambridge awarded its degrees.

2. [J. J.], 'Desultory remarks on academic and non-academic mathematics and mathematicians', *Phil. Mag.* **25** (1844), 83.

3. E. H. Neville, 'Andrew Russell Forsyth 1858–1942', *J. London Math. Soc.* **17** (1942), 239.

4. The Board of Mathematical Studies was composed of the mathematical professors, with moderators and examiners of the current year and the two preceding years.

5. Macmillan's sharp commercial sense extended to publishing textbooks with mass circulation *and* availing themselves of the services of the University Press to have them printed. It was not until the 1870s that the University Press fully realized their own potential by acting as publisher themselves. For Cambridge publishing see David McKitterick, *A History of Cambridge University Press, Vol. 2, Scholarship and Commerce 1698–1872* (Cambridge, 1998).

6. Baron Rayleigh [R. J. Strutt], *John William Strutt, Third Baron Rayleigh* (Edward Arnold, 1924), p. 29.

7. W. P. Turnbull, *An Introduction to Analytical Plane Geometry* (Deighton Bell, 1867).

8. F. J. A. Hort to J. L. Davies, [1857–59], Cambridge Local Library, Cambridge Collection, A10.1208.

9. L. Stephen, *Life of Henry Fawcett* (Smith, Elder, and Co., 1885), p. 25.

10. J. R. Seeley (ed.), *The Student's Guide to the University of Cambridge* (Deighton Bell, 1863), p. 15.

11. Winstanley (1947), p. 146.

12. H. M. Harrison, *The Life of John Couch Adams: Voyager in Time and Space* (Book Guild, 1994), p. 77.

13. W. Everett, *On the Cam. Lectures on the University of Cambridge in England* (Beeton, 1866), p. 43.

14. W. Airy (ed.), *Autobiography of Sir George Biddell Airy* (Cambridge, 1896), p. 278.

15. Airy (note 14), p. 279.

16. K. Pearson, 'Old Tripos days at Cambridge, as seen from another viewpoint', *Math. Gazette* **20** (1936), 27–36. Compare this account with a reminiscence by A. R. Forsyth, 'Old Tripos days at Cambridge', *Math. Gazette* **19** (1935), 162–79.

17. The University Syndicate was appointed on 17 May 1877; *University Reporter*, 22 May 1877.

18. As a result of the reforms the new system of examinations was still daunting. The old examination system was replaced by eight days of examination held in the June of each year. It consisted of Part I (the 'first four days') with seven elementary examinations and, after a short interval, Part II (the 'second four days') with another seven examinations in advanced subjects. The seventh examination in each part consisted of 'problems'.

19. E. F. B. MacAlister, *Sir Donald MacAlister of Tarbet* (Macmillan, 1935), pp. 18–19.

20. T. Muir, 'The promotion of research with special reference to the present state of the Scottish universities and secondary schools', Mathematical Society of Edinburgh, 8 Feb. 1884, (Alexander Gardner, 1884), p. 11. The Ph.D. was introduced to Britain at the end of the First World War.

21. J. W. L. Glaisher, 'Presidential address—Section A, British Association for the Advancement of Science', *British Association Report*, 1890, p. 724.

22. G. H. Hardy, 'The case against the mathematical tripos', *Math. Gazette* **13** (1926), 61–71, on p. 64.

23. S. P. Thompson, *The Life of William Thomson, Baron Kelvin of Largs*, Vol. 1 (Macmillan, 1910), p. 433.

24. It became evident that six months was not enough for the preparation of students for Part III of the mathematical tripos, and from 1886 the Part III advanced course was lengthened to one year and the whole process renamed to become Part I and Part II.

25. George Bernard Shaw's *Mrs Warren's Profession*, written in 1894 and performed in 1902.

26. J. Barrow-Green, '"A Corrective to the Spirit of too Exclusively Pure Mathematics": Robert Smith (1689–1768) and his prizes at Cambridge University', *Annals of Science* **56** (1999), 271–316.

27. Hardy (note 22).

28. W. M. Coates, the 3rd wrangler in 1886, was coached by E. J. Routh and had thirteen students himself. For the private coaches, the proposed changes signalled the curtailment of their own livelihoods.

29. The last senior wrangler was P. J. Daniell, a versatile mathematician who worked in both pure and applied mathematics and whose memory is kept alive by the Daniell integral.

30. The mathematician H. R. Hassé interpreted the abolition of the traditional mathematical tripos as the passing of a 'problem age' which tested technique or manipulation of symbols in distinction to the 'example age' which tested an understanding of the theory or theorems in terms of examples. For him it was the difference between 'learning mathematics' and 'reading mathematics'. (See H. R. Hassé, 'My fifty years of mathematics', *Math. Gazette* **35** (1951), 153–64.) But regulation cannot easily erase a culture, and many features of this traditional Cambridge degree were passed on to the 'new universities' that were founded after the First World War.

## CHAPTER 2 (KEITH HANNABUSS)

General background information about Oxford and the University during the Victorian period can be found in W. R. Ward's *Victorian Oxford* (Frank Cass, 1965); A. J. Engel's *From Clergyman to Don* (Clarendon Press, Oxford, 1983), and T. Hinchliffe's *North Oxford* (Yale Univ. Press, 1992). A more immediate view is provided in the Revd. W. Tuckwell's *Reminiscences of Oxford* (Cassell, 1901).

Victorian mathematics is covered in Chapter 11 ('The mid-nineteenth century'), Chapter 12 ('Henry Smith'), and Chapter 13 ('James Joseph Sylvester') of *Oxford Figures: 800 Years of the Mathematical Sciences* (ed. J. Fauvel, R. Flood, and R. Wilson) (Oxford, 2000).

A good biography of Baden Powell, though with more emphasis on theology than mathematics, is P. Corsi's *Religion and Science* (Cambridge, 1988). There is also a useful chapter on Powell in W. Tuckwell's *Pre-Tractarian Oxford* (Cassell, 1909).

There is no full-length biography of Henry Smith, although there are good biographical essays in the reprints of his *Report on the Theory of Numbers* and *The Collected Mathematical Papers of Henry John Stephen Smith* (Oxford, 1894, repr. Chelsea, 1965). Further background material about Smith and science in mid-Victorian Oxford can be found in V. Morton's *Oxford Rebels* (Alan Sutton, 1987).

Sylvester is well covered by a biography and a selection of his letters by Karen Hunger Parshall, *James Joseph Sylvester: Life and Work in Letters* (Oxford, 1998), and *James Joseph Sylvester: A Jewish Mathematician in the Victorian World* (Johns Hopkins Univ. Press, 2006).

Far more has been written about Charles Dodgson than any other Oxford mathematician, although few books give more than a cursory discussion of his mathematics. Exceptions include Robin Wilson, *Lewis Carroll in Numberland: His*

*Fantastical, Mathematical, Logical Life* (Allen Lane, 2008), Martin Gardner, *The Universe in a Handkerchief* (Copernicus, 1996), W. W. Bartley III, *Lewis Carroll's Symbolic Logic* (Harvester Press, 1977), and F. Abeles, *The Mathematical Pamphlets of Charles Lutwidge Dodgson* (Lewis Carroll Society of North America, 1994).

**NOTES**

1. The Christ Church mathematics lecturer Charles Dodgson, better known to non-mathematicians as Lewis Carroll, was granted special permission to remain a deacon.

2. Tuckwell (1901), p. 4.

3. Tuckwell (1909), p. 166.

4. V. Brittain, *The Women at Oxford* (Harrap, 1960), p. 37.

5. Ward (1965), p. 198.

6. Ward (1965), p. 166.

7. See Corsi (1988) or Anthony Hyman, *Charles Babbage, Pioneer of the Computer* (Oxford, 1982).

8. Fauvel, Flood, and Wilson (2000), p. 236.

9. Tuckwell (1901), p. 204.

10. Wilson (2008), p. 48.

11. Engel (1983), Chapter II, Section D.

12. P. Sutcliffe, *The Oxford University Press*, Part II (Clarendon Press, Oxford, 1978), Section 4.

13. M. B. Hall, *All Scientists Now* (Cambridge, 1984), pp. 117–18.

14. D. MacHale, *George Boole. His Life and Work* (Boole Press, 1985), p. 168; G. C. Smith, *The Boole–De Morgan Correspondence 1842–1864* (Clarendon Press, Oxford, 1982), p. 83.

15. C. H. Pearson, Biographical Sketch, *H. J. S. Smith Mathematical Papers*, p. xv.

16. T. Smith, 'The Balliol–Trinity laboratories', *Balliol Studies* (ed. J. Prest) (Leopard's Head Press, 1982), p. 190.

17. J. W. L. Glaisher, 'Introduction', *H. J. S. Smith Collected Mathematical Papers*, p. xci.

18. N. Kurti, 'Opportunity lost in 1865?', *Nature* **308** (1984), 313–14.

19. This paper also introduced the Smith–Minkowski–Siegel mass formula, from which he was able to deduce the existence of a new integral lattice in eight dimensions, later identified as the root lattice of the exceptional Lie algebra $e_8$.

20. Glaisher (note 17), pp. lxv–lxxi.

21. See Gert Sabidussi, 'Correspondence between Sylvester, Petersen, Hilbert and Klein on invariants and the factorisation of graphs 1889–1891', *Discrete Math.* **100** (1992), 99–155.

22. L. J. Rogers, 'Second memoir on the expansion of certain infinite products', *Proc. London Math. Soc.* (1) **25** (1894), 318–43.

23. G. H. Hardy, *Ramanujan* (Cambridge, 1940, repr. Chelsea, 1978), Section 6.8.

24. Robin Wilson, *Four Colours Suffice: How the Map Problem was Solved* (Allen Lane, 2002), Chapter 7.

25. Personal communication, J. Campbell.

26. J. Campbell, 'On a law of combination of operators bearing on the theory of continuous transformation groups', *Proc. London Math. Soc.* **28** (1897), 381–90 and **29** (1898), 14–32.

27. J. M. Keynes, 'Francis Ysidro Edgeworth', *Dictionary of National Biography*; also *Essays in Biography* (Macmillan, 1933).

28. G. Faber, *Jowett* (Faber and Faber, 1957), p. 310.

29. I. B. Cohen, 'Florence Nightingale', *Scientific American* (March 1984), 98–107; M. Diamond and M. Stone, 'Nightingale on Quetelet', *J. Royal Statistical Soc.* **A144** (1981), 66–79, 176–213, 332–51.

30. E. Abbott and L. Campbell, *The Life and Letters of Benjamin Jowett* (John Murray, 1897), p. 378.

31. E. V. Quinn and J. M. Prest, *Dear Miss Nightingale* (Clarendon Press, 1987), p. 314.

32. Diamond and Stone (note 29).

33. Abbott and Campbell (note 30), p. 478.

34. E. B. Elliott, *Why and how the Society began and kept going*, Lecture to the Oxford Mathematical and Physical Society (16 May 1925), 8–9.

## CHAPTER 3 (ADRIAN RICE)

Several of the institutions featured in this chapter have published commemorative histories—in particular, the constituent colleges of the University of London, including H. Hale Bellot's *University College London 1826–1926* (Univ. of London Press, 1929), and, more recently, Negley Harte and John North, *The World of UCL 1828–1990* (University College, London, 1991). See also Fossey J. C. Hearnshaw's *The Centenary History of King's College London 1828–1928* (G. G. Harrap & Co., 1929); Margaret J. Tuke's *A History of Bedford College for Women 1849–1937* (Oxford, 1939); Linna Bentley's *Educating Women: A Pictorial History of Bedford College, University of London, 1849–1985* (Alma, 1991); J. Mordaunt Crook (ed.), *Bedford College, University of London: Memories of 150 Years* (Royal Holloway and Bedford New College, 2001); A. Rupert Hall's *Science for Industry: A Short History of the Imperial College of Science and Technology and Its Antecedents* (Imperial College, London, 1982); Hannah Gay's *The History of Imperial College London, 1907–2007: Higher Education and Research in Science, Technology and Medicine* (Imperial College Press, 2007); and

Cecil Delisle Burns's *A Short History of Birkbeck College* (Univ. of London Press, 1924). For an overview of the history of the University of London as a whole, see Negley Harte's *The University of London 1836–1986* (Athlone Press, 1986).

Informative, though somewhat dated, accounts of the history of the Royal Military Academy in Woolwich and the Regent Street Polytechnic may be found in Henry Donald Buchanan-Dunlop (ed.), *Records of the Royal Military Academy 1741–1892* (F. J. Cattermole, 1895), and Joshua G. Fitch and William Garnett's 'Polytechnic', in *Encyclopædia Britannica*, 11th edition, Vol. 22 (Cambridge, 1911), pp. 38–42, respectively.

A variety of contextual and social histories of Victorian London exist, including Donald J. Olsen's *The Growth of Victorian London* (Holmes & Meier, 1976); Liza Picard's *Victorian London: The Life of a City, 1840–1870* (Macmillan, 2006), which focuses on the first half of the Victorian period; Stella Margetson's *Fifty Years of Victorian London, from the Great Exhibition to the Queen's Death* (Macdonald, 1969), which deals with the latter half of the 19th century; and Lynda Nead's *Victorian Babylon: People, Streets and Images in Nineteenth-Century London* (Yale Univ. Press, 2005), which concentrates on the mid-Victorian period.

Finally, in addition to published memoirs and biographies of many of the mathematicians mentioned in this chapter, such as De Morgan, Sylvester, and Pearson, there are other sources that shed additional light on the activities of certain London-based figures. These include Cyril Domb's article 'James Clerk Maxwell in London 1860–1865', *Notes & Records of the Royal Society* 35 (1980), 67–103, which concerns Maxwell's period as a professor at King's College, and the copious diaries of Thomas Archer Hirst, compiled and edited by William H. Brock & Roy M. MacLeod, *Natural Knowledge in Social Context: The Journals of Thomas Archer Hirst, F.R.S.* (Mansell, 1980), which include substantial portions concerning his work at University College and the Royal Naval College. Edited highlights from these diaries were later published in six complementary articles by J. Helen Gardner and Robin J. Wilson, 'Thomas Archer Hirst—Mathematician Xtravagant, I–VI,' *American Math. Monthly* 100 (1993), 435–41, 531–38, 619–25, 723–31, 827–34, 907–15.

This chapter is an abridged and updated version of a paper first published in 1996; for further details of much of the information covered in this chapter, see Adrian Rice, 'Mathematics in the Metropolis: a survey of Victorian London,' *Historia Math.* 23 (1996), 376–417.

## NOTES

1. Emilie I. Barrington, *Life of Walter Bagehot* (Longmans, 1914), p. 118.

2. Harriet Ann Jevons (ed.), *Letters and Journal of W. Stanley Jevons* (Macmillan, 1886), p. 150.

3. Bellot (1929), p. 263.

4. James Joseph Sylvester, Presidential address: Mathematics and Physics Section, *Report of the Thirty-ninth Meeting of the B.A.A.S. held at Exeter in August, 1869* (John Murray, 1870), pp.1–9; reprinted as 'A plea for the mathematician', *Nature* 1 (1869–70), 237–39, 261–63, quote on p. 261.

5. Micaiah J. M. Hill, 'Obituary notice of Olaus Henrici', *Proc. London Math. Soc. (2)* 17 (1918), xlii–xlix, on p. xlv.

6. Hill (note 5), p. xlix.

7. Bellot (1929), p. 322.

8. J. W. L. Glaisher, 'Notes on the early history of the Society', *J. London Math. Soc.* 1 (1926), 51–64, on p. 62.

9. Bellot (1929), p. 390.

10. Louis N. G. Filon, 'Obituary Notice of M. J. M. Hill', *J. London Math. Soc.* 4 (1929), 313–18, on p. 317.

11. *Dictionary of National Biography 1931–1940* (Oxford, 1949), p. 683.

12. (As note 11).

13. Gresham College is unusual for a higher-level educational institution because it has no students and awards no degrees. It was established in the City of London in 1596 under the will of financier Sir Thomas Gresham, with seven endowed professors, in divinity, music, astronomy, geometry, physic (= medicine), law, and rhetoric, offering free lectures to the general public. Famous Gresham professors of geometry have included Henry Briggs, Isaac Barrow, Robert Hooke, and Karl Pearson; recent professors have included Sir Christopher Zeeman, Ian Stewart, Sir Roger Penrose, Robin Wilson, and John Barrow. For more on the history of the college, see Richard Chartres and David Vermont, *A Brief History of Gresham College 1597–1997* (Gresham College, 1998).

14. *DNB 1931–1940* (note 11), p. 682.

15. Harte and North (1991), p. 31.

16. See Adrian Rice, 'Inspiration or Desperation? Augustus De Morgan's appointment to the Chair of Mathematics at London University in 1828', *British Journal for the History of Science* 30 (1997), 257–74.

17. Hearnshaw (1929), p. 89.

18. Hearnshaw (1929), p. 305.

19. Hearnshaw (1929), p. 247.

20. Hearnshaw (1929), p. 247.

21. Hearnshaw (1929), p. 305.

22. Elaine Kaye, *A History of Queen's College, London, 1848–1972* (Chatto and Windus, 1972).

23. Tuke (1939).

24. Sophia E. De Morgan, *Memoir of Augustus De Morgan* (Longmans, Green, & Co., 1882), p. 174. Despite its professors being male, Queen's College also had women on its staff, one of the most notable being Mary Everest Boole (1832–1916), wife of the mathematician George Boole, who taught there for some time after her husband's death in 1864.

25. Clive W. Kilmister, 'The Teaching of Mathematics in the University of London', *Bull. London Math. Soc.* **18** (1986), 321–37, on p. 324.

26. Harte (1986), p. 131.

27. Harte (1986), p. 128.

28. William P. Ker (ed.), *Notes and Materials for the History of University College, London. Faculties of Arts and Science* (H. K. Lewis, 1898), pp. 57, 63.

29. Kilmister (note 25), p. 332.

30. S. Gordon Wilson, *The University of London and Its Colleges* (University Tutorial Press, 1923), p. 61.

31. Harte (1986), p. 134; Wilson (note 30), pp. 55–56.

32. Niccolò Guicciardini, *The Development of Newtonian Calculus in Britain 1700–1800* (Cambridge, 1989), p. 108.

33. Guicciardini (note 32), p. 112.

34. Guicciardini (note 32), p. 110.

35. Charles Hutton, *A Course of Mathematics*, Vol. 2 (ed. O. Gregory), 11th ed. (Longman, Rees & Co., 1837), p. 203, note.

36. Buchanan-Dunlop (1895), p. 102.

37. Buchanan-Dunlop (1895), p. 120.

38. Buchanan-Dunlop (1895), p. 120.

39. Buchanan-Dunlop (1895), p. 129.

40. Percy A. MacMahon, 'Obituary notice of James Joseph Sylvester', *Proc. Royal Society* **63** (1898), ix–xxv, on p. xviii.

41. Joseph Larmor, 'Obituary Notice of Morgan William Crofton', *Proc. London Math. Soc. (2)* **14** (1915), xxix–xxx, on p. xxx.

42. Buchanan-Dunlop (1895), pp. 135–36.

43. Charles Hutton, *A Philosophical and Mathematical Dictionary*, Vol. 1 (F. C. & J. Rivington, 1815), p. 16.

44. Andrew Russell Forsyth, 'Obituary notice of William Burnside', *J. London Math. Soc.* **3** (1928), 64–80, on p. 66.

45. Burns (1924), p. 23.

46. See Joan L. Richards, *Mathematical Visions: The Pursuit of Geometry in Victorian England* (Academic Press, 1988), pp. 196–98; William H. Brock, 'Geometry and the universities: Euclid and his modern rivals 1860–1901', *History of Education* **4** (1975), 21–35; and Florian Cajori, 'Attempts made during the eighteenth and nineteenth centuries to reform the teaching of geometry, *American Math. Monthly* **17** (1910), 181–201.

47. Hall (1982).

48. *A Short History of the City and Guilds of London Institute* (City and Guilds of London Institute, 1896), p. 7.

49. Joyce Brown (ed.), *A Hundred Years of Civil Engineering at South Kensington* (Civil Engineering Department, Imperial College London, 1985), p. 18.

50. (As note 48), p. 6.

51. Fitch and Garnett (1911), p. 40.

52. Wilson (note 30), p. 66.

53. Fitch and Garnett (1911), p. 41.

54. Fitch and Garnett (1911), p. 41.

## CHAPTER 4 (A. J. S. MANN AND A. D. D. CRAIK)

The Scottish universities in Victorian times are discussed by R. D. Anderson, *Education & Opportunity in Victorian Scotland* (Oxford, 1983, repr. with corrections by Edinburgh Univ. Press, 1989) and by G. E. Davie, *The Democratic Intellect: Scotland and her Universities in the Nineteenth Century* (Edinburgh Univ. Press, 1961). Mathematics in Scotland during this period, and the growing inter-relationship with Cambridge, are covered in Alex D. D. Craik's *Mr Hopkins' Men: Cambridge Reform and British Mathematics in the Nineteenth Century* (Springer, 2007), Chapters 9 and 11.

Biographies of some of the mathematicians discussed, and much additional information, can be found in the references cited below, and also on-line at the St Andrews MacTutor History of Mathematics website, www-groups.dcs.st-and.ac.uk/~history [accessed 12 Jan 2011].

### NOTES

1. Graham Richards, 'Bain, Alexander (1818–1903)', *Oxford Dictionary of National Biography*, (Oxford, 2004). Our account is based on Bain's posthumously published autobiography, Alexander Bain, *Autobiography* (Longmans, Green & Co., 1904), pp. 6–7.

2. Bain (note 1), pp. 6–7.

3. Bain (note 1), p. 7.

4. 'With great pomp and ceremony', Innes allowed Bain to examine, for half an hour, his copy of Newton's *Principia*, in Motte's English translation. (Bain (note 1), p. 20.)

5. Some of his friends made a telescope through which Bain saw the satellites of Jupiter and watched eclipses of the moon.

6. John Cruickshank was professor of mathematics at Marischal College from 1829 to 1860, and retired when Marischal and King's Colleges merged.

7. Bain (note 1), p. 51.

8. Bain (note 1), p. 74.

9. Bain (note 1), p. 78.

10. Bain (note 1), p. 130.

11. John Butt, *John Anderson's Legacy: the University of Strathclyde and its Antecedents 1796–1996* (East Linton, Scotland, Tuckwell Press and Strathclyde Univ., 1996).

12. Regarding the latter, Bain recalled that 'I was, however, saved from it by the jealousy of James Thomson, the mathematical professor, who was looking forward to having his son appointed as soon as the Chair became vacant, and was consequently in dread of rivalry, even of the most improbable candidates' (Bain (note 1), p. 178.) In 1846, William Thomson was duly appointed in preference to David Gray. On this and most other matters relating to William Thomson, see the major biography by Crosbie Smith and M. Norton Wise, *Energy and Empire. A Biographical Study of Lord Kelvin* (Cambridge, 1989), or the recent collection *Kelvin: Life, Labours and Legacy* (ed. R. Flood, M. McCartney, and A. Whitaker) (Oxford, 2008).

13. See Alex D. D. Craik's 'Geometry, analysis and the baptism of slaves: John West in Scotland and Jamaica', *Historia Math.* **25** (1998), 29–74, and 'James Ivory, mathematician: 'the Most Unlucky Person that Ever Existed'', *Notes & Records of the Royal Society of London* **54** (2000), 223–47, and Kenneth J. Cameron's *The Schoolmaster Engineer: Adam Anderson of Perth & St Andrews* (Abertay Historical Society, 1988/2007).

14. Political, religious and family patronage were endemic at all the Scottish universities at this and earlier times: see R. L. Emerson's *Professors, Patronage and Politics: The Aberdeen Universities in the Eighteenth Century* (Aberdeen Univ. Press, 1992), and *Academic Patronage in the Scottish Enlightenment: Glasgow, Edinburgh and St Andrews Universities* (Edinburgh Univ. Press, 2008).

15. Anderson (1983/1989).

16. Davie (1961).

17. Davie (1961), p. 106.

18. Anderson (1983/1989), p. 1.

19. Hester Lynch Piozzi, *Anecdotes of Samuel Johnson*, consulted on-line: http://infomotions.com/etexts/gutenberg/dirs/etext00/andsj10.htm [accessed 12 Jan 2011].

20. Anderson (1983/1989), p. 76.

21. Ronald M. Birse, 'Chrystal, George (1851–1911)', *Dictionary of National Biography* (Oxford, 2004).

22. The *pons asinorum*, or 'bridge of asses', was Proposition 5 of the first book of Euclid's *Elements*, concerning the equality of the base angles and opposite sides of an isosceles triangle.

23. George Chrystal, 'Promoter's address to graduates of arts and science, University of Edinburgh', *The Scotsman*, Edinburgh, 22 April 1885, also on the MacTutor website (Chrystal). Frederick Fuller, a Cambridge graduate, was

the next Englishman after Kelland to hold a Scottish mathematics chair. David Thomson had studied at Glasgow and Cambridge, and was for a time assistant in natural philosophy at Glasgow. Further information on mathematics teachers at the Aberdeen colleges is in Betty Ponting, 'Mathematics at Aberdeen', *Aberdeen University Review* **48** (1979–80), 26–35, 162–76 (Aberdeen Univ. Press for Aberdeen University Alumnus Association).

24. Alex D. D. Craik (2007), and Andrew Warwick, *Masters of Theory: Cambridge and the Rise of Mathematical Physics* (Univ. of Chicago Press, 2003).

25. Alex D. D. Craik, 'Calculus and analysis in early 19th century Britain: the work of William Wallace', *Historia Math.* **26** (1999), 239–67, on p. 260.

26. Robert Louis Stevenson, *Fleeming Jenkin* (Charles Scribner's Sons, 1887), 233–35.

27. Stevenson (note 26).

28. George Chrystal, 'Promoter's address to graduates of arts and science, University of Edinburgh', *The Scotsman*, Edinburgh, 15 April 1892, also on the MacTutor website (Chrystal).

29. Leslie wrote to the publisher Archibald Constable that Wallace had imputed this motive to him, though 'this scarcely enters my mind': see Alex D.D. Craik, 'Geometry versus analysis in early 19th-century Scotland: John Leslie, William Wallace, and Thomas Carlyle', *Historia Math.* **27** (2000), 133–63, on p. 141.

30. Philip Kelland, *How to Improve the Scottish Universities* (Adam and Charles Black, 1855).

31. Quoted by Anderson (1983/1989), p. 14.

32. Quoted by Anderson (1983/1989), pp. 35–36. Playfair's name is now attached to the cipher system, invented by Charles Wheatsone, which he promoted.

33. Bound with the National Library of Scotland's copy of *Testimonials in favour of the Rev. Philip Kelland, M. A....as candidate for the chair of mathematics in the University of Edinburgh* (1838), NLS Shelfmark S.155.f(2).

34. Davie (1961), pp. 132–33.

35. See Craik (2007), p. 235, which however omits the name of John Cruickshank (Marischal College, 1829–60) from the list of professors.

36. In Edinburgh, John Playfair and John Leslie had taught a course on differential calculus only in alternate years, and to rather few students. William Wallace did so each year in his advanced classes of 1820–27 to a total of just one hundred and seven students, of whom only seventy-eight paid fees: see Craik (note 29), on p. 158.

37. Thomas Muir, 'The Promotion of Research; with Special Reference to the Present State of the Scottish Universities and Secondary Schools', *Address delivered before the Mathematical Society of Edinburgh, 8th February, 1884* (Alexander Gardner, 1884).

38. See Craik (2007), Chapters 9 and 11.

39. Anderson (1983/1989), p. 43.

40. Anderson (1983/1989), p. 266, quoting *Hansard*.

41. Quoted in Anderson (1983/1989), p. 259.

42. This was a less severe condition than at Oxford and Cambridge. At Oxford, all entrant students, as well as academic staff, had to subscribe to the thirty-nine articles of the Church of England, while, at Cambridge, this was a condition of graduation.

43. Jack Morrell, 'Leslie, Sir John (1766–1832)', *Dictionary of National Biography* (Oxford, 2004).

44. Stewart J. Brown, *Thomas Chalmers and the Godly Commonwealth in Scotland* (Oxford, 1982).

45. R. N. Smart, 'Literate ladies—a fifty year experiment', *The Alumnus Chronicle* (Alumnus Association of the University of St Andrews), No. 59 (1968), 21–31. See also A. D. D. Craik, 'Review of Robert N. Smart, *Biographical Register of the University of St Andrews 1747–1897*', *Bulletin of the British Society for the History of Mathematics* **5** (Summer 2005), 42–45.

46. A. E. L. Davis, The Davis archive: 'Mathematical Women in the British Isles, 1878–1940', MacTutor website.

47. Chrystal (note 28).

48. Quoted by Craik ('James Ivory', note 13), and Davie (1961).

49. Tony Crilly, 'The *Cambridge Mathematical Journal* and its descendants: the linchpin of a research community in the early and mid-Victorian age', *Historia Math.* **26** (2004), 455–97.

50. *Testimonials* (note 33).

51. These were published as *The Scottish University System Suited to the People* (Adam and Charles Black, 1854) and *How to Improve the Scottish Universities* (note 30) in 1855. When reform was no longer the issue of the day, he devoted his introductory lecture of 1858 to an account of his recent travels in Canada and the United States, published as *Transatlantic Sketches* (Adam and Charles Black, 1858).

52. Robert Louis Stevenson, 'Some College Memories', *Memories and Portraits* (Chatto & Windus, 1906).

53. Anderson (1983/1989), p. 267.

54. Tenures of the St Andrews chair by John Couch Adams (1858–59) and George Chrystal (1877–79) were too brief to make their mark. W. L. F. Fischer (professor 1859–77, and previously professor of natural philosophy 1847–59), although of a scholarly disposition, seems not to have been a stimulating presence.

55. There is an often-told story that William sent his scout to the announcement of the tripos results, and on the scout's return asked him who was *2nd* wrangler, to be shocked by the reply 'You, Sir'. But this is unauthenticated, and seems uncharacteristic of Thomson. See Mark McCartney, 'William Thomson: An Introductory Biography', in Flood, McCartney, and Whitaker (note 12).

56. Letter from Thomson to Stokes, quoted in David Lindley, *Degrees Kelvin: The Genius and Tragedy of William Thomson* (Aurum, 2004).

57. R. A. Rankin, 'Hugh Blackburn: a little-known mathematical friend of Lord Kelvin', *Newsletter of the British Society for the History of Mathematics* **43** (2001), 7–14.

58. See Smith and Wise (note 12), Flood, McCartney, and Whitaker (note 12), and Lindley (note 56).

59. See references in note 58.

60. Cargill Gilston Knott, *Life and Scientific Work of Peter Guthrie Tait* (Cambridge, 1911).

61. J. M. Barrie, 'Professor Tait', *An Edinburgh Eleven* (Hodder & Stoughton, 1924).

62. Letter to Chrystal, 1901, quoted by Knott (note 60).

63. This part of Tait's work is the theme of one of Edwin Morgan's *Sonnets from Scotland* (1984):

PETER GUTHRIE TAIT, TOPOLOGIST
Leith dock's lashed spars roped the young
   heart of Tait.
What made gales tighten, not undo, each knot?
Nothing's more dazzling than a ravelling plot.
Stubby crisscrossing fingers fixed the freight
so fast he started sketching on the spot.
The mathematics of the twisted state
uncoiling its waiting elegances, straight.
Old liquid chains that strung the gorgeous tot
God spliced the mainbrace with, put on the
   slate,
and sent creation reeling from, clutched hot
as caustic on Tait's brain when he strolled late
along the links and saw the stars had got
such gouts and knots of well-tied fire the mate
must sail out whistling to his stormy lot.

From Edwin Morgan, *New Selected Poems* (Carcanet, 2000); we are grateful to Iain Orr for bringing this poem to our attention.

64. Letter from Helmholtz to his wife, quoted by Lindley (note 56).

65. There is an old story that Tait calculated the longest distance to which a golf ball could possibly be driven, only to see it exceeded by Freddie within a fortnight; this is recorded in a cutting from the *Evening Dispatch* which Tait included in his scrapbook, now preserved in the

James Clerk Maxwell Foundation, Edinburgh. But Tait's calculated distance assumed no 'underspin'. The cutting is quoted by J. J. O'Connor and E. F. Robertson, 'Peter Guthrie Tait and the Scrapbook', MacTutor website (Tait). The story can also be found in, for example, R. A. Durran, 'Tait, Frederick Guthrie (1870–1900)', *Dictionary of National Biography* (Oxford, 2004).

66. J. M. Barrie (note 61).

67. *The Student* (1906–07) **409** (8 February 1907), 486–488, quoted by J. J. O'Connor and E. F. Robertson, 'George Chrystal', MacTutor website (Chrystal).

68. Reproduced in Neil Campbell and R. Martin S. Smellie, *The Royal Society of Edinburgh (1783–1983): The First Two Hundred Years* (Royal Society of Edinburgh, 1983).

69. G. Chrystal (note 28).

70. J. J. O'Connor and E. F. Robertson, 'The setting up of the Scottish Leaving Certificate', MacTutor website (Education).

71. Birse (note 21).

72. Hilary Mason, 'J E A Steggall: teaching mathematics 1880–1933', *Bulletin of the British Society for the History of Mathematics* **1** (2004), 27–39.

73. J. E. A. Steggall, *Picturesque Perthshire* (Valentine, Dundee), n.d. [1906].

74. Alex D. D. Craik, 'Science and technology in 19th century Japan: the Scottish connection', *Fluid Dynamics Research* **39** (2007), 24–48.

75. They, and several other Scots at Cambridge, are discussed at greater length by Craik (2007), Chapters 9 and 11, and by Andrew Warwick, *Masters of Theory: Cambridge and the Rise of Mathematical Physics* (Univ. of Chicago Press, 2003).

76. Craik (note 74).

77. Alex D. D. Craik's 'The logarithmic tables of Edward Sang and his daughters', *Historia Math.* **30** (2003), 47–84, and 'Edward Sang (1805–1890): calculator extraordinary', *Newsletter of the British Society for the History of Mathematics* **45** (Spring 2002), 32–43.

78. Craik (Logarithmic tables, note 77) gives a full list of Sang's publications.

79. David K. Brown, 'Russell, John Scott (1808–1882)', *Dictionary of National Biography*, Oxford, 2004; George S. Emmerson, *John Scott Russell, a Great Victorian Engineer and Naval Architect* (John Murray, 1977).

80. David Gavine, 'Henderson, Thomas (1798–1844)', *Dictionary of National Biography* (Oxford, 2004).

81. Hermann A. Brück, 'Smyth, Charles Piazzi (1819–1900)', *Dictionary of National Biography* (Oxford, 2004).

82. Mary R. S. Creese, with contributions by Thomas M. Creese, *Ladies in the Laboratory?* (Scarecrow,

1998–2004); Grace Wyndham Goldie, 'Somerville, Mary (1897–1963)', *Dictionary of National Biography*, Oxford, 2004.

83. Campbell and Smellie (note 68), and on the Royal Society of Edinburgh's website.

84. J. J. O'Connor and E. F. Robertson, 'The Royal Society of Edinburgh and the purchase of 22–24 George Street', MacTutor Archive (Chrystal).

85. Knott (note 60), p. 31.

86. Craik (Logarithmic tables) (note 77).

87. Philip Kelland, *Professor Kelland's address as President at opening session, 1853–54* (NLS shelfmark ABS.2.97.38 (12)).

88. This account is based on Robert A. Rankin, 'The first hundred years (1883–1983)', *Proc. Edinburgh Math. Soc.* **26** (1983), 135–50.

89. This circular is quoted in full by Rankin (note 86), p. 136; see also E. F. Robertson and J. J. O'Connor, *Edinburgh Mathematical Society 125th Anniversary booklet* (Edinburgh Math. Society, 2008).

90. George A. Gibson, John Sturgeon Mackay, M.A., LL.D., *Proc. Edinburgh Math. Soc.* **32** (1913–14), 151–59 (also available on line).

91. Although Volume 2 appeared in 1884, Volume 1 was delayed until 1894 because funds were not at first sufficient.

92. Rankin (note 88), p. 140.

93. Personal communication.

## CHAPTER 5 (RAYMOND FLOOD)

A good starting place are the books *Science in Ireland 1800–1930: Tradition and Reform* (ed. J. Nudds, N. McMillan, D. Weaire, and S. McK. Lawlor) (Dublin, 1988) and *Science and Society in Ireland: 1800–1950* (ed. P. J. Bowler and N. Whyte) (Institute of Irish studies, Queen's University, Belfast, 1997).

There are useful essays on many of the people mentioned in this chapter in Ken Houston (ed.), *Creators of Mathematics: The Irish Connection* (University College Dublin Press, 2000) and *Physicists of Ireland: Passion and Precision* (ed. M. McCartney and A. Whitaker) (Institute of Physics Press, 2003).

Three informative biographies of Hamilton are R. P. Graves's three-volume *Life of Sir William Rowan Hamilton* (Dublin, 1882–1889), T. L. Hankins, *Sir William Rowan Hamilton* (Johns Hopkins Univ. Press, 1980), and S. O'Donnell, *William Rowan Hamilton* (Boole Press, Dublin, 1983). For MacCullogh see the B. K. P. Scaife's article 'James MacCullagh', *Proc. Royal Irish Academy* **90C** (3) (1990), 67–106, as well as the *Collected Works of James MacCullagh* (ed. B. D. Jellett and S. Haughton) (Hodges, Figgis, and Co., 1880). Rod Gow has a fine article on George Salmon available on his website, http://maths.ucd.ie/~rodgow/ [accessed 12 Jan 2011]. Further discussion of Kelvin can be found in *Kelvin and Ireland* (ed. R. Flood, M. McCartney, and A. Whitaker) (Journal of Physics Conference Series **158**, 2009).

## NOTES

1. Charles H. Murray, 'The founding of a university', *University Review* **2** (Nos. 3/4) (1960), 13.

2. I. Grattan-Guinness, 'Mathematical research and instruction in Ireland, 1782–1840', in Nudds *et al.* (1988), p. 17.

3. O'Donnell (1983), p. 25.

4. Quoted in Hankins (1980), p. 19.

5. Grattan-Guinness (note 2), p. 18.

6. Article by T. L. Hankins in the *Dictionary of Scientific Biography*, Vol. 6 (ed. C. G. Gillespie) (Charles Scribner's Sons, 1972), 90.

7. In the *North British Review* **45** (1866), 37–74.

8. From Sir Robert Stawell Ball, *Great Astronomers*, Isbister (1895), pp. 303–34.

9. The law of the modulus holds if the norm of the product equals the product of the norms:

$$\left(a_1^2 + a_2^2 + \cdots + a_n^2\right) \times \left(b_1^2 + b_2^2 + \cdots + b_n^2\right)$$
$$= c_1^2 + c_2^2 + \cdots + c_n^2.$$

10. Michael Crowe, *A History of Vector Analysis* (Notre Dame, 1967), p. 185

11. W. R. Hamilton, 'Memorandum respecting a new system of roots of unity', *Phil. Mag.* **XII** (1856), 446.

12. Augustus De Morgan, An obituary published in the *Gentleman's Magazine and Historical Review* **I** (new series) (1866), 128–34.

13. Hankins (1980), p. 1.

14. Jellett and Haughton (1880), pp. 17–19.

15. Hankins (1980), p. 166.

16. Graves (1882–89), Vol. 2, p. 391.

17. Graves (1882–89), Vol. 3, pp. 331–32.

18. W. R. Hamilton, *Proc. Royal Irish Academy* **1** (1837–38), 212.

19. T. D. Spearman, 'James MacCullagh' in Nudds *et al.* (1988), p. 54. From *A Full Report of the Proceedings at the Election of Two Members to Serve in Parliament for the University of Dublin* (Dublin, 1847), incorporated in Vol. 2012 of the Haliday collection of pamphlets in the library of the Royal Irish Academy.

20. Spearman (note 19), p. 56.

21. Jellett and Haughton (1880), preface.

22. Graves (1882–1889), Vol. 3, p. 334.

23. Graves (1882–1889), Vol. 3, p. 335.

24. Ivor Grattan-Guinness, *History of the Mathematical Sciences* (Fontana, 1997), p. 581.

25. See *Obituary Notices of Fellows of the Royal Society* **11** (Nov. 1942), p. 197.

26. As note 25.

27. For further discussion, see Andrew Warwick, *Masters of Theory: Cambridge and the Rise of Mathematical Physics* (Univ. of Chicago Press, 2003) and Raymond Flood's article on Larmor in Mark McCartney and Andrew Whitaker (2003).

28. Horace Lamb, 'Presidential address', Section A (British Association Meeting, 1904), p. 421.

29. *Nature* **63** (No. 1636) (7 March 1901), p. 445.

30. As note 29, p. 447.

31. A. Warwick, 'Frequency, theorem and formula: remembering Joseph Larmor in electromagnetic theory', *Notes and Records Roy. Soc. (London)* **47** (1) (1993), 49–60.

32. 'Letter from Lewes Jail', *Eamon de Valera Centenary* (Dublin Institute for Advanced Studies, 1982), pp. 21–22.

33. *Eamon de Valera Centenary* (Dublin Institute for Advanced Studies, 1982), p. 23.

## CHAPTER 6 (JUNE BARROW-GREEN)

Little general work has so far been done on mathematics and the British Empire, and indeed this chapter is only a preliminary survey, but there are a number of general studies on science and the British Empire. For a useful overview of literature on the latter, see Mark Harrison's article 'Science and the British Empire', *Isis* **96** (2005), 56–63. Another useful, if at times patchy, source on science and empire is the collection of essays, *Nature and Empire: Science and the Colonial Enterprise* (ed. R. MacLeod) (Osiris, 2000).

On the growth of mathematics in different locales, see G. Cohen, 'Counting Australia', in *The People, Organisations and Institutions of Australian Mathematics* (Halstead Press, in association with the Australian Mathematical Society, 2007), T. Archibald and L. Charbonneau, 'Mathematics in Canada before 1945: a preliminary survey', *Mathematics in Canada 1945-1995* (Canadian Mathematical Society, 1995), pp. 1–43, and D. A. Nield, 'University mathematics in Auckland. A historical essay', *Mathematical Chronicle* **12** (1983), 1–33. For the Observatory at Cape Town, see B. Warner, *Astronomers at the Royal Observatory, Cape of Good Hope: a History, with Emphasis on the Nineteenth Century* (Balkema, Cape

Town, 1979). For the British Association, see *The Parliament of Science. The British Association for the Advancement of Science 1831–1981* (ed. R. MacLeod and P. Collins) (Science Reviews Ltd., Northwood, 1981). For Haileybury and Addiscombe, see F. C. Danvers, *Memorials of Old Haileybury College* and H. M. Vibart, *Addiscombe: its Heroes and Men of Note* (both Archibald Constable, 1894), and for the Royal Indian Engineering College, see J. G. P. Cameron, *A Short History of the Royal Indian Engineering College, Coopers Hill* (private circulation, 1960).

For general sources of information on the lives of individuals, see the *Complete Dictionary of Scientific Biography*; the *Oxford Dictionary of National Biography* (both also on line by subscription), and P. C. Fenton, *Dictionary of New Zealand Biography*, http://www.dnzb.govt [accessed 12 Jan 2011]. For a study of an individual within a colonial context, see the two articles by J. G. Jenkin, 'William Henry Bragg in Adelaide', *Isis* **95** (2004), 58–90, and 'The Appointment of W. H. Bragg, F.R.S., to the University of Adelaide', *Notes & Records of the Royal Society of London* **40** (1985), 75–99.

For general Cambridge context, see Andrew Warwick *Masters of Theory: Cambridge and the Rise of Mathematical Physics* (Univ. of Chicago Press, 2003), and Alex D. D. Craik, *Mr. Hopkins' Men: Cambridge Reform and British Mathematics in the 19th Century* (Springer, 2007).

## NOTES

1. The 16th century mathematician, astrologer and antiquary, John Dee was the first to use the term 'Brytish Impire' in his *General and Rare Memorials Pertayning to the Perfect Arte of Navigation* (1577), p. 3.

2. Wranglers were students in the top division of the mathematical tripos (see Chapter 1). For information about the origin of the term 'tripos' and the structure of the tripos examination in the 19th century, see W. W. Rouse Ball, *A History of the Study of Mathematics at Cambridge* (Cambridge, 1889). For a discussion of the Smith's prize examination, which was deemed a better measure of mathematical originality than the tripos examination, see J. E. Barrow-Green, '"A Corrective to the Spirit of Too Exclusively Pure Mathematics": Robert Smith (1689–1768) and his prizes at Cambridge University', *Annals of Science* **56** (1999), 271–316.

3. C. D. Burns, *A Short History of Birkbeck College* (Univ. of London Press), 1924.

4. I. S. Turner, 'The first hundred years of mathematics in the University of Sydney', *Royal Australian Historical Society Journal and Proceedings* **41** (1955), 245–66, on p. 246.

5. Sir William Charles Windemeyer (1834–1897) went on to have a distinguished, if controversial, legal career. He was chancellor of Sydney University from 1895 to 1896.

6. Turner (note 4), p. 245.

7. Cohen (2007), p. 48.

8. Turner (1955), p. 248.

9. Cohen (2007), pp. 48–49.

10. Turner (1955), p. 250.

11. T. W. Moody and J. C. Beckett, *Queen's Belfast 1845–1949. The History of a University* (Faber & Faber, 1959), pp. 159–66.

12. G. Cohen, 'The appointment of the first four professors of mathematics in the University of Melbourne', *Gazette of the Australian Math. Society* **33** (2006), 14–21, on p. 14.

13. G. Blainey, *A Centenary History of the University of Melbourne* (Melbourne Univ. Press, 1957), pp. 39–40.

14. Blainey (note 13), p. 11.

15. Cohen (note 12), p. 15.

16. K. H. Parshall, *James Joseph Sylvester. Life and Work in Letters* (Clarendon Press, Oxford, 1998), p. 144.

17. K. H. Parshall, *James Joseph Sylvester. Jewish Mathematician in a Victorian World* (Johns Hopkins Univ. Press, 2006), p. 223.

18. *The Times*, 19 February 1875.

19. G. C. Fendley, 'Nanson, Edward John (1850–1936)', *Australian Dictionary of Biography* (Melbourne Univ. Press, 1973), p. 663.

20. Since 1885 the Smith's prizes have been awarded on the basis of an essay, rather than on an examination performance. The title of Michell's essay was 'The vibrations of curved rods and shells'.

21. Cohen (note 12), p. 15.

22. R. T. Glazebrook and A. E. H. Love, 'Horace Lamb. 1849–1934', *Obituary Notices of Fellows of the Royal Society* **1** (1935), 374–92, on p. 378.

23. E. N. da C. Andrade and K. Lonsdale, 'William Henry Bragg. 1862–1942', *Obituary Notices of Fellows of the Royal Society* **4** (1943), 276–300, on p. 279.

24. For a discussion of Bragg's research in Adelaide, see Jenkin (2004).

25. Jenkin (1985), p. 93.

26. Jenkin (1985), p. 94.

27. Glazebrook and Love (note 22), p. 377.

28. G. E. Thompson, *A History of the University of Otago (1869–1919)* (J. Wilkie & Co, Dunedin, 1921), p. 12.

29. P. C. Fenton, 'John Shand', *Dictionary of New Zealand Biography*, http://www.dnzb.govt.nz [accessed 12 Jan 2011].

30. Sir Robert Stout, 'A tribute to the late Dr. Shand', *The Otago University Review* **29** (1915), 18.

31. W. P. Morrell, *The University of Otago. A Centennial History* (University of Otago Press, Dunedin, 1969), p. 77.

32. Nield (1983), p. 3.

33. For the contents of the syllabuses, see Nield (1983), p. 6.

34. Warwick (2003), p. 232.

35. T. S. Aldis was 2nd wrangler in 1866; the third brother, A. J. Aldis, was 6th wrangler in 1863.

36. W. S. Aldis, *A Textbook of Algebra* (Clarendon Press, Oxford, 1887), Preface.

37. Nield (1983), p. 8.

38. Nield (1983), p. 9.

39. K. Sinclair, *A History of the University of Auckland 1883–1983* (Auckland Univ. Press, 1983), pp. 131–32.

40. R. Narasimhan, 'The coming of age of mathematics in India', *Miscellanaea Mathematica* (ed. P. Hilton, F. Hirzebruch, and R. Remmert) (Springer, 1991), pp. 236–58, on p. 236.

41. The Presidency College, Calcutta, founded in 1817 as the Hindu College, was the first college to be established in India for the delivery of English education.

42. N. K. Sinha, *Asutosh Mookerjee: A Biographical Study* (Asutosh Mookerjee Centenary Committee, Calcutta, 1966), p. 176.

43. Sir A. Mookerjee, *A Diary of Sir Asutosh Mookerjee, with his Life Sketch by Syama Prasad Mookerjee* (Asutosh Mookerjee Memorial Institute, Calcutta, 1998), p. 167.

44. A full list of the texts referred to by Mookerjee in his diary (1883–86) is given in Mookerjee (note 43).

45. Mookerjee (note 43), 171.

46. J. T. Hathornthwaite, *A Manual of Elementary Algebra* (Bell & Sons, 1894), p. 139.

47. For a discussion of Todhunter as a textbook writer, see J. E. Barrow-Green, "The advantage of proceeding from an author of some scientific reputation': Isaac Todhunter and his mathematical textbooks', *Teaching and Learning in Nineteenth-Century Cambridge* (ed. J. Smith and C. Stray) (Boydell Press, 2001), pp. 177–203.

48. English editions of Todhunter's texts also appeared in the United States, Australia, and China, while the translations were similarly widespread with editions also appearing in Italian, Chinese, and Japanese.

49. E. J. Routh, 'Isaac Todhunter', *Proceedings of the Royal Society* 37 (1884), 27–32, on p. 29.

50. A. McConnell, 'John Henry Pratt', *Oxford Dictionary of National Biography* (Oxford, 2004).

51. G. F. Childe, *Investigations in the Theory of Reflected Ray-surfaces and their Relation to Plane Reflected Caustics* (J. C. Juta, Cape Town, 1857), pp. vii–viii.

52. G. F. Childe, *Singular Properties of the Ellipsoid and Associated Surfaces of the Nth Degree* (Macmillan and J. C. Juta, Cape Town, 1861).

53. W. Ritchie, *The History of the SA College: 1829–1918*, 2 vols. (Maskew Miller, Cape Town, 1918), p. 716.

54. Smith was not the only wrangler in South Africa to have a career outside mathematics. Henry Cotterell, senior wrangler of 1835, became Bishop of Grahamstown in 1857, having previously been chaplain for the East India Company. John Colenso, 2nd wrangler in 1836 and author of popular textbooks, notably *Elements of Algebra* (1841) and *Arithmetic* (1843), was appointed the first Bishop of Natal in 1853; he later gained notoriety for holding unorthodox and liberal views.

55. T. Muir, 'The promotion of research: with special reference to the present state of the Scottish universities and secondary schools', An address delivered before the Mathematical Society of Edinburgh, 8 February 1884 (privately printed, A. Gardner, Paisley, 1884).

56. For further information about Beattie, Crawford, and Brown, see L. Crawford, 'Sir John Carruthers Beattie, Kt., D.Sc., LL.D.', *Royal Society of Edinburgh Year Book 1947*, pp. 10–11, S. Skewes, 'Lawrence Crawford, M. A. (Cantab.), D.Sc. (Glas.), LL.D. (Wit.)', *Royal Society of Edinburgh Year Book 1952*, pp. 14–15, and L. Crawford, 'Alexander Brown, M.A., B.Sc.', *Royal Society of Edinburgh Year Book 1949*, pp. 12–13.

57. For a detailed history of the Cape Observatory during the Victorian period, see Sir David Gill, *A History and Description of the Royal Observatory, Cape of Good Hope* (H.M.S.O., London, 1913).

58. Warner (1979), p. 73.

59. Warner (1979), p. 77.

60. It has so far proved impossible to ascertain the identity of Bard, since no-one of that name appears in any of the Cambridge lists for the period.

61. To quote Gill himself: '...to many it seemed that Clerk Maxwell was not a very good professor. But to those who could catch a few of the sparks that flashed as he thought aloud at the blackboard in lecture, or when he twinkled with wit and suggestion in after-lecture conversation, Maxwell was supreme as an inspiration'. See Gill (note 57), p. xxxi.

62. Anon., 'Sir David Gill', *The Observatory* 37 (1914), 115–17, on p. 115.

63. We are grateful to Tom Archibald for supplying us with much of the information about university life in Canada in the 19th century. See Archibald and Charbonneau (1995), pp. 1–43.

64. One man who almost contributed to the development of mathematics in Canada was Duncan Gregory, a notable

figure in the context of mid 19th-century British algebra. Gregory was offered a position at King's College, Toronto, in 1841 but declined it on grounds of poor health; he died three years later.

65. R. C. Wallace, *Some Great Men of Queen's* (Ayr Publishing, Manchester, New Hampshire, 1969), p. 58.

66. Wallace (note 65), p. 59.

67. For further information about these meetings, and specifically their cultural context, see M. Worboys, 'The British Association and Empire: science and social imperialism, 1880–1940', in MacLeod and Collins (1981), pp. 170–87.

68. Lord Rayleigh, 'Recent progress in physics', *Science* **4** (1884), 179–84.

69. Anon., 'Lord Rayleigh', *Science* **4** (1884), 161–63, on p. 163.

70. Anon., 'The Toronto meeting of the British Association', *Science* **6** (1897), 333–38, on p. 334.

71. H. Callendar, 'Mathematics and physics at the British Association', *Science* **6** (1897), 464–72, on p. 465. Callendar was professor of physics at McGill University, Montreal. He had graduated from Cambridge as 16th wrangler in 1885 and was professor of physics at Royal Holloway College from 1888 to 1893. He returned to England in 1902 to take up the chair of physics at the Royal College of Science, later Imperial College.

72. *The Times*, 21 August 1897, p. 6.

73. C. Babbage, *Passages from the Life of a Philosopher* (Longman & Co., 1864), p. 473.

74. A. D. D. Craik, 'James Ivory, mathematician: 'The Most Unlucky Person that Ever Existed'', *Notes & Records of the Royal Society* **54** (2000), 223–47, on p. 244, note 34.

75. Danvers (1894), p. 68.

76. Danvers (1894), p. 189.

77. Vibart (1894), pp. 225–26.

78. J. Cape, *A Course of Mathematics Principally Designed for the Use of Students in the East India Company's Military College at Addiscombe* (Longman & Co, 1839/1844), p. vii.

79. Cape (note 78), p. vii.

80. For further information about the relationship between Wrigley and Charles Darwin, see the Darwin Correspondence Project, http://www.darwinproject.ac.uk [accessed 12 Jan 2011].

81. J. A. Venn, *Alumni Cantabrigienses: A Biographical List of All Known Students, Graduates and Holders of Office at the University of Cambridge from the Earliest Times to 1900*, Part 2, Vol. 6 (Cambridge, 1954), p. 600.

82. Vibart (1894), p. 206.

83. The College buildings are now part of Brunel University.

84. Cameron (1960), p. 11.

85. A. R. Forsyth, 'Joseph Wolstenholme', *Dictionary of National Biography* (Smith, Elder & Co., 1901).

86. Forsyth (note 85).

87. Letter from General Sir Alexander Taylor to the Under Secretary of the India Office, 1 May 1889, British Library, Oriental and India Office Collections.

88. L. Stephen, *Sir Leslie Stephen's Mausoleum Book* (*with an Introduction by Alan Bell*) (Clarendon Press, Oxford, 1977), p. 79.

89. Mr Alfred Lodge, *The Times*, 6 December 1937.

90. British Library, India Office Records, J & K.

91. O. Lodge, 'George Minchin Minchin', *Obituary Notices of Fellows Deceased. Proceedings of the Royal Society* (A) **92** (1916), xlvi–l, on p. xlvi.

92. Lodge (note 91), p. l.

## CHAPTER 7 (SLOAN EVANS DESPEAUX)

A thorough analysis of the mathematics and mathematicians of 19th-century British scientific journals can be found in Sloan E. Despeaux, *The Development of a Publication Community: Nineteenth-Century Mathematics in British Scientific Journals* (Ph.D. thesis, Univ. of Virginia, 2002).

Excellent studies on individual journals or societies include Tony Crilly, 'The *Cambridge Mathematical Journal* and its descendants: The linchpin of a research community in the early and mid-Victorian Age', *Historia Math.* **31** (2004), 455–97; Teri Perl, *The Ladies' Diary* or *Woman's Almanack*, 1704–1841, *Historia Math.* **6** (1979), 36–53; Janet Burt (Delve), *The Development of the Mathematical Department of the Educational Times from 1847 to 1862* (Ph.D. thesis, Middlesex Univ., 1998); Adrian C. Rice, Robin J. Wilson, and J. Helen Gardner, 'From student club to national society: the founding of the London Mathematical Society in 1865', *Historia Math.* **22** (1995), 402–21; Adrian C. Rice and Robin J. Wilson, 'From national to international society: the London Mathematical Society, 1867–1900', *Historia Math.* **25** (1998), 185–217; Michael H. Price's *Mathematics for the Multitude* (Mathematical Association, 1994); and Joe Albree and Scott H. Brown, 'A valuable monument of mathematical genius: *The Ladies Diary* (1704–1840)', *Historia Math.* **36** (2009), 10–47.

For more on British mathematical contributions abroad during this period, see Sloan Evans Despeaux, 'Mathematics sent across the Channel and the Atlantic: nineteenth-century British mathematical contributions to international scientific journals', *Annals of Science* **65** (2008), 73–99. More information about Continental mathematical journals can be found in Wolfgang Eccarius, 'August Leopold Crelle als Herausgeber wissenschaftlicher Fachzeitschriften', *Annals of Science* **33** (1976), 229–61; Silvina Duvina, 'Le *Journal de Mathématiques Pures et Appliquées* sous la férule de J. Liouville

(1836–1874)', *Sciences et Techniques en Perspective* **28** (1994), 179–217; Laura Martini, 'The politics of unification: Barnaba Tortolini and the publication of research mathematics in Italy, 1850–1865', *Il Sogno di Galois: Scritti di Storia della Matematica Dedicati a Laura Toti Rigatelli per il suo 60° Compleanno* (ed. R. Franci, P. Pagli, and A. Simi) (Studi della Matematica Medioevale, Univ. di Siena, 2003), 171–98; and June E. Barrow-Green, 'Gösta Mittag-Leffler and the foundation and administration of *Acta Mathematica*', *Mathematics Unbound: The Evolution of an International Mathematical Community, 1800–1945* (ed. K. H. Parshall and A. C. Rice) (American and London Mathematical Societies, 2002), pp. 138–64; Karen Hunger Parshall and David E. Rowe, *The Emergence of the American Mathematical Research Community, 1876–1900: J. J. Sylvester, Felix Klein, and E. H. Moore* (American and London Mathematical Societies, 1994), pp. 88–94; Elena Aussejo and Mariano Hormigon (ed.), *Messengers of Mathematics: European Mathematical Journals 1810–1939* (Siglo XXI, 1993); and Heinrich Behnke, 'Rückblick auf die Geschichte der Mathematischen Annalen', *Math. Annalen* **200** (1973), i–vii. For more on mathematical societies in Europe, see Danny J. Beckers, 'Untiring labour overcomes all! The Dutch Mathematical Society in European perspective', *Historia Math.* **28** (2001), 31–47.

## NOTES

1. J. W. L. Glaisher, 'Mathematical journals', *Nature* **22** (1880), 73–75, on p. 74.

2. Raymond Archibald, 'Notes on some minor English mathematical serials', *Math. Gazette* **14** (1929), 379–400.

3. Perl (1979), p. 37.

4. Ruth Wallis and Peter Wallis, 'Female philomaths', *Historia Math.* **7** (1980), 57–64, on p. 58.

5. Thomas Turner Wilkinson, 'Mathematical periodicals', *Mechanics Magazine* **49** (1848), 5–7, on p. 6.

6. Thomas Stephens Davies, William Rutherford, and Stephen Fenwick, 'Prospectus', *The Mathematician* **1** (1843–44), 1–3, on p. 2.

7. Hugh Godfray, 'Approximate rectification of the circle', *The Mathematician* **3** (1849–50), 121–22, on p. 122.

8. 'Solutions of mathematical exercises—CLXVI. Dr. Burns, Rochester', *The Mathematician* **3** (1849–50), 281.

9. William Rutherford and Stephen Fenwick, 'Preface', *The Mathematician* **3** (1849–50).

10. J. S. Mackay, 'Notice sur le journalisme mathématique en Angleterre', *Comptes Rendu de l'Association Française pour l'Avancement des Sciences* **2** (1893), 303–308, on p. 308.

11. Burt (Delve) (1998), pp. 74, 113–14.

12. Burt (Delve) (1998), p. 131.

13. Burt (Delve) (1998), pp. 322, 325.

14. Letter from James Joseph Sylvester to William Miller, 16 Oct. 1865, in Karen Hunger Parshall, *James Joseph Sylvester: Life and Work in Letters* (Clarendon Press, Oxford, 1998), p. 127.

15. W. K. Clifford, *Richmond and Twickenham Times* (17 August 1889), quoted in B. F. Finkel, 'Biography: W. J. C. Miller', *American Math. Monthly* **3** (1896), 159–163, on p. 162.

16. *Mathematical Questions and Solutions from the 'Educational Times'* **43** (1885), 34, 112, 127, 131, 132, 141.

17. J. W. L. Glaisher, 'Mathematical journals', *Nature* **22** (1880), 73–75, on p. 74.

18. For more on the Cambridge educational system during the Victorian era, see Chapter 1.

19. Patricia Allaire, *The Development of British Symbolical Algebra as a Response to "The Problem of Negatives" with an Emphasis on the Contribution of Duncan Farquharson Gregory* (UMI Dissertation Services, Ann Arbor, 1997), p. 76.

20. [Duncan F. Gregory], 'Preface', *Cambridge Math. Journal* **1** (1837), 1–2, on p. 1.

21. Despeaux (2002), p. 163.

22. Elaine Koppelman, 'The calculus of operations and the rise of abstract algebra', *Archive for History of Exact Sciences* **8** (1971), 155–242, on p. 157.

23. Augustus De Morgan, quoted in Koppelman (note 22), p. 189. For more on the calculus of operations in the *Cambridge Mathematical Journal*, see Sloan E. Despeaux, '"Very full of symbols": Duncan F. Gregory, the calculus of operations, and the *Cambridge Mathematical Journal*', *Episodes in the History of Modern Algebra (1800–1950)* (ed. J. Gray and K. H. Parshall) (American and London Mathematical Societies, 2007), pp. 49–72.

24. George Boole, 'Exposition of a general theory of linear transformations', *Cambridge Mathematical Journal* **3** (1841–43), 1–20. For more on the influence of this article on the British approach to invariant theory, see Karen Hunger Parshall, 'Toward a history of nineteenth-century invariant theory', *The History of Modern Mathematics*, Vol. 1 (ed. D. E. Rowe and J. McCleary) (Academic Press, 1989), pp. 157–206; and Paul R. Wolfson, 'George Boole and the origins of invariant theory', *Historia Math.* **35** (2008), 37–46.

25. Crilly (2004), p. 487.

26. S. P. Thompson, *The Life of William Thomson, Baron Kelvin of Largs*, Vol. 1 (Macmillan, 1910), pp. 113–20.

27. For more on mathematics in Dublin during the Victorian era, see Chapter 5.

28. Despeaux (2002), p. 168.

29. Tony Crilly, *Arthur Cayley: Mathematician Laureate of the Victorian Age* (Johns Hopkins Univ. Press, 2006), pp. 145–46.

30. A surface of degree $n$ is defined by a homogeneous polynomial of degree $n$.

31. Arthur Cayley, 'On the triple tangent planes of surfaces of the third order', *Cambridge and Dublin Math. J.* **4** (1849), 118–32.

32. Rod Gow, 'George Salmon 1819–1904', *Creators of Mathematics: The Irish Connection* (ed. K. Houston) (University College Dublin Press, 2000), pp. 37–45, on pp. 42–43.

33. For more on Dickson, see Della Fenster, *Leonard Eugene Dickson and his Work in the Theory of Algebras* (Ph. D. thesis, Univ. of Virginia, 1994).

34. Leonard E. Dickson, 'A triply infinite system of simple groups', 'The first hypoabelian group generalized', 'Simplicity of the Abelian group on two pairs of indices in the Galois field of order $2n$, $n > 1$', and 'A class of linear groups including the Abelian group', *Quarterly Journal of Pure and Applied Mathematics* **29** (1898), 169–78; **30** (1899), 1–16; **30** (1899), 383–84; and **31** (1900), 60–66.

35. Parshall and Rowe (1994), p. 381.

36. 'Introduction', *Oxford, Cambridge, and Dublin Messenger of Mathematics* **1** (1862), 1–4.

37. Advertisement, *Messenger of Mathematics* **1** (1871), iii–iv, on p. iii.

38. After Glaisher's death, *The Messenger of Mathematics* was absorbed into a new series of *The Quarterly Journal*, centred at Oxford. The first volume of this new series appeared in 1930, and its editors were the Oxford mathematicians Theodore Chaundy, William Ferrar, and Edgar Poole, with the assistance of Arthur Dixon, Edwin Bailey Elliott, G. H. Hardy, Augustus Love, Edward Milne, F. B. Pidduck, and E. C. Titchmarsh, all from Oxford. With this team, the journal 'went from strength to strength, bolstering along the way Oxford's research reputation within the international mathematics community'—John Fauvel, '800 Years of Mathematical Traditions', *Oxford Figures: 800 Years of the Mathematical Sciences* (ed. J. Fauvel, R. Flood, and R. Wilson) (Oxford, 2000), pp. 1–27, on p. 16.

39. Godfrey Harold Hardy, 'Dr. Glaisher and the "Messenger of Mathematics"', *Messenger of Math.* **58** (1929), 159–60, on p. 159.

40. See Despeaux (2008).

41. This table is taken from Robert Mortimer Gascoigne, *A Historical Catalogue of Scientific Periodicals, 1665–1900: With a Survey of their Development* (Garland Publishing, 1985).

42. Michel Chasles, *Rapport sur les Progrès de la Géométrie* (Imprimerie nationale, Paris, 1870), pp. 378–79, quoted in Hélène Gispert, 'The effects of war on France's international role in mathematics, 1870–1914', *Mathematics Unbound: The Evolution of an International Mathematical Community, 1800–1945* (ed. K. H. Parshall and A. C. Rice), American and London Mathematical Societies, 2002), pp. 105–21, on pp. 106–7.

43. J. W. S. Cassels, 'The Spitalfields Mathematical Society', *Bulletin London Math. Soc.* **11** (1979), 241–58, on pp. 242, 245–51; Spitalfields is an area in the east end of London.

44. Philip C. Enros, 'The Analytical Society (1812–1813): precursor of the renewal of Cambridge Mathematics', *Historia Math.* **10** (1983), 26–37, on p. 37.

45. Rice, Wilson, and Gardner (1995), pp. 404, 407, 410, 415.

46. Rice, Wilson, and Gardner (1995), p. 411.

47. The societies and dates in Box 7.6 are from Beckers (2001), pp. 40–44.

48. James W. L. Glaisher, 'Notes on the early history of the Society', *J. London Math. Soc.* **1** (1926), 51–64, on p. 60. Glaisher was also Secretary of the Royal Astronomical Society from 1877–84, and from 1881 until the end of his secretaryship was the editor of its publications.

49. Glaisher (note 48), p. 63.

50. Rice and Wilson (1998), p. 205.

51. Rice and Wilson (1998), p. 207.

52. Rice and Wilson (1998), p. 197.

53. Thomas Archer Hirst, quoted in Rice and Wilson (1998).

54. For more on geometry and the debate over Euclid, see Chapter 14.

55. James Maurice Wilson, quoted in Price (1994), p. 22.

56. James M. Wilson, 'The early history of the Association', *Math. Gazette* **10** (1921), 239–46, on p. 241.

57. Price (1994), p. 23.

58. Joan L. Richards, *Mathematical Visions: The Pursuit of Geometry in Victorian England* (Academic Press, 1988), pp. 173–74.

59. Price (1994), pp. 30–31.

60. T. A. A. Broadbent, '*The Mathematical Gazette*: our history and aims', *Math. Gazette* **186** (October 1946), pp. 186–94, on p. 186.

61. Price (1994), p. 38.

62. For a listing of these papers, see 'Publications issued by the Association for the Improvement of Geometrical Teaching', reproduced in Price (1994), p. 37.

63. E. M. Langley, 'Origin of the *Mathematical Gazette*', *Math. Gazette* **1** (1894), quoted in Price (1994), p. 40.

64. Price (1994), p. 64. Taking educational contributions to include correlation with other subjects the examination system, teacher supply and education, and educational research, Price calculated that 12 per cent of the *Gazette* was educational for 1894–95, 16 per cent for 1896–97, 10 per cent for 1898–99, and 3 per cent for 1900. Questions for answer in the *Gazette* 'became virtually extinct' after 1908. Price (1994), p. 65.

65. Cargill G. Knott, Andrew J. G. Barclay, and Alexander Y. Fraser, 'Circular', January 23, 1883; quoted in Robert A. Rankin, 'The first hundred years', *Proc. Edinburgh Math. Soc.* **26** (1983), 135–50, on p. 136.

66. Rankin (note 65), pp. 135, 137.

67. Glaisher (note 17), p. 75.

68. Adrian Rice, Robin Wilson, and Helen Gardner cite November 1866 as a date by which the LMS had reached national proportions. Rice, Wilson, and Gardner (1995), p. 415.

## CHAPTER 8 (A. D. D. CRAIK)

Much of this chapter is a reworking of material in Alex D. D. Craik's *Mr. Hopkins' Men: Cambridge Reform and British Mathematics in the 19th Century* (Springer, 2007).

Other studies of mathematical and scientific education at Cambridge during the 19th century, and the research achievements of its graduates, are Peter M. Harman's *Wranglers and Physicists: Studies on Cambridge Physics in the Nineteenth Century* (Manchester Univ. Press, 1985), and Andrew Warwick's *Masters of Theory: Cambridge and the Rise of Mathematical Physics* (Univ. of Chicago Press, 2003).

A useful general source, with surveys of many scientific topics, is Ivor Grattan-Guinness's *Companion Encyclopedia of the History and Philosophy of the Mathematical Sciences*, 2 vols. (Routledge, 1994). Much information on individuals is in the multi-volume *Oxford Dictionary of National Biography*, also available electronically, and in the St Andrews University electronic *MacTutor History of Mathematics Archive*, http://www-history.mcs.st-andrews.ac.uk/history/ [accessed 12 Jan 2011].

The lives and work of Stokes, Kelvin, Tait, and Maxwell are extensively treated in Silvanus P. Thompson's *The Life of William Thomson Baron Kelvin of Largs*, 2 vols. (Macmillan, 1910); Cargill G. Knott's *Life and Scientific Work of Peter Guthrie Tait* (Cambridge, 1911); David B. Wilson's *Kelvin & Stokes, A Comparative Study in Victorian Physics* (Adam Hilger, 1987); Crosbie Smith & M. Norton Wise's *Energy and Empire. A Biographical Study of Lord Kelvin* (Cambridge, 1989); and Peter M. Harman's *The Natural Philosophy of James Clerk Maxwell* (Cambridge, 1998).

On 'aether' theories and electromagnetism, see Sir Edmund Whittaker's *A History of the Theories of Aether and Electricity. Vol. 1: The Classical Theories*, 2nd enlarged ed. (Nelson, 1951), and Bruce J. Hunt's *The Maxwellians* (Cornell, 1991). On the history of fluid mechanics, see Olivier Darrigol's *Worlds of Flow: A History of Hydrodynamics from the Bernoullis to Prandtl* (Oxford, 2005). On elasticity theory, see the brief historical introduction to A. E. H. Love's *A Treatise on the Mathematical Theory of Elasticity*, 4th ed. (Cambridge, 1927, also a Dover reprint), or Isaac Todhunter and Karl Pearson's monumental *A History of the Theory of Elasticity and of the Strength of Materials from Galilei to the Present Time* (Cambridge, 1886–93).

### NOTES

1. Augustus De Morgan, *A Budget of Paradoxes* (Longman Green, 1872), p. 145.

2. Charles Hutton, *A Mathematical and Philosophical Dictionary*, Vol. 2 (J. Johnson and G. G. & J. Robinson, 1795), pp. 81–82.

3. Walter W. Rouse Ball, *A Short Account of the History of Mathematics* (Macmillan, 1888), p. 411.

4. John Playfair, 'Review of Laplace, *Traité de Méchanique Céleste*', *Edinburgh Review* **11** (1808), 249–84.

5. Ivor Grattan-Guinness, *Convolutions in French Mathematics, 1800–1840*, 3 vols. (Birkhäuser, 1990).

6. Alex D. D. Craik, 'James Ivory, mathematician: 'the most unlucky person that ever existed'', *Notes & Records of the Royal Society* **54** (2000), 223–47, and 'James Ivory's last papers on the "Figure of the earth" (with biographical additions)', *Notes & Records of the Royal Society* **56** (2002), 187–204, and Mary D. Cannell, *George Green: Mathematician and Physicist 1793–1841*, 2nd ed. (S.I.A.M., 2001).

7. There is a huge literature on Cambridge University in the 19th century: see Warwick (2003), Craik (2007), and the references therein.

8. See John Fauvel, Raymond Flood, and Robin Wilson, *Oxford Figures: 800 Years of the Mathematical Sciences* (Oxford, 2000).

9. See, for example, Craik (2007).

10. A 'popular' account of the controversy is Tom Standage's *The Neptune File. Planet Detectives and the Discovery of Worlds Unseen* (Allen Lane, 2000). A recent scholarly reassessment of the evidence is Nicholas Kollerstrom, 'An hiatus in history: the British claim for Neptune's co-prediction, 1845–1846', *History of Science* **44** (2006), 1–28, 349–71.

11. George Biddell Airy, *Mathematical Tracts* (J. Smith, 1826, and later enlarged editions).

12. John Henry Pratt successfully combined his scientific interests with the ecclesiastical post of Archdeacon of Calcutta.

13. Isaac Todhunter, *A History of the Mathematical Theories of Attraction and the Figure of the Earth*, 2 vols. (Constable, 1873, Dover reprint, 1962).

14. Thomas L. Hankins, *Sir William Rowan Hamilton* (Johns Hopkins Univ. Press, 1980).

15. The best biography of Tait is still that of Knott (1911). The collaboration of Thomson and Tait is treated by Thompson (1910) and by Smith and Wise (1989).

16. Alex D. D. Craik, 'The logarithmic tables of Edward Sang and his daughters', *Historia Math.* **30** (2003), 47–84.

17. David E. Cartwright, *Tides: A Scientific History* (Cambridge, 1999).

18. See Alex D. D. Craik, 'The origins of water wave theory', *Ann. Rev. Fluid Mech.* **36** (2004), 1–28, and 'George Gabriel Stokes on water wave theory', *Ann. Rev. Fluid Mech.* **37** (2005), 23–42. See also Olivier Darrigol, 'The spirited horse, the engineer, and the mathematician: water waves in nineteenth-century hydrodynamics', *Arch. Hist. Exact Sci.* **58** (2003), 21–95, and Chapter 2 of Darrigol (2005).

19. Olivier Darrigol, 'Between hydrodynamics and elasticity theory: the first five births of the Navier–Stokes equation', *Arch. Hist. Exact Sci.* **56** (2002), 95–150; see also Chapter 3 of Darrigol (2005).

20. See Darrigol (2005).

21. An early precursor was 'Hooke's law', proposed by Robert Hooke in 1678; this is equivalent to the statement that the extension of a spring is proportional to the force applied at its end. A later, but pre-Victorian, British contribution was that of Thomas Young (1807), after whom 'Young's modulus' is named.

22. See Alfred B. Basset, *A Treatise on Hydrodynamics*, 2 vols. (Deighton Bell, 1888, Dover reprint, 1961), Horace Lamb, *Hydrodynamics* (Cambridge, 1895), John William Strutt (Baron Rayleigh), *The Theory of Sound* (Macmillan, 1877–78), and Augustus E. H. Love, *A Treatise on the Mathematical Theory of Elasticity* (Cambridge, 1892–93), all available in Dover reprints.

23. These institutions amalgamated in 1886 and thereafter employed mainly Japanese staff. Sekiya Seikei, the world's first professor of seismology, was appointed to a newly created chair in the Imperial University in 1889.

24. See Alex D. D. Craik, 'Science and technology in 19th century Japan: the Scottish connection', *Fluid Dynamics Research* **39** (2007), 24–48, and references therein.

25. As an antidote to uncritical eulogies of Victorian engineering and free enterprise, see Charles McKean, *Battle for the North. The Tay and Forth Bridges and the 19th-Century Railway Wars* (Granta, 2006).

26. However, in 1879 and 1893 (respectively), MacCullagh's aether model was re-examined, and its merits recognized, by his fellow-Irishmen George Francis Fitzgerald and Joseph Larmor.

27. Whittaker (1951) remains authoritative. More recent accounts, focusing on particular individuals, are Wilson (1987), Smith and Wise (1989), and Harman (1998).

28. Recently, knot theory has found a new physical application in the modern theory of superstrings.

29. More precisely, Thomson had earned his degree but did not in fact graduate, perhaps because he feared that this might debar him from enrolling as an undergraduate at Cambridge.

30. On the age of the earth and related matters, see Smith and Wise (1989) and Craik (2007).

31. See, for example, Knott (1911) and Smith and Wise (1989).

32. See Harman (1998).

33. See Smith and Wise (1989) and Harman (1998).

34. See Hunt (1991), Harman (1998), and Warwick (2003).

35. See Hankins (note 14) and M. J. Crowe, *A History of Vector Analysis*, 2nd ed. (Dover, 1985).

36. See Craik (note 24).

37. See Smith and Wise (1989).

38. Gillian Cookson, *The Cable: the Wire that Changed the World* (Tempus, 2003), gives a comprehensive account of the telegraph cables saga. Thomson's involvement is also described in Thompson (1910) and Smith and Wise (1989).

39. See Smith and Wise (1989) and Thompson (1910).

40. The remarkable fact that all of these six (and others mentioned in this chapter) are Scottish or Irish raises questions that cannot be examined here, but see Craik (2007), particularly Chapter 11.

41. See Love (1927), p. 31.

## CHAPTER 9 (ALLAN CHAPMAN)

There is a rich body of accessible sources on Victorian astronomy.

For the social and institutional history, see Mary Brück's *Women in Early British and Irish Astronomy: Stars and Satellites* (Springer, 2009), which is the best study of women in British astronomy known to the author.

For specific observatory studies, there is A. J. Meadows, *Greenwich Observatory, Vol. 2: Recent History (1836–1875)* (Taylor & Francis, 1975), and J. A. Bennett's *Church, State, and Astronomy in Ireland: 200 Years of the Armagh Observatory* (Armagh Observatory and the Institute of Irish Studies, Queen's University, Belfast, 1990). For British university observatories, see Roger Hutchins, *British University Observatories 1772–1939* (Ashgate, 2008).

For the 'Grand Amateur' tradition, see A. Chapman's *The Victorian Amateur Astronomer: Independent Astronomical Research in Britain 1820–1920* (Wiley–Praxis, 1998). For the pivotal role that Sir William and Caroline Herschel played in framing the cosmological landscape within which the Victorians would work, see Michael A. Hoskin's *William Herschel and the Construction of the Heavens* (Oldbourne, 1963), and *The Herschel Partnership as Viewed by Caroline* (Science History Publications, Cambridge, 2003).

Excellent studies of particular areas of Victorian research from a technical perspective by modern scholars include J. B. Hearnshaw's *The Analysis of Starlight: One Hundred and Fifty Years of Astronomical Spectroscopy* (Cambridge, 1986) and *The Measurement of Starlight: Two Centuries of Astronomical Photometry* (Cambridge, 1996), and Owen Gingerich, 'Unlocking the chemical secrets of the Cosmos', in *The Great Copernicus Chase and Other Adventures in Astronomical History* (Sky Publishing Co., USA, and Cambridge, 1992).

For developments in instrumentation see I. S. Glass, *Victorian Telescope Makers: The Lives and Letters of Thomas and Howard Grubb* (Institute of Physics, Bristol, 1997), Anita McConnell's *Instrument Makers to the World: A History of Cooke, Troughton and Simms* (York, 1992), and H. C. King's *The History of the Telescope* (Griffin, 1955). See also Derek Howse's *Greenwich Observatory, Vol. 3: The Buildings and the Instruments* (Taylor and Francis, 1975), and A. Chapman's *Dividing the Circle: The Development of Critical Angular Measurement in Astronomy 1500–1850* (Wiley–Praxis, 1990/1995).

For major works by Victorian authors themselves, we cannot overestimate the significance of Agnes Clerke's *A Popular History of Astronomy in the Nineteenth Century* (1885/1893) (from a time when 'popular' meant without mathematical equations, rather than simplistic), and Sir Robert Stawell Ball's *The Story of the Heavens* (1886) and *The Story of the Sun* (1910).

For an excellent and recently published study of Victorian popular astronomy (along with various other sciences) see Bernard Lightman's *Victorian Popularisers of Science: Designing Nature for New Audiences* (Univ. of Chicago Press, 2007).

## NOTES

1. For contemporary accounts of this development, see Ball (1886/1910) and Clerke (1893).

2. See Michael Hoskin (ed.), *The Cambridge Illustrated History of Astronomy* (Cambridge, 1997), Christopher Walker (ed.), *Astronomy Before the Telescope* (British Museum, London, 1996), and Allan Chapman, *Gods in the Sky: Astronomy, Religion, and Culture from the Ancients to the Renaissance* (Macmillan, Channel 4, 2001).

3. See Chapman (1998).

4. Sir George Biddell Airy, 'The endowment of research', letter to the *English Mechanic*, no. 831 (25 February 1881), 587.

5. See Meadows (1975), Chapter 1, and Edwin Dunkin, 'A far off vision: a Cornishman at Greenwich Observatory', from Dunkin's *Auto-Biographical Notes* (Royal Institution, Cornwall, 1999), Chapter 5, for a human computer's life.

6. See Roger Hutchins (2008), Chapters 1 and 4, I. S. Glass, *Victorian Telescope Makers: The Lives and Letters of Thomas and Howard Grubb*, pp. 29–32 for Dunsink, A. Chapman, 'Sir Robert Stawell Ball (1840–1913): Royal Astronomer in Ireland and astronomy's public voice', *J. Astronomical History and Heritage* **10** (2007), 198–210, and Hermann A. Brück, *The Story of Astronomy in Edinburgh from its Beginnings until 1975* (Edinburgh University Press, 1983). See also Bennett (1990), and Hermann A. Brück, 'Lord Crawford's observatory at Dun Echt, 1872–1892', *Vistas in Astronomy* **35** (1992), 81–138.

7. See Thomas Sprat, *History of the Royal Society* (London, 1667, 2nd ed. 1702), p. 67; J. L. E. Dreyer, *History of the Royal Astronomical Society*, (R.A.S., 1923, 1987), Chapter 1, and Chapman (1998), Chapter 2.

8. Mrs Sarah Challis to Mrs Rebecca Hale, Cambridge, 23 July 1847, in D. W. Dewhirst, 'A royal look through the Northumberland telescope', *Pulsar* (Journal of the Cambridge University Astronomical Society), 10 December 1989, original MS letter preserved in Cambridge University Library.

9. Sir G. B. Airy, *Autobiography* (Cambridge, 1896), pension, pp. 105–8, knighthood offers, pp. 111–13, 187, 254–56, 296.

10. See Michael A. Hoskin (1963 and 2003) and M. A. Hoskin (ed.), *Caroline Herschel's Autobiographies* (Science History Publications, Cambridge, 2003).

11. For the authoritative contemporary treatment of these techniques see William Pearson, *Practical Astronomy* (London, 1829), especially Vol. II. See also A. Chapman (1990), Chapter 7.

12. See Anita McConnell, *Instrument Makers to the World*, Gilbert Satterthwaite, 'Airy's transit circle', *J. Astronomical History and Heritage* **4** (2001), 115–41, and H. C. King (1955), Chapter XI.

13. Chapman (1998), Chapters 2 and 3.

14. Derek Howse (1975), Chapters 2 and 3, and *Nevil Maskelyne. The Seaman's Astronomer*, (Cambridge, 1989), pp. 85–96.

15. See Mary Somerville, *On the Connexion of the Physical Sciences*, 3rd ed. (London, 1836), p. 357, and A. Chapman, *Mary Somerville and the World of Science* (Canopus, Bristol, 2004), pp. 56–57.

16. Edmond Halley, 'Considerations of the changes of the latitudes of some of the principal fix't stars', *Phil. Trans.* **30** (1718), 736–38.

17. See Joseph Ashbrook, 'The story of Groombridge 1830', in J. Ashbrook, *The Astronomical Scrapbook. Skywatchers, Pioneers, and Seekers in Astronomy* (Sky Publishing Co., USA, and Cambridge, 1984), pp. 352–59, and Sir John Herschel, *Outlines of Astronomy*, 2nd ed. (London, 1849), for proper motions.

18. See William Henry Smyth, *Cycle of Celestial Objects: 2, The Bedford Catalogue* (London, 1844), for binary star positions, and Chapman (1998), pp. 42–43, 58, etc.

19. Sir William Herschel, 'Catalogue of a second thousand new nebulae and clusters of stars', *Phil. Trans.* **79** (1789), 212–55.

20. See Sir John Herschel, *Results of Astronomical Observations during the years 1834–8 at the Cape of Good Hope* (London, 1847), numerical catalogue of double stars, pp. 171–242, and J. F. Herschel, 'Catalogue of nebulae and clusters of stars', *Phil. Trans.* **154** (1864), 1–137.

21. See Thomas Woods, *The Monster Telescope* (1845), p. 4, and A. Chapman (1998), p. 344, n. 34.

22. See William Parsons, Lord Rosse, 'Observations on the nebulae', *Phil. Trans.* **140** (1850), 499–514, and D. W. Dewhirst and Michael Hoskin, 'The Rosse spirals', *J. History of Astronomy* **22** (1991), 258–66.

23. Auguste Comte, *Cours de Philosophie Positive*, II, 'Astronomie' (Paris, 1835), pp. 8–9.

24. For a contemporary discussion of the 'calorific' (infra-red) and 'invisible' (ultra-violet) rays from the sun, see Somerville, (note 15), pp. 225–40.

25. For a masterly and accessible account of Victorian spectroscopy, see the astrophysicist Sir Joseph Norman Lockyer, *Stargazing, Past and Present* (London, 1878), Chapter 27. See also A. Chapman, 'The [German] astronomical revolution', in *Möbius and his Band: Mathematics and Astronomy in Nineteenth-Century Germany*, (ed. J. Fauvel, R. Flood, and R. Wilson) (Oxford, 1993), pp. 34–77.

26. Sir G. B. Airy to Arthur Biddell [his uncle], 17–18 November 1842, describing the Turin solar eclipse. Manuscript letter, in possession of Church Wardens of St. Mary's Church, Playford, Suffolk; transcription in possession of the author.

27. William Lassell, 'Trollhätten Falls. Observations by W. Lassell Esq.', *Memoirs of the Royal Astronomical Society* **XXI** (1852), 44–50, on p. 45. A. Chapman, 'William Lassell (1799–1880): Practitioner, patron, and "Grand Amateur" of Victorian astronomy', *Vistas in Astronomy* **32** (1989), 341–370; reprinted in Chapman, *Astronomical Instruments and their Users: Tycho Brahe to William Lassell* (Variorum, Ashgate, 1996).

28. Airy (note 9), pp. 241–42.

29. Clerke (1893), pp. 178–206.

30. Stephen Peter Rigaud, 'Account of Harriot's astronomical papers', *Miscellaneous Works of Revd Dr James Bradley* (Oxford, 1832, 1833), p. 32. John North, 'Thomas Harriot and the first telescopic observations of sunspots', John W. Shirley (ed.), *Thomas Harriot. Renaissance Scientist* (Oxford, 1972), pp. 129–65.

31. Clerke (1893), p. 184.

32. Clerke (1893), pp. 198–200. Stuart Clark, *The Sun Kings. The Unexpected Tragedy of Richard Carrington and the Tale of How Modern Astronomy Began* (Princeton, 2007), pp. 9–24.

33. Samuel Smiles (ed.), *James Nasmyth, Engineer. An Autobiography* (London, 1889), pp. 370–74.

34. Lockyer (note 25), Chapters 28 and 29, and Gingerich (1992).

35. Clerke (1893), pp. 209–11.

36. Clerke (1893), pp. 212–15.

37. Charles E. Mills and C. F. Brooke, *A Sketch of the Life of Sir William Huggins* (London, 1936), p. 23. Chapman (1998), pp. 113–17. Two major modern reference works in the history of astrophysics are J. B. Hearnshaw (1986 and 1996).

38. Sir William Huggins, 'On the spectra of some of the chemical elements', *Phil. Trans.* **154** (1864), 139–60, and W. Huggins and W. A. Miller, 'On the spectra of some of the fixed stars', *Phil. Trans.* **154** (1864), 413–35.

39. Mary T. Brück, 'The family background of Lady Huggins (Margaret Lindsay Murray)', *Irish Astronomical J.* **20** (30 March 1992), 210–11. Barbara J. Becker, 'Eclecticism, opportunism, and the evolution of a new research agenda: William and Margaret Huggins and the origins of astrophysics' (unpublished Ph.D. thesis, Johns Hopkins Univ., USA, 1993).

40. 'Prof. Henry Draper', obituary by William H. M. Christie, *The Observatory* (6 January 1883), 23–24. For a Draper bibliography, see Chapman (1998), pp. 354–55.

41. Angelo Secchi, 'Note sur les spectres prismatiques de corps celestes', *Comptes Rendus* **57** (1863), 71–75. Hearnshaw (1986), pp. 57–66.

42. Kevin Krisciunas, *Astronomical Centers of the World* (Cambridge, 1988), pp. 130–33.

43. Glass (1997), pp. 13–16, and McConnell (1992), pp. 50–78.

44. King (1955), pp. 242–44. Deborah Jean Warner, *Alvan Clark & Sons. Artists in Optics* (Smithsonian Institution Press, Washington DC, 1968), pp. 3–37.

45. W. Herschel, 'Description of a forty-feet reflecting telescope', *Phil. Trans.* **85** (1795), 347–409. J. A. Bennett, '"On

the power of penetrating space", the telescopes of William Herschel', *J. History of Astronomy* 7 (1976), 75–108.

46. See Chapman, 'William Lassell...' (note 27), pp. 349–62, and Chapman (1998), pp. 102–4.

47. Lord Rosse (William Parsons), 'On the construction of specula of six-feet aperture; and a selection of observations of nebulae made with them', *Phil. Trans.* **151** (1861), 681–747.

48. W. Lassell, 'Description of a machine for polishing specula...', *Memoirs of the Royal Astronomical Society* **18** (1849), 1–20.

49. W. Lassell, 'Observations of planets and nebulae at Malta', *Memoirs of the Royal Astronomical Society* **36** (1867), 1–32. Glass (1997), Chapter 2.

50. King (1955), pp. 261–62.

51. Henry Draper, 'On a reflecting telescope for celestial photography, erected at Hastings near New York', *Report... British Association for the Advancement of Science, 1860* (London, 1861), Sections report 63–64. Chapman (1998), pp. 118–21, 136–37.

52. Janet and Mark Robinson (ed.), *The Stargazer of Hardwicke. The Life and Work of Thomas William Webb* (Gracewing, 2006).

53. Barbara Slater, *The Astronomer of Rousdon: Charles Grover, 1842–1921* (Steam Mill, 2005).

54. Alex Smith, 'A working-man astronomer and his telescopes [John Glass]', *English Mechanic and World of Science* **1810** (1 Dec. 1899), 360–61, and Chapman (1998), pp. 161–218.

55. Samuel Smiles, *Men of Invention and Industry* (London, 1884), pp. 361–69, and Chapman (1998), pp. 209–13.

56. See Chapman (1998), pp. 243–93, for foundation histories of several amateur astronomical societies after 1858, and women's membership.

## CHAPTER 10 (DORON D. SWADE)

A comprehensive account of mechanical aids, devices, and machines is Michael Williams, *A History of Computing Technology* (Prentice Hall, 1985). There is a large and growing literature on Charles Babbage and his engines. An accessible account of Babbage's efforts, and those of others, to build calculating engines, is the author's *The Cogwheel Brain: Charles Babbage and the Quest to Build the First Computer* (Little, Brown, 2000), published in the US as *The Difference Engine* (Viking–Penguin, 2001). This includes an account of the construction of the first complete Babbage engine built to original designs. Anthony Hyman's biography of Babbage, *Charles Babbage: Pioneer of the Computer* (Oxford, 1982), is the standard reference on Babbage's life and is especially good on the context of Babbage's times.

There are three major monographs on 19th-century calculating engines. Bruce Collier's *The Little Engines that Could've* (Garland, 1970/1990) remains the most detailed and authoritative account of the historical development of Babbage's ideas; Michael Lindgren's *Glory and Failure: The Difference Engines of Johann Muller, Charles Babbage, and Georg and Edvard Scheutz* (MIT Press, 1990), is an authoritative technical study framed in the cultural and economic context of the times; finally, there is the author's study of the utility of 19th-century calculating engines, *Calculation and Tabulation in the Nineteenth Century: George Biddell Airy versus Charles Babbage* (Ph.D. thesis, University College London, 2003). The standard reference source for Babbage's collected published writings is *The Works of Charles Babbage*, 11 vols. (ed. M. Campbell-Kelly) (William Pickering, 1989); this is referred to below as '*Works*'. *The IEEE Annals of the History of Computing* is a rich source of material on computing prehistory. Allan Bromley's uniquely authoritative accounts of Babbage's engine designs are featured in various issues of the *IEEE Annals*.

## NOTES

1. Thomas Carlyle, 'Signs of the times', *Edinburgh Review*, June 1829.

2. For an illustrated compendium of 19th-century inventions (1865–1900), reported in all seriousness in scientific journals, see Leonard de Vries (ed.), *Victorian Inventions* (John Murray, 1971). For a selection of serious and quirky inventions patented during Queen Victoria's reign, see Stephen van Dulken, *Inventing the 19th Century: The Great Age of Victorian Inventions* (British Library, 2001).

3. De Vries (note 2), p. 156 (italics by the author). The advertisement was published in 1873.

4. The term 'industrial revolution' continues to hold its own against challenges to its *bona fides* as an historical movement. For a defence of the term and discussion, see Neil Cossons, 'Industrial archaeology: the challenge of the evidence', *Antiquaries Journal* **87** (2007), 7 ff.

5. Translation from the Latin quoted in Ernst Martin, *The Calculating Machines* (Die Rechensmashinen): *Their History and Development* (transl. and ed. P. A. Kidwell and M. R. Williams), Charles Babbage Institute Reprint Series for the History of Computing (MIT Press, 1992). The date 1685 is taken from an object label in the 'Computing and mathematics' gallery at the Science Museum, which features a looser translation.

6. The association of elevated social class with abstraction and analysis, and the role of machines in deskilling calculation, recur in Charles Babbage's reference to the social organization of labour in de Prony's great cadastral tables project in the late 18th century. See Charles Babbage, *On the Economy of Machinery and Manufactures*, 4th ed. (Charles Knight, 1835). For a reprint, see *Works*; see Vol.

8, pp. 138, 141, for Babbage's reference to classes and the hierarchy of skills.

7.  Sources in order: Babbage (1822), *Works*, Vol. 2, pp. 6, 15; Baily (1823), *Works*, Vol. 2, p. 45; Juris, Judex (1861), *Works*, Vol. 1, p. 3.

8.  Menabrea (1842), *Works*, Vol. 3, p. 113.

9.  For a technical history of mechanical calculators, see Michael R. Williams, *A History of Computing Technology* (Prentice-Hall, 1985). For an uneven collectors' compendium of mechanical calculators from 1642 to 1925, see Martin (note 5). A rough sketch of Schickard's 'calculating clock' in 1623 was found by Franz Hammer in 1935, although details of it were not published until 1957. Histories of calculation predating 1957 invariably credit Pascal with the invention of the first mechanical calculator. For an accessible account of the reconstruction of Schickard's calculator, see Stan Augarten, *Bit by Bit: An Illustrated History of Computers* (Allen & Unwin, 1984), pp. 15–22.

10.  For a brief authoritative account of the origins and development of the slide rule, see D. Baxandall, *Catalogue of the Collections in the Science Museum: Mathematics: Calculating Machines and Instruments* (HMSO, 1926). For a summary history of the slide rule, see Mary Croarken, *Early Scientific Computing in Britain* (Oxford, 1990), pp. 7, 8. For a chronology of slide rule development from 1620, and a classification of slide rules, their mathematical principles, and physical examples, see G. D. C. Stokes, 'Slide rules', *Napier Tercentenary Celebration: Handbook of the Exhibition of Napier Relics and of Books, Instruments, and Devices for Facilitating Calculation* (ed. E. M. Horsburgh) (Royal Society of Edinburgh, 1914), pp. 155–80.

11.  Quoted in Williams (1985), p. 114.

12.  See D. Baxandall, *Catalogue of the Collections in the Science Museum: Calculating Machines and Instruments.* Second edition revised and updated by Jane Pugh (ed.) (Science Museum, London, 1975). The second edition has a narrower and less exotic range of examples.

13.  See Baxandall (note 10), pp. 53, 54, and Williams (1985), pp. 117, 118. Fuller's rule is usually referred to as a 'spiral rule', but since the scales are inscribed on a cylinder, rather than a cone, the form is strictly helical, not spiral.

14.  Stokes (note 10), p. 174.

15.  L. J. Comrie, *Chamber's Six-figure Mathematical Tables*, Vol. 1 (W. & R. Chambers, 1948), p. v.

16.  See Stephen Johnston, 'Making the arithmometer count', *Bull. Scientific Instrument Society* **52** (March 1997), 12–21; see pp. 12–14 for separate statements on the introduction of the arithmometer.

17.  National Archive, Kew. Graham to Treasury, 29 January 1870, RG 29-2, Vol. 2, f. 111. Further requests were made in 1873 and 1877: Graham to Treasury, 31 March 1873, RG 29-2, Vol. 2, f. 149, and 23 February 1877, f. 249.

18.  Johnston (note 16), p. 16

19.  Graham to Treasury, 28 July 1873, RG 29-2, Vol. 2, f. 162.

20.  Graham to Treasury, 16 March 1877, f. 250.

21.  Quoted in Johnston (1997), p. 17.

22.  For the relative merits of arithmometers, including life expectancy, usage, reliability, and costs of repair, see the report by Brydges Henniker, Henniker to Treasury, 2 February 1893, RG 29-3, Vol. 3, f. 129.

23.  For the Buxton memoir, see Anthony Hyman (ed.), *Memoir of the Life and Labours of the Late Charles Babbage Esq. F.R.S. by H. W. Buxton*, Vol. 13 (Tomash, 1988), pp. 48–49 (emphasis original). Dionysus Lardner, 'Babbage's calculating engine', *Edinburgh Review* **59** (1834), 263–327, reprinted in *Works*, Vol. 2, p. 119. Letter from Lady Byron to Dr. King, 21 June 1833, *Ada, the Enchantress of Numbers: A Selection from the Letters of Lord Byron's Daughter and her Description of the First Computer* (ed. B. A. Toole) (Strawberry Press, 1992), p. 51 (emphasis original).

24.  See Swade (2000). For a description of the construction of Babbage's Difference Engine No. 2, see Doron Swade, 'The construction of Charles Babbage's Difference Engine No. 2', *IEEE Annals of the History of Computing* **27** (3) (2005), 70–88.

25.  For background into why the Astronomical Society commissioned the tables, see William J. Ashworth, 'The calculating eye: Baily, Herschel, Babbage and the business of astronomy', *British J. History of Science* **27** (1994), 409–41.

26.  Peter Ackroyd, *Dan Leno and the Limehouse Golem* (Minerva, 1994), p. 116.

27.  There are three known accounts of this episode written by Babbage: 1822, 1834, and 1839: see Collier (1990), pp. 14–18. The quotation cited is taken from the third account which appears in the Buxton memoir (see Hyman (note 23), p. 46). The first account leaves open whether it was Babbage or Herschel who made the suggestion. In the second and third accounts, Babbage claimed ownership for himself. All three accounts refer to steam. The third account is the most dramatized and is the only one to include direct speech.

28.  Anthony Hyman, *Charles Babbage: Pioneer of the Computer* (Oxford, 1982), pp. 147, 148.

29.  J. N. Hays, 'The rise and fall of Dionysius Lardner', *Annals of Science* **38** (1981), 527–42.

30.  Lardner (note 23), reprinted in *Works*, Vol. 2, pp. 118–86.

31.  Lardner (note 23), *Works*, Vol. 2, p. 138 (emphasis original).

32.  Lardner to Babbage, 2 June 1830, BL ADD MS 37185, f. 206.

33. For a listing of the papers and discussion, see Swade (2003), p. 116.

34. Charles Babbage, *Passages from the Life of a Philosopher* (Longman, 1864), p. 112.

35. Babbage, A letter to Sir Humphrey Davy, Bart., President of the Royal Society, on the application of machinery to the purpose of calculating and printing mathematical tables (Cradock and Joy, 1822), reprinted in *Works*, Vol. 2, pp. 5–14.

36. Ada Lovelace, 'Sketch of the Analytical Engine', *Scientific Memoirs* 3 (1843), 666–731, reprinted in *Works*, Vol. 3, 663–731. For Menabrea (1842), see *Works*, Vol. 3, 109.

37. For details of the Difference Engine No. 2 halting mechanism, see Doron Swade, 'Charles Babbage's Difference Engine No. 2: technical description', *Science Museum Papers in the History of Technology* (1996), 44–45.

38. *Works*, Vol. 2, p. 43. Babbage makes a similar statement to Davy, see *Works*, Vol. 2, 7.

39. *Works*, Vol. 2, pp. 38–40.

40. Babbage (1826), *Works*, Vol. 2, p. 62.

41. *Works*, Vol. 2, p. 34; read 13 December 1822.

42. For a thematic treatment see Ivor Grattan-Guinness, 'Charles Babbage as an algorithmic thinker', *IEEE Annals of the History of Computing* 14 (3) (1992), 34–48. Babbage wrote that the engine's stimulus for his enquiries was 'singular in the history of mathematical science'; see Babbage (1826), *Works*, Vol. 2, p. 61.

43. In a manuscript on the analytical engine, dated 26 December 1837, unpublished in his lifetime; see Hyman (note 23). Reprinted in *Works*, Vol. 3, pp. 60–61.

44. *Works*, Vol. 2, 43.

45. For detailed analysis of Lardner's motivations in giving false emphasis to the role of errors, and for the background to his article, see Swade (2003), Chapter 3.

46. See Michael Lindgren, *Glory and Failure: The Difference Engines of Johann Muller, Charles Babbage, and Georg and Edvard Scheutz* (transl. C. G. McKay), 2nd ed. (MIT Press, 1990).

47. For a detailed analysis of utility, see Swade (2003), Chapter 3.

48. Babbage did not assume that machines were infallible by virtue of being mechanical. He went to elaborate lengths to include security devices that ensured correct working and the integrity of results. For Babbage, machines were not inherently infallible, but could be made so. For details of 19th-century table-making, see Doron Swade, 'The unerring certainty of mechanical agency: machines and table making in the nineteenth century', *The History of Mathematical Tables: From Sumer to Spreadsheets* (ed. M. Campbell-Kelly *et al.*) (Oxford, 2003), pp. 143–74.

49. Lardner (note 23), *Works*, Vol. 2, p. 138; B. H. Babbage (1872), *Works*, Vol. 2, p. 226.

50. For further details of 'modern' features evidenced in Babbage's engines, see Doron Swade, 'Automatic computation: the origins of computational method', *Rutherford Journal* (*The New Zealand Journal for the History and Philosophy of Science and Technology*) (ed. J. Copeland), 2009. For von Neumann's seminal paper, see John von Neumann, *First Draft of a Report on the Edvac* (Moore School of Electrical Engineering, Univ. of Pennsylvania, 1945), reprinted in *The Origins of Digital Computers: Selected Papers* (ed. B. Randell) (Springer-Verlag, 1982), pp. 375–82.

51. Babbage devised a symbolic descriptive language that he called the 'mechanical notation'. For further details and examples, see note 53.

52. Between 1847 and 1849, Babbage intermitted work on the analytical engine to design Difference Engine No. 2.

53. See Swade (2003), Chapter 5.

54. See Michael R. Williams, 'Difference Engines: from Muller to Comrie', *The History of Mathematical Tables: From Sumer to Spreadsheets* (ed. M. Campbell-Kelly *et al.*) (Oxford, 2003).

55. See Mary Croarken, *Early Scientific Computing in Britain* (Clarendon Press, 1990).

## CHAPTER 11 (M. EILEEN MAGNELLO)

John Ramsay McCulloch's two-volume *A Statistical Account of the British Empire* (London, 1839) is a useful introduction to the way in which data collection, enumeration, and tabulation became a part of the statistical landscape in early Victorian Britain; he examined, for example, the statistics of the population and geography of the British Empire, the growth of industrialization, and provisions for the poor; a digital copy of this book can be downloaded via Google books.

Complementing this work are the seminal publications of the historians M. J. Cullen, *The Statistical Movement in early Victorian Britain: The Foundation of Empirical Social Research* (Harvester Press, 1975) and Lawrence Goldman, 'Statistics and the science of society in early Victorian Britain: An intellectual context for the General Register Office', *Journal for the Social History of Medicine* 5 (1991), 415–35, which provide richly nuanced accounts of the historical context of statistics in the early Victorian period.

The prodigious writing of the literary critic Elaine Freedgood exemplified the way in which statistical ideas and thinking were interwoven into the canons of Victorian literature. Her *Victorian Writing about Risk: Imagining a Safe England in a Dangerous World* (Cambridge, 2000) is an exemplary account of the Victorians' attempt to reduce and manage statistical risk (or 'banishing panic', in the words of William Farr). She also addressed the role of sanitation and political economy in the emergence of vital statistics, and argued that statistics became the empirical arm of political economy in early- to mid-Victorian Britain. Theodore Porter's book *The Rise of Statistical Thinking, 1820–1900* (Princeton Univ. Press, 1986) examines Adolphe Quetelet's influential role in the development of vital statistics for the early-to mid-Victorians.

On the history of statistics in the Victorian period, see also: Ian Hacking, *The Taming of Chance* (Cambridge, 1990); M. Eileen Magnello, 'Victorian vital and mathematical statistics', *Bulletin of the British Society for the History of Mathematics* **21**(3) (November 2006), 219–29, and 'Eminent Victorians and early statistical societies', *Significance: Statistics Making Sense* (a journal of the Royal Statistical Society) **6**(2) (July 2009), 86–88; M. Eileen Magnello and Borin Van Loon, *Introducing Statistics: A Graphic Guide* (Icon Press, 2009); Stephen Stigler, *History of Statistics: The Measurement of Uncertainty before 1900* (Belknap Press of Harvard Univ. Press, 1986) and *Statistics on the Table: The History of Statistical Concepts and Methods* (Harvard Univ. Press, 1999). The *Probabilistic Revolution Vol. 1: Ideas in History* (ed. L. Krüger, L. J. Daston, and M. Heidelberger) and *Vol. 2: Ideas in the Sciences* (ed. L. Krüger, G. Gigerenzer, and M. S. Morgan), both from MIT Press (1987), contain a number of chapters that address Victorian probability and statistics.

For the emergence of medical statistics and the development of vital statistics in the Victorian period, see John Eyler, *Victorian Social Medicine: The Ideas and Methods of William Farr* (Johns Hopkins Univ. Press, 1979), which contains a chapter on the statistical work of Florence Nightingale, and 'Constructing vital statistics: Thomas Rowe Edmunds and William Farr, 1835–1845', *History of Epidemiology* **47** (2002), 9; Edward Higgs, *Making Sense of the Census. Census records for England and Wales, 1801–1901: A Handbook for Historical Researchers* (National Archives and Institute of Historical Research, 2005) and *Life, Death and Statistics; Civil Registration, Censuses and the work of the General Register Office, 1837–1952* (Local Population Studies, Hatfield, 2004); M. Eileen Magnello and Anne Hardy (ed.), *The Road to Medical Statistics* (Rodopi, 2002), pp. 95–124; and J. Rosser Matthews, *Quantification and the Quest for Medical Certainty* (Princeton Univ. Press, 1995).

On the lives and work of Babbage, Bentham, Chadwick, Edgeworth, Jevons, and Malthus, see Doron Swade's *The Cogwheel Brain: Charles Babbage and the Quest to Build the First Computer* (Little, Brown, 2000); Charles Milner Atkinson, *Jeremy Bentham: His Life and Work* (Univ. Press of the Pacific, Honolulu, 2004); David Gladstone, *Setting the Agenda: Edwin Chadwick and Nineteenth-Century Reform* (Routledge, 1997); Lluis Barbe, *Francis Ysidro Edgeworth: A Portrait With Family and Friends* (Edward Elgar Publ., 2010); Margaret Schabas, *A World Ruled by Number: William Stanley Jevons and the Rise of Mathematical Economics* (Princeton Univ. Press, 1990); Patricia James, *Population Malthus: His Life and Times* (Routledge and Kegan Paul, 1979, and Routledge, 2006).

For Nightingale and Farr, see: Monica Baly, *Florence Nightingale and the Nursing Legacy*, 2nd ed. (Whurr, 1997); Anna Stickler, *Florence Nightingale Curriculum Vitae with Information about Florence Nightingale and Kaiserswerth* (Diakoniewerk, 1965). For the role of statistics in Nightingale's life, see Lynn MacDonald (ed.), *Florence Nightingale on Public Health Care, Collected Works of Florence Nightingale, Vol. 6* (Wilfrid Laurier Univ. Press, 2003); John Eyler's chapter on Nightingale in his book on Farr (see below) and M. Eileen Magnello, 'The passionate statistician', *Navigating Nightingale* (ed. S. Nelson and A. M. Rafferty) (Cornell Univ. Press, 2010); and John Eyler, *Victorian Social Medicine: The Ideas and Methods of William Farr* (Johns Hopkins Univ. Press, 1979).

## NOTES

1. Graunt used ten thousand parish records in England and Wales that contained information on sex, age, and cause of death (e.g., burnt, drowned, murdered, 'poysoned', shot, and starved), in his *Natural and Political Observations upon the London Bills of Mortality* (1662). From the mortality rates derived from the Bills of Mortality, the mathematician James Dodson (*c.* 1710–57), Master of the Royal Mathematical School, Christ's Hospital (Sussex), calculated a set of risks in the 1750s that led to the development of a mutual system of life assurance; this formed the basis of the Equitable Society, established in 1762. Dodson introduced the word *actuary* into the assurance world.

2. Karl Metz, 'Paupers and numbers: the statistical argument for social reform in Britain during the period of industrialization' (Krüger, Daston, and Heidelberger, 1987). See also Philip Kreager, 'Death and method: the rhetorical space of seventeenth-century vital measurement' (Magnello and Hardy, 2002).

3. Sir John Sinclair, *The Statistical Account of Scotland drawn up from Communications of the Ministers of Different Parishes* (Edinburgh, 1791). The Dutch espousal of the Achenwallian *staaten-kunde*, as a method of structuring noteworthy facts concerning the state and as a quantified description of the nation and region, led to their introduction of the term *statistiek* in 1807 as 'the discipline which studies the various manners in which each separate state is governed'. See *The Statistical Mind in a Pre-Statistical Era: The Netherlands 1750–1850* (ed. P. M. M. Klep and I. H. Stamhuis) (Askant, 2002), pp. 22–23.

4. Metz (note 2).

5. Sinclair (note 3).

6. It was during this time when the monarchical Bourbon state began to implement administrative statistics in France. In 1800 the *Statistique Générale* was established to supervise this work. See Alain Desrosières, 'Official statistics in 19th-century France: the SGF as a case study', *J. Social History of Medicine* **5** (1991), 517.

7. Francis Bisset Hawkins, *The Elements of Medical Statistics* (Rees, Orme, Brown, and Green, 1829), p. 2. Hawkins was at Exeter College, Oxford, a fellow of the Royal College of Physicians, and a fellow and council member of the Statistical Society of London, when the society was established in 1834.

8. Ernst Mayr, *Population, Species and Evolution* (Belknap Press of Harvard University, 1963), p. 4.

9. M. Eileen Magnello, 'The introduction of mathematical statistics into medical research: the roles of Karl Pearson, Major Greenwood and Austin Bradford Hill' (Magnello and Hardy, 2002), especially pp. 96–99.

10. Gregory King made one of the first systematic attempts to estimate the population of England and Wales in 1665, positing the number of houses as 1.3 million and the number of people as 5.5 million. The Dutch astronomer and political arithmetician, Nicolaas Struyck (1687–1769) built on Edmond Halley's work on comets and research into determining the size of various populations. Struyck's greatest ambition was to make a reasoned estimate of the total number of people on earth; he wanted to know if the population was increasing, stable, or decreasing. For more on Struyck, see J. Zuideervaart Huib, 'Early quantification of scientific knowledge: Nicolaas Struyck (1686–1796) as a collector of empirical data', (Klep and Stamhuis, note 3), pp. 125–48.

11. Metz (note 2), p. 343.

12. In his *Essay on the Principle of Population*, Malthus divided the world into two parts—modern Europe and the rest of the world. He noticed that the relatively late marriage customs in Western Europe produced lower birth rates and consequently a smaller rate of increase of populations, whether or not epidemics intervened. In 1965 the historical demographer John Hajnal (b. 1924) corroborated Malthus's views when he discovered the existence of a European demographic dividing line from Trieste to Leningrad (now St Petersburg). Hajnal found that from 1500 to 1900 the people to the north and west married relatively late, with a substantial proportion never marrying at all: the rest of the world married early and hardly anyone remained single beyond 40. This pattern of European marriage developed over the course of the early modern period in the 16th century and extended to the modern era at the beginning of the 20th century.

13. *Anon*, 'Introduction', *J. Statistical Society of London* (1 May 1838), p. 4.

14. It was not until 1851, when a more comprehensive census was undertaken in England and Wales, that provisions were made to include age, sex, occupation, and birthplace, as well as counting the blind and the deaf.

15. Quetelet also played a pivotal role in launching the International Statistical Congress in 1853. This event heralded the beginning of international statistical cooperation, leading to the establishment of the International Statistics Institute (ISI) in the Netherlands in 1885.

16. I. D. Hill, 'Statistical Society of London—Royal Statistical Society: the first 100 years: 1834–1934', *J. Royal Statistical Society* 147 (2) (1984), 130.

17. In 1856 the Statistical section of the British Association was changed to the Section of economic science and statistics, which became the Economics section of the British Association in 1948.

18. [Adolphe Quetelet], 'Report of the Fourth Meeting of the British Association, London, (1835)', *British Association for the Advancement of Science* (1835), xxxix.

19. Anon (note 13).

20. Lawrence Goldman (1991).

21. Goldman (1991), p. 434.

22. Victoria Coven, 'A history of statistics in the social sciences', *Gateway: An Academic Journal on the Web* (Spring 2003), 1.

23. Quetelet borrowed the phrase 'social physics' from the French philosopher Auguste Comte (1778–1857), who coined the word 'sociology' after he learnt that Quetelet had begun to use his original terms.

24. M. Eileen Magnello, *A Centenary History of the Christie Hospital Manchester* (Alden Press, 2001).

25. Roy Porter, *The Greatest Benefit to Mankind: A Medical History of Humanity from Antiquity to the Present* (Harper Collins, 1997).

26. As Edward Higgs has noted, the manner of Lister's appointment was not unusual, since 'political patronage was the typical means of appointing senior civil servants in the early 19th century'. See Edward Higgs, *Some Forgotten Men: The Registrars General of England and Wales and the History of State Demographic and Medical Statistics, 1837–1920*.

27. Edward J. Higgs, personal communication, 16 May 2007.

28. Higgs (note 27).

29. See Higgs (2004).

30. Farr was first appointed as Compiler of abstracts and subsequently promoted to Statistical Superintendent of the GRO. For further elaboration about the medico-demographic programme, see Higgs (2004).

31. We thank Roger Thatcher, Registrar General from 1976–1986, for bringing this to our attention, personal correspondence, 22 January 2007.

32. John M. Eyler, 'Constructing vital statistics: Thomas Rowe Edmunds and William Farr, 1835–1845', *History of Epidemiology* 47 (2002), 9.

33. Goldman (1991), p. 415.

34. [Major Greenwood], 'Medical statistics', *Lancet* (7 May 1921), 986.

35. For the French usage of vital statistics in medicine, see Bernard-Pierre Lécuyer, 'Probability in vital and social statistics: Quetelet, Farr and the Bertillions' (Krüger, Daston, and Heidelberger, 1987), pp. 317–35.

36. Eyler (note 32), p. 6.

37. Snow was also a leading pioneer in the development of anaesthesia in Britain. He was Queen Victoria's obstetrician and administered chloroform during the births of Prince Leopold (7 April 1853) and Princess Beatrice (14 April 1857). See also Peter Vinten-Johansen, Howard Brody, Nigel Paneth, and Stephen Rachman, *Cholera, Chloroform and the Science of Medicine: A Life of John Snow* (Oxford, 2003).

38. William Coleman, 'Experimental physiology and statistical inference: the therapeutic trial in 19th-century Germany' (Krüger, Gigerenzer, and Morgan, 1987), pp. 201–28.

39. Gosset, William Sealy [Student], 'The probable error of a mean', *Biometrika* 6 (1908), 1–8. Fisher used the *t* symbol and described Student's distribution (and others based on the normal distribution) and the role of degrees of freedom in *Proc. International Congress of Mathematics*, Toronto, Vol. 2, pp. 805–13. Although the paper was presented in 1924, it was not published until 1928.

40. The 'passionate statistician' was the sobriquet given to Nightingale by her first biographer Sir Edward Cook, in *The Life of Florence Nightingale* (Macmillan, 1913).

41. For a well-developed and highly contextualized study of the Nightingale family, which places them within the social, religious, and political milieu of Victorian society, see Gillian Gill, *Nightingales: Florence and Her Family* (Hodder and Stoughton, 2004).

42. Gillian Gill made this observation when Florence put together a small table of 'vegetables' and 'fruits' held in her kitchen and larder playhouse, see Gill (note 41), p. 40.

43. Karl Pearson, *The Life, Letters and Labours of Francis Galton*, Vol II (Cambridge, 1924), p. 250.

44. Cited in Lytton Strachey, *Eminent Victorians: Cardinal Manning, Florence Nightingale, Dr. Arnold and General Gordon* (Chatto & Windus, 1918). For more on William Derham's religious ideas about statistics, see Egon Pearson (ed.), *The History of Statistics in the 17th and 18th Centuries Against the Changing Background of Intellectual, Scientific and Religious Thought. Lectures given by Karl Pearson at University College London during the Academic Sessions 1921–1923* (Griffin, 1978).

45. Lytton Strachey (note 44).

46. See Gill (note 41). Gill further remarked that Florence was aware that 'hospitals were crowded, filthy, meagerly funded and badly run. All too often the staff consisted of brutal doctors, drunken orderlies and women down on their luck', p.191.

47. See I. Bernard Cohen, 'Florence Nightingale', *Scientific American* 250 (3) (March 1984), 250.

48. Florence Nightingale, Letter to Adolphe Quetelet (8 November 1872), reposited in the Archival collection at the Wellcome Trust Library, London; also cited in John Bibby, *Notes Towards a History of Teaching Statistics*, (John Bibby Books, 1986), p. 113.

49. Gill (note 41), p. 128.

50. Edgeworth, a student at Balliol College, Oxford, received a first class degree in *Literae Humaniores* in 1869. His first set of statistical lectures was delivered at UCL in his Newmarch Lectures in 1884–85 and 1890–91.

51. Pearson (note 43), pp. 419–20.

52. I. Castles, 'William Stanley Jevons', *Statisticians of the Centuries* (ed. C. C. Heyde and E. Seneta) (Springer-Verlag, 2001), p. 201.

53. Schabas (1990).

54. For Bentham's influence on Jevons, see Nathalie Sigot, 'Jevons' debt to Bentham: mathematical economy, morals and psychology', *Manchester School* 70 (2002), 262. See also Stephen M. Stigler, 'Jevons as statistician', *Manchester School* 50 (1982), 354–65.

55. Stigler (1999), see especially Chapter 5.

56. Francis Ysidro Edgeworth, 'The hedonical calculus', *Mind* 4 (1879), 394–408.

57. Stigler (1999).

## CHAPTER 12 (M. EILEEN MAGNELLO)

Much of the material in this chapter arises from the author's extensive publications on Pearson. See, for example: 'Karl Pearson's Gresham Lectures: W. F. R. Weldon, speciation and the origins of Pearsonian statistics', *British Journal for the History of Science* 29 (1996), 43–63; 'Karl Pearson's mathematisation of inheritance: From Galton's ancestral heredity to Mendelian genetics, 1895–1909', *Annals of Science* (1998), 35–94; 'The non-correlation of biometrics and eugenics: Rival forms of laboratory work in Karl Pearson's career at University College London', *History of Science* 37 (1999), 79–106, and 38 (2000), 123–150; 'The reception of Mendelism by the biometricians and the early Mendelians (1899–1909)', *A Century of Mendelism in Human Genetics* (ed. J. M. Keynes, A. W. F. Edwards, and J. Peel) (CRC Press, April 2004); 'Karl Pearson and the origins of modern statistics: An elastician becomes a statistician', *The Rutherford Journal: New Zealand Journal for the History and Philosophy of Science and Technology* 1 (December 2005); 'Karl Pearson and the establishment of mathematical statistics', *International Statistical Review:*

Papers Honouring Karl Pearson (1857–1936), 77 (April 2009), 3–29; and M. Eileen Magnello and Borin Van Loon, *Introducing Statistics: A Graphic Guide* (Icon Press, 2009). For Pearson's statistical writings, see the author's 'Karl Pearson', *Statisticians of the Centuries* (ed. C. C. Heyde and E. Seneta) (Springer-Verlag, 2001), pp. 248–56, 'Karl Pearson', *International Encyclopaedia of the Social Sciences* 6, 2nd ed., (ed. W. A. Darity, Jr.) (Macmillan Reference USA, 2008), pp. 190–94, and also *Encyclopedia.com* http://www.encyclopedia.com [accessed 12 Jan 2011].

For other articles on Pearson's statistics, see the nine Pearson articles in the above-mentioned *Papers Honouring Karl Pearson (1857–1936)*, and John Aldrich's articles 'Correlations genuine and spurious in Pearson and Yule', *Statistical Science* 10(4) (1994), 364–76, 'Doing least squares: Perspectives from Gauss and Yule', *International Statistical Review* 66(1) (1998), 61–81, and 'The language of the English biometric school', *International Statistical Review* 71 (2003), 109–29. Aldrich's prodigious website *Karl Pearson, A Reader's Guide*: http://www.economics.soton.ac.uk/staff/aldrich/kpreader.htm [accessed 16 Jan 2011] provides an annotated overview of nearly all published material on Pearson, as well as related material on the history of statistics and of mathematics. See also Stephen Stigler's publications: *The History of Statistics. The Measurement of Uncertainty before 1900* (Belknap Press of Harvard Univ. Press, 1986), *Statistics on the Table.* (Harvard Univ. Press, 1999), 'Francis Galton's account of the invention of correlation', *Statistical Science* 4 (May 1989), 73–86, 'Karl Pearson and quasi-independence', *Biometrika* 79 (1992), 563–75 (revised in *Statistics on the Table*), 'Karl Pearson's theoretical errors and the advances they inspired', *Statistical Science* 23(2) (May 2008), 261–71, and 'Remembering Karl Pearson after 150 years', *STATS: The Magazine for Student of Statistics*, 49 (2008), 3–4.

For background reading on Darwin and evolutionary biology, especially how these ideas influenced Galton, Pearson, and Weldon, see Peter Bowler's *Evolution: The History of an Idea* (Univ. of California Press, 1989, 4th ed., 2009) and *Charles Darwin: The Man and his Influence* (Cambridge Science Biographies, 1990), and Robert Olby, 'The dimensions of scientific controversy: The biometric–Mendelian debate', *British Journal for the History of Science* 22 (1988), 299–320.

The way in which the Victorian mathematical tripos examinations at Cambridge University influenced and shaped the life and work of the Victorian wranglers, including Karl Pearson (3rd wrangler), has been elucidated by Andrew Warwick in 'The World of Cambridge Physics', *The Physics of Empire: Public Lectures* (ed. Richard Staley) (Whipple Museum of History of Science, Cambridge, 1994); 'Exercising the student body. Mathematics and athleticism in Victorian Cambridge', *Science Incarnate: Historical Embodiments of Natural Knowledge* (ed. S. Shapin and C. Lawrence) (Univ. of Chicago Press, 1998), and *Masters of Theory: Cambridge and the Rise of Mathematical Physics* (Univ. of Chicago Press, 2003). The development of mathematics physics laboratories in 19th-century Britain is examined by Graeme Gooday, 'Precision measurement and the genesis of physics teaching laboratories in Victorian Britain', *British Journal for the History of Science* 23 (1990), 25–51.

For biographical material on Galton, see Francis Galton's *Memories of My Life* (E. P. Dutton, 1909), and Karl Pearson's three-volume monumental biography of Galton in *The Life, Letters and Labours of Francis Galton*, (Cambridge, 1914–30), which contains many letters written by Galton and his contemporaries. Derek Forrest's book, *Francis Galton: The Life and Work of a Victorian Genius* (Taplinger Press, 1974) presents a compact and readable account of Galton's life. Michael Bulmer's *Francis Galton: Pioneer of Heredity and Biometry* (Johns Hopkins Univ. Press, 2003) is not a biography in the traditional sense; instead, Bulmer is more interested in imagining how Galton would have dealt with problems if he were alive today.

The two-part biographical article of Karl Pearson by his son Egon gives some insight from a more intimate and contemporary view; he covers nearly all aspects of his father's life, thus providing the only biographical book on Karl Pearson: 'Karl Pearson. An appreciation of some aspects of his life and work, Part 1, 1857–1905, and Part 2, 1906–1936', *Biometrika* (1936), 193–257, and (1938), 161–248 (reprinted by Cambridge, 1938). See also J. B. S. Haldane's 'Karl Pearson', *UCL Magazine* (June 1936), 252–53, and 'Karl Pearson', *Speeches Delivered at a Dinner held in University College* (1958); Churchill Eisenhart, 'Karl Pearson', *Dictionary of Scientific Biography* 10 (Charles Scribner's Sons, 1974), pp. 447–73. Theodore Porter's book, *Karl Pearson: The Scientific Life in a Statistical Age* (Princeton Univ. Press, 2004) provides an illuminating and erudite account of the younger Pearson, and shows how his literary talents emerged from his comprehensive undertaking of medieval German folklore, literature, and history. However, Porter says little about Pearson's creation of the discipline of mathematical statistics, and hardly even acknowledges the quintessential roles that Darwinian variation and Weldon played in the creation and development of Pearson's statistics.

For Weldon, see Karl Pearson, 'Walter Frank Raphael Weldon, 1860–1906', *Biometrika* (1906), 1–52 (reprinted in E. S. Pearson and Maurice Kendall, *Studies in the History of Statistics and Probability, Vol. 1* (Charles Griffin, 1970), pp. 265–322), the author's 'W. F. R. Weldon' in *Statisticians of the Centuries* (ed. C. C. Heyde and E. Seneta) (Springer-Verlag, 2001), pp. 261–64, and 'W. F. R. Weldon' in *The Oxford Dictionary of National Biography* (ed. C. Matthew) (Oxford, 2004). For Edgeworth, see Stephen Stigler's 'Francis Ysidro Edgeworth, statistician' (with discussion), *Journal of the Royal Statistical Society A* 141 (1978), 287–322; reprinted (without the discussion) in *Alfred Marshall (1842–1924) and Francis Edgeworth (1845–1926)* (ed. M. Blaug), *Pioneers of Economics* 29 (Elgar Reference Collection, 1992), revised in *Statistics on the Table*.

The author wishes to thank University College London for permission to quote material from the manuscript collection of Karl Pearson (KP:UCL).

## NOTES

1. Charles Darwin, *Variation of Plants and Animals under Domestication*, 2 vols. (John Murray, 1875), pp. 313–17, 347.

2. Karl Pearson, 'Editorial (II). The spirit of *Biometrika*', *Biometrika* (1 October 1901), 4.

3. Christopher Pritchard, *The normal curve of evolutionary biology, 1869–1877, with special reference to the support given to Francis Galton and George Darwin* (Ph.D. thesis, Open University, 2005), p. 147.

4. Francis Galton, 'Co-relations and their measurement, chiefly from anthropometric data', *Proceedings of the Royal Society* **45** (1888), 135, reprinted in *Nature* **39** (1889), 238; W. F. R. Weldon, 'Certain correlated variations in *Crangon vulgaris*', *Proc. Royal Society of London* **51** (1892), 2–19.

5. For further elaboration of Weldon's work, see M. Eileen Magnello (1996).

6. W. F. R. Weldon, 'The variations occurring in certain *Decapod crustacean*: I. *Crangon vulgaris*', *Proc. Royal Society* **47** (1890), 446.

7. Weldon (note 6).

8. This work is discussed more fully in Chapter 3 of M. Eileen Magnello, *Karl Pearson: Evolutionary Biology and the Emergence of a Modern Theory of Statistics, 1884–1936* (D.Phil. thesis, Oxford University, 1993).

9. Magnello (note 8).

10. W. F. R. Weldon (note 4), p. 11.

11. Edgeworth introduced the term 'coefficient of correlation' in 1892 in his Newmarch lecture, and then wrote 'Correlated averages', *Phil. Mag.* (5th Series) **34** (1892), 190–204, and 'The law of error and correlated averages', *Phil. Mag.* (5th Series) **34** (1892), 429–38, 518–26. Stephen Stigler has argued that Edgeworth wanted to develop a mathematical apparatus to handle multivariate normal distributions; see Stigler (1999), p. 107.

12. Karl Pearson, 'Applications of geometry to practical life', *Nature* (21 January 1891), 275. Free public lectures (in various fields) continue to be delivered at Gresham College to this day.

13. Karl Pearson, Gresham lecture on 'The geometry of statistics' (17 November 1891), 11; this is held in the University College London manuscript collection of Pearson papers/49 (KP:UCL).

14. Pearson (note 13).

15. Pearson (note 13).

16. Pearson (note 13).

17. Karl Pearson, Application: To the electors of natural philosophy in the University of Edinburgh, 1901; KP:UCL/11/9.

18. Karl Pearson, Letter to W. F. R. Weldon, 28 February 1899; KP:UCL/266/8.

19. Karl Pearson, Gresham lecture on 'Normal curve of frequency' (22 November 1893); KP:UCL/49.

20. W. F. R. Weldon, 'On certain correlated variations in *Carcinus mœnas*', *Proc. Royal Society* **54** (1893), 329.

21. Karl Pearson, Gresham lecture on 'Compound curves' (24 November 1893), 15; KP:UCL/49.

22. Karl Pearson, Gresham lecture on 'Skew curves' (13 November 1893); KP:UCL/49.

23. Pearson (note 22).

24. Alfred Marshall, 'On the graphical method of statistics', *Jubilee Volume of the Royal Statistical Society* (1889), 251–260.

25. Karl Pearson, Gresham lecture on 'Maps and chartograms' (20 November 1891); KP:UCL/49/21.

26. For further elaboration on Pearson's work on the method of moments and his work on the chi-square, see M. Eileen Magnello, 'Karl Pearson, Paper on the chi-square goodness of fit test', *Landmark Writings in Western Mathematics: Case Studies, 1640–1940* (ed. I. Grattan-Guinness) (Elsevier, 2005), Chapter 56, pp. 724–31.

27. Pearson introduced the method of moments in his Gresham lecture on 'Skew curves' (23 November 1893); KP:UCL/ 49.

28. Student, 'Errors of routine analysis', *Biometrika* **15** (1927), 151–64.

29. John Venn, 'Law of error', *Nature* (1 September 1887), 411.

30. Francis Ysidro Edgeworth, 'Law of error and the elimination of chance', *Phil. Mag.* (5th series) **21** (1886), 308–24.

31. Karl Pearson, 'Asymmetrical frequency curves', *Nature* (26 October 1893), 615.

32. Eisenhart (1974), p. 461.

33. See Richard J. Smith, 'Alternative semi-parametric likelihood approaches to generalised method of moments estimation', *Economic Journal* **107**, No. 441 (March 1997), 503–19.

34. Karl Pearson, Gresham lecture on 'Normal curves' (22 November 1893); KP:UCL/49, 3.

35. Magnello (1996).

36. These correlational methods include Pearson's product-moment correlation coefficient (1896), multiple correlation and part correlation (1896), phi-coefficient (1899), tetrachoric correlation (1899), chi-square test of association for contingency tables (1904), contingency coefficient (1904), correlation ratio (1905), biserial correlation (1909), triserial correlation (1914) and polychoric

correlation (1922). (Yule introduced the partial correlation in 1897.) For further elaboration of Pearson's correlational and regression methods, see Magnello (1998).

37. Karl Pearson, *Grammar of Science*, 3rd ed. (Charles and Black, 1911), p. 173.

38. This was first published in an abstract, Karl Pearson, 'Mathematical contributions to the theory of evolution. III. Regression, heredity and panmixia (Abstract)', *Proc. Royal Society of London* (1895), 69–71, and then the complete paper, Karl Pearson, 'Mathematical contributions to the theory of evolution. III. Regression, heredity and panmixia', *Phil. Trans. A* **187** (1896), 253–318.

39. Stigler (1986), pp. 16–17.

40. The experimental psychologist Stanley Smith Stevenson (1906–1973) popularized Pearson's classificatory system when he introduced a methodical typology with his 'scale of measurements', which consisted of nominal, ordinal, interval, and ratio scales. Stevenson introduced the idea of ratio scales (the type of variable that Pearson used) and 'interval' scales in his paper, S. S. Stevens, 'On the theory of scales of measurement', *Science* **103**, no. 2684 (1946), 677–80. For further elaboration on Pearson's classificatory system, see Magnello (1998) and Magnello and Van Loon (2009).

41. For more details, see Magnello (note 26).

42. This historiographical problem is examined in considerable detail in Magnello (1999), pp. 79–106, 123–150; a more updated account can be found in Magnello (2007). For those scholars who espoused the belief that eugenics and biometrics were the same, see Daniel Kevles, *In the Name of Eugenics* (New York, 1985), p. 39; Donald Mackenzie, *Statistics in Britain: The Social Construction of Scientific Knowledge, 1856–1930* (Edinburgh, 1980), p. 180; Theodore Porter, *The Rise of Statistical Thinking* (Princeton Univ. Press, 1986), p. 305; Richard Solloway, *Demography and Degeneration* (Chapel Hill, 1990), pp. 117–18; Tukufu Zuberi, *Thicker than Blood: How Racial Statistics Lie* (Univ. of Minnesota Press, 2001).

43. M. Eileen Magnello, 'The non-correlation of eugenics and biometrics', *History of Science* **37** (1999), 79–106, 123–50.

44. Magnello (note 43).

45. Magnello (note 43), p. 84.

46. Magnello (note 43), p. 83.

47. For a recent account, see Michael Bulmer (2003).

48. This view was espoused initially by Bernard Norton, 'Biology and philosophy: the methodological foundation of biometry', *J. History of Biology* **7** (1975), 85–93, and took on an even greater role in the work of Theodore Porter, especially in his book on Pearson (2004).

## CHAPTER 13 (I. GRATTAN-GUINNESS)

Useful sources for this chapter are P. M. Harman (ed.), *Wranglers and Physicists* (Manchester Univ. Press, 1985); and especially the late Maria Panteki's Ph.D. thesis, *Relationships between Algebra, Differential Equations and Logic in England: 1800–1860* (Middlesex Univ., 1992), Chapters 2–5, from which Grattan-Guinness has written 'D company: the rise and fall of differential operator theory in Britain, 1810s–1870s', *Archives Internationales d'Histoire des Sciences* **60** (2010, publ. 2011), 477–528.

### NOTES

1. N. Guicciardini, *The Development of Newtonian Calculus in Britain, 1700–1800*, Part 3 (Cambridge, 1989).

2. J. Caramalho Domingues, *The Calculus and Lacroix* (Birkhäuser, 2008).

3. On Woodhouse, see C. Philips, 'Robert Woodhouse and the evolution of Cambridge mathematics', *History of Science* **44** (2000), 69–91. At the time when Newtonian calculus still ruled, Woodhouse produced the history, *A Treatise of Isoperimetrical Problems, and the Calculus of Variations* (1810)—a remarkable novelty.

4. On these developments, see E. Koppelman, 'The calculus of operations and the rise of abstract algebra', *Archive for History of Exact Sciences* **8** (1971), 155–242; S. E. Despeaux, '"Very full of symbols": Duncan F. Gregory, the calculus of operations and the *Cambridge Mathematical Journal*', *Episodes in the History of Modern Algebra (1800–1950)* (ed. J. J. Gray and K. H. Parshall), (American and London Mathematical Societies, 2007), pp. 49–72, and especially Panteki (1992), Chapters 2–5.

5. In 1970 the author visited the Institut Mittag-Leffler, near Stockholm (see I. Grattan-Guinness, 'Materials for the history of mathematics in the Institut Mittag-Leffler', *Isis* **62** (1971), 363–74), and found about fifty volumes of versions of Weierstrass's lecture courses on various subjects prepared by students, and lists of students who attended the courses. There does not appear to have been a single Briton involved in either source.

6. Significant British contributions to the foundations of mathematical analysis in the period 1830–1890 seem to comprise a small number of works: William Thomson on Fourier series (1840–1845); I. Grattan-Guinness, 'On the early work of William Thomson: mathematical physics and methodology in the 1840s', *Lord Kelvin: Life, Labours and Legacy* (ed. R. G. Flood, M. McCartney, and A. Whitaker) (Oxford, 2008), pp. 44–55, on pp. 45–47; Augustus De Morgan, on summing divergent series, and also on iterative tests for the convergence of infinite series, in his textbook (1842) (see I. Grattan-Guinness, *The Development of the Foundations of Mathematical Analysis*

from *Euler to Riemann* (MIT Press, 1970), Appendix; G. G. Stokes on a mode of convergence of an infinite series of functions that corresponds to uniform convergence and is often misidentified with it (1847) (see Grattan-Guinness (1970), pp. 113–16); Henry Wilbraham on the 'Gibbs phenomenon' (1848) (see E. Hewitt and R. E. Hewitt, 'The Gibbs–Wilbraham phenomenon: an episode in Fourier analysis', *Archive for History of Exact Sciences* 21 (1981), 129–60, on pp. 147–50); and H. J. S. Smith on the integration of discontinuous functions and a set similar to Cantor's ternary set, and some other aspects of point-set topology (1875) (see T. W. Hawkins, *Lebesgue's Theory of Integration: Its Origins and Development*, Univ. of Wisconsin Press, 1970, pp. 37–41). The copious historical footnotes in Whittaker and Watson's textbook on 'modern analysis' (1927) suggest that no further British stones await turning.

7. See A. Warwick, *Masters of Theory. Cambridge and the Rise of Mathematical Physics* (Univ. of Chicago Press, 2003); A. D. D. Craik, *Mr Hopkins' Men. Cambridge Reform and British Mathematics in the 19th Century* (Springer, 2007); and J. Barrow-Green, '"A correction to the spirit of too exclusively pure mathematics": Robert Smith (1689–1768) and his prizes at Cambridge University', *Annals of Science* 56 (1999), 271–316.

8. See Harman (1985).

9. See Panteki (1992), Chapters 4–5. At that time a new operator theory was developing with the mathematician and electrical engineer Oliver Heaviside. After an international effort, his wild methods were finally reined in early in the new century, especially with the systematization of the so-called 'Laplace transform'; see J. Lützen, 'Heaviside's operational calculus and the attempts to rigorise it', *Archive for History of Exact Sciences* 21 (1979), 161–200.

10. G. Boole, 'On the integration of linear differential equations with constant coefficients', *Cambridge Math. Journal* 2 (1840), 114–19.

11. Surprisingly, this book was translated into German in 1857.

12. Guicciardini (note 1), p. 138.

13. See S. E. Despeaux, *The Development of a Publication Community: Nineteenth-Century Mathematics in British Scientific Journals*, Ph.D. thesis, Univ. of Virginia, 2002, as well as Chapter 7 of this book.

14. These data are based upon the somewhat crude technique of searching by title keywords in the catalogue of the British Library, and so are to be taken as underestimates. A book that covers more than one topic is counted multiply; not counted in are new editions, or books produced in the USA or the British Empire. In the period 1810–1840 there were very few textbooks or substantial encyclopedia articles on the calculus, but there were many of both in

mechanics; see the author's lists in Harman (1985), 88–89, 98–99.

15. J. Barrow-Green, 'The advantage of proceeding from an author of some scientific reputation: Isaac Todhunter and his mathematics textbooks', *Teaching and Learning in Nineteenth-Century Cambridge* (ed. J. Smith and C. Stray) (Boydell and Brewer, 2001), pp. 177–203.

16. A. Rice, 'A gradual innovation: the introduction of Cauchian calculus into mid-nineteenth-century Britain', *Proc. Canadian Society for the History and Philosophy of Math.* 13 (2001), 48–63.

17. A. De Morgan, *The Elements of Algebra Preliminary to the Differential Calculus* (John Taylor, 1835), p. 155.

18. A. De Morgan, *The Differential and Integral Calculus* (Baldwin and Cradock, 1842), pp. 3, 8–11, 35, 49–51.

19. De Morgan (note 18), Chapter 6.

20. De Morgan (note 18), pp. 633–40.

21. De Morgan (note 18), Chapter 19.

22. An Italian translation was published in Naples in 1880.

23. I. Todhunter, *A Treatise on the Differential Calculus, and the Elements of the Integral Calculus*, 5th ed. (Macmillan, 1871), p. vi.

24. For this, Panteki (1992) is still our principal guide, with an introduction in Grattan-Guinness (to appear).

25. E. T. Whittaker, 'Andrew Russell Forsyth', *Obituary Notices of Fellows of the Royal Society* 4 (1942), 209–27, on p. 217.

26. Boole held the notion of singular solution in especial esteem: on the one hand it was a solution, but on the other hand it lay outside the normal range of solutions. This was exactly an example of $x$ and its complement $1 - x$ that lay at the heart of his algebra of logic: see I. Grattan-Guinness, *The Search for Mathematical Roots, 1870–1940. Logics, Set Theories and the Foundations of Mathematics from Cantor through Russell to Gödel* (Princeton Univ. Press, 2000), p. 52.

27. L. Roth, Old Cambridge days, *American Math. Monthly* 78 (1971), 223–36, on p. 230.

28. Roth (note 27).

29. Whittaker (note 25), p. 218.

30. E. H. Neville, 'Andrew Russell Forsyth', *J. London Math. Society* 17 (1942), 237–56, on p. 245.

31. W. F. Osgood, 'The theory of functions', *Bull. American Math. Society* 1 (1895), 142–54, on p. 143.

32. Roth (note 27), p. 231.

33. Whittaker (note 25), p. 217.

34. Neville (note 30), p. 245.

35. Whittaker (note 25), p. 218.

36. Whittaker (note 25).

37. I note here the 'competition' between British and American mathematicians in developing real and complex analysis between 1890 and the First World War. The old country was given a close ride by its former colony: for example, W. F. Osgood's book on infinite series (1897) much preceded Bromwich's, and the first volume of James Pierpont's textbook on real-variable functions (1905) was slightly ahead of Hobson's. Some Americans had attended Weierstrass's lecture courses.

38. See Grattan-Guinness (note 26), Chap. 6.

39. E. T. Whittaker and G. N. Watson, *A Course of Modern Analysis* (2nd revised ed., Cambridge, 1915; 3rd ed., 1920; 4th ed., 1927).

## CHAPTER 14 (AMIROUCHE MOKTEFI)

The standard edition of Euclid's *Elements* is Thomas L. Heath's *The Thirteen Books of Euclid's Elements*, translated from Heiberg's text, with introduction and commentary, (Cambridge, 1908, 2nd ed. revised with additions, 1926). A useful and nicely illustrated history of English editions of Euclid is to be found in June Barrow-Green's '"Much necessary of all sorts of men": 450 years of Euclid's *Elements* in English', *Bull. British Society for the History of Mathematics* 21 (2006), 2–25.

For a general history of geometry, both Euclidean and non-Euclidean, see Jeremy Gray's *Ideas of Space: Euclidean, Non-Euclidean, and Relativistic* (Oxford, 1989), and Robin Hartshorne's *Geometry: Euclid and Beyond* (Springer, 2000). The standard work on British geometry in the 19th century is Joan L. Richards, *Mathematical Visions: The Pursuit of Geometry in Victorian England* (Academic Press, 1988).

This chapter also uses material from A. Moktefi's 'How to supersede Euclid: geometrical teaching and the mathematical community in nineteenth-century Britain', *(Re)Creating Science in Nineteenth-Century Britain* (ed. A. M. Caleb) (Cambridge Scholars Publishing, 2007), pp. 216–29. Other accounts of the Euclid debate in Britain can be found in W. H. Brock, 'Geometry and the universities: Euclid and his modern rivals, 1860–1901', *History of Education* 4 (2) (1975), 21–35, and Robin Wilson, 'Geometry teaching in England in the 1860s and 1870s: two case studies', *HPM 2004 & ESU4: Proceedings of the ICME10 Satellite Meeting of the HPM Group & The Fourth European Summer University on the History and Epistemology in Mathematics Education* (ed. F. Furinghetti, S. Kaijser, and C. Tzanakis) (2006), pp. 167–73. See also Chapter 4 of Joan Richards (1988).

For a detailed history of The Association for the Improvement of Geometrical Teaching, and later The Mathematical Association, see Michael H. Price's *Mathematics for the Multitude? A History of the Mathematical Association* (Math. Association, Leicester, 1994).

For a comparison with geometrical teaching in other western countries (France, Germany, Italy, and the United States of America), see F. Cajori, 'Attempts made during the eighteenth and nineteenth centuries to reform the teaching of geometry', *American Math. Monthly* 17 (10) (1910), 181–201.

## NOTES

1. D. Lardner, *The First Six Books of the Elements of Euclid*, 12th ed. (Henry G. Bohn, 1861), p. ix.

2. W. Whewell, *Of a Liberal Education* (J. W. Parker, 1845), p. 29.

3. See R. C. Archibald, 'The first translation of Euclid's Elements into English and its source', *American Math. Monthly* 57 (1950), 443–52.

4. See Barrow-Green (2006) and Heath, Vol. 1 (1926), pp. 109–12.

5. Montagu Burrows, *Pass and Class: An Oxford Guide-Book, through the Courses of Literae Humaniores, Mathematics, Natural Science, and Law and Modern History*, 2nd ed. (J. H. and J. Parker, 1861), pp. 183–85.

6. R. Potts, *Euclid's Elements of Geometry* (Cambridge, 1845).

7. See W. Johnson, 'Isaac Todhunter (1820–1884): textbook writer, scholar, coach and historian of science', *International Journal of Mechanical Sciences* 38 (1996), 1231–70; J. Barrow-Green, "The advantage of proceeding from an author of some scientific reputation': Isaac Todhunter and his mathematical textbooks', *Teaching and Learning in Nineteenth-Century Cambridge* (ed. J. Smith and C. Stray) (Boydell Press, 2001), pp. 177–203.

8. Barrow-Green (note 7), p. 189.

9. Barrow-Green (2006), p. 15.

10. On H. MacColl, see M. Astroh, I. Grattan-Guinness, and S. Read, 'A survey of the life and work of Hugh MacColl', *History and Philosophy of Logic* 22 (2001), 81–98.

11. H. MacColl, 'Review of A. E. Layng's *Euclid's Elements of Geometry* (Blackie & Son, 1890)', *Athenaeum* 3310 (4 April 1891), 443.

12. J. J. Sylvester, 'A plea for the mathematician II', *Nature* (6 January 1870), 261–62.

13. See J. M. Wilson, *An Autobiography: 1836–1931* (Sidgwick & Jackson, 1932), and G. Howson, *A History of Mathematics Education in England* (Cambridge, 1982), pp. 123–40.

14. J. M. Wilson, *Elementary Geometry: Part I* (Macmillan, 1868), pp. v–xii.

15. J. M. Wilson (note 14), p. xii.

16. See J. M. Wilson, 'Euclid as text-book of elementary geometry', *Educational Times* 21 (1868), 126–28, and

J. M. Wilson, *A Lecture on Mathematical Teaching* (W. Billington, Rugby, 1870).

17. On Levett, see C. H. P. Mayo and C. Godfrey, 'A great schoolmaster', *Math. Gazette* **11**, no. 165 (July 1923), 325–29.

18. R. Levett, 'Euclid as a textbook', *Nature* **2** (26 May 1870), 65.

19. On Hirst, see J. H. Gardner and R. J. Wilson, 'Thomas Archer Hirst—Mathematician Xtravagant' (in six parts), *American Math. Monthly* **100** (1993), 435–41, 531–38, 619–25, 723–31, 827–34, 907–15.

20. The Association for the Improvement of Geometrical Teaching, *First Annual Meeting Report* (J. Allen, Birmingham, 1871), p. 14.

21. AIGT (note 20), pp. 18–19.

22. A. W. Siddons, 'Progress', *Math. Gazette* **20**, no. 237 (February 1936), 18.

23. Siddons (note 22).

24. A. De Morgan, 'Review of J. M. Wilson's *Elementary Geometry*', *Athenaeum* **2125** (18 July 1868), 72.

25. I. Todhunter, *The Conflict of Studies and Other Essays* (Macmillan, 1873); see also W. Johnson, 'Todhunter's *The Conflict of Studies* and other essays', *International J. Mechanical Sciences* **38** (1996), 1367–77.

26. C. L. Dodgson, *Euclid and his Modern Rivals*, 2nd ed. (Macmillan, 1885), p. 225.

27. C. L. Dodgson, *Euclid: Books I, II* (Macmillan, 1882).

28. C. L. Dodgson, *A New Theory of Parallels* (Macmillan, 1888, 4th ed., 1895). See also W. B. Frankland, *Theories of Parallelism* (Cambridge, 1910), p. 52; F. F. Abeles, 'Avoiding 'the bewildering region of infinites and infinitesimals': C. L. Dodgson's Euclidean parallel axiom', *Proc. Canadian Society for the History and Philosophy of Math.* **8** (1995), 143–51.

29. David E. Smith, *The Teaching of Elementary Mathematics* (Macmillan, 1900), p. 229.

30. See M. Dampier, 'The *Mathematical Gazette*: a brief history', *Math. Gazette* **80**, no 487 (March 1996), 5–12.

31. A. Cayley, 'Note on Lobatchewsky's imaginary geometry', *Phil. Mag.* **29** (1865), 231–33.

32. See J. Richards, 'The reception of a mathematical theory: non-Euclidean geometry in England, 1868–1883', *Natural Order: Historical Studies of Scientific Culture* (ed. B. Barnes and S. Shapin) (Sage, 1979), pp. 143–66.

33. B. Russell, *An Essay on the Foundations of Geometry* (Cambridge, 1897). See also J. Richards, 'Bertrand Russell's essay on the foundations of geometry and the Cambridge mathematical tradition', *Russell: the Journal of Bertrand Russell Studies* **8**, no. 1 (1988), 59–80.

34. B. Russell, 'The teaching of Euclid', *Math. Gazette* **2**, no 33 (May 1902), 165.

35. On Perry, see G. J. N. Gooday, 'Perry, John (1850–1920)', *Dictionary of National Biography*, Oxford, 2004.

36. J. Perry, 'The teaching of mathematics', *Nature* **62**, no. 1605 (2 August 1900), 317–18.

37. R. F. Muirhead, 'The teaching of mathematics', *Math. Gazette* **2**, no. 29 (October 1901), 81–83.

38. E. M. Langley, 'The teaching of mathematics', *Math. Gazette* **2**, no. 30 (December 1901), 105–6; C. Godfrey, 'The teaching of mathematics—a compromise', *Math. Gazette* **2**, no. 30 (December 1901), 106–8; A. W. Siddons, 'From a public school point of view', *Math. Gazette* **2**, no 30 (December 1901), 108–11.

39. A. R. Forsyth, 'Report of the British Association Committee on the teaching of mathematics', *Math. Gazette* **2**, no. 35 (October 1902), 198–99.

40. 'The M. A. Committee on Geometry', Report of the M. A. Committee on Geometry, *Math. Gazette* **2**, no. 33 (May 1902), 168–172; C. G. (probably Charles Godfrey), 'Report on the teaching of geometry', *Nature* **66**, no. 1704 (26 June 1902), 201–2. On the history of the Teaching Committee, see A. W. Siddons, 'The first twenty years of the teaching committee', *Math. Gazette* **36**, no. 317 (September 1952), 153–57.

41. For an account of the history of geometry in Cambridge in the second half of the 19th century and beyond, see J. Barrow-Green and J. Gray, 'Geometry at Cambridge, 1863–1940', *Historia Math.* **33** (2006), 315–56.

42. A. W. Siddons, 'Progress', *Math. Gazette* **20**, no. 237 (February 1936), 20–21.

43. See Anonymous, 'Mathematical reform at Cambridge', *Nature* **68**, no. 1756 (25 June 1903), 178–79.

44. Anonymous, 'The Mathematical Association', *Math. Gazette* **3**, no. 50 (March 1905), 152.

45. A particularly successful modern manual was C. Godfrey and A. W. Siddons, *Elementary Geometry: Practical and Theoretical* (Cambridge, 1903).

46. On the uniformity or freedom of sequences, see for instance, Anonymous, 'The Mathematical Association', *Math. Gazette* **3**, no. 50 (March 1905), 145–57; J. M. Child, 'The need of a sequence in geometry', *Math. Gazette* **4**, no. 64 (May 1907), 80–81; C. Godfrey, 'Is there need of a recognized sequence in geometry?', *Math. Gazette* **4**, no. 65 (July 1907), 100–1.

## CHAPTER 15 (KAREN HUNGER PARSHALL)

Certain aspects of the history of algebra in 19th-century Britain have received extensive historical analysis; for example, much work has been done on the rise of symbolical algebra and the associated development of the calculus of operations in the first half of the century.

Among works to consult on the former are: Marie-José Durand-Richard, 'L'école algébrique anglaise: les conditions conceptuelles et institutionnelles d'un calcul symbolique comme fondement de la connaissance', *L'Europe Mathématique—Mythes, Histoires, Identités/Mathematical Europe—Myth, History, Identity* (ed. C. Goldstein, J. Gray, and J. Ritter) (Éditions de la Maison des Sciences de l'Homme, 1996), pp. 447–77; Menachem Fisch, "The emergency which has arrived': the problematic history of nineteenth-century British algebra—a programmatic outline', *British Journal for the History of Science* 27 (1994), 247–76; and 'The making of Peacock's *Treatise on Algebra*: a case of creative indecision', *Archive for History of Exact Sciences* 54 (1999), 137–179; Helena Pycior, 'George Peacock and the British origins of symbolical algebra', *Historia Math.* 8 (1981), 23–45, and 'Early criticism of the symbolical algebra', *Historia Math.* 9 (1982), 392–412; Joan L. Richards, 'The art and perception of British algebra: a study in the perception of mathematical truth', *Historia Math.* 7 (1980), 343–65, and 'Augustus De Morgan, the history of mathematics, and the foundations of algebra', *ISIS* 78 (1987), 7–30.

On the latter, see Patricia Allaire and Robert E. Bradley, 'Symbolical algebra as a foundation for calculus: D. F. Gregory's contribution', *Historia Math.* 29 (2002), 395–426; and Elaine Koppelman, 'The calculus of operations and the rise of abstract algebra', *Archive for History of Exact Sciences* 8 (1971–72), 155–242.

On the development of algebras, see Michael J. Crowe's *A History of Vector Analysis: The Evolution of the Idea of a Vectorial System* (Univ. of Notre Dame Press, 1967); Thomas Hawkins, 'Hypercomplex numbers, Lie groups, and the creation of group representation theory', *Archive for History of Exact Sciences* 8 (1972), 243–87, and 'Another look at Cayley and the theory of matrices', *Archives Internationales d'Histoire des Sciences* 26 (1977), 82–112; and Karen Hunger Parshall, 'Joseph H. M. Wedderburn and the structure theory of algebras', *Archive for History of Exact Sciences* 32 (1985), 223–349.

On invariant theory, consult especially Tony Crilly, 'The rise of Cayley's invariant theory (1841–1862)', *Historia Math.* 13 (1986), 241–54, and 'The decline of Cayley's invariant theory (1863–1895)', *Historia Math.* 15 (1988), 332–47; and Karen Hunger Parshall, 'Toward a history of nineteenth-century invariant theory', *The History of Modern Mathematics*, Vol. 1 (ed. David E. Rowe and John McCleary) (Academic Press, 1989), pp. 157–206.

These, as well as other aspects of the development of algebra in Victorian Britain, are also addressed in the following works: Tony Crilly, *Arthur Cayley: Mathematician Laureate of the Victorian Age*, (Johns Hopkins Univ. Press, 2006); Thomas Hankins, *Sir William Rowan Hamilton*, (Johns Hopkins Univ. Press, 1980); Karen Hunger Parshall, *James Joseph Sylvester: Life and Work in Letters*, (Clarendon Press, Oxford, 1998), and *James Joseph Sylvester: Jewish Mathematician in a Victorian World*, (Johns Hopkins Univ. Press, 2006). See also Sloan Evans Despeaux, *The Development of a*

*Publication Community: Nineteenth-century Mathematics in British Scientific Journals* (Ph.D. thesis, Univ. of Virginia, 2002).

The collected works of several figures central to the history of algebra in Victorian Britain have also been published: William Burnside, *The Collected Papers of William Burnside* (ed. P. M. Neumann, A. J. S. Mann, and J. C. Tompson) (Oxford, 2004); *The Collected Mathematical Papers of Arthur Cayley*, 14 vols. (ed. A. Cayley and A. R. Forsyth) (Cambridge, 1889–98); *Mathematical Papers by William Kingdon Clifford* (ed. R. Tucker) (Macmillan, 1882, reprinted by Chelsea, 1968); Sir William Rowan Hamilton, *The Mathematical Papers of Sir William Rowan Hamilton*, 3 vols. (ed. H. Halberstam and R. E. Ingram) (Cambridge, 1967); *Percy Alexander MacMahon, Collected Papers*, 2 vols. (ed. G. E. Andrews) (MIT Press, 1978–86); and *The Collected Mathematical Papers of James Joseph Sylvester*, 4 vols. (ed. H. F. Baker), (Cambridge, 1904–12) (reprinted by Chelsea, 1973).

For general overviews of the history of algebra that incorporate discussions of developments in 19th-century Britain, see Isabella Bashmakova and Galina Smirnova, *The Beginnings and Evolution of Algebra* (Math. Association of America, 2000); Bartel L. van der Waerden, *A History of Algebra from al-Khwārizmī to Emmy Noether* (Springer-Verlag, 1985); and Hans Wussing, *The Genesis of the Abstract Group Concept: A Contribution to the History of the Origin of Abstract Group Theory* (transl. Abe Shenitzer) (MIT Press, 1984).

## NOTES

1. In her masterful study of what she terms the 19th-century British 'publication community', Sloan Despeaux has documented that algebraic topics accounted for somewhat more than 10 per cent of the research output of British mathematicians over the course of the 19th century, with analytic geometry comprising between 15 and 20 per cent of British research output and mathematical physics averaging some 20 per cent over that same hundred-year period. See Despeaux (2002), pp. 311–12.

2. George Peacock, *Treatise on Algebra* (J. & J. J. Deighton, 1830).

3. Fisch (1999), p. 155 (his emphasis).

4. Peacock, *Treatise*, xi, as quoted in Fisch (1999), pp. 166–67.

5. George Peacock, 'Report on the recent progress and present states of certain branches of analysis', *Report on the Third Meeting of the British Association for the Advancement of Science* (John Murray, 1834), p. 195, as quoted in Fisch (1999), p. 168.

6. Fisch (1994), p. 264 (his emphasis).

7. See William Rowan Hamilton, 'Theory of conjugate functions, or algebraic couples; with a preliminary and elementary essay on algebra as the science of pure time', *Trans.*

*Royal Irish Academy* **17** (1837), 293–422 = *Math. Papers*, Vol. 3, pp. 3–96.

8. Hamilton, *Math. Papers*, Vol. 3, p. 3.

9. Augustus De Morgan, 'On the foundations of algebra, no. II', *Trans. Cambridge Philosophical Society* **7** (1842), 287–300, on p. 287.

10. De Morgan (note 9), pp. 289–90; see also Richards (1987) for a historical contextualization of De Morgan's position.

11. De Morgan (note 9), p. 287.

12. Augustus De Morgan, *Elements of Algebra Preliminary to the Differential Calculus*, 2nd ed. (Taylor and Walton, 1837), p. 4, from the preface to the first edition.

13. Duncan Gregory, 'On the real nature of symbolical algebra', *Trans. Royal Society of Edinburgh* **14** (1840), 208–16; see 209–10 for the two classes discussed here. For more on Gregory's contributions, see Allaire and Bradley (2002).

14. Gregory (note 13), p. 208.

15. On the development of the calculus of operations in Britain and on the numerous contributors to it, see Koppelman (1971–72), and pp. 155–57, 187–89 therein for these examples: compare also Parshall (2006), pp. 97–98. On *The Cambridge Mathematical Journal*, see Tony Crilly, 'The Cambridge Mathematical Journal and its descendants: 1830–1870', *Historia Math.* **31** (2004), 255–97.

16. For a fascinating discussion of Hamilton's philosophical views and the influences on them, see Hankins (1980), pp. 247–79.

17. See Hamilton, 'Theory of conjugate functions', and compare the discussion in Crowe (1967), 27–33. This has been a tale oft-told.

18. Shortly after its discovery, Hamilton announced his result to James MacCullagh and William Sadlier at a meeting of the council of the Royal Irish Academy on 16 October 1843, and in a letter to his friend John Graves, the next day. He first published it in 'On a new species of imaginary quantities connected with the theory of quaternions', *Proc. Royal Irish Academy* **2** (1844), 424–34 = *Math. Papers*, Vol. 3, pp. 111–16. Graves wrote this in his reply to Hamilton on 31 October 1843; see Hankins (1980), p. 300. It is well known that Hamilton himself vigorously pursued the theory of quaternions after his discovery of them, but not from a purely algebraic point of view. Hamilton, and subsequently his many followers (the so-called 'quaternionists') saw the quaternions as a key mathematical tool for use in physics. On the development of the quaternions as a vectorial system and on the history of vectorial systems in general, see Crowe (1967).

19. Graves made his discovery known in a letter to Hamilton dated 26 December 1843, while Cayley published his in 'On Jacobi's elliptic functions, in reply to the Revd. B. Bronwin: and on quaternions', *Phil. Mag.* **26** (1845),

208–11 = *Math. Papers*, Vol. 1, p. 127. Hamilton introduced the biquaternions in his *Lectures on Quaternions* (Hodges and Smith, 1853); see, in particular, articles 637 and 669.

20. See Augustus De Morgan, 'On triple algebra', *Trans. Cambridge Philosophical Society* **8** [1844], 241–54; Charles Graves, 'On algebraic triplets', *Proc. Royal Irish Academy* **3** (1847), 51–54, 57–64, 80–84, and 105–8; Thomas Kirkman, 'On pluquaternions, and homoid products of sums of *n* squares', *Phil. Mag.* **33** (1848), 447–59 and 494–509; and James Cockle, 'On the symbols of algebra, and on the theory of tessarines', *Phil. Mag.* **34** (1849), 406–10. Kirkman and Cockle both pursued this line of thought into 1850 with papers in the *Philosophical Magazine*. For more on Kirkman's work on pluquaternions, in particular, see Norman L. Biggs, 'T. P. Kirkman, mathematician', *Bull. London Math. Society* **13** (1981), 97–120, on pp. 100–101; Biggs also provides a complete bibliography of Kirkman's publications.

21. William Kingdon Clifford, 'Applications of Grassmann's extensive algebra', *American J. Math.* **1** (1878), 350–58 = *Math. Papers*, pp. 266–76. Peirce's 'Linear associative algebra' was distributed privately in lithograph in 1870 but published as 'Linear associative algebra with notes and addenda by C. S. Peirce, son of the author', in *American J. Math.* **4** (1881), 97–229. As early as 1873, Clifford had already considered an algebra which he termed the 'biquaternions' but which, unlike Hamilton's biquaternions, was (in modern terms) the tensor product of the quaternions and the complex field. See William Kingdon Clifford, 'Preliminary sketch of biquaternions', *Proc. London Math. Society* **4** (1873), 381–95 = *Math. Papers*, pp. 181–200.

22. See James Joseph Sylvester, 'A word on nonions', *Johns Hopkins Univ. Circulars* **1** (August 1882), 241, 242 = *Math. Papers*, Vol. III, pp. 647–50. This work is also discussed in Parshall (1985), 245–47 and in Parshall (2006), 265–66; in the latter discussion, a file conversion resulted in misplaced negative signs in the matrices *u* and *w*.

23. James Joseph Sylvester, 'Lectures on the principles of universal algebra', *American J. Math.* **6** (1884), 270–86 = *Math. Papers*, Vol. IV, pp. 208–24. The historical import of this work was discussed in Parshall (1985), 249–50 and 261–63.

24. Sylvester coined the term 'matrix' in 'Addition to the articles "On a new class of theorems" and "On Pascal's theorem"', *Phil. Mag.* **37** (1850), 363–70 = *Math. Papers*, Vol. I, pp. 145–51, on p. 150. For the quotation presented here, see James Joseph Sylvester, 'On the relation between the minor determinants of linearly equivalent quadratic functions', *Phil. Mag.* **1** (1851), 295–305, or *Math. Papers*, Vol. I, pp. 241–50; see p. 247 for the quotation. Compare also the discussion in Parshall (2006), 101–2.

25. For an exhaustive treatment of the contributors (British and otherwise) to the theory of determinants, see Thomas Muir, *The Theory of Determinants in the Historical Order of Development*, 4 vols. (Macmillan, 1906–23, reprinted by Dover, 1960). As early as 1851, Spottiswoode published *Elementary Theorems Relating to Determinants* (Longman, Brown, Green, and Longman, 1851), while Dodgson made his most significant statement on the subject in *An Elementary Treatise on Determinants with Their Application to Simultaneous Linear Equations and Algebraical Geometry* (Macmillan, 1867).

26. Arthur Cayley, 'A memoir on the theory of matrices', *Phil. Trans. Royal Society of London* **148** (1858), 17–37 = *Math. Papers*, Vol. 2, pp. 475–96, on pp. 475–76. Thomas Hawkins initially highlighted the significance of Cayley's work here in Hawkins (1977); see also the discussion in Parshall (1985), pp. 233–39, and Crilly (2006), pp. 226–30.

27. Cayley (note 26), p. 476.

28. On these late 19th-century developments in general, and on Wedderburn's contributions in particular, see Parshall (1985).

29. Arthur Cayley, 'On the theory of groups, as depending on the symbolic equation $\theta^n = 1$', *Phil. Mag.* **7** (1854), 40–47 and 408–9 = *Math. Papers*, Vol. 2, pp. 123–32; for the quotations in this paragraph and the next, see pp. 123–24. Crilly contextualizes Cayley's 1854 work in group theory in Crilly (2006), pp. 185–88; see also Wussing (1984), pp. 230–33, van der Waerden (1985), pp. 150–52, and Bashmakova and Smirnova (2000), pp. 124–25.

30. Arthur Cayley, 'On the theory of groups, as depending on the symbolic equation $\theta^n = 1$. Third part', *Phil. Mag.* **18** (1859), 34–37 = *Math. Papers*, Vol. 4, pp. 88–91.

31. For a historical account of Kirkman's contributions to group theory, see Biggs (note 20), pp. 105–9.

32. Thomas P. Kirkman, 'On the theory of groups and many valued functions', *Memoirs of the Manchester Literary and Philosophical Society* **1** (1862), 274–398.

33. Thomas P. Kirkman, 'Hints on the theory of groups', *Messenger of Math.* **1** (1862), 58–68, 187–92; for an assessment of this work, see Biggs (note 20), p. 108.

34. For the references and a discussion of this work, see Parshall (1998), 97–101.

35. For this and the next quote, see Arthur Cayley, 'The theory of groups', *American J. Math.* **1** (1878), 50–52 = *Math. Papers*, Vol. 10, pp. 401–3. For a sense of the context of some of the late 19th-century American work, see Karen Hunger Parshall and David E. Rowe, *The Emergence of the American Mathematical Research Community, 1876–1900: J. J. Sylvester, Felix Klein, and E. H. Moore* (American and London Mathematical Societies 1994), pp. 321–22,

372–81. On this, and Cayley's other group-theoretic work in the late 1870s, see Crilly (2006), pp. 348–54.

36. See Arthur Cayley, The theory of groups: graphical representation, *American J. Math.* **1** (1878), 174–76 = *Math. Papers*, Vol. 10, pp. 403–5.

37. William Burnside, *Theory of Groups of Finite Order* (Cambridge, 1897, reprinted by Dover, 1955), p. viii of the reprint edition.

38. George Boole, 'Exposition of a general theory of linear transformations', *Cambridge Mathematical J.* **3** (1841–42), 1–20, 106–19. I have altered Boole's original notation so that it conforms with that later adopted by Sylvester and Cayley. A more detailed version of this section appeared in Karen Hunger Parshall, 'The British development of the theory of invariants (1841–1895)', *Bulletin of the British Society for the History of Mathematics* **21** (2006), 186–99.

39. Notice that the notation indicates a linear change of the variables $x$ and $y$ to the variables $x'$ and $y'$. Boole only implicitly assumed that the determinant of the transformation is non-zero.

40. Boole (note 38), pp. 6, 11, 19.

41. Cayley presented this sequence of results in 'On the theory of linear transformations', *Cambridge Math. J.* **4** (1845), 193–209 = *Math. Papers*, Vol. 1, pp. 80–94, especially pp. 93–94.

42. Arthur Cayley, 'On linear transformations', *Cambridge and Dublin Math. J.* **1** (1846), 104–22 = *Math. Papers*, Vol. 1, pp. 95–112; see p. 95 for the quotation. For the broader historical context of Cayley's work in invariant theory, see Crilly (1986) and (1988). Compare the discussion of these matters in Parshall and Rowe (1994), pp. 67–68.

43. Letter from James Joseph Sylvester to Arthur Cayley, 3 February 1850, Sylvester Papers, St. John's College, Cambridge, Box 9.

44. James Joseph Sylvester, 'On the intersections, contacts, and other correlations of two conics expressed by indeterminate coordinates', *Cambridge and Dublin Math. J.* **5** (1850), 262–82 = *Math. Papers*, Vol. I, pp. 119–37.

45. I have adapted Sylvester's notation here, using the variables $x$, $y$, $z$ instead of his $\xi$, $\eta$, $\zeta$. Sylvester also considered both $\square(U + \lambda V)$ and the more general$)\square$ $(\lambda U + \mu V)$; see the commentary in Parshall (1998), pp. 28–29.

46. Letter from Arthur Cayley to James Joseph Sylvester, 5 December 1851, Sylvester Papers, St. John's College, Cambridge, Box 2, or Parshall (1998), p. 37; here $U$ is an invariant and $a, b, \ldots, k$ are the coefficients of the underlying homogeneous polynomial of degree $n$ in two unknowns.

47. James Joseph Sylvester, 'On the principles of the calculus of forms', *Cambridge and Dublin Math. Journal* 7 (1852), 52–97, 179–217 = *Math. Papers*, Vol. I, pp. 284–327, 328–63, on p. 284.

48. Sylvester (note 47), p. 352. The mathematical exposition here follows that given in Karen Hunger Parshall, 'The mathematical legacy of James Joseph Sylvester', *Nieuw Archief voor Wiskunde* 17 (1999), 247–67, on pp. 255–56.

49. James Joseph Sylvester, 'On a theory of the syzygetic relations of two rational integral functions, comprising an application to the theory of Sturm's functions, and that of the greatest algebraical measure', *Phil. Trans. Royal Society of London* 143 (1853), 407–548 = *Math. Papers*, Vol. I, pp. 429–586.

50. Arthur Cayley, 'An introductory memoir upon quantics', *Phil. Trans. Royal Society of London* 144 (1854), 244–58 = *Math. Papers*, Vol. 2, pp. 221–34; and 'A second memoir upon quantics', *Phil. Trans. Royal Society of London* 146 (1856), 101–26 = *Math. Papers*, Vol. 2, pp. 250–81.

51. George Salmon, *Lessons Introductory to the Modern Higher Algebra* (Hodges & Smith, 1859).

52. For more details on some of the later partition-theoretic work sparked by these invariant-theoretic investigations, see Karen Hunger Parshall, 'America's first school of mathematical research: James Joseph Sylvester at The Johns Hopkins University', *Archive for History of Exact Sciences* 38 (1988), 153–96, on pp. 172–88.

53. See the anonymous review, 'The Rev. G. Salmon's *Lessons Introductory to the Modern Higher Algebra*', *Phil. Mag.* 18 (1859), 67–68, as quoted in Despeaux (2002), p. 389.

54. Paul Gordan, 'Beweis, dass jede Covariante und Invariante einer binären Form eine ganze Function mit numerische Coefficienten einer endlichen Anzahl solchen Formen ist', *Journal für die reine und angewandte Math.* 69 (1868), 323–54.

55. Arthur Cayley, 'A ninth memoir on quantics', *Phil. Trans. Royal Society of London* 161 (1871), 17–50 = *Math. Papers*, Vol. 7, pp. 334–53, on p. 353. Tony Crilly discusses the ninth memoir at some length in Crilly (2006), pp. 301–2.

56. James Joseph Sylvester (assisted by Fabian Franklin), 'Tables of the generating functions and groundforms for the binary quantics of the first ten orders', *American J. Math.* 2 (1879), 223–251 = *Math. Papers*, Vol. 3, pp. 283–311.

57. David Hilbert, 'Über die Endlichkeit des Invariantensystems für binäre Grundformen', *Math. Annalen* 34 (1889), 223–26.

58. Arthur Cayley, 'On the finite number of covariants of a binary form', *Math. Annalen* 34 (1889), 319–20 (not in *Math. Papers*).

59. David Hilbert, 'Ueber die Theorie des algebraischen Formen', *Math. Annalen* 36 (1890), 473–534, and 'Ueber die vollen Invariantensysteme', *Math. Annalen* 42 (1893), 313–73.

60. Edwin Bailey Elliott, *An Introduction to the Algebra of Quantics* (Oxford, 1895).

61. Letter from William Thomson to Hermann von Helmholtz, 31 July 1864, as quoted in Crilly (2006), p. 276.

## CHAPTER 16 (I. GRATTAN-GUINNESS)

The most ample historical survey of British algebraic logic is *British Logic in the Nineteenth Century* (ed. D. M. Gabbay and J. Woods), Vol. 4 of the *Handbook of the History of Logic* (Elsevier, 2007). Still valuable is L. Liard's *Les Logiciens Anglais Contemporains* (Germer Baillère, 1878), which goes up to its dedicatee, Jevons; it also covers inductive logic. Maria Panteki provides a very detailed account of both algebraic logic and the attendant mathematical theories, such as differential operators and functional equations, in her doctoral dissertation, *Relationships between Algebra, Differential Equations and Logic in England: 1800–1860*, Ph.D. thesis, Univ. of Middlesex, 1992.

A short survey of the algebraic tradition is given in I. Grattan-Guinness, *The Search for Mathematical Roots, 1870–1940. Logics, Set Theories and the Foundations of Mathematics from Cantor through Russell to Gödel* (Princeton Univ. Press, 2000), Chapter 2 and parts of Chapter 4; set theory is treated in detail in Chapter 3, and mathematical logic in Chapters 5–7.

Among general histories of logic, W. Kneale and M. Kneale, *The Development of Logic* (Clarendon Press, 1962), Chapters 6, 7, and 11, is quite serviceable, though astonishingly they omit Whately entirely. The same comments apply also to I. M. Bochenski's *A History of Formal Logic* (Univ. of Notre Dame Press, 1961, reprinted by Chelsea, 1970, parts 4 and 5).

For comments on an earlier draft of this chapter, the author is indebted to Amirouche Moktefi and Adrian Rice.

### NOTES

1. For a general review see, for example, Kneale and Kneale (1962), Chapter 5. A short survey appears in J. Maat, 'The status of logic in the seventeenth century', *The History of the Concept of the Formal Sciences* (ed. B. Löwe et al.) (College Publications, King's College, 2006), pp. 157–67. For a review of the status of syllogistic logic during that period, see J. van Evra, 'The development of logic as reflected in the fate of the syllogism', *History and Philosophy of Logic* 21 (2000), 115–34. This journal is cited below as *HPL*.

2. An edition of these papers is included in A. De Morgan, *On the Syllogism and Other Logical Writings* (ed. P. Heath) (Routledge and Kegan Paul, 1966).

3. See De Morgan (note 2), p. 163. This design appears on the De Morgan Medal that is awarded to distinguished mathematicians by the London Mathematical Society.

4. See De Morgan (note 2), p. 119.

5. See D. D. Merrill, *Augustus De Morgan and the Logic of Relations* (Kluwer, 1990).

6. A. De Morgan, 'Calculus of functions', *Encyclopaedia Metropolitana* **2** (1836), 305–92. (This is the date of the offprint version; the volume carries '1845'.)

7. A. De Morgan, 'Calculus of probabilities', *Encyclopaedia Metropolitana* **2** (1837), 393–490. (This is the date of the offprint version, which carries the title 'A treatise on the theory of probabilities'; the volume carries '1845'). See also De Morgan's *An Essay on Probabilities* (Longman, 1838).

8. See T. Hailperin, 'The development of probability logic from Leibniz to MacColl', *HPL* **9** (1988), 131–91, and A. Rice, '"Everybody makes errors": the intersection of De Morgan's logic and probability, 1837–1847', *HPL* **24** (2003), 289–305.

9. We note here an interesting but uninfluential pioneer of imitating some notions from common algebra in syllogistic logic, and also in analyzing the quantification of the predicate: see M. Panteki, 'Thomas Solly (1816–1875): an unknown pioneer of the mathematicization of logic in England, 1839', *HPL* **14** (1993), 133–69. De Morgan added a note on Solly at the end of his *Formal Logic* (1847) as it completed production, and mentioned him in passing twice in his suite of papers (see De Morgan (note 2), pp. 42, 140).

10. The best biography of Boole is D. MacHale, *George Boole—His Life and Work* (Boole Press, 1985); see also the Boole editions, *Studies in Logic and Probability* (ed. R. Rhees) (Open Court, 1952), and *Selected Manuscripts on Logic and its Philosophy* (ed. I. Grattan-Guinness and G. Bornet) (Birkhäuser, 1997). A good collection of commentaries ancient and modern is included in J. Gasser (ed.), *A Boole Anthology* (Kluwer, 2000).

11. See Chapter 13.

12. The names 'distributive' and 'commutative' were introduced in 1814 in the context of functional equations by the French mathematician F.-J. Servois (1767–1847) (see Grattan-Guinness (2000), p. 19).

13. For more on Babbage and his calculating engines, see Chapter 10.

14. G. C. Smith (ed.), *The Boole–De Morgan Correspondence*, (Clarendon Press, Oxford, 1982).

15. I. Grattan-Guinness, 'The correspondence between George Boole and Stanley Jevons, 1863–1864', *HPL* **12** (1991), 15–35. On Jevons as a logician, see W. Mays and D. P. Henry, 'Jevons and logic', *Mind* (new ser.) **62** (1953), 484–505.

16. The theory of multisets was not so long in coming, though not from logic but in an extraordinary long paper of 1886 by A. B. Kempe (1849–1922), of which Peirce was the only serious reader. On this bizarre piece of history, see I. Grattan-Guinness, 'Re-interpreting "λ": Kempe on multisets and Peirce on graphs, 1886–1905', *Trans. C. S. Peirce Society* **38** (2002), 327–50.

17. W. S. Jevons, 'On the mechanical performance of logical inference', *Phil. Trans. Royal Society of London* **160** (1870), 497–518 (reprinted in *Pure Logic and Other Minor Works* (ed. R. Adamson and H. A. Jevons), Macmillan, 1890, reprinted in Thoemmes, 1991, pp. 137–72).

18. Space prevents an account here of these important but geographically extra-Victorian developments. See *Studies in the Logic of Charles S. Peirce* (ed. N. Houser *et al.*) (Indiana Univ. Press, 1997), and G. Brady (ed.), *From Peirce to Skolem. A Neglected Chapter in the History of Logic* (Elsevier, 2000). A short account is given in Grattan-Guinness (2000), Chapter 4; see also V. Peckhaus, *Logik, Mathesis Universalis und allgemeine Wissenschaft. Leibniz und die Wiederentdeckung der formalen Logik im 19. Jahrhundert* (Akademie Verlag, 1997), Chapter 6.

19. See W. H. Brock (ed.), *The Atomic Debates* (Leicester Univ. Press, 1967).

20. On the gradual growth of awareness in Britain of Leibniz's work in logic during the second half of the 19th century, see Peckhaus (note 18), Chapter 5. For reasons of chronology, Leibniz became a corroborating angel of the endeavours of these logicians, rather than an active inspiration of them.

21. These manuscripts by A. J. Ellis are held in the Archives of the Royal Society, which published only short summaries in its *Proceedings*.

22. See the survey of MacColl's work, mostly the second phase, in the *Nordic J. of Philosophical Logic* **3** (1998–1999), 1–234. On his life, see M. Astroh, I. Grattan-Guinness, and S. Read, 'A survey of the life and work of Hugh MacColl', *HPL* **22** (2001–02), 81–98, and S. Rahman, 'Hugh MacColl: eine bibliographische Erschliessung seiner Hauptwerke', *HPL* **18** (1997), 165–83.

23. In the mid-1880s, MacFarlane moved to the USA and became interested in the history of mathematics; in 1916 there appeared posthumously his *Lectures on Ten British Mathematicians of the Nineteenth Century*, which included both De Morgan and Boole.

24. This diagram appears in John Venn's *Symbolic Logic*, 1st ed. (1881), p. 122. For the history of such diagrams, and modern extensions of them, see A. W. F. Edwards, *Cogwheels of the Mind* (Johns Hopkins Univ. Press, 2004).

25. On the differences, see I. Grattan-Guinness, 'The Gergonne relations and the intuitive use of Euler and Venn

diagrams', *International J. of Mathematical Education in Science and Technology* **8** (1977), 23–30.

26. W. E. Johnson, 'The logical calculus', *Mind* (new ser.) **1** (1892), 3–30, 235–50, 340–57.

27. See E. Shen, 'The Ladd-Franklin formula in logic: the antilogism', *Mind* (new ser.) **36** (1927), 54–60.

28. See *Lewis Carroll's Symbolic Logic* (ed. W. W. Bartley, III) (Harvester Press, 1977). For commentaries, see F. Abeles, 'Lewis Carroll's formal logic', *HPL* **26** (2005), 33–46, and 'Lewis Carroll's visual logic', *HPL* **28** (2007), 1–19, and especially A. Moktefi, 'Lewis Carroll's logic', in Gabbay and Woods (2007), pp. 457–505.

29. This example comes from Bartley (note 28), 98.

30. MacFarlane had tried a diagrammatic method similar to Carroll's boxes in the 1880s. Carroll's method of trees is similar in logical form to the 'antilogism' method of handling syllogisms that Ladd-Franklin had published in 1883, which he might have seen (see note 27).

31. L. Carroll, 'What the tortoise said to Achilles', *Mind* (new ser.) **4** (1895), 278–80, reprinted various times, including in Bartley (note 28), pp. 431–34. On the possible bearing of his argument upon the purported paradox itself, including notice of a Carroll manuscript, see I. Grattan-Guinness, 'Achilles is still running', *Trans. C. S. Peirce Society* **10** (1974), 8–16.

32. On the distinction between parts and moments, where Edmund Husserl (1859–1928) was an important pioneer, see B. Smith (ed.), *Parts and Moments. Studies in Logic and Formal Ontology* (Philosophia, 1982).

33. P. E. B. Jourdain, *The Philosophy of Mr. B\*tr\*nd R\*ss\*ll* (Allen and Unwin, 1918), reprinted in *Selected Essays on the History of Set Theory and Logics (1906–1918)* (ed. I. Grattan-Guinness) (CLUEB, Bologna, 1991), pp. 245–342; some chapters were published earlier as articles. The design of the book imitated that of Russell's own book *A Critical Exposition of the Philosophy of Leibniz with an Appendix of Leading Passages* (Cambridge, 1900). See also the editorial material in M. Gardner (ed.), *The Annotated Alice. Alice's Adventures in Wonderland and Through the Looking-Glass* (C. N. Potter, 1960), and in P. Heath (ed.), *The Philosopher's Alice: Alice's Adventures in Wonderland; & Through the Looking-Glass* (St Martin's Press, 1974) (although on pp. 64–65 Heath does not really capture the grin).

34. The influence of Cauchy upon A. De Morgan, *The Differential and Integral Calculus* (Taylor and Walton, 1842) is evident in Chapters 0–2 (limits and the derivative), intermittently in Chapters 6 and 20 (the integral), but not at all in, for example, Chapter 19 (formal summation of series). De Morgan mentioned Cauchy rarely. See also Chapter 13 of this book.

35. The history of mathematical analysis has been written up in a fair amount of detail. An introductory account that covers also the material of this section is I. Grattan-Guinness (ed.), *From the Calculus to Set Theory, 1630–1910: An Introductory History* (Duckworth, 1980, reprinted by Princeton Univ. Press, 2000).

36. See various chapters in U. Cassina, *Critica dei Principî della Matematica e Questione di Logica* (Cremonese, 1961) and *Dalla Geometria Egiziana alla Matematica Moderna* (Cremonese, 1961), and M. Borga, P. Freguglia, and B. Palladino, *I Contributi Fondazionale della Scuola di Peano* (Franco Angeli, 1985). A short survey is given in Grattan-Guinness (2000), Chapter 5.

37. See N. Griffin, *Russell's Idealist Apprenticeship* (Clarendon Press, Oxford, 1991).

38. On all the matters discussed in this section, see Grattan-Guinness (2000), especially Chapters 6 and 7. The historical literature on this story is very philosophical, but often without grasp of the mathematics. In addition, the influence upon Russell of the remarkable but little-read German logicist for arithmetic, Gottlob Frege (1848–1925), is greatly exaggerated, while Peano and Cantor are largely or even wholly passed over. An exception to these characteristics is F. A. Rodriguez-Consuegra, *The Mathematical Philosophy of Bertrand Russell: Origins and Development* (Birkhäuser, 1991).

39. On this and other paradoxes of the time, see A. Garciadiego, *Bertrand Russell and the Origins of the Set-Theoretic "Paradoxes"* (Birkhäuser, 1992).

40. Russell's own recollection of his encounter with Peano in Paris and the writing of *The Principles of Mathematics* is very faulty (*Autobiography*, Vol. 1 (Allen & Unwin, 1967) pp. 144–45 and elsewhere). For a detailed account see I. Grattan-Guinness, 'How did Bertrand Russell write *The Principles of Mathematics* (1903)?', *Russell* (new ser.) **16** (1996–97), 101–27. On the book, see J. Vuillemin, *Leçons sur la Première Philosophie de Russell* (Colin, 1968).

41. While vast, *Principia Mathematica* is also incomplete. Whitehead was due to add his own fourth volume on aspects of geometry, and during the 1910s he wrote much of it, but then he abandoned it and no manuscript survives. See M. Harrell, 'Extension to geometry of *Principia Mathematica* and related systems II', *Russell* (new ser.) **8** (1988), 140–60; and I. Grattan-Guinness, 'Algebras, projective geometry, mathematical logic, and constructing the world: intersections in the philosophy of mathematics of A. N. Whitehead', *Historia Math.* **29** (2002), 427–62 (printing correction, **30** (2003), 96).

42. The differences between algebraic and mathematical logic are often not well recognized. They are emphasized in I. Grattan-Guinness, 'Living together and living apart: on the interactions between mathematics and logics from the

French Revolution to the First World War', *South African Journal of Philosophy* **7** (2) (1988), 73–82, and a shorter version in 'The mathematical turns in logic', *The Rise of Modern Logic from Leibniz to Frege* (ed. D. Gabbay and J. Woods), Vol. 3 of the *Handbook of the History of Logic* (Elsevier, 2004), pp. 545–56.

43. The posher versions of part–whole theory are sometimes called 'mereology': see Smith (note 32), pp. 56–59, 113–21. De Morgan made quite a detailed use of a more informal version, in connection with syllogisms, in the last (1862) of his suite of papers on logic; see De Morgan (note 2), pp. 306–11.

44. See De Morgan (note 2), p. 78. Even such a keen bibliophile as Peano may not have noticed this passage in a long footnote, and his own later use of 'mathematical logic' is presumably independent.

45. See Chapter 15. Scottish mathematics was noticeably more geometrical than algebraic in character, partly under the influence of the country's 'common-sense' philosophy. See R. Olsen, *Scottish Philosophy and British Physics, 1750–1880. A Study in the Foundations of the Victorian Scientific Style* (Princeton Univ. Press, 1975).

## CHAPTER 17 (ROBIN WILSON)

Little has been written on the history of graph theory and combinatorics. In graph theory, three topics (polyhedra, trees, and the four-colour problem) are discussed at length in Norman Biggs, Keith Lloyd, and Robin Wilson's book *Graph Theory, 1736–1936* (Oxford, 1976/1998); this book includes extracts from several of the papers discussed here. For an account of the combinatorial topics (triple systems and partitions), the same authors have written a survey, 'The history of combinatorics' in the *Handbook of Combinatorics* (ed. R. Graham, M. Grötschel, and L. Lovász) (Elsevier Science, 1995). For more information on the four-colour problem, see Robin Wilson's *Four Colours Suffice* (Allen Lane, 2002).

## NOTES

1. Thomas Pynchon, *Gravity's Rainbow* (Vintage Books, 2000), p. 90.

2. J. J. Sylvester, 'Elementary researches in the analysis of combinatorial aggregation', *Phil. Mag. (3)* **24** (1844), 285–96 = *Math. Papers*, Vol. I, pp. 91–102.

3. *The Lady's and Gentleman's Diary* (1844), p. 84.

4. *The Lady's and Gentlemen's Diary* (1846), p. 76.

5. T. P. Kirkman, 'On a problem in combinations', *Cambridge and Dublin Math. J.* **2** (1847), 191–204.

6. T. P. Kirkman, 'Query VI', *Lady's and Gentleman's Diary* (1850), p. 48.

7. Editorial note by W. J. C. Miller, Solution to Problem 2905, *Math. Questions and their Solutions from the Educational Times* **13** (1870), 79.

8. T. P. Kirkman, 'Theorems in the doctrine of combinations', *Phil. Mag. (4)* **24** (1852), 209.

9. R. R. Anstice, 'On a problem in combinatorics', *Cambridge and Dublin Math. J.* **7** (1852), 279–92.

10. J. J. Sylvester, 'Note on the historical origin of the unsymmetrical six-valued functions of six letters', *Phil. Mag. (4)* **21** (1861), 369–77, on p. 371 = *Math. Papers*, Vol. II, 264–71, on p. 266.

11. T. P. Kirkman, On the puzzle of the fifteen young ladies', *London, Edinburgh & Dublin Philos. Mag. and J. Sci.* **23** (1862), 198–204, on p. 198.

12. A. Cayley, 'On the triadic arrangements of seven and fifteen things', *Phil. Mag. (3)* **37** (1850), 50–53, on p. 52 = *Math. Papers*, Vol. I, pp. 481–84, on p. 483.

13. R. H. F. Denniston, 'Sylvester's problem of the 15 schoolgirls', *Discrete Math.* **9** (1974), 229–33.

14. D. K. Ray-Chaudhuri and R. M. Wilson, 'Solution of Kirkman's schoolgirl problem', *Combinatorics, American Math. Soc. Proc. Symp. Pure Math.* **9** (1971), 187–203.

15. J. Steiner, 'Combinatorische Aufgabe', *J. für die reine und ungewandte Mathematik* **45** (1853), 181–82.

16. M. Reiss, 'Ueber eine Steinersche combinatorische Aufgabe, welche im 45sten Bande dieses Jornals, Seite 181, gestellt worden ist', *J. für die reine und ungewandte Mathematik* **56** (1859), 326–44.

17. T. P. Kirkman, 'Appendix I: Solution of Questions 8886 and 9009', *Math. Questions and Solutions from the Educational Times* **47** (1887), 125–27.

18. T. P. Kirkman, 'On the representation of polyedra', *Phil. Trans Roy. Soc. London* **146** (1856), 413–18.

19. A. Cayley, 'On the theory of the analytical forms called trees—part II', *Phil. Mag. (4)* **18** (1859), 374–78.

20. J. J. Sylvester, 'On an application of the new atomic theory to the graphical representation of the invariants and covariants of binary quantics,—with three appendices', *Amer. J. Math.* **1** (1878), 64–125, on p. 109 = *Math. Papers*, Vol. III, pp. 148–206, on p. 190.

21. Sylvester (note 20), p. 64 = *Math. Papers*, Vol. III, p. 148.

22. J. J. Sylvester, 'Chemistry and algebra', *Nature* **17** (1877–78), 284 = *Math. Papers*, Vol. 3, pp. 103–4.

23. A. Cayley, 'A theorem on trees', *Quart. J. Pure Appl. Math.* **23** (1889), 376–78 = *Math. Papers*, Vol. XII, pp. 639–56.

24. G. W. Leibniz, *Math. Schriften*, Vol. IV.2, Specimen de divulsionibus aequationum. Letter 3 dated 2 September, 1674; see D. Mahnke, 'Leibniz auf der Suche nach einer allgemeinen Primzahlgleichung', *Bibliotheca Math.* **13** (1912–13), 29–61, on p. 37.

25. P. J. MacMahon, 'Combinatory analysis: a review of the present state of knowledge', *Proc. London Math. Soc.* **28** (1896–97), 5–32.

26. J. J. Sylvester, 'On the partitions of numbers', *Quart. J. Pure Appl. Math.* **1** (1855–57), 141–52 = *Math. Papers*, Vol. II, pp. 90–99.

27. G. H. Hardy and S. Ramanujan, 'Asymptotic formulae in combinatory analysis', *Proc. London Math. Soc. (2)* **17** (1918), 75–115 = *Math. Papers (Hardy)*, Vol. 1, pp. 306–39. MacMahon's table of partition numbers is on pp. 114–15 (= p. 338).

28. For a reproduction of this letter, see Wilson (2002).

29. [A. De Morgan], 'Review of W. Whewell's *The Philosophy of Discovery*', *Athenaeum*, No. 1694 (14 April 1860), 501–3.

30. [A. Cayley], *Proc. London Math. Soc.* **9** (1877–78), 148.

31. A. B. Kempe, 'On the geographical problem of the four colours', *American J. Math.* **2** (1879), 193–200.

32. P. J. Heawood, 'Map-colour theorem', *Quart. J. Pure Appl. Math.* **24** (1890), 332–38.

## CHAPTER 18 (JEREMY GRAY)

The books on Cayley and Sylvester by Tony Crilly and Karen Parshall—T. Crilly, *Arthur Cayley: Mathematician Laureate of the Victorian Age*, and Karen Parshall, *James Joseph Sylvester: Jewish Mathematician in a Victorian World* (both from Johns Hopkins Univ. Press, 2006)—will surely and rightly be the first things to be consulted by anyone who wants to know about British mathematics in the 19th century, and readers will find in them a wealth of relevant material. Crilly is particularly informative on the people involved and on the Cambridge and London of the day. Parshall also discusses the situation of Jews in Britain and the world of American mathematics, especially around Johns Hopkins University where Sylvester taught successfully for many years. There is more to be said about the mathematics than Crilly provides, but these books set a high standard for anyone following them.

If similar scholarly attention to British applied mathematicians was now forthcoming, which brought out the originality and importance of their mathematics in the pursuit of science, and which helped to narrow the gap between history of mathematics and history of science, we would all benefit.

## NOTES

1. See Crilly (2006) and Parshall (2006).

2. A. Hyman, *Charles Babbage: Pioneer of the Computer* (Oxford, 1982); D. Swade, *The Cogwheel Brain: Charles Babbage and the Quest to Build the First Computer* (Little, Brown, 2000).

3. W. Whewell, *Of a Liberal Education in General, and with Particular Reference to the Leading Studies of the University of Cambridge*, in 3 parts (John W. Parker, 1845–52).

4. Crilly (2006), pp. 469–471.

5. A. Cayley, 'On the intersection of curves', *Cambridge Math. Journal* **3** (1843), 211–13 = *Collected Math. Papers*, Vol. 1, pp. 25–27.

6. See I. Bacharach, Über den Cayley'schen Schnittpunktsatz', *Math. Annalen* **26** (1886), 275–99, and, for a history of the Cayley–Bacharach theorem, see D. E. Eisenbud, M. Green and J. Harris, 'Cayley-Bacharach theorems and conjectures', *Bull. American Math. Society* **33** (1996), 295–324.

7. See T. Hawkins, 'Another look at Cayley and the theory of matrices', *Archives Internationals d'Histoire des Sciences* **26** (1977), 82–112.

8. See A. Cayley, 'On the triple tangent planes of surfaces of the third order', *Cambridge and Dublin Math. Journal* **4** (1849), 118–32 = *Math. Papers*, Vol. 1, pp. 445–56.

9. Parshall (2006), p. 120.

10. Crilly (2006), p. 120.

11. J. L. Richards, *Mathematical Visions: The Pursuit of Geometry in Victorian England* (Academic Press, 1988).

12. W. K. Clifford, *Mathematical Papers* (Macmillan, 1882).

13. G. H. Hardy, 'Ernest William Hobson', *Obituary Notices of Fellows of the Royal Society of London* **3** (1934), pp. 237–49.

14. E. T. Whittaker, 'Andrew Russell Forsyth', *Obituary Notices of Fellows of the Royal Society of London* **4** (1942), 209–27.

15. W. F. Osgood, 'The theory of functions', *Bull. American Math. Society* **1** (1895), 142–54.

16. W. Burnside, *The collected papers of William Burnside* (ed. P. M. Neumann, A. J. S. Mann, and J. C. Tompson) (Oxford, 2004).

17. D. Wilson, *Kelvin and Stokes: A Comparative Study in Victorian Physics* (Adam Hilger, 1987).

18. J. Heard, *The Rise of the Pure Mathematical Disciplines in Britain, ca. 1870–1939*, Ph.D. thesis, Imperial College, London, 2004.

19. A. C. Rice and R. J. Wilson, 'From national to international society: The London Mathematical Society, 1867–1900', *Historia Math.* **25** (1998), 185–217, and 'The rise of British analysis in the early 20th century: the role of G. H. Hardy and the London Mathematical Society', *Historia Math.* **30** (2003), 173–94.

20. The often valuable historical work of Jean Dieudonné has this limitation; see, for example, J. Dieudonné, *A History of Algebraic and Differential Topology, 1900–1960* (Birkhäuser, 1988).

21. This is a failing criticized recently by Goldstein writing about French work on number theory; see C. Goldstein, 'The Hermitian form of reading the *Disquisitiones Arithmeticae*', in *The Shaping of Arithmetic after C. F. Gauss's Disquisitiones Arithmeticae*, (ed. C. Goldstein, N. Schappacher, and J. Schwermer) (Springer, 2007), pp. 377–410.

22. L. Koenigsberger, *Herman von Helmholtz*, 3 vols. (Braunschweig, 1902–03).

23. S. P. Thomson, *The Life of William Thomson, Baron Kelvin of Largs*, 2 vols. (Macmillan, 1910).

24. L. Campbell and W. Garnett, *The Life of James Clerk Maxwell* (Macmillan, 1882).

25. See Wilson (note 17).

26. See, for example, U. Bottazzini, *Il Calcolo Sublime: Storia dell' Analisi Matematica da Euler a Weierstrass* (Editore Boringhieri, 1981), English transl., *The Higher Calculus* (Springer, 1986).

27. O. Darrigol, *Worlds of Flow: A History of Hydrodynamics from the Bernoullis to Prandtl* (Oxford, 2005).

28. M. V. Berry, 'Uniform asymptotic smoothing of Stokes' discontinuities', *Proc. Roy. Soc.* (A) **422** (1989), 7–21, and 'Stokes' phenomenon: smoothing a Victorian discontinuity', *Inst. Hautes Études Sci. Publ. Math.* **68** (1989), 211–21.

29. H. I. and T. Sharlin, *Lord Kelvin: The Dynamic Victorian* (Penn State Univ. Press, 1979).

30. C. Smith and M. N. Wise, *Energy and Empire: A Biographical Study of Lord Kelvin* (Cambridge, 1989).

31. See, for example, J. Buchwald, *From Maxwell to Microphysics: Aspects of Electromagnetic Theory in the Last Quarter of the Nineteenth Century* (Univ. of Chicago Press, 1985), and B. J. Hunt, *The Maxwellians* (Cornell Univ. Press, 1991).

32. E. Scholz, *Symmetrie, Gruppe, Dualität: zur Beziehung zwischen theoretischer Mathematik und Anwendungen in Kristallographie und Baustatik des 19. Jahrhunderts* (Birkhäuser Verlag, 1989).

33. It is described in the forthcoming book by Bottazzini and Gray on the history of complex function theory.

34. A. Warwick, *Masters of Theory: Cambridge and the Rise of Mathematical Physics* (Univ. of Chicago Press, 2003).

35. J. E. Barrow-Green and J. J. Gray, 'Geometry at Cambridge, 1863–1940', *Historia Math.* **33** (2006), 315–56.

36. See Richards (note 11).

37. R. Taton, *L'Oeuvre Scientifique de Monge* (Paris, 1951).

38. B. Belhoste, *Cauchy, 1789–1857, Un Mathématicien Légitimiste au XIXe Siècle. Un Savant, une Époque* (Librairie Classique Eugène Belin, 1985), English transl., *Augustin-Louis Cauchy: A Biography* (Springer, 1991).

39. J. Lützen, *Joseph Liouville, 1809–1882: Master of Pure and Applied Mathematics* (Springer, 1990).

40. V. Maz'ya, and T. Shaposhnikova, *Jacques Hadamard, A Universal Mathematician* (American and London Math. Societies, 1998).

41. A. Stubhaug, *Niels Henrik Abel and His Times. Called Too Soon by Flames Afar* (transl. by R. H. Daly from the 1996 second Norwegian edition, Springer, 2000); A. Stubhaug, *The Mathematician Sophus Lie: It was the audacity of my thinking* (transl. by R. H. Daly from the 2000 Norwegian original, Springer, 2002).

42. J. W. Dauben, *Georg Cantor: His Mathematics and Philosophy of the Infinite* (Harvard Univ. Press, 1979); W. Purkert and H. J. Ilgauds, *Georg Cantor* (Biographienhervorragender Naturwissen-schaftler, Techniker und Mediziner, Vol. 79, Teubner, 1985).

43. G. W. Dunnington, *Gauss—Titan of Science* (1956, republished with a new introduction and appendices by J. J. Gray, Math. Assoc. of America, 2003); W. K. Bühler, *Gauss, a Biographical Study* (Springer, 1981); H. Wussing, *Carl Friedrich Gauss* (Biographienhervorragender Naturwissenschaftler, Techniker und Mediziner, Vol. 15, Teubner, 1989).

44. D. Laugwitz, *Bernhard Riemann 1826–1866. Wendepunkte in der Auffassung der Mathematik, Vita Mathematica*, Vol. 10 (Birkhäuser Verlag, 1996); English transl. by Abe Shenitzer (with the editorial assistance of the author, H. Grant, and S. Shenitzer), *Riemann 1826–1866: Turning Points in the Conception of Mathematics* (Birkhäuser).

45. R. Tobies, *Felix Klein* (Biographienhervorragender Naturwissenschaftler, Techniker und Mediziner, Vol. 50, Teubner, 1981).

# NOTES ON CONTRIBUTORS

**June Barrow-Green** is a senior lecturer in the history of mathematics at the Open University, and is a former President of the British Society for the History of Mathematics. She is an editor of *Historia Mathematica*, and was an associate editor of *The Princeton Companion to Mathematics*; other publications include *Poincaré and the Three Body Problem*. Her research concentrates on the history of late 19th- and early 20th-century mathematics and she has recently completed a study of the British mathematical community during the First World War. In 2010 she was a visiting research professor at the University of Nancy. She is currently working on G. D. Birkhoff's contributions to dynamical systems theory.

**Allan Chapman** teaches at the University of Oxford, where he is associated with the Faculty of History and Wadham College, and was a Visiting Professor in the history of science at Gresham College, London, from 2003 to 2004. His interests lie mainly in the history of astronomy, with particular emphasis on the development of astronomical instruments and observatories. He is an enthusiastic popularizer of his subject, and has written biographies of many scientists, including Robert Hooke and Mary Somerville.

**Alex D. D. Craik** is a graduate of St Andrews and Cambridge Universities and a Fellow of the Royal Society of Edinburgh. He taught applied mathematics for many years at St Andrews University, where he is currently an emeritus professor. He has published extensively on fluid mechanics and on the history of mathematics, most recently *Mr Hopkins' Men: Cambridge Reform and British Mathematics in the 19th Century*.

**Tony Crilly** is emeritus reader in mathematical sciences at Middlesex University. Current interests are the history of 19th-century pure mathematics and the popularization of mathematics through its history. A former secretary and treasurer of the British Society for the History of Mathematics, he is the author of *Arthur Cayley: Mathematical Laureate of the Victorian Age* and, more recently, *50 Mathematical Ideas You Really Need to Know*.

**Sloan Evans Despeaux** is an associate professor of mathematics at Western Carolina University in Cullowhee, North Carolina. Her research interests include 19th-century mathematics, mathematicians, and scientific journals in Britain. She has recently published papers in *Historia Mathematica* and *Annals of Science* and completed a four-year term as abstracts editor of *Historia Mathematica*.

**Raymond Flood** is a fellow and former vice president of Kellogg College, Oxford University, and previously was university lecturer in computing studies and mathematics at the Department for Continuing Education, Oxford University. His main research interests lie in statistics and the history of mathematics, and he was formerly president of the British Society for the History of Mathematics. He is a co-editor of *The Nature of Time, Let Newton be!, Möbius and his Band, Oxford Figures, Music and Mathematics: from Pythagoras to Fractals*, and *Kelvin: Life, Labours and Legacy*.

**Ivor Grattan-Guinness** is emeritus professor of the history of mathematics and logic at Middlesex University, England. He was editor of the history of science journal *Annals of Science* from 1974 to 1981. In 1979 he founded the journal *History and Philosophy of Logic*, and edited it until 1992. He also edited a substantial *Companion Encyclopedia of the History and Philosophy of the Mathematical Sciences*, and published *The Fontana History of the Mathematical Sciences. The Rainbow of Mathematics* and *The Search for Mathematical Roots, 1870–1940. Logics, Set Theories and the Foundations of Mathematics from Cantor through Russell to Gödel*. He was the associate editor for mathematicians and statisticians for the *Oxford Dictionary of National Biography*, and edited a large collection of essays on *Landmark Writings in Western Mathematics, 1640–1940*. In August 2009 the International Commission for the History of Mathematics awarded him the Kenneth O. May Medal in the History of Mathematics.

**Jeremy Gray** is a professor of the history of mathematics at the Open University, and an honorary professor at the University of Warwick, where he lectures on the history of mathematics. In 2009 he was awarded the Albert Leon Whiteman Memorial Prize by the American Mathematical Society for his work in the history of mathematics. He has written many books on the history of modern mathematics, the most recent being *Plato's Ghost: The Modernist Transformation of Mathematics*.

**Keith Hannabuss** is Billmeir fellow and tutor in mathematics at Balliol College, Oxford University, and CUF lecturer in the Mathematical Institute. His main research interests are in quantum field theory and operator algebras, combined with a long-standing interest in the history of mathematics and theoretical physics. He is the author of *An Introduction to Quantum Theory*, and contributed to *Oxford Figures* and to the *History of the University of Oxford*.

**Eileen Magnello** is a research associate in the department of science and technology studies at University College London. She trained as a statistician in the mid-1970s and did her doctorate in the history of science at St Antony's College, Oxford University. Her main research lies in the history of statistics, especially that of Karl Pearson. She has published a series of papers on Pearson, and is editing Pearson's sesquicentenary volume. Her recent book, *Introducing Statistics*, places statistical methods and probability within their historical context. She is in the process of setting up a History Group of the Royal Statistical Society.

**Tony Mann** teaches at the University of Greenwich, where he was formerly Head of the Department of Mathematical Sciences. He has held a number of positions within the British Society for the History of Mathematics, of which he is currently President. He is a co-editor, with Peter M. Neumann and Julia Tompson, of *The Collected Papers of William Burnside*, published by Oxford University Press in 1994.

**Amirouche Moktefi** is a research fellow in history and philosophy of science at Strasbourg University. His recent research was on the mathematics and logic of Charles L. Dodgson (alias Lewis Carroll). He currently works broadly on Victorian mathematics and logic, in addition to some other history of science interests, notably on visual reasoning.

**Karen Hunger Parshall** is professor of history and mathematics at the University of Virginia and past chair of the International Commission for the History of Mathematics. Her research interests lie in the history of 19th- and 20th-century mathematics, with particular emphasis on the historical development of modern algebra. She has written *The Emergence of the American Mathematical Research*

*Community, 1876–1900: J. J. Sylvester, Felix Klein, and E. H. Moore* (with David E. Rowe), *James Joseph Sylvester: Life and Work in Letters*, and *James Joseph Sylvester: Jewish Mathematician in a Victorian World*. She is a co-editor of *Experiencing Nature, Mathematics Unbound: The Evolution of an International Mathematical Research Community, 1800–1945*, and *Episodes in the History of Modern Algebra (1800–1950)*.

**Adrian Rice** is professor of mathematics at Randolph-Macon College in Ashland, Virginia, where his research focuses on 19th-century and early 20th-century British mathematics. His publications include *Mathematics Unbound: The Evolution of an International Mathematical Research Community, 1800–1945* (edited with Karen Hunger Parshall) and *The London Mathematical Society Book of Presidents, 1865–1965* (written with Susan Oakes and Alan Pears). He is a two-time recipient of the Mathematical Association of America's Trevor Evans Award for outstanding expository writing.

**Doron D. Swade** is an engineer, historian, and a museum professional. He has published widely on history of computing, curatorship, and museology. He is a leading authority on Charles Babbage, and he masterminded the construction of the first complete Babbage engine built to original 19th-century designs. He was Senior Curator of computing at the Science Museum, London, and later Assistant Director & Head of Collections. He is currently visiting professor (history of computing) at Portsmouth University and honorary research fellow (computer science) at Royal Holloway University of London. He was awarded an MBE in 2009 for services to the history of computing.

**Robin Wilson** is emeritus professor of pure mathematics at the Open University, emeritus professor of geometry at Gresham College, London, a former fellow of Keble College, Oxford University, and President-elect of the British Society for the History of Mathematics. He has written and edited a number of books on graph theory, including *Introduction to Graph Theory* and *Four Colours Suffice*, and on the history of mathematics, including *Lewis Carroll in Numberland*. He is involved with the popularization and communication of mathematics and its history, and in 2005 was awarded a Pólya prize by the Mathematical Association of America for 'outstanding expository writing'.

# ACKNOWLEDGEMENTS

Shorter versions of the Introduction (Adrian Rice) and Chapters 5 (Raymond Flood), 11 and 12 (M. Eileen Magnello), 15 (Karen Hunger Parshall) and 18 (Jeremy Gray) appeared in the *BSHM* (British Society for the History of Mathematics) *Bulletin*, Vol. 21, No. 3, 2006.

## INTRODUCTION

University Museum and Hamilton – courtesy of Robin Wilson.
Carroll – courtesy of the Alfred C. Berol collection, Fales Library, New York University.
Babbage's Difference Engine No. 1, demonstration assembly, 1832 – courtesy of the Science Museum, London.
Thomson – courtesy of the National Portrait Gallery, London. NPG 1708(f). 'William Thomson, Baron Kelvin' by Elizabeth King (née Thomson), pencil, 1840 © National Portrait Gallery, London.
De Morgan medal – courtesy of the London Mathematical Society.

## CHAPTER 1

Hopkins – courtesy of Peterhouse, Cambridge.
Routh, Adams, and Glaisher – courtesy of the London Mathematical Society.
Cayley – courtesy of Robin Wilson.

## CHAPTER 2

Oxford viva – courtesy of Robin Wilson.
Smith, Sylvester, and Campbell – courtesy of the London Mathematical Society.

## CHAPTER 3

De Morgan, Hirst, Henrici, Clifford, and Bryant – courtesy of the London Mathematical Society.

## CHAPTER 5

Hamilton's note on quaternions and Humphrey Lloyd – courtesy of Trinity College, Dublin.

Casey, Salmon – courtesy of the London Mathematical Society.
Sketch of Hamilton on bridge – courtesy of Robin Wilson.
MacCullagh – courtesy of the National Gallery of Ireland.
Cubic surface – courtesy of the John Fauvel collection, Open University.
Torchlight procession – courtesy of Raymond Flood.

## CHAPTER **6**

Nanson – courtesy of the University of Melbourne archives.
Lamb – courtesy of the University of Adelaide archives.
Aldis – courtesy of the New Zealand Mathematical Society.
Mookerjee – courtesy of the University of Calcutta.
Royal Observatory at Cape of Good Hope – courtesy of the South African Library.

## CHAPTER **7**

All images – courtesy of Sloan Despeaux.

## CHAPTER **8**

BA lecture – courtesy of Robin Wilson.
Thomson and Tait title page – courtesy of Raymond Flood.
Stokes and Love – courtesy of the London Mathematical Society.

## CHAPTER **9**

Greenwich meridian transit – courtesy of Graham Dolan.

## CHAPTER **10**

Johan Müller's 'Universal Calculator', 1794 – courtesy of Hessisches Landesmuseum Darmstadt.
Thomas de Colmar Arithmometer, Babbage, Difference Engine No. 1 Design Drawing, Analytical Engine 1871, Analytical Engine 1840, Scheutzes' Difference Engine, Difference engine prototype – courtesy of the Science Museum, London.
Difference Engine No. 2, 2004 and Output Apparatus for Difference Engine No. 2 – courtesy of Doron Swade.

## CHAPTER **13**

Forsyth, Whittaker, Hobson – courtesy of the London Mathematical Society.
Preface to Hardy – courtesy of the John Fauvel collection, Open University.

## CHAPTER 14

Byrne's *Euclid* – courtesy of the John Fauvel collection, Open University.
Todhunter, Hirst – courtesy of the London Mathematical Society.
Syllabus – courtesy of Robin Wilson.
Dodgson – courtesy of Edward Wakeling.


## CHAPTER 15

Binary quantics – courtesy of Robin Wilson.
Burnside, Elliott – courtesy of the London Mathematical Society.


## CHAPTER 16

De Morgan – courtesy of the London Mathematical Society.


## CHAPTER 17

Woolhouse, MacMahon, Kempe – courtesy of the London Mathematical Society.
Kirkman, Heawood – courtesy of Robin Wilson.

While every effort has been made to secure copyright, anyone who feels that their copyright has been infringed is invited to contact the publishers, who will endeavour to correct the situation at the earliest possible opportunity.

# INDEX OF NAMES

An asterisk indicates the appearance of a picture.

Crofton, Morgan William, 65, 73
Cruickshank, John, 78–9

## D

D'Alembert, Jean le Rond, 179, 181, 185
Darwin, Charles Robert, 2, 31, 93, 146–7, 192, 232, 262–3, 279–80, 283, 284*, 285–6, 292, 299
Darwin, George Howard, 23, 27, 31, 143, 147, 184, 285, 295, 306
Davie, G. E., 80, 85
Davies, Thomas Stephens, 64, 158, 159
Davy, Humphry, 248, 251
De Colmar, Thomas, 244
De la Rue, Warren, 217, 228, 231
De Moivre, Abraham, 264, 284
De Morgan, Augustus, 7–8, 13*, 14, 21, 53–54, 55*, 56–61, 64, 72, 90, 104, 111–12, 114–15, 138, 160, 162, 168, 177, 264, 302, 309–11, 324, 328, 331, 339, 342–3, 345, 360, 361*, 362, 364–70, 373–4, 377, 391–2, 398–9, 408–9, 413
De Valera, Eamon, 118*, 119
Dixon, Andrew Cardew, 105–6
Dodgson, Charles Lutwidge (Lewis Carroll), 1, 6, 7*, 41, 170, 331*, 332, 346, 358, 368–9
Donkin, William Fishburn, 40, 180
Draper, Henry, 225–6, 231
Drew, William Henry, 24, 60, 72
Duncan, Thomas, 80, 82, 91, 97, 100
Dupuis, Nathan Fellowes, 142, 143*

## E

Edgeworth, Francis Ysidro, 48–9, 278*, 279, 284, 286, 288, 294
Eisenstein, Ferdinand Gotthold Max, 44, 354
Elliott, Edwin Bailey, 14, 47, 48, 50, 353, 355, 356*
Ellis, Alexander John, 20, 366
Ellis, Robert Leslie, 21, 23, 342–3, 366
Esson, William, 46, 48–50
Euclid, 26, 56, 64, 69, 85, 124, 136, 142–3, 170, 196, 320–36, 360–1, 410, 412–3
Euler, Leonhard, 7, 179, 181, 183, 185–7, 303–6, 309, 314–5, 361, 367, 388–90, 392, 394
Ewing, James Alfred, 27, 96, 101, 180, 187

## F

Faraday, Michael, 189, 194

Farr, William, 49, 263, 269, 270*, 272, 275, 280, 283
Fawcett, Philippa, 31*, 62, 133–4
Fenwick, Stephen, 64, 158–9
Ferrers, Norman MacLeod, 23, 24, 164, 391
Fisch, Menachem, 340–1
Fischer, Wilhelm Franz Ludwig, 23, 79, 95, 329
Fisher, Ronald Aylmer, 272, 292, 295
FitzGerald, George Francis, 103–4, 115*, 116–18, 180, 306
Forbes, James David, 84–5, 91, 191
Forsyth, Andrew Russell, 13, 23, 27, 29, 67, 69, 135, 144, 148, 186, 311*, 312–15, 335–6, 405, 408–9, 411
Fourier, Joseph, 90, 135, 185, 187, 191, 196, 223, 267, 316–7, 407, 411
Fraser, Alexander Yule, 99, 172
Fraunhofer, Josef von, 212, 215–16, 219–22, 224–7
Frend, William, 4, 8, 339
Fresnel, Augustin Jean, 187–8
Fuller, Frederick, 23, 82, 91, 100, 243

## G

Galton, Francis, 12, 14, 23, 41, 49, 263, 273, 277–8, 280, 283–285, 286*, 289–90, 295, 297–300
Gauss, Carl Friedrich, 2, 13, 284, 296, 333, 347
George II (King), 62, 207
Gibbs, Josiah Willard, 14, 195
Gibson, George, 313–14
Glaisher, James Whitbread Lee, 23, 25*, 29, 44–5, 141, 143, 156, 161, 165, 168, 172
Godfray, Hugh, 125, 158–9
Goodeve, Thomas Minchin, 60, 69, 72
Gordan, Paul Albert, 48, 354, 355*, 388, 403–5, 413
Gosset, William Sealy ('Student'), 272, 293, 295
Grassmann, Hermann Günther, 14, 345
Grattan-Guinness, Ivor, 107–8, 110, 116, 409
Graunt, John, 261, 264, 279
Graves, Charles, 114, 345
Graves, John, 5, 345
Green, George, 9, 10, 20, 124, 162–3, 179, 185, 187–8, 193–4, 306, 317, 407, 409, 411
Greenhill, Alfred George, 143–4, 148, 150, 152, 187
Gregory, David, 86, 89
Gregory, Duncan Farquharson, 4–5, 20, 85, 90, 161–2, 306, 342*, 343
Gregory, Olinthus, 63*, 157

Grubb, Howard, 208, 225, 227*
Grubb, Thomas, 206, 227–9
Guicciardini, Niccolò, 62, 307
Gurney, Theodore, 124–5
Guthrie, Francis, 138*, 139, 391*
Guthrie, Frederick, 68, 138, 391

## H

Hall, Thomas Grainger, 59–60, 72
Halley, Edmond, 210, 216, 264
Hamilton, William Rowan, 5*, 13, 94, 103–4, 107–109, 110*, 111–14, 116–19, 163, 180, 184, 195, 198, 306, 313, 341, 344–5, 347, 377–8, 383–5, 391, 400–1, 404, 406
Hamilton, William Stirling, 7, 84, 90, 111, 361–2
Hankins, Thomas, 109, 111
Hardy, Godfrey Harold, 13, 15, 29–30, 32, 50, 165, 317–18, 391, 395, 405, 408–9
Harrison, John, 207, 247
Hart, Harry, 65, 67
Hawkins, Francis Bisset, 262, 271
Hayward, Robert Baldwin, 171, 330
Heaviside, Oliver, 11*, 194–5
Heawood, Percy, 46–7, 394*
Helmholtz, Hermann Ludwig Ferdinand von, 30, 44, 94, 117, 186, 189, 191–2, 333, 411
Henrici, Olaus Magnus Friedrich Erdmann, 56, 57*, 60–1, 69–70, 72–3, 143–4, 331, 399
Hermite, Charles, 44–5, 164–5, 404, 413
Herschel, John Frederick William, 3–4, 24, 123, 125, 140, 162, 188, 201, 203, 208, 211–13, 217, 219, 224, 234, 246, 255, 304, 340, 342
Herschel, William, 207–8, 210, 212–13, 215, 224
Hertz, Heinrich, 188, 194–5
Hilbert, David, 2, 13–14, 48, 50, 355*, 388, 403, 413
Hill, Micaiah John Muller, 57, 187
Hirst, Thomas Archer, 56*, 66, 73, 160, 168, 170, 328*, 399
Hobson, Ernest William, 29, 96, 144, 317*, 405
Hogg, Quintin, 70–1
Hopkins, William, 23*, 92–3, 179, 183, 191
Houël, Jules, 332, 412
Hudson, William Henry Hoar, 60, 144
Huggins, Margaret, *see* Murray, Margaret Lindsay
Huggins, William, 203, 223, 224*, 225–6, 231, 236